American Interceptor: US Navy Convoy Fighter Projects

US NAVY CONVOY FIGHTER PROJECTS

Jared A. Zichek

Published in Great Britain by Tempest Books
an imprint of Mortons Books Ltd.
Media Centre
Morton Way
Horncastle LN9 6JR
www.mortonsbooks.co.uk
© 2022 by Tempest Books

All rights reserved. No part of this publication may be reproduced or transmitted in any form or by any means, electronic or mechanical including photocopying, recording, or any information storage retrieval system without prior permission in writing from the publisher.
ISBN 978-1-911658-94-8

The right of Jared A. Zichek to be identified as the author of this work has been asserted in accordance with the Copyright, Designs and Patents Act 1988.

Typeset by Jayne Clements (jayne@hinoki.co.uk), Hinoki Design and Typesetting
Original cover artwork by Piotr Forkasiewicz

Contents

Introduction	7
Chapter 1: Convoy Fighter Origins	8
Chapter 2: Convair Model 5 Convoy Fighter	24
Chapter 3: Goodyear GA-28A and GA-28B	49
Chapter 4: Lockheed L-200	81
Chapter 5: Martin Model 262 and 262P	174
Chapter 6: Northrop N-63 and N-63A	220
Chapter 7: Winners and Losers	262
Chapter 8: Related Projects and Studies	328
Sources and Bibliography	346
Image credits	348
Index	349

Introduction

THIS BOOK is about the US Navy's Convoy Fighter competition of November 1950, a contest to produce a turboprop-powered 'tailsitter' single-place aircraft capable of vertical unassisted takeoff from, and landings on, small platform areas of convoy vessels. The principal mission of this fighter was to protect convoys of merchant ships from enemy air attack, fending off the hostile aircraft until more capable carrier-based fighters could arrive and neutralize the threat.

The aircraft which resulted from this competition, the Convair XFY-1 Pogo and Lockheed XFV-1, did not meet the expectations of the Navy, revealing serious problems in the initial conception of the vertical takeoff and landing (VTOL) turboprop tailsitter concept. These strange machines were unceremoniously cancelled in 1955 after relatively short periods of experimental flight testing, becoming museum pieces and occasional objects of derision among historical aviation writers.

The main focus of the book is on 'paper projects', the unbuilt studies submitted to the Navy and rejected for various reasons. It is a compilation of the monographs I previously self-published on the individual entrants in the Convoy Fighter completion, a handsome hardcover edition with superior print quality, supplemented by new material on American interwar VTOL tailsitter projects and more information on the early preliminary development of the XFY-1. The book is not a detailed history of the XFV-1 and XFY-1 as built and tested, as my original research did not uncover much in the way of new information in this regard.

If readers want to find out more about the actual XFY-1 and XFV-1 aircraft, I encourage them to seek out the Ginter monographs on each type as well as the various articles which have been published about them over the decades. There is undoubtedly a substantial amount of primary documents on both aircraft waiting to be discovered in the US National Archives and other repositories, but I leave it to other researchers to undertake the challenging task of uncovering this material and assembling the ultimate histories of these unusual VTOL fighters.

In the meantime, I hope readers enjoy this book, which is packed full of interesting information and exciting imagery from an era when the US military was willing to take big gambles on risky concepts. In the case of the VTOL turboprop tailsitter, it did not pay off, but we are left with these fascinating artifacts of what might have been from a golden age of aerospace history unlikely to be equalled or exceeded.

Jared A. Zichek, Editor
Retromechanix.com
Hauser, Idaho, USA
March 2022

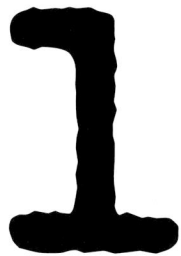

Convoy Fighter Origins

THE DREAM of combining the best aspects of the airplane with those of the helicopter into one vehicle, often referred to as a convertible aircraft, is an old one, likely extending back to the earliest days of flight. Among the many convertible aircraft layouts studied, the vertical takeoff and landing (VTOL) tailsitter configuration was among the more popular, with numerous proposals and patents appearing throughout the first half of the twentieth century. The majority of these resembled a conventional aircraft with a strengthened aft fuselage and tail to enable the vehicle to operate standing on its end.

These designs often had oversize propellers, larger than a conventional aircraft but smaller than a helicopter rotor. They were designed to rise vertically, transition to horizontal flight after reaching the desired altitude, travel at the speed of a conventional aircraft, then transition back to the vertical orientation and descend tail first to land in the upright position, with no traditional runway being required.

The engineering challenges inherent in the VTOL tailsitter, along with the lack of a formal requirement from the military, deterred most of the major aircraft manufacturers from seriously pursuing the concept, and it was left mostly to eccentric individual inventors with varying levels of formal engineering education to work out how such a machine would function. The proposals and patents produced by these inventors were often reviewed with great skepticism by engineers within the National Advisory Committee for Aeronautics (NACA), the Navy and Army Air Corps; they were almost always rejected due to concerns about stability, control, engineering feasibility, and so forth.

However, there were a few personnel within both the Navy Bureau of Aeronautics (BuAer) and US Army Air Corps (USAAC) Materiel Command who did not dismiss the concept out of hand; in the early postwar period, they would gradually recognize the potential of the VTOL tailsitter configuration as technology advanced and missions emerged that, in theory at least, these strange aircraft were well suited to perform.

EARLY AMERICAN VTOL TAILSITTER CONCEPTS
The following patents and proposals, dating from 1929-41, represent a small sampling of the various VTOL tailsitter designs developed in the United States prior to the Second World War. There are many more which can be found in online patent databases; there are likely still more waiting to be discovered in various aerospace archives and museums. While none of these studies made it beyond the drawing board, they may have helped plant a seed in the minds of Army and Navy planners that would grow in the early postwar period, leading to formal research into the feasibility of the concept

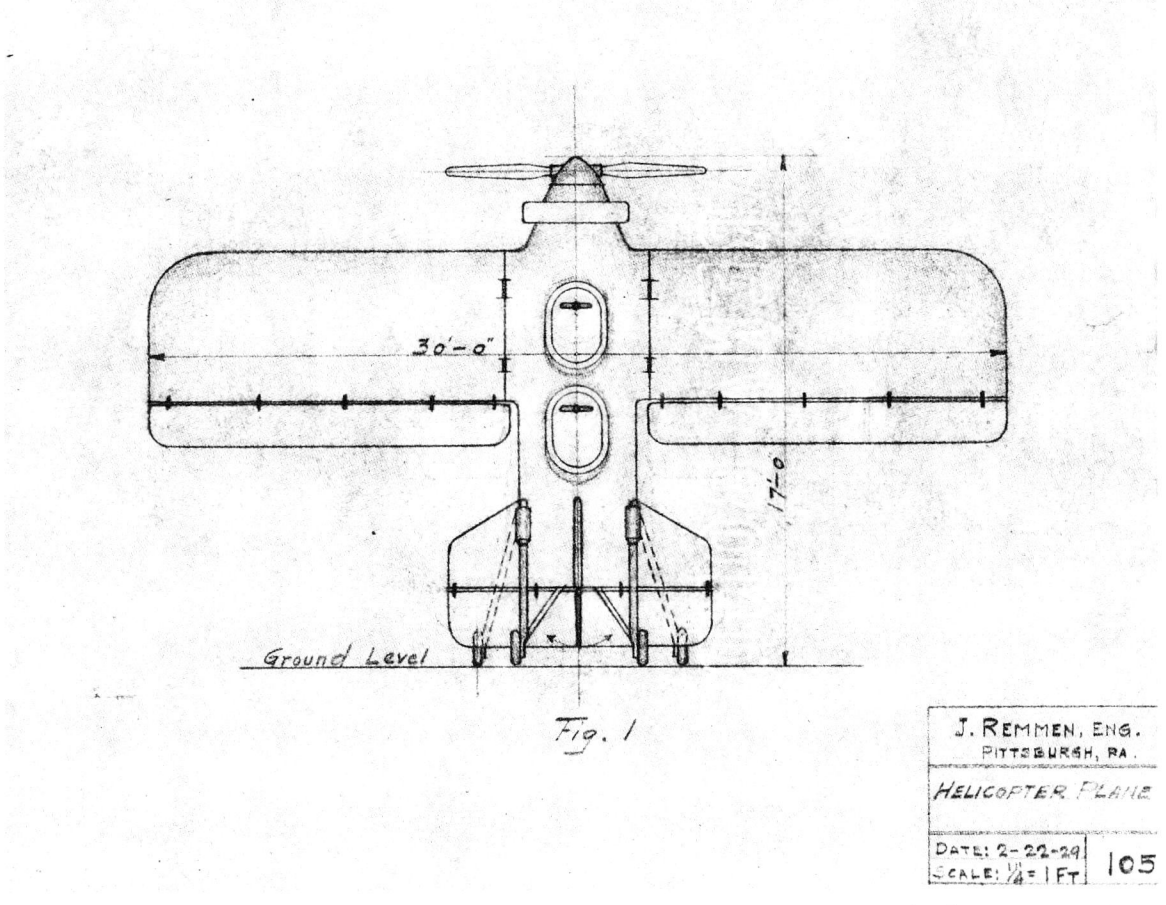

A general arrangement of the Remmen Helicopter Plane of 1929, a VTOL tailsitter concept submitted to the US government's Aeronautical and Patents Design Board, evaluated by NACA, and ultimately rejected due to perceived controllability issues.

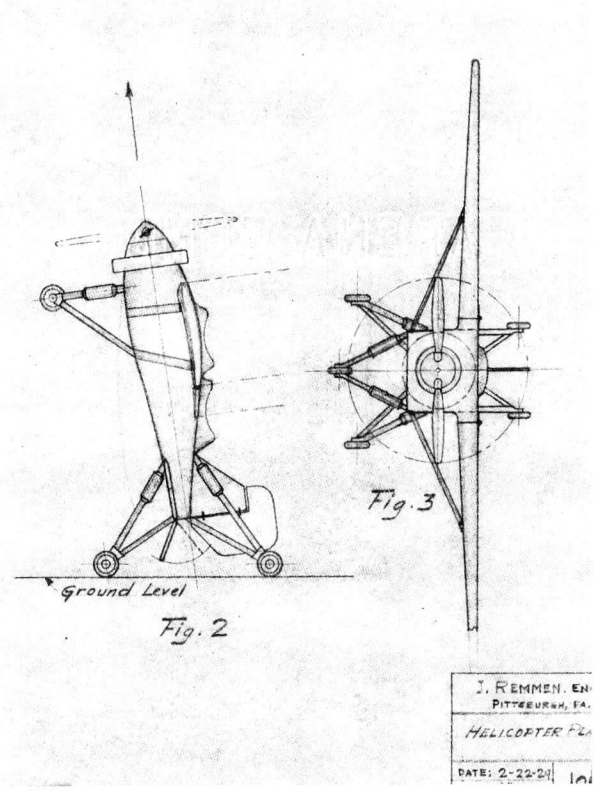

and eventually to the construction of actual VTOL tailsitter aircraft.

REMMEN HELICOPTER PLANE (1929)

In a letter to the Aeronautical and Patents Design Board dated February 26, 1929, engineer J. Remmen of Pittsburgh, Pennsylvania proposed a "Helicopter Plane" tailsitter concept which was, according to the inventor, able to take off the ground vertically to travel horizontally at high speed, or remain suspended in the air without horizontal motion, and finally to descend vertically under full control of the pilot and any fairly level ground of restricted area, say the size of a city lot.

Remmen considered his concept primarily as an airplane, designed along standard lines with some slight modifications; nevertheless, it also embodied some of the characteristics of the helicopter. The fuselage of the Remmen Helicopter Plane was short with the empennage ending just above the ground line; it was equipped with a special landing gear to enable takeoffs and landings in the vertical position.

The machine was designed along conventional lines with supporting planes for horizontal flight, ailerons, stabilizers, elevator, fin and a standard control.

Remmen assumed that the plane was equipped with a 200hp engine and a propeller giving a lifting force of 15lb/hp, the total lifting force was 3,000lb. With the plane fully loaded, it weighed 2,000lb; thus, there was ample reserve lifting force for taking off from the ground vertically. The Helicopter Plane could then proceed vertically until the ceiling was reached or it could level out by setting the elevator in the down position so that the propeller slipstream would bring the ship into the desired angle of climb or a horizontal flight position. In this position it was supported entirely on the wings with the ailerons, elevator and rudder functioning normally. The ceiling for a vertical climb would have likely been lower than the ceiling attained by spiralling.

To achieve a hovering position, the plane was brought 'nose-up' by putting the elevator in the up position. Remmen claimed the Helicopter Plane had no critical stalling angle. Due to the propeller torque, the Helicopter Plane had the tendency to turn around its longitudinal axis in this position; the wings, however, would resist this tendency. Along with proper manipulation of the ailerons, the airplane could be made to turn one way or the other or be held stationary as the propeller slipstream would strike that part of the ailerons nearest the fuselage. This was the reason the ailerons were extended close to the fuselage, the outer ends in this case being ineffective. The rudder had no function in the hovering position.

The ascent or descent of the airplane, in this position, was governed by the pitch and speed of the propeller, provided of course that it was below the ceiling for a vertical climb. By manipulating the elevator, the Remmen Helicopter Plane was slightly inclined in the hovering position and in conjunction with the ailerons made to move slowly in any desired direction. When a landing spot was selected the speed or pitch of the propeller was changed to suit the desired landing speed. The question of how to keep the right side up when in a hovering position required no explanation; as long as the power plant was working, the airplane would hang on the propeller like it was tied to a rope.

In case of engine failure the Remmen Helicopter Plane was, of course, in the same predicament as any other plane under similar circumstances; by necessity, it would have to glide to the ground and land as best as it could. It was for such an emergency that the front landing gear was provided.

Remmen believed his design could meet the key requirements of a successful helicopter design, as set forth by Major Victor W. Pagé, an authority on the subject. These were the ability of the machine to get off the ground; safely landing in the event of an engine failure; keeping the right side up; and the ability to travel horizontally in any desired direction. Remmen summed up the novel features of his design as follows: "Set the plane on end, give it ample motor power and a propeller with sufficient thrust to lift the total weight and then 'let 'er go'; the rest is nothing new."

Remmen asserted that the advantages of this airplane were so self-evident as to need no particular mention. It could take off from most any fairly level spot on the ground, a house top or a ship's deck and land in the same places with equal ease. As the airplane had excess power it was able to dart swiftly from place to place and assume a hovering position in the air whenever desired.

Remmen foresaw the need for a variable pitch propeller, as a smaller pitch was required to provide the maximum lift for vertical travel, while greater pitch was required for fast travel in a horizontal direction. It might have been possible to strike a compromise between the two, i.e., great thrust and slow speed for vertical travel and less thrust and greater speed for horizontal travel, and in that way use a fixed pitch propeller. However, Remmen designed a variable pitch propeller for his Helicopter Plane which fulfilled this purpose better than the fixed pitch alternative. A geared propeller might have been necessary, and two or more motors might have been required for heavier aircraft.

Remmen acknowledged that there were several minor details to be taken care of in the construction of his design, but was confident that the general outline of the main principle would be sufficient to convince the government of the merit of his invention.

NACA evaluated all proposals sent to the Aeronautical Patent and Design Board, which was set up by Congress to promote aeronautical innovation through monetary awards, and the Helicopter Plane proposal was evaluated on March 11, 1929 by G. W. Lewis, NACA's director of aeronautical research. He considered Remmen's device to be a normal airplane

1/48 Scale

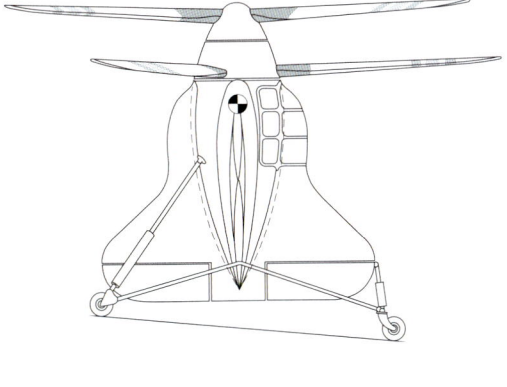

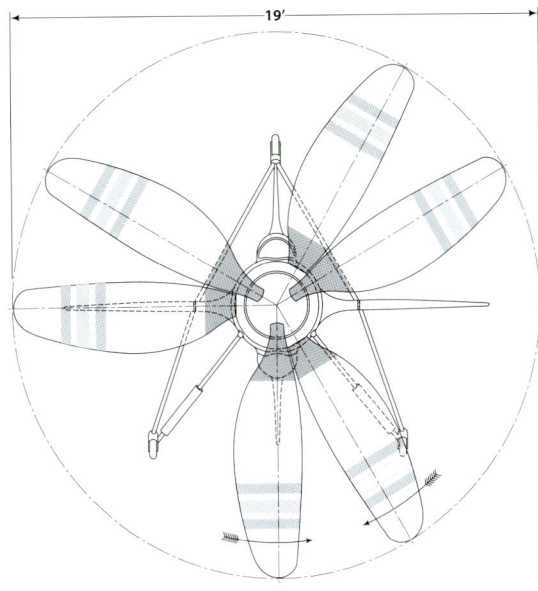

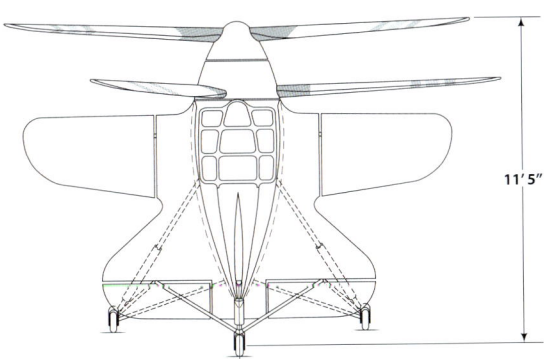

GAMBY VERTIGO PLANE MODEL-A
July 22, 1933

Original title block to the Gamby Vertigo Plane blueprint. The general arrangement has been traced to improve clarity. NACA considered the Vertigo completely unworkable.

with a landing gear placed at the tail. The airplane was to rest on the ground with the nose upward and the thrust propeller to act as the direct lift propeller.

Lewis noted that identical arrangements had been proposed before. Remmen's arrangement, however, provided no means whereby control of altitude, attitude, and position could be effected. Below flying speed as an airplane the arrangement would be unstable and uncontrollable, hence, it was unable to accomplish the purpose for which it was intended.

Lewis believed that the Helicopter Plane was not new and not better than others which had been suggested to accomplish the same purpose; no award was recommended. The lack of financial support from the government seems to have precluded further development of the Remmen Helicopter Plane, as there is no evidence in the historical record that it made it beyond the drawing board.

GAMBY VERTIGO PLANE (1934)

Georges Henri Gamby of Santa Monica, California submitted his Vertigo Plane concept to the chairman of the Aeronautical Patent and Design Board at the War Department, Joseph S. Ames, on March 12, 1934. There is scant information available on Gamby, who appears to have been an amateur inventor of French ancestry; English was definitely not his first language, judging by the spelling and grammatical errors in his letter to the government. He had some basic knowledge of aircraft construction and blueprints though, perhaps having worked on the factory floor of a local manufacturer such as Lockheed or Douglas.

According to his proposal, the Vertigo was designed to rise straight up and assume horizontal

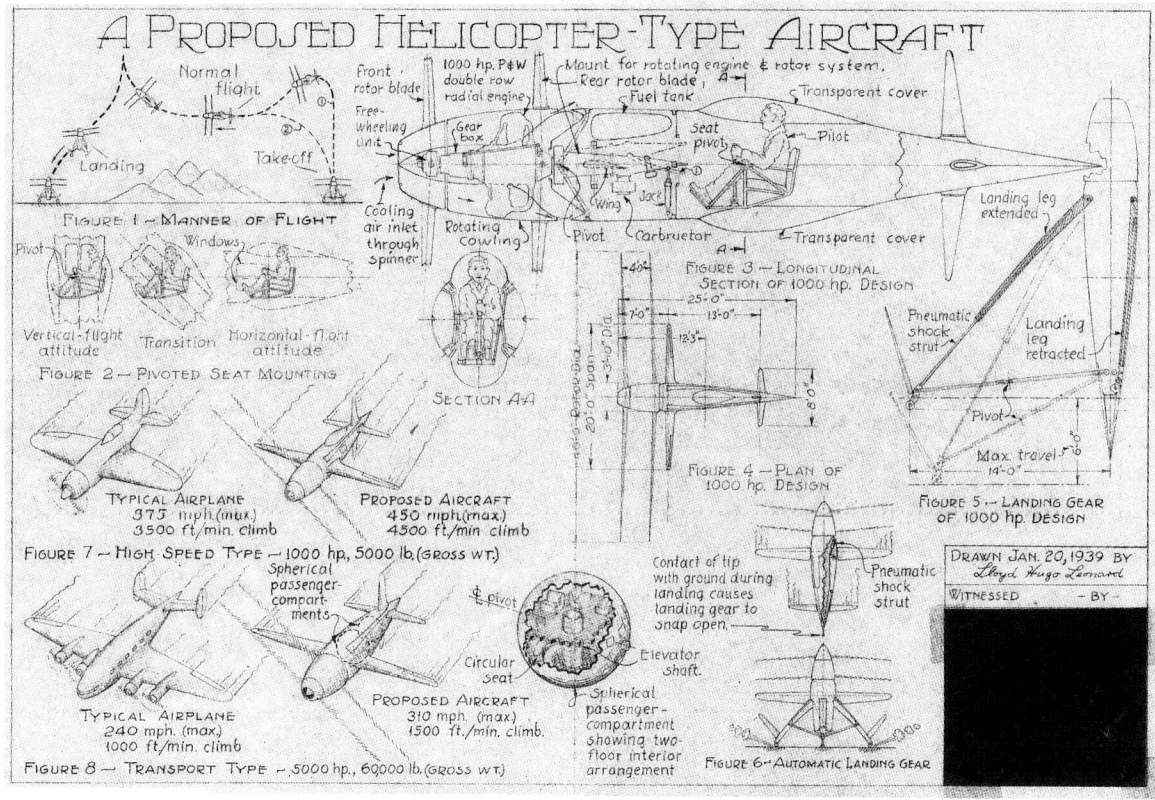

A drawing of L.H. Leonard's extraordinary 1,000 Horsepower Helicopter-Type Aircraft proposal of 1939, a convertible aircraft which featured a large diameter contra-rotating propeller, each unit having two blades, as well as a stalky landing gear which, when extended, provided a wide footprint to support the aircraft in the upright position. The drawing shows numerous details of the design, as well as a concept of a 1,000 hp 5,000lb transport with two spherical rotating passenger compartments (shown in the lower left corner).

flight after clearing obstacles, the lift and forward propulsion being maintained by two large contra-rotating propellers mounted on the nose of the plane. The propellers were geared to a lower speed than that of the engine, a Warner 110hp, and revolved in opposite directions to equalize the torque. The lower propeller or "rotator" turned with an oscillation motion (not shown in the accompanying drawing); Gamby believed this movement to be essential for the aircraft to achieve décollage (takeoff) from the ground.

The controls were the same as those of a conventional airplane, with only the pilot seat pivoting according to the position of the vehicle. The assembly and riveting were of Gamby's own design, permitting the use of rivet squeezers for almost the entire construction, eliminating the use of bucking bars.

Additional details of the design can be gleaned from the blueprint. The huge broad chord propellers/rotors had a 19ft diameter; in contrast, the bullet-shaped fuselage was short and compact, with an overall length (or height, depending on the attitude) of only 11ft 5in. The huge propeller disc would have restricted the maximum speed of the aircraft in horizontal flight. The large canopy featured ample transparencies for the pilot to view the ground during landing, though no ventral window was provided. No provision was shown for channelling cooling air to the radial engine and the wings were all-moving with no separate ailerons or flaps. The large tail was of cruciform configuration and the fixed landing gear featured prominent bracing struts, which would have created significant drag.

Lewis, at NACA, evaluated Gamby's design but found little merit in it. In his letter dated March 28, 1934, Lewis noted that identical arrangements had been proposed and Gamby's airplane was neither newer nor better than its predecessors. Just like Remmen's design five years earlier, the Vertigo provided no means whereby control of the altitude, attitude, and position could be effected below flying speed. As an airplane, the arrangement would be unstable and uncontrollable; hence, it was unable to

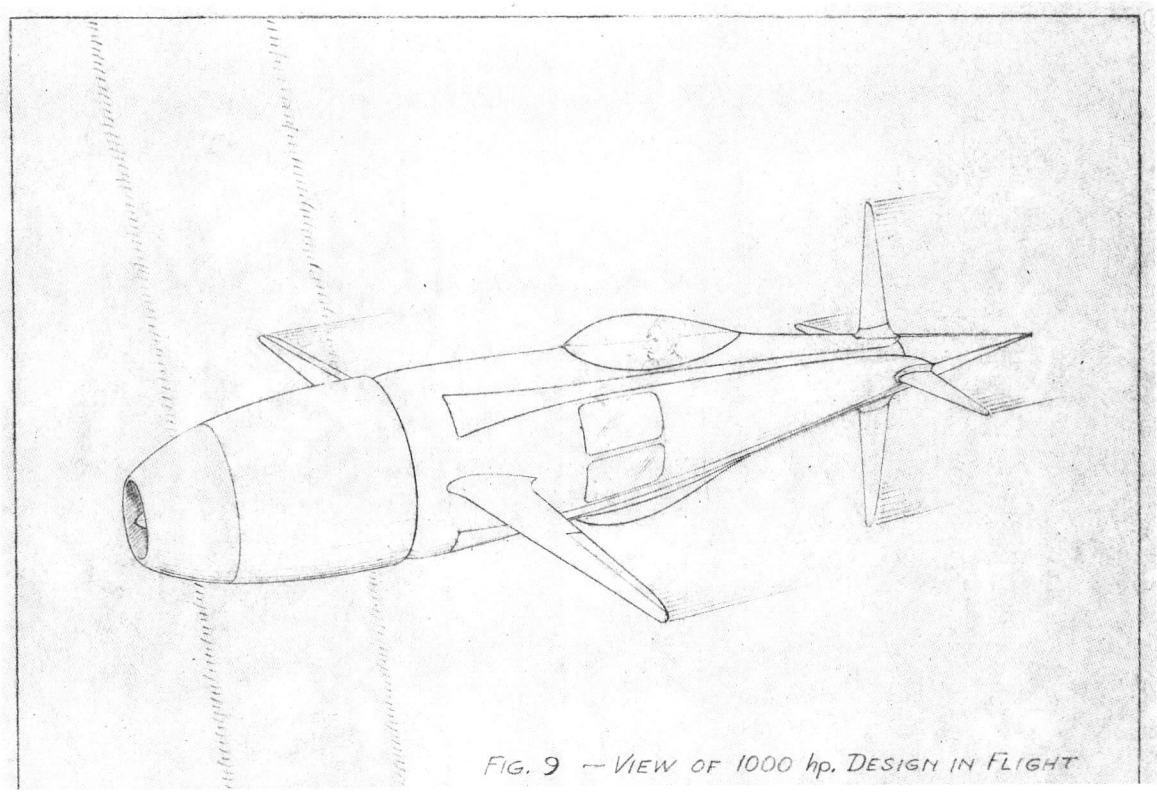

An artist's impression of the 1,000 Horsepower Helicopter-Type Aircraft in flight, emphasizing its clean lines, diminutive, high aspect ratio flying surfaces, huge propellers, and ducted spinner. The cockpit featured extensive glazing, including a secondary ventral teardrop canopy, to aid the pilot in vertical flight and landing.

accomplish the purpose for which it was intended. Once again, no award was recommended.

Likely discouraged by this response, Gamby never formally patented his design nor attempted to build a prototype, as far as the historical record shows.

LEONARD 1,000 HORSEPOWER HELICOPTER-TYPE AIRCRAFT (1939)

Lloyd H. Leonard was a NACA engineer employed at the Langley Memorial Aeronautical Laboratory who devoted a good portion of his life to the development of a practical VTOL tailsitter which combined the best aspects of the helicopter with those of the aircraft. His Helicopter-Type Aircraft proposal of 1939 was capable of rising vertically, leveling off in a substantially horizontal position for normal flight, after having reached the desired altitude, and descending endwise to land again in the upright position, with no runway being required. This VTOL capability permitted the lifting surface required, being employed only to sustain the craft at flying speeds, to be reduced to a minimum. The aircraft was designed to operate at high wing loading and maximum lifting efficiency.

To enable vertical takeoffs and landings, Leonard devised an improved landing gear comprising four units extending longitudinally of the fuselage at one end during flight, and capable of being spread or expanded to provide a structure with a wide base adapted to support the fuselage in the upright position. Viewing the aircraft from the bottom, the extension of the gear was not unlike the blossoming of flower petals. The landing gear also incorporated a resilient hydraulic system for cushioning the impact and absorbing the shock when coming into contact with the ground upon landing.

The Leonard Helicopter-Type Aircraft descended normally under power, using its large diameter contra-rotating propellers, controlling the speed such that the descent could be moderated as much as desired to effect a gentle landing. In the event of engine failure, the large propellers themselves, when disengaged from the engine, could slow the descent sufficiently to prevent damage or injury to the craft or its pilot. A speed reduction gearing between the engine and propeller was provided to allow the aircraft to fly at high speed horizontally and lower speeds in vertical flight.

The aircraft also had an improved manually controlled means for automatically changing the

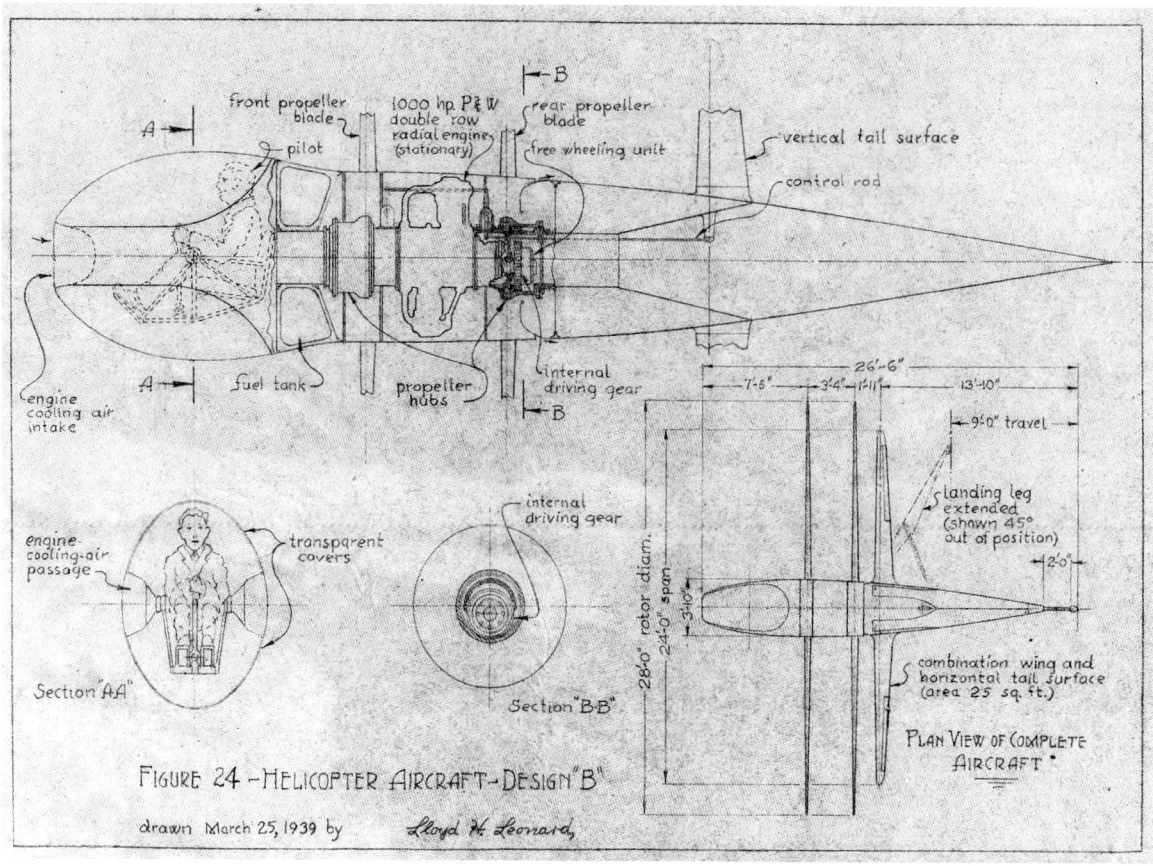

Leonard also developed a Design "B" of his Helicopter-Type Aircraft, which placed the pilot's compartment at the nose of the aircraft with an extensively glazed canopy for maximum full vision, the radial engine and propellers being located behind the pilot at mid-fuselage. Neither version of Leonard's radical VTOL tailsitter concept found favour with the Navy or the Army Air Force.

pitch of the blades at different points in the revolution of the propeller, as desired by the pilot. In this way, the blades could be caused to assume a steeper pitch at one side of the machine, for example, than at the other, thus exerting a turning moment on the plane of the rotor at the will of the pilot. Leonard envisioned a mechanism whereby both pitch and speed could be simultaneously changed by the movement of a single control member.

One configuration proposed by Leonard placed the radial engine at the front of the aircraft with a ducted spinner, the pilot's compartment being located about mid-fuselage. The cockpit had extensive glazing, with a teardrop canopy both above and below the fuselage, as well as side windows to aid the pilot during descent and landing. It had small wing and a cruciform tail, both being of high aspect ratio.

The alternate Design B placed the pilot's compartment at the nose of the aircraft with an extensively glazed canopy, the engine and propellers being located behind this compartment at mid-fuselage. This gave the pilot maximum vision at all times and when landing tail first the weight of the engine was below, rather than above, the pilot. This particular configuration only had a pair of vertical tail surfaces and a compact wing, both sets of surfaces being of high aspect ratio. It also had a nose inlet which split into two ducts that curved around the pilot and extended aft to provide cooling air to the buried radial engine. Both configurations were powered by a Pratt & Whitney radial engine of 1,000hp, likely the R-2800 Double Wasp.

Leonard's proposal was submitted to the Navy around April 1939, accompanied by an in-depth 22-page engineering analysis report. The official response letter has not been located, but surviving BuAer notes evaluating the design suggest significant skepticism. According to one BuAer analyst, the profile wing drag of contemporary navy fighter airplanes was approximately 25% of the total parasite drag. On this basis, the analyst did not see how there could be a gain in maximum speed over 10% by going to wing loadings of 200lb/sq ft.

Leonard used a low drag coefficient and made no allowances for protuberances, roughness, guns, radio, etc. There also appeared to be an error in

```
ENGINEERING ANALYSIS OF PROPOSED        EA-1
 1000 hp. HELICOPTER-TYPE AIRCRAFT
I. SPECIFICATIONS:
   Gross weight ——————————————— W = 5000 lb
   Engine power (Twin Row P.&W.) ——— P = 1000 hp.
   Power loading ——————————————— W/P = 5.0 lb/hp.
   Rotor diameter ——————————————— D = 40 ft.
   Disc area ——————————————————— S_D = 1257 sq.ft.
   Disc loading ————————————————— W/S_D = 3.98 lb/ft²
   No. of rotors ————————————————— Two in tandem
   No. rotor blades ————————————— Four (2 per rotor)
   Fuselage length ———————————— l_F = 25.0 ft.
   Fuselage max. diam. ———————— D_F = 46 inches
   Fuselage basic form ——————————— NACA 111(modified)
   Fuselage frontal area (incl. windshields) S_F = 13.5 sq.ft.
   Wing span ———————————————————— b = 20.0 ft.
   Wing area (incl. that covered by fuselage) S_W = 25.0 sq.ft.
   Wing loading ——————————————— W/S_W = 200 lb/ft²
   Aspect ratio ——————————————————— A = 16
   Taper ratio ——————————————————— c_t/c_0 = 3
   Tail area (total exposed area) = S_W/2 — S_T = 12.5 sq. ft.
   Landing gear travel ——————————— d = 6.0 ft.
   Rotor solidity (both rotors as unit) — σ = .0278

II. NOTATION:
   a, acceleration, ft.per sec. per sec.
   C_Dp, drag coeff. abs. of rotor based on disc area
   C_Dw,    "     "   "   "    "    of wing
   C_DW,T,  "     "   "   "    "    & tail as a unit
   C_DT,    "     "   "   "    tail (based on total tail area)
   C_DF,    "     "   "   "    of fuselage (based on frontal area)
   C_T = T/(ρ n² D⁴), rotor thrust coeff. (both rotors)
   J = V/nD
   P_E, maximum engine horsepower for take-off
   ρ, air density
   ρ_0 = .002378, std density at sea level
   η, rotor propulsive efficiency
   n, rotor rotational speed, rev. per sec.
   V, velocity (in ft/sec. unless otherwise stated)
```

his maximum speed calculation in the engineering analysis report. Adverse effects of increased weight, structural and mechanical complications, and stability and control problems appeared out of proportion to the maximum probable gain in speed. Another analyst believed that the propellers and pitch changing device would prove almost impossible to handle. "This is like trying to run before we can walk," he wrote.

Based on surviving correspondence, it appears that Leonard submitted the same proposal (or a variant thereof) to the USAAC on June 7 of the same year. His proposal was evaluated by Army Air Corps analysts and he received a letter back from Major F. O. Carroll, chief of the experimental engineering section of the Materiel Division of Wright Field on July 31, 1939. It was Carroll's opinion that although the type of aircraft proposed was undoubtedly in the realm of possibility, its realization required the solving of many problems before a successful aircraft could be developed.

The problem of the helicopter was, by itself, difficult to solve. Whether the proposed design offered this solution could not be properly estimated due to the lack of mechanical details furnished and due to insufficient experience with satisfactory helicopter control. Considering the Leonard proposal as a high speed aircraft flying horizontally, the question of propeller efficiency entered into consideration. It was held that the diameter of propellers sufficient to operate the aircraft as a helicopter at low air stream velocities was too large for efficient operation at high speed.

The gyroscopic effect of such large propellers would have interfered with manoeuvrability, which for a pursuit aircraft had to be exceptionally good. The effect of slipstream on the wings would have been asymmetrical and, therefore, large aileron correction and possible asymmetrical angle of incidence of the wings would have been necessary. The difficulty during the transition period from the helicopter to airplane stage could not be underestimated and could have required the development of a special control for the transition.

Special attention also had to be paid to the installation of the engine, requiring that its functioning would not be adversely affected by the transition from vertical to horizontal. The proposed combination of landing gear with tail surface mounting did not appear to be desirable either. Despite the negative evaluation of the design, Leonard's proposal documents and data were retained by the USAAC Materiel Division for future reference.

Though Leonard's Helicopter-Type Aircraft of 1939 did not find favour with the military, he continued to further refine and develop the concept, receiving various patents related to it into the 1950s. He eventually switched from a piston engine power plant to jet propulsion, specifically tip jets driving rotating wings in the manner of the Focke-Wulf Triebflügel fighter project devised during the Second World War. None of these concepts made it further than the patent stage and Leonard appears to have abandoned his pursuit of the ideal convertible aircraft configuration after 1955.

BARLING AIRCRAFT PATENT (1940)

A patent application for a VTOL tailsitter aircraft design was filed by English expat Walter H. Barling on June 18, 1940, and it was awarded on January 19, 1943, receiving no. 2,308,802.

Barling was a legitimate aeronautical engineer, his greatest claim to fame being the design of the Witteman-Lewis XNBL-1 'Barling Bomber', the largest American aircraft of the 1920s. His 1940 patent concept was much smaller but arguably more innovative, being adapted for high speed travel in normal horizontal fight and capable of taking off and landing vertically.

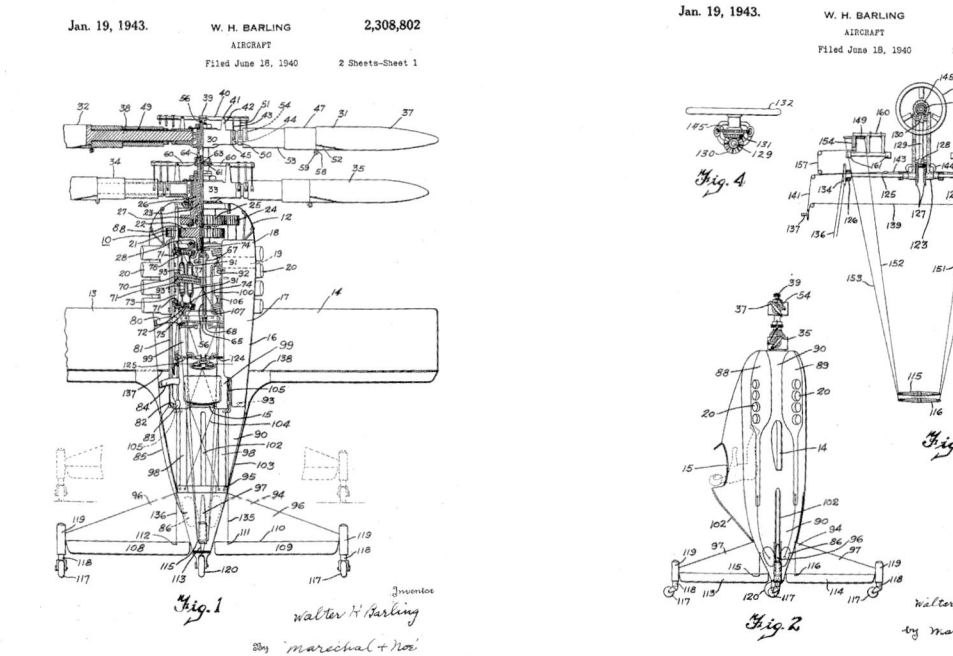

Drawings taken from the 1940 patent for a VTOL tailsitter concept designed by Walter H. Barling. This small aircraft was a departure for Barling, who was better known for designing huge bomber aircraft, specifically the Tarrant Tabor of 1919 and the Witteman-Lewis XNBL-1 of 1923.

The Barling Aircraft had a cruciform tail structure adapted to carry the main frame in a substantially upright position; it was movable axially relative to the fuselage, with a hydraulic shock-absorbing telescopic connection between tail structure and the fuselage for effectively cushioning the shock of landing.

The vehicle had compact wings with narrow chord, the span being slightly wider than the large propellers. The slipstream of the propellers flowed over almost the entire wing, which was equipped with wide ailerons to improve controllability during hovering and transitionary flight.

The aircraft was equipped with a propeller capable of efficient operation at various speeds of travel through the air, and having blade sections which were relatively adjustable to assume different pitch angles at different distances along the blade axis. It should be noted that Barling's huge contra-rotating controllable-pitch propellers were more cumbersome and less aerodynamic than similar units of the period, having a prominent drag-inducing lever arm assembly attached to the front of each unit. As with earlier designs, the contra-rotating propellers were adopted to counteract torque and improve the controllability of the aircraft during hovering flight. The diameter of the propellers, relative to the size of the aircraft, was much larger than a typical propeller but smaller than a typical helicopter rotor.

Any suitable number of engines could be employed for driving the propeller or propellers; Barling's design was powered by four engines. On the crankshaft of one of the engines was a drive gear which meshed in normal operation with a gear fixed on one of the propeller shafts; a second engine was similarly connected to the same propeller shaft, while the third and fourth engines were provided with drive gears meshing with a gear on the other propeller shaft. The blades of the upper propeller could be pivotally connected to the hub for movement about hinge axes for auto-rotational operation.

In case of failure of one of the engines, the construction was such as to permit the gears on the engine crankshafts to be moved out of mesh with the gears and/or to permit the entire release of an engine or engines which could descend to the ground independently of the body. Each engine would move out far enough by means of a bar and guide assembly so that it would fall between the adjacent tail surfaces. The end of the bar was connected by a cable to a parachute which was packed in a compartment in the empennage, and which could be pulled out automatically by the weight of the engine so that it would descend under the support of the parachute, which opened automatically.

If all the engines had been ejected, the gasoline would be dumped from the gas tanks. The propeller blade sections were then adjusted to their most efficient blade angles for auto-rotational operation, the rotation of the freely revolving propellers during descent providing adequate sustaining force for the main body, and the pilot, in view of the greatly reduced load with the engines separated from the aircraft.

Viewed in profile, the fuselage had a hunchback appearance where the pilot sat in an open cockpit well above the centre line, the general outline being reminiscent of the later XFY-1 Pogo. While Barling's proposed mechanism for ejecting the engines was elaborate, the pilot still had to bail out in the usual way — a hazardous undertaking given the tall vertical tail behind him and huge propeller in front of him. Perhaps Barling should have developed an ejection seat as well.

It seems likely that Barling tried to interest the USAAC in his unusual VTOL tailsitter, given his prior association with them in the 1920s. The US Navy was likely approached too, though no documentary evidence of his attempts to sell his design to the military has thus far been found. In any case, no financial backing was apparently secured, and the design never left the drawing board.

BELL AIRCRAFT PATENT (1941)
On January 8, 1941, Arthur M. Young of Bell Aircraft Corporation filed a patent application for an innovative VTOL tailsitter design; it was awarded on August 14, 1945, receiving patent no. 2,382,460.

The Bell Aircraft consisted of an abbreviated fuselage containing a buried engine behind the pilot. There was a drive shaft extending upwardly from the engine directly under the pilot, as with the P-39, and a gear engaging the drive shaft and two large counter-rotating coaxial propellers. It had a short wing mounted mid-fuselage beneath the propellers and lacked a traditional empennage. The retractable landing gear comprised three shock absorbing units adapted for resisting impact from the rear of the aircraft in the vertical position.

There was a means for feathering the blades of the upper propeller and a flywheel incorporated into the control system for automatically stabilizing the aircraft at any inclination selected by the pilot. A gear shift could change the propeller drive ratio, or a special gear could be used which was arranged to drive the propellers with equal torque in opposite directions, but having differential properties which permitted the speed of each propeller to vary while the algebraic sum was a constant.

There was also a pitch-changing mechanism which increased the pitch of the forward or upper propeller, while decreasing it for the rear propeller. This meant that during takeoff the forward propeller rotated at high speed and the rear at low speed, while for high forward speed both propellers had a high pitch setting and rotated at medium speed in opposite directions.

The Bell Aircraft was able to take off vertically, rising into the air as a helicopter; and having risen to sufficient height, be manoeuvred into an inclined position, in which position it would proceed, without loss of control, in part as a helicopter, in part as an airplane, in a generally horizontal direction at moderate speed; and would finally be manoeuvred into a more greatly inclined position until the main axis of the aircraft was horizontal and it was flying at high speed as an airplane, the thrust supplied by what were the lifting propellers and lift supplied by the wing.

The ability of the aircraft to rise vertically enabled the employment of a small wing, the minimum needed to maintain lift at high speed. It was able to dispense with some of the unnecessary parasitic drag with which a typical airplane was encumbered, as it was a known fact that the conventional airplane, at high speed, had more wing than was needed to supply lift, but the wing size could not be reduced due to the necessity of taking off and landing the vehicle at a reasonable speed.

The pilot's seat was inclinable to a significant degree to better position the pilot for both horizontal and vertical flight. A ventral window below the cockpit was provided for vertical takeoffs, hovering and landings. In addition, the control stick could be conveniently reoriented for operation from either position of the seat.

The Bell Aircraft featured a special gear and a novel method of changing the pitch of the propellers so that no gear shifting was necessary and pitch change was achieved with a minimum of effort. It departed from the classic method of control with conventional control surfaces, instead obtaining control from the rotating propellers, which were at all times rotating, whether in rise or descent as a helicopter, in horizontal flight as an airplane, or in a dead engine descent as an autogyro. In all of

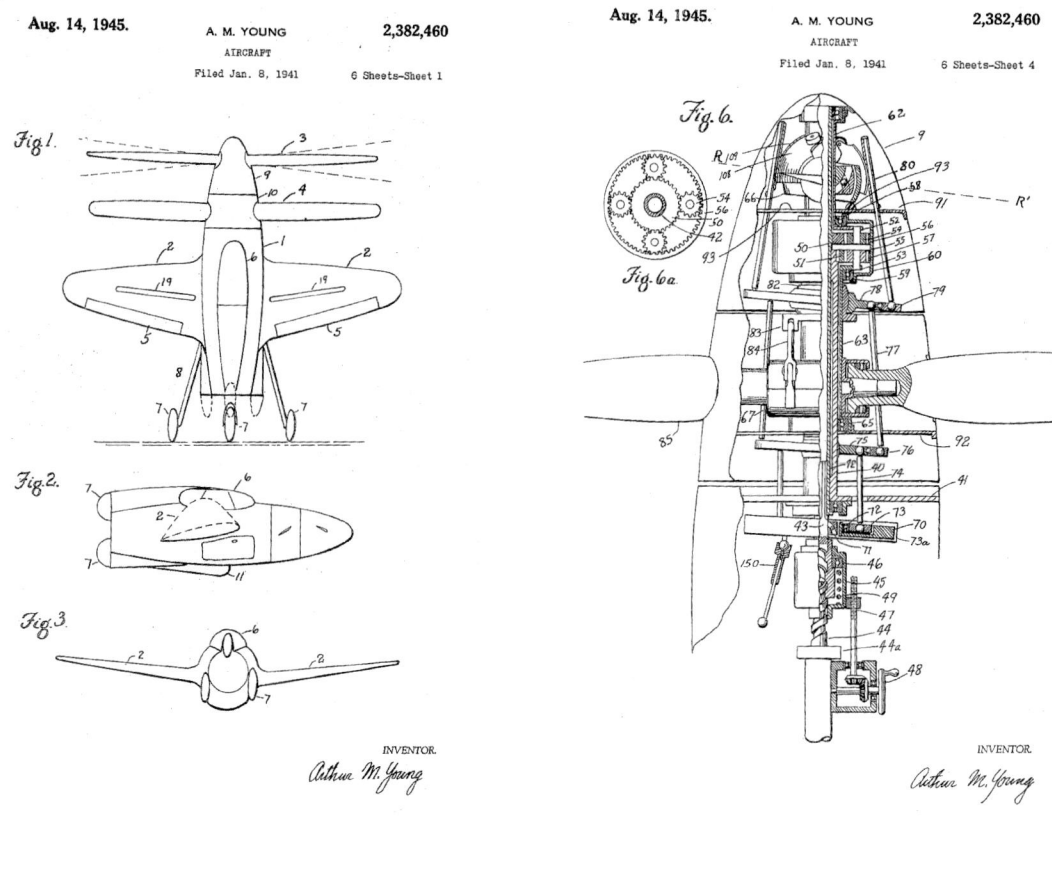

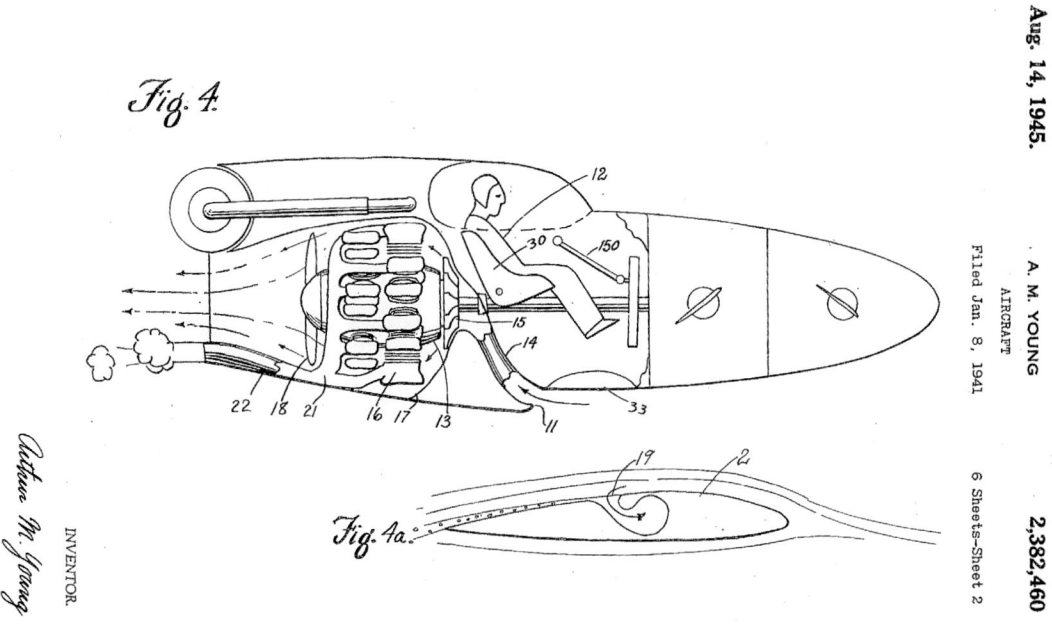

Drawings which accompanied the 1941 patent for a tailless VTOL tailsitter aircraft designed by Arthur M. Young of Bell Aircraft Corporation. A key feature of the design was the large coaxial contra-rotating propeller, where the upper unit was like a helicopter rotor, capable of feathering and inclination in any direction; it could also auto rotate in the event of an engine failure, with the vehicle descending as an autogyro. The lower unit was a more conventional controllable pitch propeller.

The patent for the Bell tailless VTOL tailsitter was awarded in 1945, with the concept being publicized in the March 1948 issue of Modern Mechanix magazine, where it was described as a "convertible jet helicopter" and depicted as a mail plane operating from the rooftops of tall buildings.

these types of flight, at least one of the coaxial rotors was maintained in a state of rapid rotation, and this lift screw provided for control of the aircraft, and, because it was always in motion with respect to the surrounding air, kept itself in condition to provide adequate control of the aircraft.

A free-wheeling device permitted the upper rotor to auto rotate in the event of engine failure. The upper rotor, being capable of feathering and inclination in any direction, was essentially a helicopter rotor. The lower one was a controllable pitch propeller used on conventional aircraft. The reason for this difference was that the forward rotor in the Bell Aircraft was used for control since it was in a position to exert greater leverage, being further from the CG of the aircraft.

The radial engine was buried behind the pilot in order to bring the aircraft's centre of gravity well back so that the wing, which was placed so that its centre of pressure was close to the centre of gravity of the entire aircraft, could be kept far back allowing clearance for the propellers and vision for the pilot. Since the machine was intended to land on its rear, it was advisable to locate the concentrated mass of the engine below the pilot, so that in the event of an unduly hard landing, the pilot would be protected from injury.

As the aircraft had no tail, it was possible to provide an exit both for the spent cooling air and the exhaust from the engine. A scoop under the fuselage took in cooling air which was conducted by a duct to a blower on to the cylinders. It then passed over the cylinders, which could be provided with baffles, then entering the chamber behind the engine which was roughly in the form of a jet, before finally being discharged into the wake of the aircraft. A fan could have been provided to further accelerate the air as it left the craft. The proportions of the various ducts, etc., were so designed that the air would leave the rear of the nacelle at a velocity somewhat greater than the maximum speed of the aircraft through the air, so that at all times this jet of air expelled from the rear of the nacelle provided some useful thrust and thereby helped offset the lost power caused by the drag of the scoop.

An alternative to the ventral scoop was an open slot located on top of each wing at approximately mid-chord, coinciding with the point at which breakdown in the airflow over the wing occurred. An opening in this position helped maintain non-turbulent flow over the surface of the wing, reducing the profile drag of the wing and increasing lift at large angles of attack.

Despite coming from an employee of a successful aircraft manufacturer, Arthur Young's innovative VTOL tailsitter never made it beyond the drawing board. The patent for the design was filed in early 1941, and with the Pearl Harbor attack occurring at the end of the year, Bell was likely compelled to concentrate on more practical designs, such as the P-39/P-63 series, in its contribution to the war effort. The concept was revived and publicized after the war in the March 1948 issue of *Modern Mechanix*, where it was described as a "convertible jet helicopter" and depicted as a mail plane by the inimitable artist Douglas Rolfe, but nothing came of it. Despite Bell's research into the VTOL tailsitter concept, as well as its pioneering work in the helicopter field, the company did not participate in the Navy's Convoy Fighter competition of 1950, perhaps due to being preoccupied with other contractual commitments.

THE BATTLE OF THE ATLANTIC AND ITS INFLUENCE ON POSTWAR NAVAL PLANNING

Switching to an entirely different topic, one of the threats which concerned US Navy planners in the early postwar period was a replay of the Battle of the Atlantic, this time against the USSR and spread out across major sea lanes extending from the Eurasian landmass.

The original Battle of the Atlantic was the longest continuous campaign of the Second World War, running from 1939 to the surrender of Nazi Germany in May 1945. It began with the Allied naval blockade of Germany, which responded with its own counter-blockade. Large convoys of Allied merchant vessels became the target of German U-boats, warships and aircraft. Over the course of the war, these convoys were protected by Allied destroyers, escort carriers, submarines and aircraft.

It took some time for the Allies to gain the upper hand, eliminating German surface raiders by the end of 1942 and largely defeating the U-boat threat by the middle of 1943. The conflict involved thousands of ships in more than 100 convoy battles and around 1,000 single-ship encounters. The final statistics at the end of the war were sobering: 2,603 Allied merchant ships were sunk, equivalent to 13.5 million tons, as well as 175 Allied naval vessels

While the majority of these losses were attributable to German surface raiders and U-boats, German aircraft claimed their fair share of merchant ships as well, especially early in the war. In particular, the Focke-Wulf Fw 200 Condor, the 'Scourge of the Atlantic', proved to be a deadly adversary. Operated by the Luftwaffe from bases in western France against British merchant ships in the Atlantic, they targeted convoy lanes west of Britain while staying outside the range of British land-based fighters. At the time, no escort carriers were available to defend the convoys against aerial attack. Not only did the Condors shadow convoys and drop bombs on numerous merchant ships, they also directed U-boat attacks on them. One unit alone, KG 40, sank more than 343,000 tons of shipping between August 1940 and February 1941.

The primary solution to the challenge posed by the Fw 200 would be sufficient escort carriers equipped with fighters such as the Grumman Martlet, but these took some time to be introduced. In the meantime, the British Admiralty came up with an innovative stopgap solution: the catapult aircraft merchant (CAM) ship, equipped with a rocket-propelled catapult launching a single Hawker Hurricane, known colloquially as the Hurricat or Catafighter. Fifty merchant ships were converted to launch these fighters.

When a Luftwaffe bomber was sighted, the Hurricane would be launched into the air with rockets and would climb to destroy or drive away the enemy aircraft. At the conclusion of the encounter, the Hurricane pilot would bail out or ditch in the ocean near the convoy for retrieval, the aircraft of course being lost in the process. During the war, there were nine combat launches of Hurricanes from CAM ships, shooting down nine German aircraft, damaging one, and chasing three away. Losses included the ditching of eight Hurricanes and the death of one pilot. CAM operations were ended in August of 1942 as sufficient numbers of escort carriers were introduced to protect the convoys travelling to Britain and the USSR.

The Battle of the Atlantic was the greatest naval campaign in history, and it is no wonder that it would weigh heavily on the minds of US Navy planners in the early postwar period. Though the Soviet Union was primarily a land power, it did have a sizable Navy equipped with surface ships, submarines, and various patrol, bomber and attack aircraft which could present a significant threat to merchant shipping. The threat became even more serious on August 29, 1949, when the USSR successfully detonated its first atomic bomb at a remote test site in Semipalatinsk, Kazakhstan. Suddenly, the prospect of a single Soviet aircraft having the capability of wiping out an entire convoy of merchant ships with one atomic bomb became a real near-term possibility. The necessity of protecting merchant shipping in the event that the cold war went hot took on a greater urgency.

The traditional solution of protecting convoys with large numbers of destroyers and escort carriers was expensive, especially given the reduced Navy budgets prior to the Korean War, and the possibility of providing merchant ships with a form of self-defence at considerably lower cost must have been appealing. Even if the self-defensive capability, such as a fighter aircraft launched from the ship, was unable to entirely destroy an enemy bomber on its own, it could provide valuable time for escort carriers to rush to the scene and launch more capable fighters to eliminate the threat.

The success of the British with their CAM ships equipped with catapult-launched Hurricane fighters showed the feasibility of the idea, though obviously the loss of the aircraft at the end of the engagement was not ideal. What was needed was an aircraft that could both launch from and land on the limited deck space of a Liberty Ship, a vehicle that functioned both as a helicopter and a fighter aircraft. Such an unconventional aircraft would be challenging to design and build, so the Navy began to research what such an aircraft would look like.

PROJECT HUMMINGBIRD
Both the Navy and Air Force funded research into the development of VTOL fighter aircraft under the Project Hummingbird designation from 1946 to 1947. The Air Force sought to develop a VTOL fighter which could be dispersed to compact and easily concealed operating bases. Recent experience during the Second World War had shown the vulnerability of traditional large airfields to aerial bombardment as well as being overrun in a Blitzkrieg-type scenario. The advent of the atomic bomb made such airfields even more vulnerable to total destruction, with the 'victor' of such a conflict likely being determined within the first 24 hours.

The Navy, on the other hand, had endured the onslaught of Japanese kamikaze attacks on its ships in the final stages of the Pacific War. It was interested in developing a practical VTOL fighter aircraft which could protect a wide variety of vessels, including LSTs, fleet oilers, tenders, transports, destroyers, large capital ships, and merchant vessels. The VTOL fighter would have been housed in a conical-type shelter aboard these ships, being deployed when an enemy aircraft was detected. The VTOL fighter would have then been guided by radar to destroy or fend off the attackers until it could be reinforced by carrier-based fighters.

Captured German reports on such VTOL projects as the Heinkel Wespe and Lerche and the Focke-Wulf Triebflügel were incorporated into this research project. A wide variety of aircraft configurations and power plants were examined, including fighters powered by rocket, turbojet and turboprop engines. Unfortunately, very few primary documents have been uncovered about Project Hummingbird, and we can only speculate about the appearance of these unusual studies. It is obvious, however, that the results of the research were positive enough for the Navy to persevere with the concept, with the VTOL turboprop tailsitter configuration seemingly having great potential.

ADVENT OF THE TURBOPROP ENGINE
The first practical turboprop engines began to emerge towards the end of the Second World War. A turboprop engine is essentially a turbine driving a propeller by means of a reduction gearbox. The first American turboprop engine was the General Electric XT31, which equipped the experimental Consolidated Vultee XP-81, the first flight of which occurred on December 21, 1945. The advantages of the turboprop engine over the piston engine are numerous; typically, the turboprop engine has a greater power-to-weight ratio and less mechanical complexity while producing greater horsepower. Compared to early turbojets, the turboprop engine

generally had lower fuel consumption, greater range, and superior performance at medium speeds and altitudes. The initial climb and takeoff performance of the turboprop is also quite remarkable.

Of particular interest to BuAer was the Allison T40 turboprop engine, the development of which began in 1944. It was composed of two Allison T38-A-1 power sections side-by-side driving a contra-rotating Aeroproducts propeller through a common reduction gearbox. The power sections were connected to the gearbox via extension shafts, each of which incorporated a clutch to allow the power sections to be run independently, permitting an aircraft to cruise on one half of the engine and only engage the second power section when it was necessary. The XT40-A-6 produced a maximum 5,525 equivalent shaft horsepower for takeoff; it weighed 2,500lb dry and had a power-to-weight ratio of 2.222hp/lb.

These characteristics were very favorable to the development of VTOL aircraft powered by a turboprop engine. The VTOL tailsitter projects presented earlier in this chapter were all powered by piston engines, which did not have the sheer horsepower and impressive power-to-weight ratio of the newer turboprop power plants. Provided the aircraft was built with lightness in mine, the takeoff and climb performance of a VTOL turboprop tailsitter could be truly impressive. The ability of the XT40 to run on just one of its two power sections also provided an extra margin of safety in case one of the power sections failed. Thus, as BuAer became more favorably disposed to the VTOL tailsitter aircraft, the Allison T40 became its preferred power plant for the concept.

It should be noted that BuAer also seriously considered the development of a vertically launched jet fighter in the early postwar period. In 1947, BuAer awarded Ryan a contract to research this type of aircraft, which was to be launched from submarines. Ryan got as far as conducting remote-controlled VTOL tethered rig tests from 1947 to 1950 as well as testing a flying rig in 1951. Around this time, the Navy lost interest in the programme, and Ryan would go on to develop the X-13 Vertijet under a USAF contract. However, this subject matter is beyond the scope of the present book.

BUREAU OF AERONAUTICS DR-72/ DR-72A VTOL TAILSITTER STUDIES

There is mention in various Convoy Fighter proposal documents of a pair of internal BuAer VTOL tailsitter studies designated as the DR-72 and DR-72A; the acronym likely stood for 'Design Research'. These mysterious designs are cited in the Goodyear and Lockheed Convoy Fighter proposal documents. A report produced by the Naval Air Development Center (NADC) Johnsville in Pennsylvania concerning these designs accompanied the Operational Specification documents which BuAer sent to those aircraft manufacturers that expressed an interest in participating in the Convoy Fighter competition.

It is likely that a powered scale model was constructed and employed in wind tunnel and possibly hovering tests. No images of these designs have thus far been discovered, but available evidence suggests that they each had a rotatable ejection seat and gun pod fairings mounted outboard of the fuselage on the wings. Hopefully further research will shed more light on these early VTOL tailsitter studies.

PREPARING THE OPERATIONAL SPECIFICATION

By the end of the 1940s, BuAer had conducted sufficient research to determine that the VTOL tailsitter fighter was a viable proposition. Turboprop engine technology had matured to the point where it seemed like an ideal power plant for such an aircraft. The type appeared ideally suited to the protection of convoys, which were quite vulnerable to enemy air attack and critical to maintaining supply lines to American allies in a future conflict with the USSR.

The desired characteristics and performance of the future VTOL turboprop tailsitter fighter were described in Operational Specification 122 (OS-122), which was distributed to various aircraft manufacturers in July 1950. The primary mission of this aircraft was to protect convoy vessels from air attack by enemy aircraft. It was a single-place aircraft capable of vertical unassisted takeoff from, and landing on, small platform areas of convoy vessels.

Some of the key requirements for the nascent 'Convoy Fighter' included a gross weight of 16,000lb; a climb time to 35,000ft of 4.5 minutes; a loiter time of two hours; a speed of 540 knots at 35,000ft; and a combat ceiling of 45,000ft. The preferred power plant was the Allison XT40-A-8 turboprop engine having 7,500 equivalent shaft horsepower and a two-speed propeller reduction gear.

OS-122 also called for the design and construction of a 0.766 reduced scale unpressurized prototype powered by the Armstrong Siddeley Double Mamba III turboprop engine. This technology demonstrator was to maintain dimensional and dynamic similarity to the full-scale Convoy Fighter aircraft as much as possible. It was to investigate vertical takeoff and landing, as well as other flight characteristics as part of the development programme for the tactical Convoy Fighter.

Five aircraft manufacturers responded favourably to BuAer's solicitation: Convair, Goodyear, Lockheed, Martin and Northrop. Their proposals are covered in the following chapters.

Convair Model 5 Convoy Fighter

Judging by the surviving documentation, Convair's Convoy Fighter proposal was relatively brief but highly effective, presenting a relatively simple delta wing design without a complicated recovery system or landing gear. The resulting light weight and reasonable cost of the design was persuasive to BuAer, landing Convair a contract that would lead to construction and testing of the XFY-1 Pogo.

It should be noted that at least two Convair Convoy Fighter preliminary designs survive, dating from the summer of 1950, preceding the official November 1, 1950 proposal by about five months. Already referred to by the 'Pogo' nickname, both studies were unconventional tailsitters which shared a three-point landing configuration that would be abandoned in favour of the four-point cruciform layout of the final proposal, designated internally as the Model 5. The following is a summary of the original proposal brochures that led to the development of one of the most unusual aircraft of the 1950s.

CLASS VF CONVOY FIGHTER PROPOSAL
Design analysis
In its proposal to BuAer, Convair noted that the design of a fighter aircraft capable of vertical takeoff and landing presented two new challenges. The first concerned the control of the airplane in vertical flight; based on BuAer model tests of this type of airplane and Convair wind tunnel tests of the proposed design, control in vertical flight was assured. A suitable autopilot was proposed for use in vertical and transition flying.

The second challenge was the two-position pilot's seat. The proposed seat was designed in accordance with BuAer fighter seat requirements and was positioned in the normal manner for the usual horizontal type of flight. For vertical takeoff and landing, the seat was rotated 45° forward in order to provide the pilot with the necessary downwards vision. Rudder controls followed the seat motion; however, the cockpit arrangement allowed the normal stick position to be used with both seat positions. The design of the cockpit provided for ejection from both seat positions.

1. The seat controls were designed to accomplish the following at the pilot's discretion: The controls could be set so that a gradual transition of the seat from the vertical takeoff position to the horizontal flight position would take place as the airplane changed attitude from vertical to horizontal flight.
2. The controls could be set so that the seat transition would be completed by the time the airplane reached a 30° attitude.
3. The controls could be set in manual which

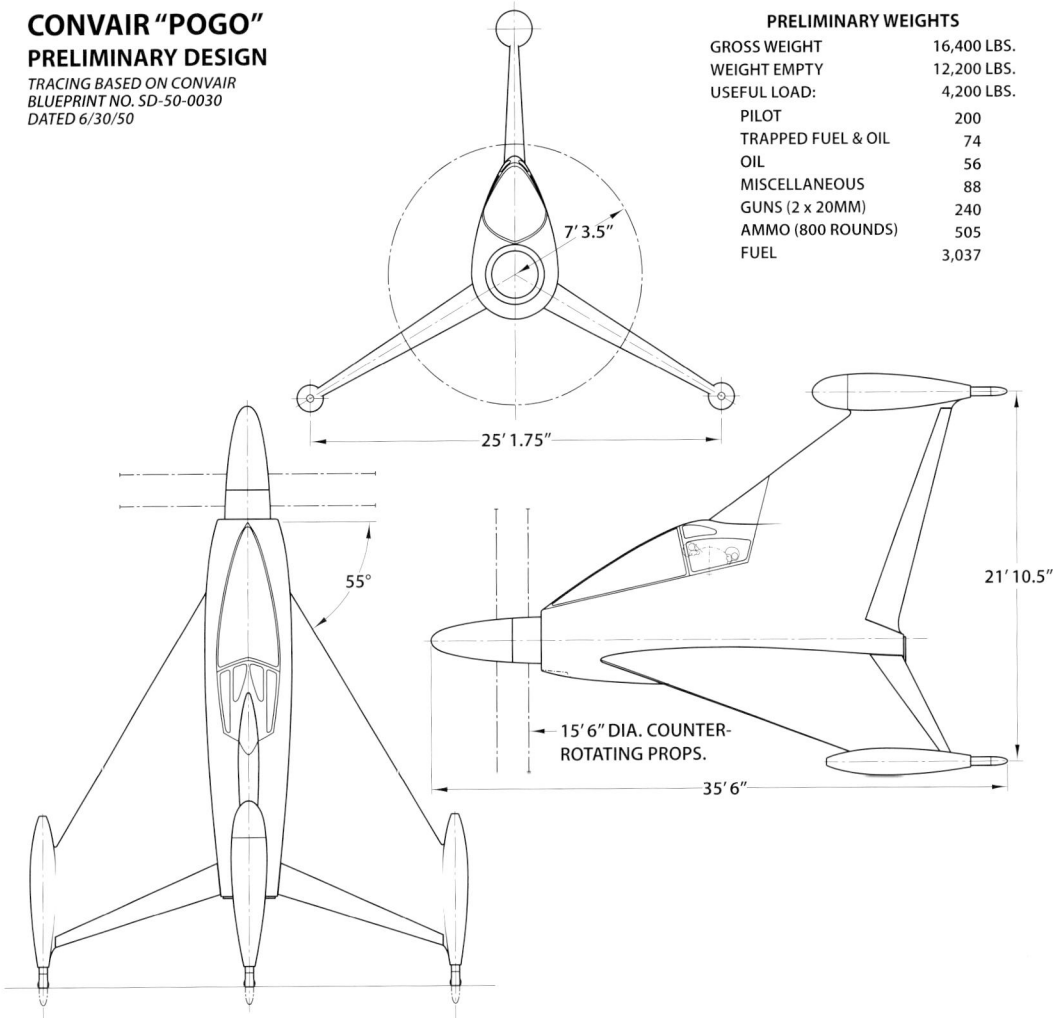

An early Convoy Fighter configuration from June 30, 1950 with three flying surfaces and substantial blending of the cockpit into the vertical fin; landings would have been especially challenging with the poor rearward visibility of this design.

allowed the pilot to place the seat in either position at any time desired.

Configuration

Preliminary studies by Convair showed that a delta wing configuration was superior in manoeuvrability and lighter in weight than a swept wing configuration when compared on an equal basis of meeting all of the performance requirements for the Convoy Fighter. The extensive work done by Convair on the XF-92A (7002) delta wing airplane including flight testing provided a reliable basis for the design as proposed.

Prototype

As requested in BuAer letter AER-AC-21 dated July 21, 1950, an unpressurized aerodynamic and dynamic scale prototype airplane (0.766 scale) using an Armstrong Siddeley Double Mamba III turboprop was designed and included in the proposal. This prototype would be used to investigate vertical takeoff and landing and other flight characteristics as part of the development programme for the tactical Convoy Fighter.

An alternative to construction of the scale prototype was to build an experimental full-size Convoy Fighter stripped of armament, electronic equipment, etc., but with the aerodynamic form maintained by mock-up armament. This experimental airplane would be powered by the existing Allison 5,525 equivalent shaft horsepower T40 engine with single speed reduction gear, modified to operate through 90°. The proposed in-service Convoy Fighter was to be powered by the advanced XT40-A-8 engine having 7,500 equivalent shaft horsepower and two-speed propeller reduction gear.

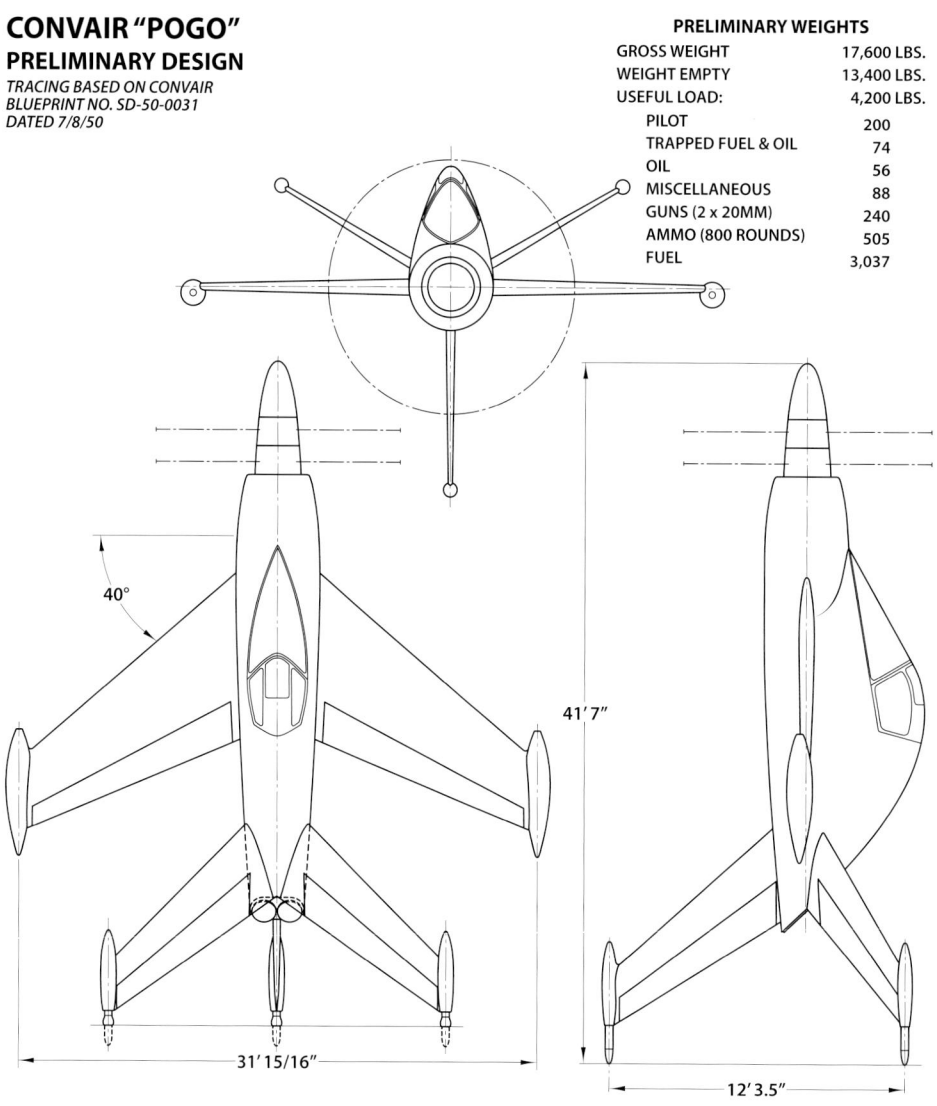

This swept wing design from July 8, 1950 featured a triple tail fin landing configuration; Convair would shift to a cruciform layout for its actual proposal, likely to improve landing stability.

The experimental Convoy Fighter had the advantage of being the same structurally and aerodynamically (except for power) as the tactical airplane and could be converted at a later date to a full tactical airplane by replacement of the power plant and installation of military equipment. It was pressurized, which allowed for testing at high altitude, thus simulating actual operation of the tactical airplane.

General description

Mission. The primary mission of this airplane was to protect convoy vessels from air attack by enemy aircraft. It was designed for vertical unassisted take-off from, and landing on, small platform areas of convoy vessels.

The Convoy Fighter was also an effective tactical weapon for marine amphibious operations. It could be used for defence of an attack force en route to the objective area followed by support of the landing operations and establishment of the beachhead using LSTs as bases. Continued support would be provided, using mobile bases ashore, as the force moved inland. This use of Convoy Fighters would release CV and CVE carriers and their aircraft for other purposes.

Configuration. This airplane had a 55° delta wing and was powered by an Allison T40-A-8 turboprop engine with a two-speed propeller reduction gear.

The one-man crew was housed in a pressurized flight compartment equipped with an ejection seat

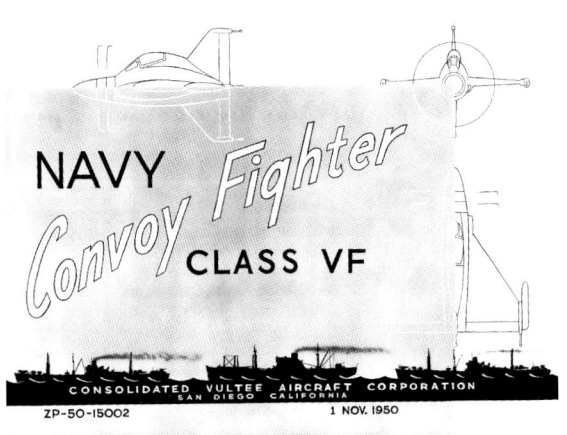

Cover to the Convair Class VF Convoy Fighter proposal brochure dated November 1, 1950.

A contemporary artist's impression of the Convair Class VF (Visual Fighter) Convoy Fighter proposal of 1950 (also known as the Model 5), which ultimately led to the XFY-1 Pogo.

built to Navy standards.

Four 20 mm fixed aircraft guns, with 150 rounds each, were installed in pairs at the wing tips. Good gun platform characteristics were achieved by the inherently high rigidity of the delta wing planform. Alternate armament installation allowed for the replacement of 20mm guns with fifty 2.75" folding fin rockets.

All fuel was contained in two integral wing tanks.

Production considerations. This design lent itself to high production assembly line methods of manufacture. Subassembly and feeder shop methods would have been used for mass production.

Landing

Landing aboard ship was accomplished from the hovering position by approaching the ship from the stern while drifting sideways with one wing pointed toward the ship. In this manoeuvre, the pilot had excellent visibility in the direction in which he was moving. If the sea was running such as to cause appreciable pitch or roll to the deck, the pilot, as he approached, unreeled a length of cable from the stern of the airplane which was caught by a man on deck, as was done when a blimp made contact. This cable was attached, by a quick connecting device on the end, to another cable that was unreeled from a winch below deck.

The winch was operated by a fluid torque converter thereby eliminating any quick jerk and, at the same time, exerting a steady pull on the cable which assured a constant rate of descent of the airplane relative to the deck. As the airplane settled down towards the deck and approached within a few feet, the pull on the cable was increased to pull the airplane down and secure it, so that neither wind nor ship movement could cause upset. Analysis showed that even with a cable pull of 1,000lb, exerted at an angle of 45° with respect to the thrust line, a rudder angle of 14.3° would balance the forces and moments in hovering flight.

Performance and aerodynamic characteristics

The basic concept of the Convoy Fighter gave rise to the following design requirements:

1. Extremely light weight.
2. Adequate stability and control at all angles of attack up to 90°.
3. Compact size and high structural rigidity.
4. Low drag at transonic speeds.
5. Freedom from buffeting at all speeds.

Extensive wind tunnel and flight experience with the XF-92A (7002) delta wing airplane led to early investigation of this type as an answer to the design requirements.

This experience demonstrated the following desirable characteristics:

1. Freedom from a sharply defined stall wherein lift and rolling control were suddenly lost at high angles of attack.
2. Low drag at transonic speed.
3. Freedom from buffeting.

In addition, the structural form of the configuration was ideal for light weight, rigidity and compactness.

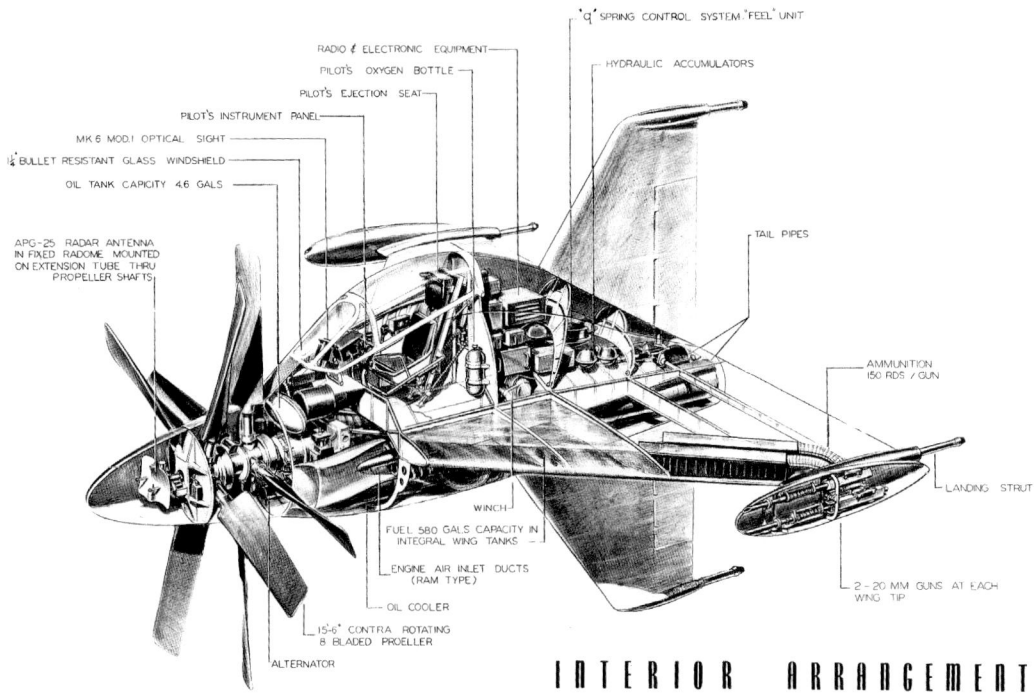

Perspective interior arrangement of the Convair Convoy Fighter revealing how its compact fuselage was tightly packed with equipment.

Design studies and wind tunnel tests of a 1/10 scale model of the Convoy Fighter in the Convair 8 x 12ft wind tunnel further confirmed this choice. It was shown that the airplane experienced uninterrupted and essentially constant positive stability in pitch up to 120° angle of attack, the maximum test angle. Directional stability to 120° angle of yaw was also retained. Positive control effectiveness in pitch and yaw was evidenced without the aid of slipstream velocity. With power on, the major part of the control surfaces remained in the slipstream and control effectiveness was not impaired by extreme attitudes.

The rigidity and compactness of the delta planform configuration for the wing and vertical tail afforded the further advantage of allowing the airplane to be flown directly to its landing platform, rather than being captured and supported by an external structure. This rigidity and the accompanying small overall dimensions facilitated handling under adverse conditions.

A 9% thick NACA 63-009 wing section (with cusp removed) was selected on the basis of good high speed performance and adequate thickness for structural rigidity and space for fuel. A leading edge sweep of 55° was selected from considerations of stability and drag at high angles of attack, and low structural weight.

Longitudinal and lateral control were obtained by the use of short-chord, full-span elevons located at the trailing edge of the wing. Conventional rudder control was provided in both the upper and lower portions of the vertical tail. All controls were power operated. For the conditions of hovering flight, an autopilot maintained stability about any given angular attitude selected by the pilot.

Wind tunnel test
Tests of the 1/10 scale powered model of the Convoy Fighter made in the Convair 8 x 12ft wind tunnel obtained basic stability and control characteristics. The model was to scale except that no airflow through the ducts was simulated and six blade propellers were used instead of eight blade as these were readily available.

As previously indicated by extensive wind tunnel and flight tests of the XF-92A (7002), it was possible to develop a configuration, after a number of changes, having satisfactory stability and control about all axes up to 90° angle of attack, thus ensuring satisfactory transition between the vertical and horizontal flight attitudes.

Test data were obtained for the proposed design through angle of attack and angle of yaw ranges from 0° to 120°. Of particular interest was the fact that pitching moment and yawing movement curves

SEAT ADJUSTMENT FOR TAKE-OFF & LANDING

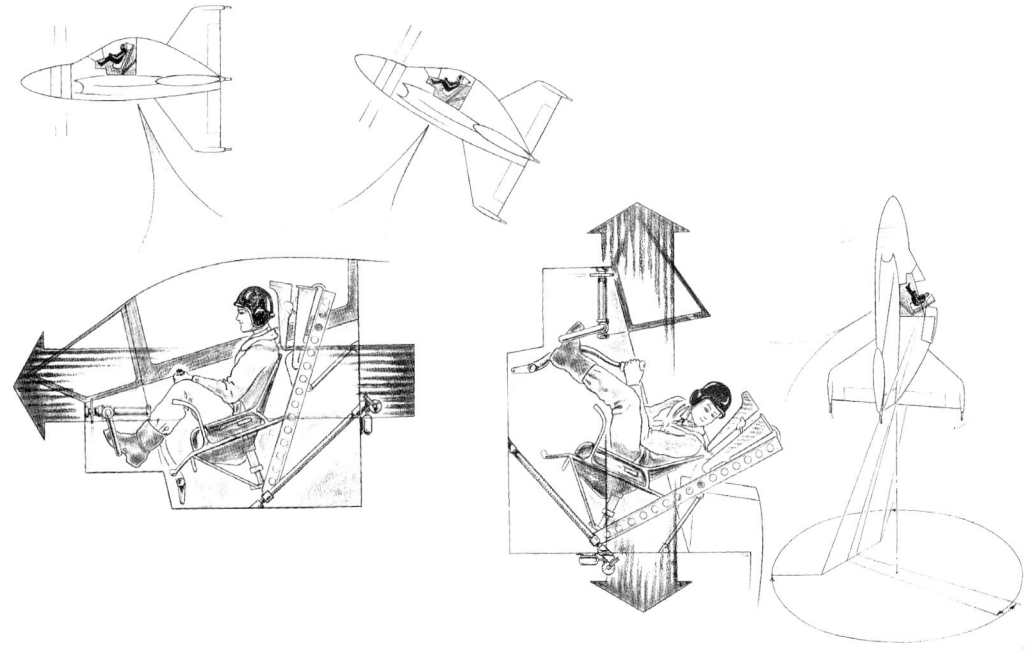

Illustration of the pilot's seat, which could rotate 45° forward during vertical takeoffs and landings to provide the pilot additional downwards visibility.

TAKE-OFF

The Convair Convoy Fighter was designed for vertical unassisted takeoffs from a small platform on a convoy vessel.

LANDING

When landing aboard ship in rough seas, the pilot could unreel a cable from the stern of the airplane and have the vehicle slowly winched downwards to achieve safe contact with the platform.

were stable and uniform with no erratic breaks throughout the entire 120° range in pitch and yaw respectively.

Stress analysis and weights

The unique structural design of the Convoy Fighter provided ease of manufacture, low weight and efficient maintenance. In this design, the delta wing carried all the major loads such as power plant, pilot's compartment, equipment and fin reactions. The fuselage was therefore essentially non-structural, which permitted the extensive use of large, quickly openable doors and removable panels.

The delta planform of the wing and vertical surfaces was particularly suitable for installation of vertical takeoff and landing gear. Four oleo struts were mounted in faired pods at the tips of the surfaces. The inherent rigidity of the delta planform resulted in a minimum of added structural weight to take landing gear loads.

CLASS VF AIRPLANE PROTOTYPE FOR CONVOY FIGHTER

As per the original OS-122 requirement, Convair designed and submitted a proposal for 0.766 scale prototype of the Convoy Fighter to serve as a technology demonstrator. This prototype was intended for use in the investigation of vertical takeoff and landing and other flight characteristics as part of the development programme for the Convoy Fighter.

The prototype airplane was a one-place, flyable, unpressurized dimensionally and dynamically similar 0.766 scale prototype of the Convoy Fighter. This airplane was capable of vertical, unassisted takeoffs from and landings on small platform areas.

The airplane had a delta wing and both an upper and lower vertical tail surface, also of delta configuration. Control surfaces were power operated. The airplane was provided with a Navy standard ejection seat which could be rotated through 45° for vision in the vertical attitude.

Alighting gear was provided on the tips of the wing and vertical tail surfaces. An auxiliary, conventional landing gear was provided for test purposes. The airplane was powered by a dual rotating propeller driven by a Armstrong Siddeley Double Mamba III turboprop engine.

Like the other manufacturers participating in the Convoy Fighter competition, Convair advised BuAer to save money by forgoing the construction of the 0.766 scale prototype and building instead a stripped version of the full-scale Convoy Fighter with a less powerful turboprop engine.

STRIPPED DOWN CONVOY FIGHTER

This experimental airplane was powered by the existing Allison 5,525 equivalent shaft horsepower T40 turboprop engine with single speed reduction gear, modified to operate through 90°. The proposed Convoy Fighter was powered by the advanced XT40-A-8 turboprop engine having 7,500 equivalent shaft horsepower and two-speed propeller reduction gear.

This experimental Convoy Fighter had the advantage of being the same structurally and aerodynamically (except for power) as the tactical airplane and

ARMING & REFUELING

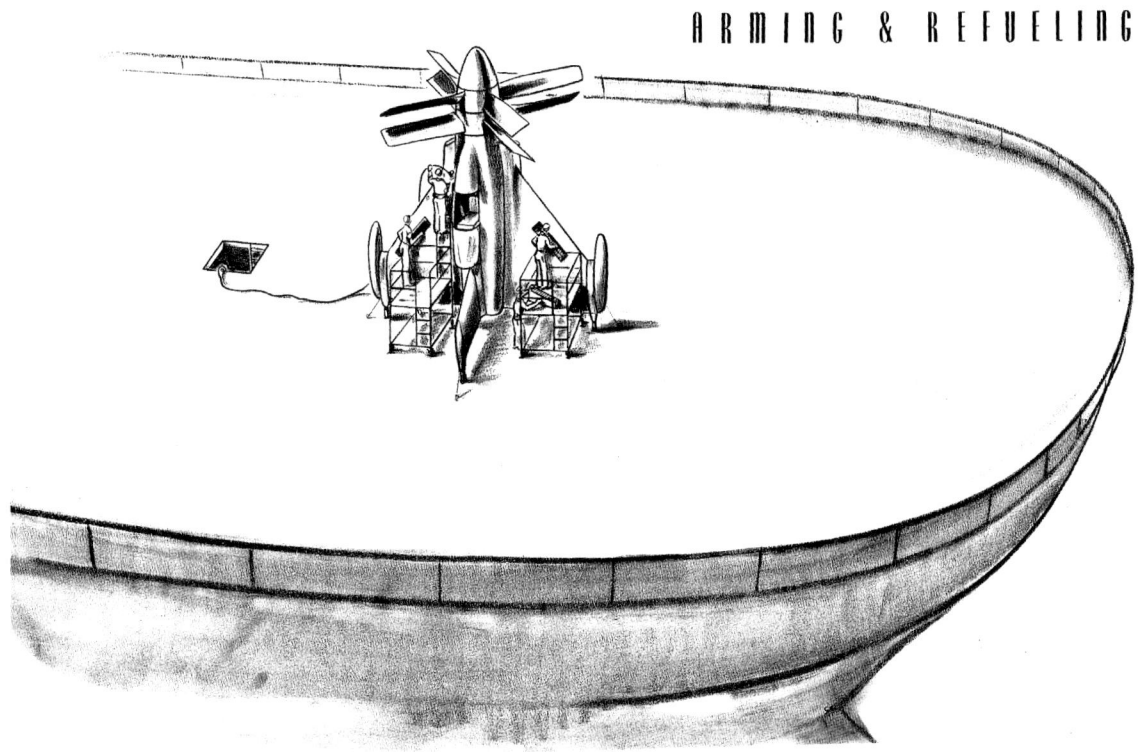

The Convoy Fighter could be refuelled and rearmed with a minimum of support equipment aboard its home vessel.

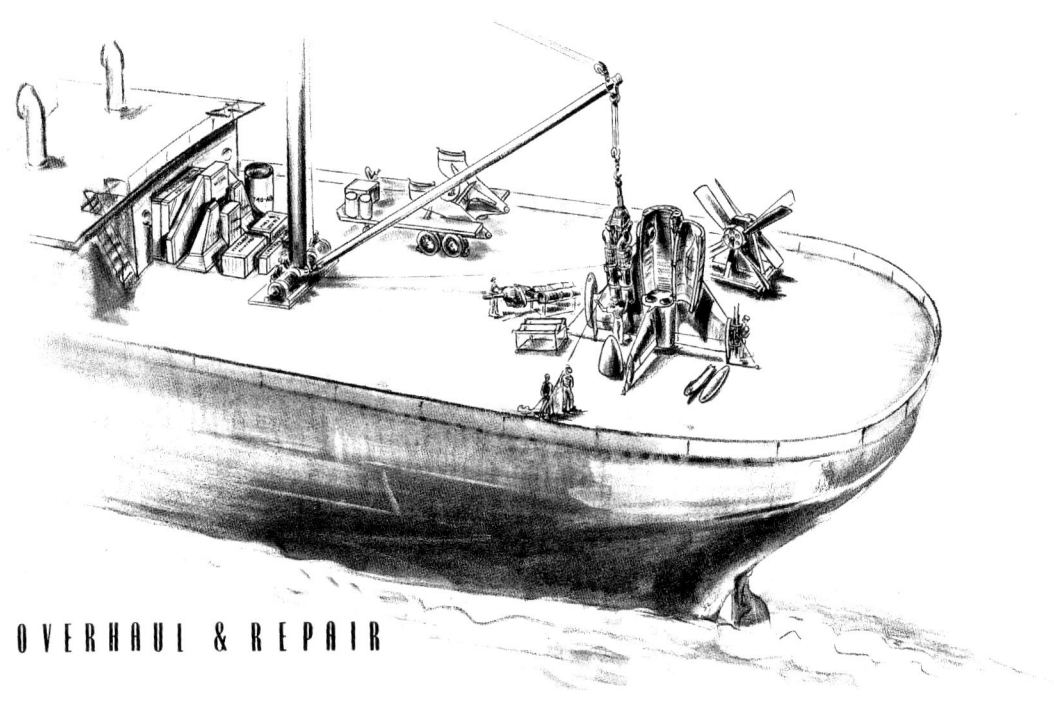

OVERHAUL & REPAIR

A shipboard cargo winch being used in the overhaul and repair of the Convair Class VF Convoy Fighter.

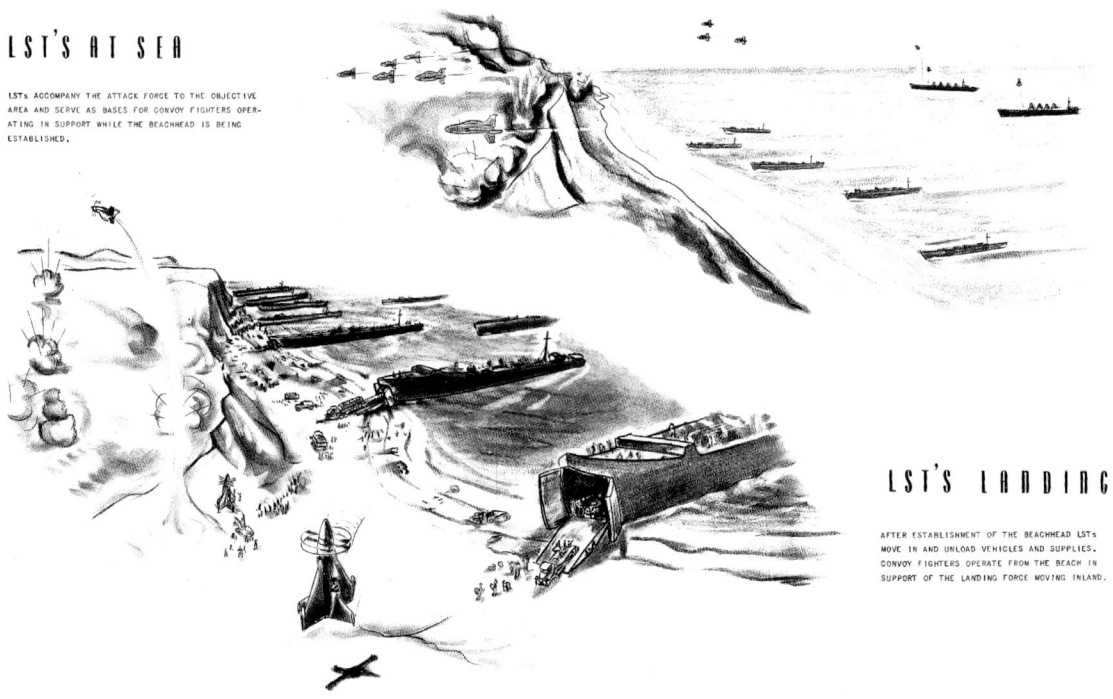

The Convoy Fighter was also foreseen as a tactical weapon in support of marine amphibious operations using LSTs as bases.

As the marines moved inland, Convoy Fighters could use mobile bases ashore to provide continued tactical support.

PERFORMANCE SUMMARY

PERFORMANCE DATA*

TAKE-OFF GROSS WEIGHT	LB	16,000
FUEL	LB	2,950
BASIC FLIGHT DESIGN GROSS WEIGHT (TAKE-OFF GROSS WEIGHT MINUS 40% FUEL)	LB	14,820
DESIGN LANDING GROSS WEIGHT (TAKE-OFF WEIGHT MINUS 60% FUEL)	LB	14,230
HIGH SPEED, BASIC FLIGHT DESIGN GROSS WEIGHT		
@ 35000 FT. ALTITUDE	KNOTS	542
@ SEA LEVEL	KNOTS	542
COMBAT RADIUS (BASIC MISSION WITH LOITER (1.53 HR)*, CRUISE OUT, COMBAT (3 MIN) AND CRUISE BACK AT 35000 FT)	N.MI.	100
AVERAGE CRUISING AIRSPEED		
OUT TO COMBAT AT MILITARY POWER (INCLUDING ACCELERATION PERIOD)	KNOTS	535
BACK FROM COMBAT ONE POWER UNIT INOPERATIVE	KNOTS	401
RATE OF CLIMB AT SEA LEVEL		
@ TAKE-OFF GROSS WEIGHT	FT/MIN	12,820
@ BASIC FLIGHT DESIGN GROSS WEIGHT	FT/MIN	14,100
TIME TO CLIMB TO 35000 FT ALTITUDE (FROM STANDING START)	MIN	4.48
RATE OF CLIMB AT 35000 FT ALTITUDE AT BASIC FLIGHT DESIGN GROSS WEIGHT	FT/MIN	5,100
COMBAT CEILING AT BASIC FLIGHT DESIGN GROSS WEIGHT (500 FT/MIN RC)	FT	46,000
FERRY RANGE (AT AVERAGE SPEED = 401 KNOTS) (16,530 LB. T.O. WT., 4480 LB. FUEL)	N.MI.	1492

*FUEL CAPACITY IS PROVIDED FOR 2 HOURS LOITER WITH A T.O. G.W. OF 16,455 LB.

** ALL PERFORMANCE AT MILITARY POWER UNLESS OTHERWISE NOTED

AIRCRAFT DIMENSIONAL DATA

WING		
TOTAL AREA	SQ.FT.	346
SPAN	FT. IN.	25'8"
ROOT CHORD	FT. IN.	22'8"
MEAN AERODYNAMIC CHORD	FT. IN.	15'7"
AIRFOIL SECTION	NACA	63-009(MOD.)
WING INCIDENCE AT ROOT	DEG.	0
AERODYNAMIC WASHOUT	DEG.	0
DIHEDRAL	DEG.	0
SWEEPBACK (LEADING EDGE)	DEG.	55
ASPECT RATIO		1.9
TAPER RATIO		5.23
VERTICAL FIN		
TOTAL AREA	SQ.FT.	150.6
SPAN	FT. IN.	19'4"
AIRFOIL SECTION	NACA	63-006.5,009(MOD.)
SWEEPBACK (LEADING EDGE)	DEG.	40
ASPECT RATIO		2.47
TAPER RATIO		3.15
FUSELAGE		
LENGTH	FT. IN.	29'5"
WIDTH (MAXIMUM)	FT. IN.	5'0"
DEPTH (MAXIMUM)	FT. IN.	8'10"

POWER PLANT

UNIT	ALLISON NAVY MODEL XT40-A-8 TURBO-PROP ENGINE
SPEC	ALLISON DIVISION OF GENERAL MOTORS CORPORATION SPEC #272-B REVISED 5-31-50
PROPELLER	8 BLADE DUAL ROTATING - 15.5 FT DIA., AF = 150 DESIGN C_{L} = 0.35 GEAR RATIOS = 13.65:1, 23.80:1

ENGINE STATIC SEA LEVEL RATINGS

CONDITION	RPM	PROP SHP	JET THRUST (LB)
TAKE-OFF	15700	6825	1685
MILITARY	14300	6955	1363
NORMAL (100%)	14000	5790	1225

Performance summary and physical characteristics of the Convair Class VF Convoy Fighter proposal.

could be converted at a later date to a full tactical airplane by replacement of power plant and installation of military equipment. It was pressurized, allowing for testing at high altitude, thus simulating actual operation of the tactical airplane.

The overall programme for development of the Convoy Fighter could be materially shortened by designing and manufacturing this stripped down version of the Convoy Fighter rather than designing and manufacturing two different airplanes.

Preliminary hovering tests
A study was made of the practicability of conducting preliminary hovering flight tests with the airplane restrained. This type of testing allowed the pilot to become familiar with hovering control prior to actual takeoff. The restraining cables were controlled by three operators, one for restraint in the vertical sense, one for pitch and one for roll. This test could be conducted in any hangar with a minimum opening of 50 x 50ft with an overhead truss strong enough to withstand a 12,000lb applied load based on suitable factors of safety. This test was based on the assumption that suitable support of the airplane could be obtained at the end of the fixed propeller shaft.

CORRESPONDENCE & COST PROPOSAL
In a letter accompanying the proposal documents dated November 14, 1950, Convair noted that emphasis was placed on experimental shop methods in order to expedite construction of the prototype Convoy Fighter. It was felt that the shortest elapsed time for manufacturing was of prime importance, in order to give maximum time for test flying and early construction of the tactical airplane.

In the design of Convoy Fighter, emphasis was placed on design for high production, thus facilitating the use of feeder shop methods for subassemblies. Special attention was given to the mating problems of major assemblies.

Extensive study was given to ease of service and maintenance, particularly as the Convoy Fighter had inherent advantages as a Marine Corps support airplane wherein it would be operated from advanced areas. Outstanding features of the proposed designs were:

1. Satisfactory stability and control in vertical flight and in the transition to horizontal flight based on powered wind tunnel tests of the proposed design conducted by Convair.
2. Complete freedom from buffet at all angles of attack based on wind tunnel tests and on flight tests of the XF-92A (7002) delta wing airplane of similar configuration.
3. Inherent structural rigidity of the delta wing planform resulting in suitable wing tip installations of armament completely

WEIGHT SUMMARY

WEIGHT EMPTY		11,785 LB
WING GROUP	1,574	
TAIL GROUP	426	
FUSELAGE	820	
ALIGHTING GEAR	300	
ENGINE	3,160	
ENGINE ACCESSORIES	290	
POWER PLANT CONTROLS	30	
PROPELLERS	2,105	
STARTING SYSTEM	30	
LUBRICATING SYSTEM	110	
FUEL SYSTEM	205	
INSTRUMENTS	50	
SURFACE CONTROLS	480	
HYDRAULIC SYSTEM	260	
ELECTRICAL SYSTEM	415	
ELECTRONICS	650	
ARMAMENT PROVISIONS	605	
FURNISHINGS	220	
AUXILIARY GEAR	55	
USEFUL LOAD		4,215 LB
PILOT	200	
FUEL	2,950	
OIL	80	
TRAPPED FUEL AND OIL	75	
GUNS (4) 20 MM	450	
AMMUNITION 600 ROUNDS	408	
EQUIPMENT	52	
GROSS WEIGHT		16,000 LB

Convair's weight summary put the design at a gross weight of 16,000lb; BuAer's analysis put it higher at 16,724lb.

outside the propeller plane.
4. Compactness of design which was ideal for shipboard operation.
5. High ground stability (against overturn) through installation of alighting gear at the vertical surface tips and at the wing tips resulting in maximum operational flexibility.

Convair emphasized its proposal for a stripped down version of the Convoy Fighter powered by the existing 5,525 equivalent shaft horsepower Allison T40 engine with single speed reduction gear and modified as required to operate through 90°. This airplane had the advantage of being the same structurally and aerodynamically (except for power) as the tactical airplane and could therefore have been converted to a tactical airplane by replacement of the power plant and installation of military equipment. Immediate design and construction of this full-scale stripped down Convoy Fighter, in lieu of a 0.766 scale prototype, effectively reduced overall cost and elapsed time of the entire Convoy Fighter programme.

As required by BuAer, Convair guaranteed the following:

1. Weight empty within 2%.
2. Takeoff gross weight as shown in the proposal, with the provision that the weight may vary, within the limiting requirement of a 5ft/sec^2 minimum acceleration during transition from vertical takeoff to normal climbing flight.
3. Performance as shown in the proposal, without tolerance.

The cost proposal followed on November 29, 1950. In its letter to BuAer, Convair noted that the Navy had requested that the Convoy Fighter proposal be prepared in two parts, Part 1 covering two scale prototype airplanes and flight test thereof and Part 2 covering two full size experimental Convoy Fighters, one static test article and flight testing.

It was understood that Part 2 would be initiated only after the flight tests of the scale prototype indicated soundness of design. Convair prepared its proposal on these bases and in addition presented a Part 3 as an alternate to Parts 1 and 2. The proposed Part 3 contemplated the construction of two airplanes and a static test article. The first of the two airplanes would be a stripped down version of the tactical article utilizing the existing Allison T-40 power plant. The airplane would be structurally and aerodynamically the same as the tactical article and would be converted later to the full tactical airplane by the addition of military equipment and replacement of the power plant.

It was believed the alternate Part 3 programme complied with the intent of the invitation and offered definite advantages from overall cost and schedule considerations.

The estimated cost for Part 1 was $2,643,511.60 and for Part 2 was $4,265,792.60, or a total of $6,909,304.20. The estimated cost for the alternate or Part 3 was $4,756,741.40. Assuming authorization to proceed was received during January 1951, the following first flight dates were established:

- Part 1 Scale Prototype: First Airplane—May 1952, Second Airplane—July 1952
- Part 2 Experimental Convoy Fighter: First Airplane—April 1954, Second Airplane—August 1954
- Part 3 (Alternate): First Airplane—August 1952, Second Airplane—December 1952

Based upon the estimated costs, Convair was prepared at any time to negotiate a fixed price or cost plus fixed fee contract. For this purpose the

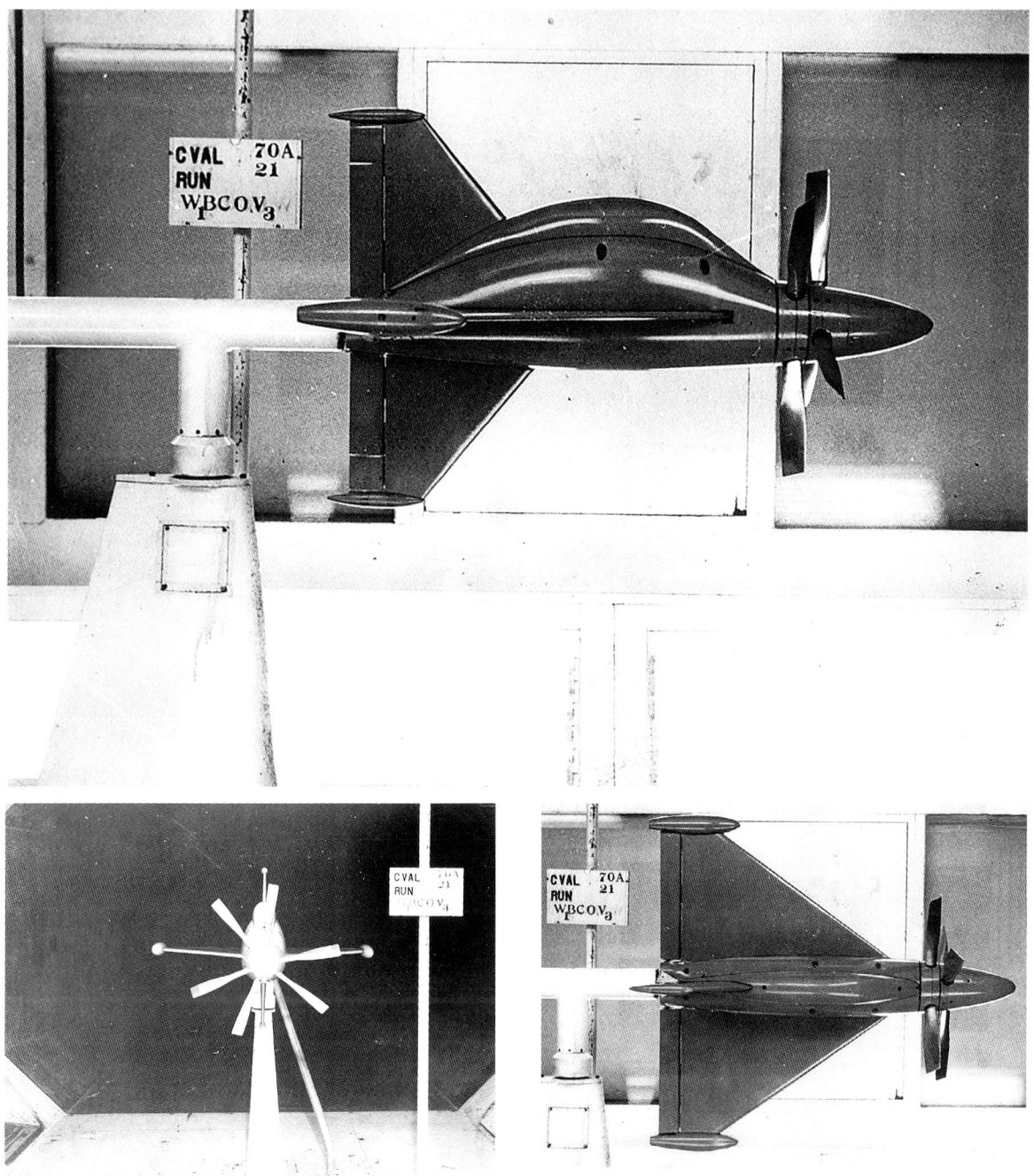

Photos of the original Convair Class VF Convoy Fighter model in the wind tunnel. Six blade propellers were substituted for eight blade units as they were more readily available.

company anticipated a 10% profit factor for a fixed price contract or a 6% fee factor for a cost plus fixed fee contract.

CONTRACT AWARD

Convair was awarded a contract in May 1951 to build and test its Convoy Fighter design, which would be designated as the XFY-1 Pogo. Lockheed and Martin were virtually tied for third, with the edge being given to Lockheed; that design would be built as the XFV-1. BuAer's likely rationale in selecting both Convair and Lockheed's proposals for further development is covered in Chapter 7, along with some details of the XFY-1's preliminary development as well as a summary of its modestly successful flight testing programme and subsequent cancellation.

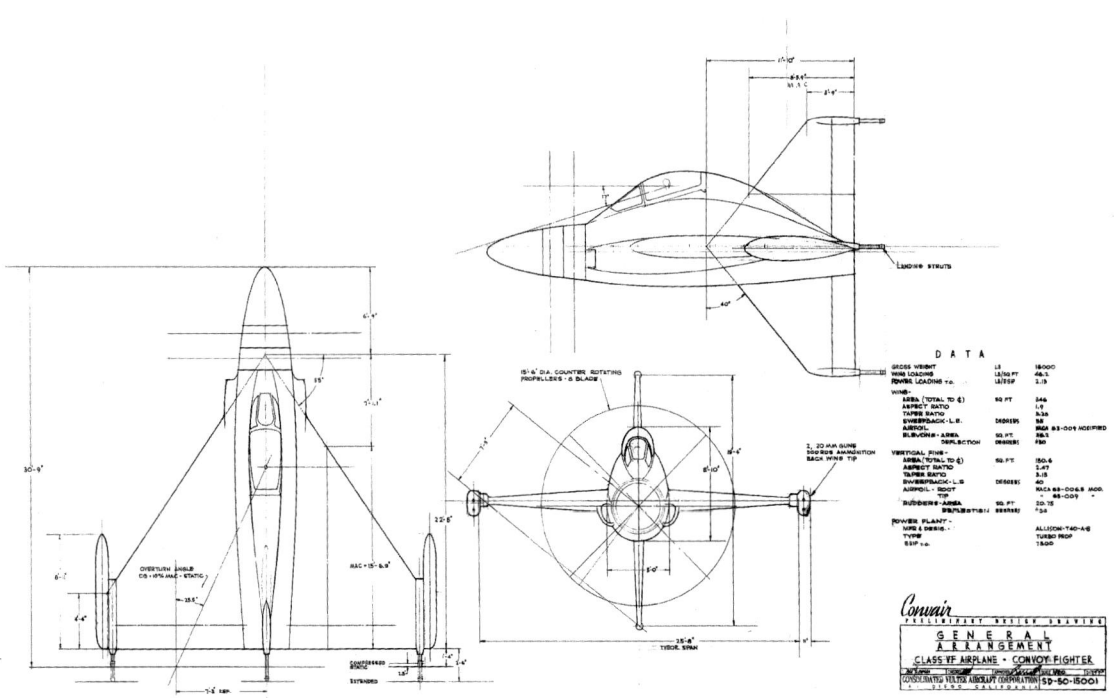

General arrangement drawing of the Convair Class VF Convoy Fighter.

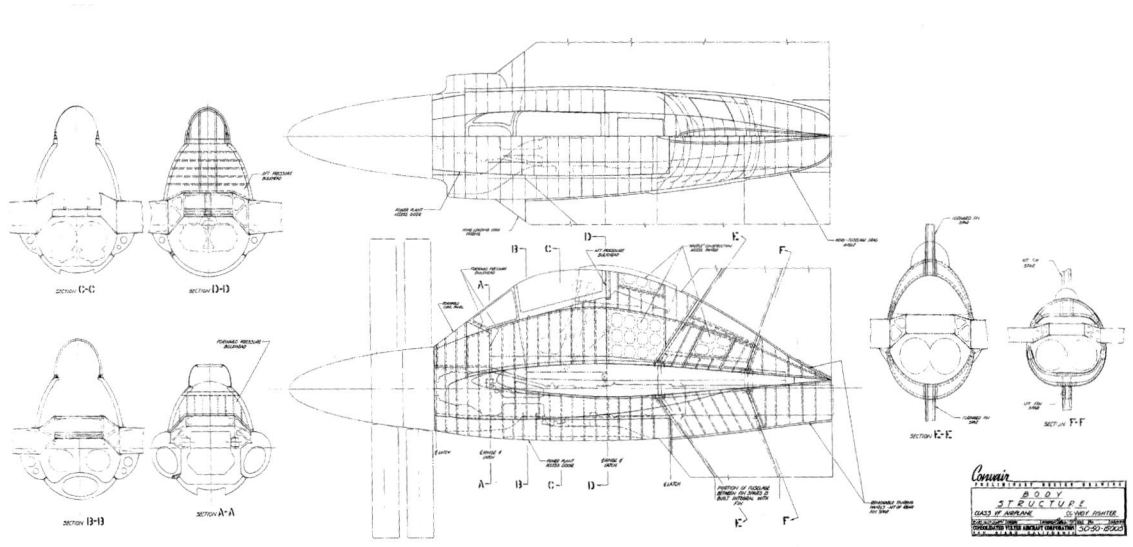

Diagram of the Convoy Fighter's body structure. With the wing carrying the major loads, the fuselage was essentially non-structural, permitting the extensive use of large, quick opening doors and removable panels.

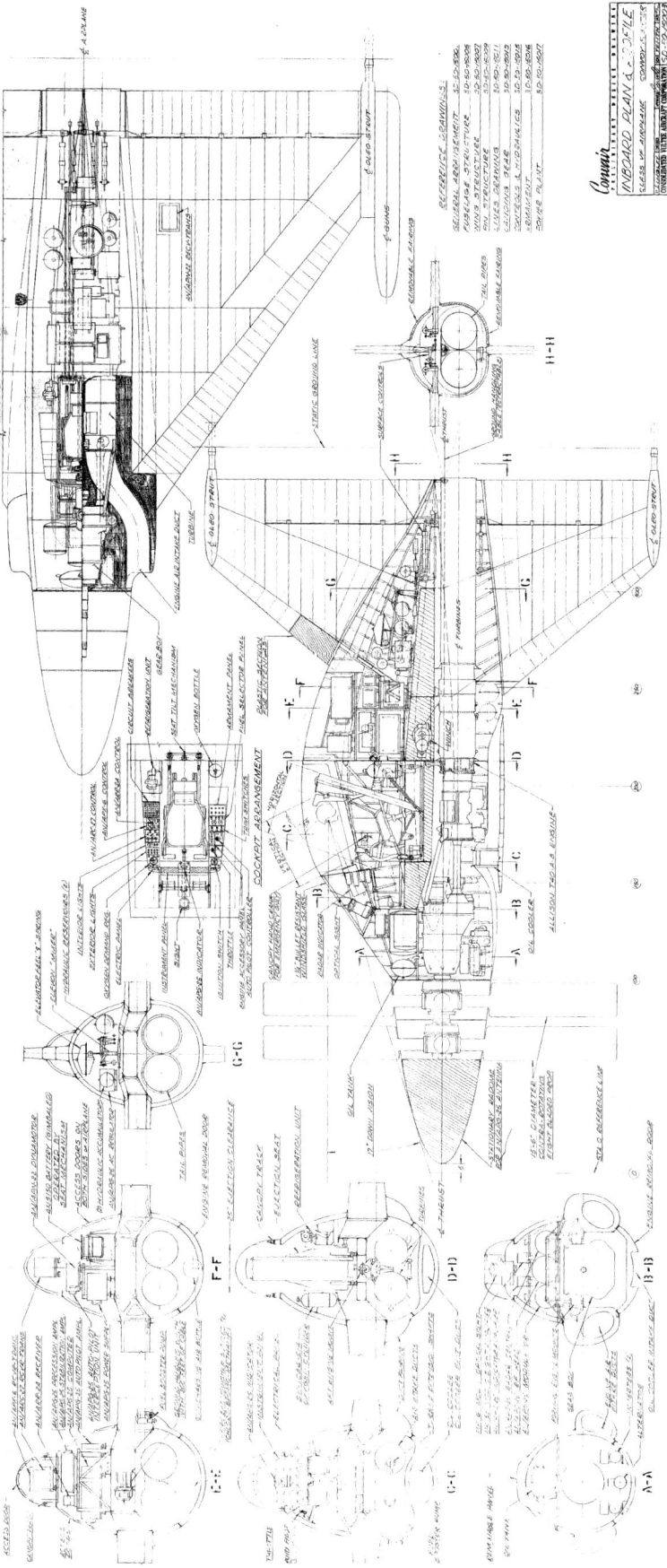

Inboard profile and cross sections of Convair's VTOL tailsitter.

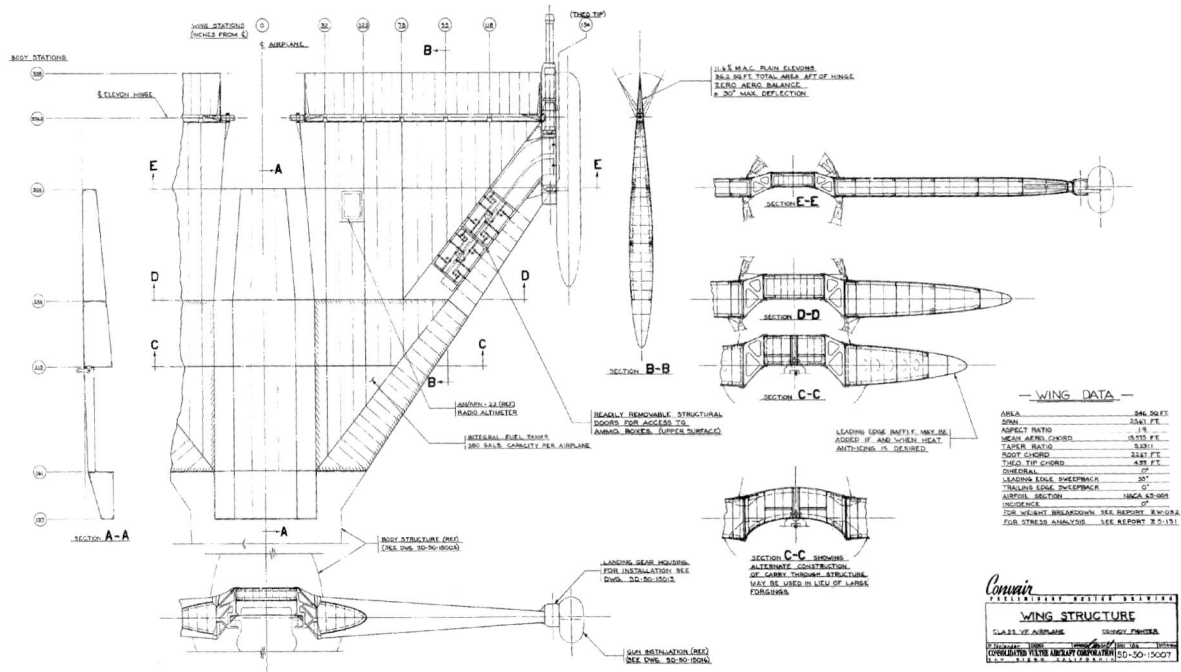

Schematic of the wing structure, which carried all the major loads, including the power plant, pilot's compartment, equipment and fin reactions.

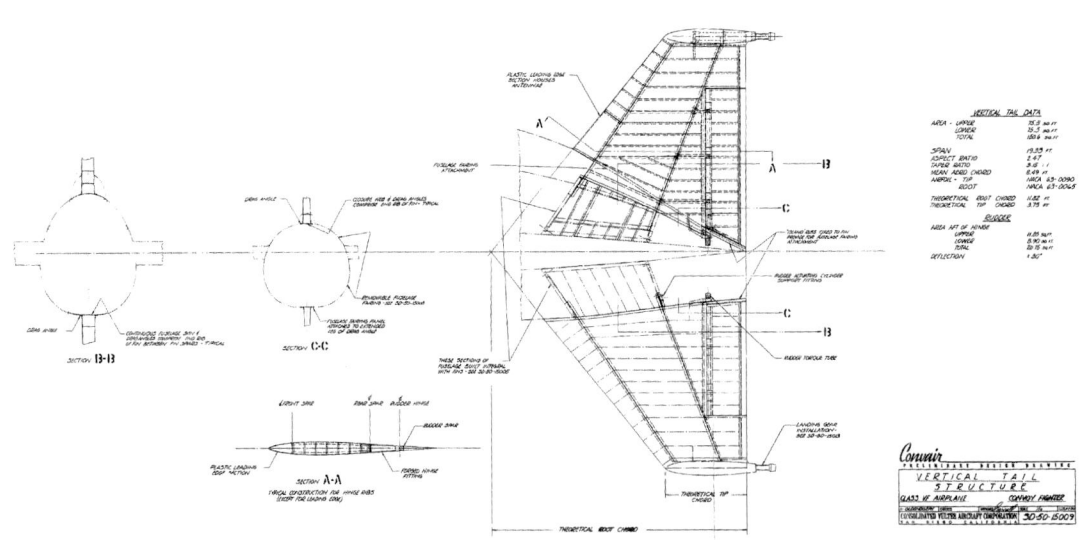

Blueprint of the vertical tail structure. Like the wing, the tail fins were of delta planform and had an inherent rigidity, resulting in a minimum of added structural weight to absorb the landing gear loads.

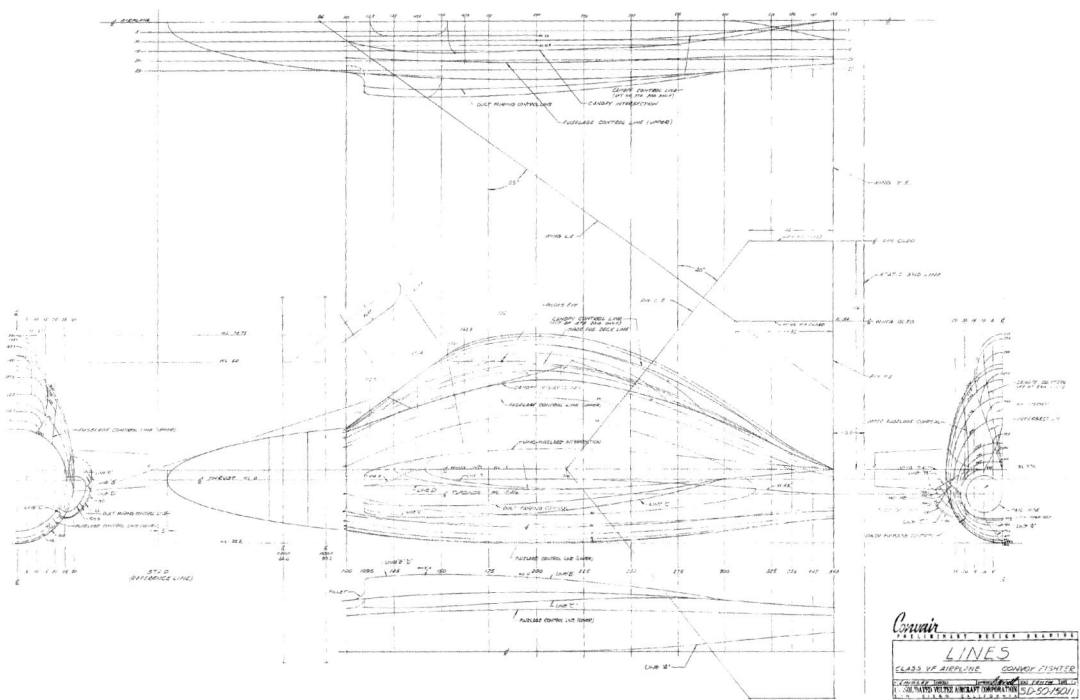

Fuselage loft lines of the Convair Class VF Convoy Fighter.

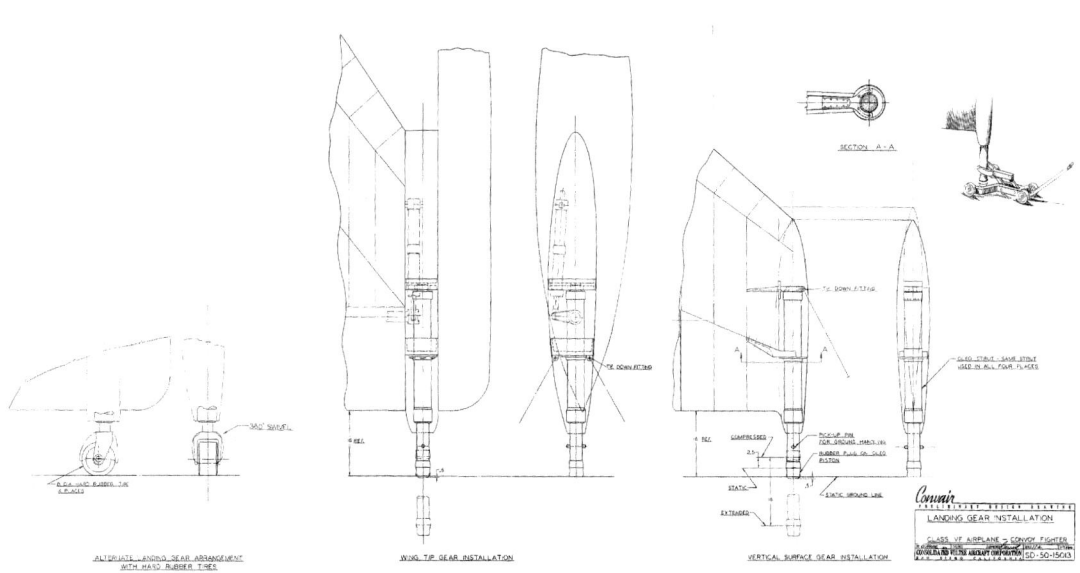

The landing gear installation consisted of four oleo struts mounted in faired pods at the tips of the wing and tail surfaces. An alternate landing gear with hard rubber tyres is shown at the far left; this was what was actually used on the prototype XFY-1.

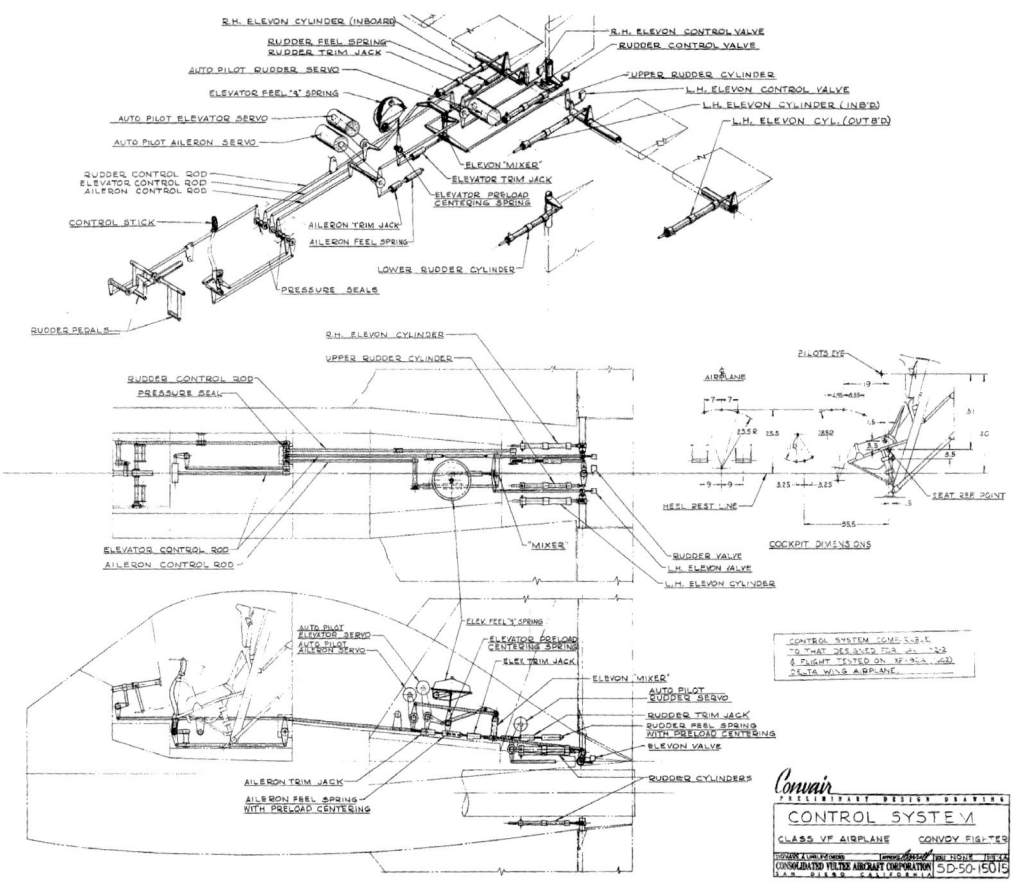

Diagram of the control system; all controls were power operated.

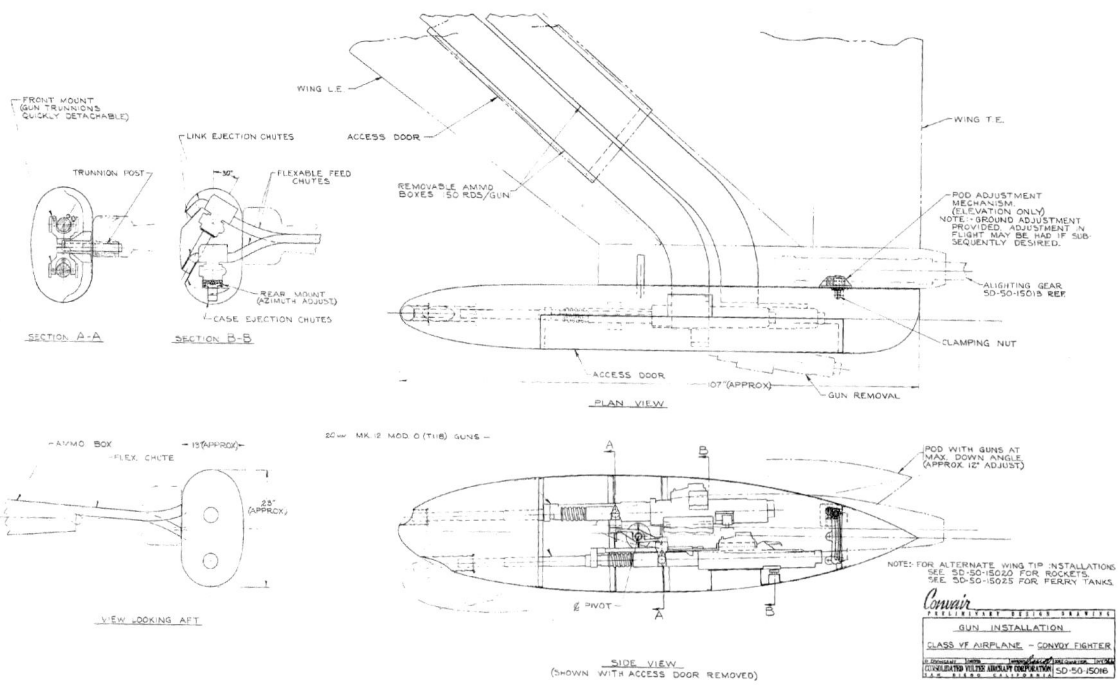

Four 20mm Mk.12 Mod. 0 (T118) fixed aircraft guns, with 150 rounds each, could be installed in pairs at the wing tips of the Convair Convoy Fighter.

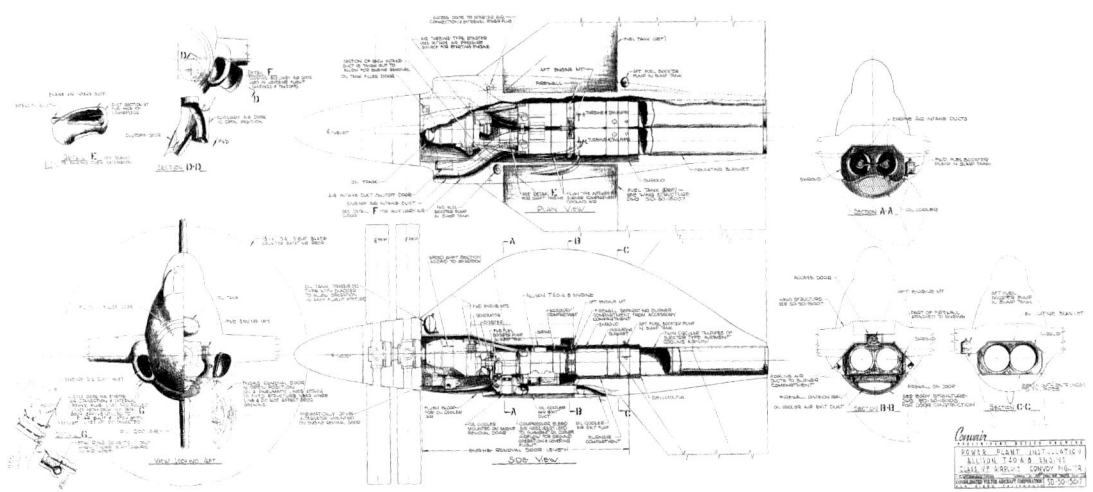

Blueprint of the Allison T40-A-8 turboprop engine installation.

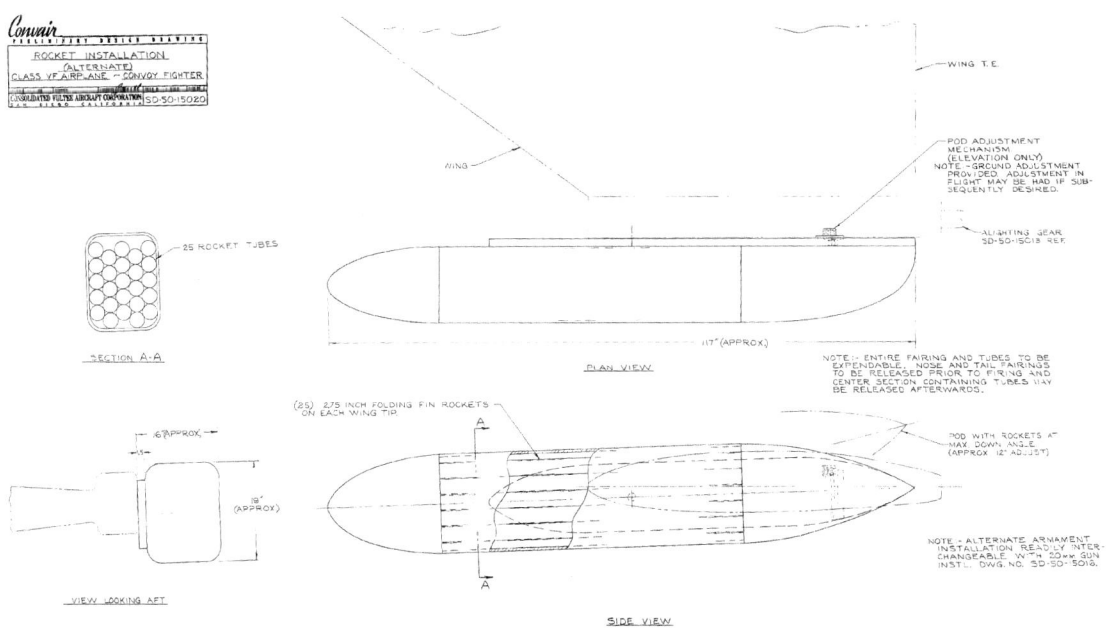

An alternate armament installation consisted of fifty 2.75in folding fin rockets—twenty-five in each wing tip fairing. Each fairing was divided into three parts, with the front and aft portions being released prior to firing, and the central section containing the tubes being released afterwards.

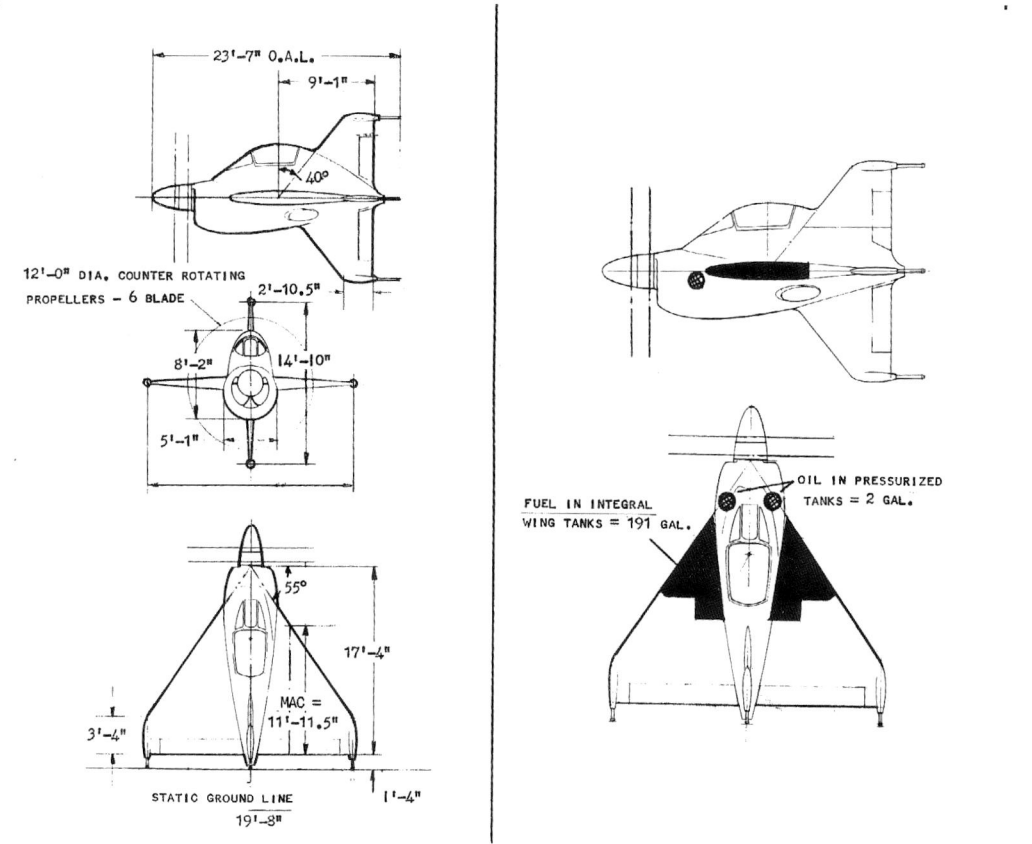

Three-view and fuel tank diagram taken from the Standard Aircraft Characteristics (SAC) charts prepared for the type.

(Left and above) Desktop model of Convair's original proposal for what became the XFY-1 Pogo.

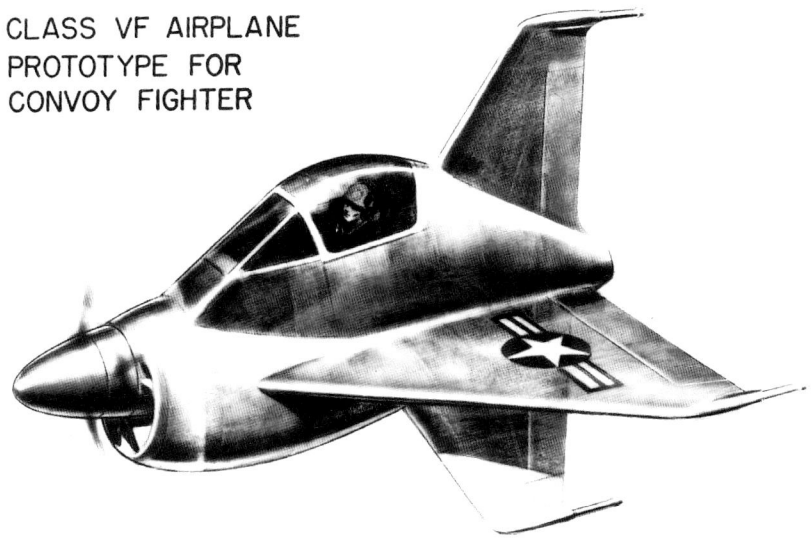

Artist's impression of the Convair's 0.766 scale prototype of the Convoy Fighter, designed to meet the original OS-122 requirement.

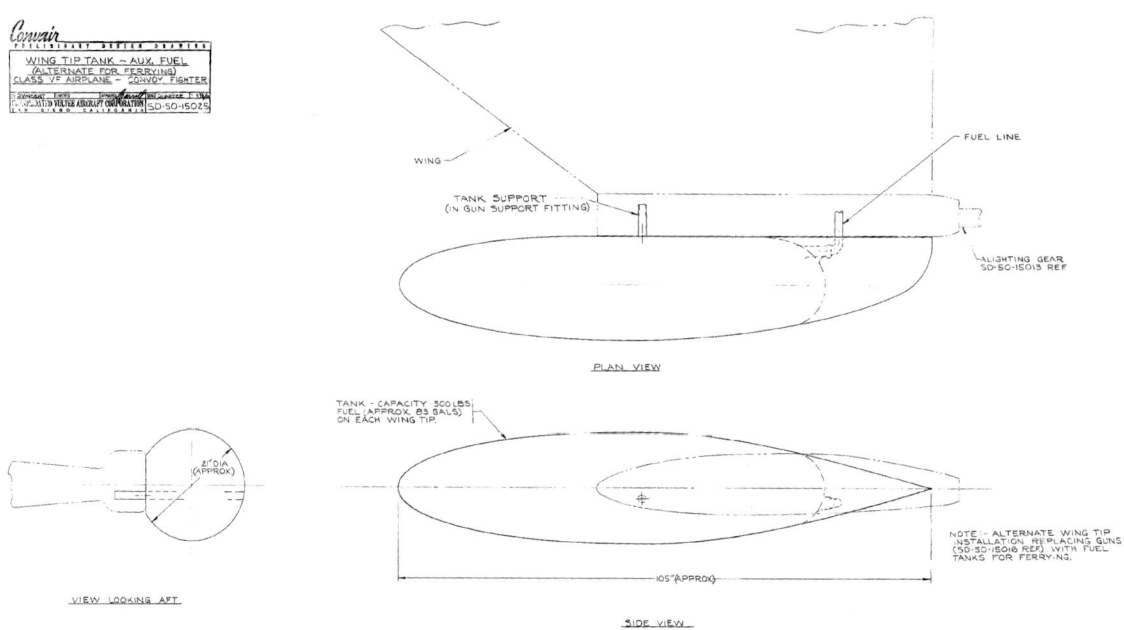

Another alternative to the gun installation was a pair of 500lb (approximately 85 gallon) fuel tanks mounted on the wing tips for ferrying purposes.

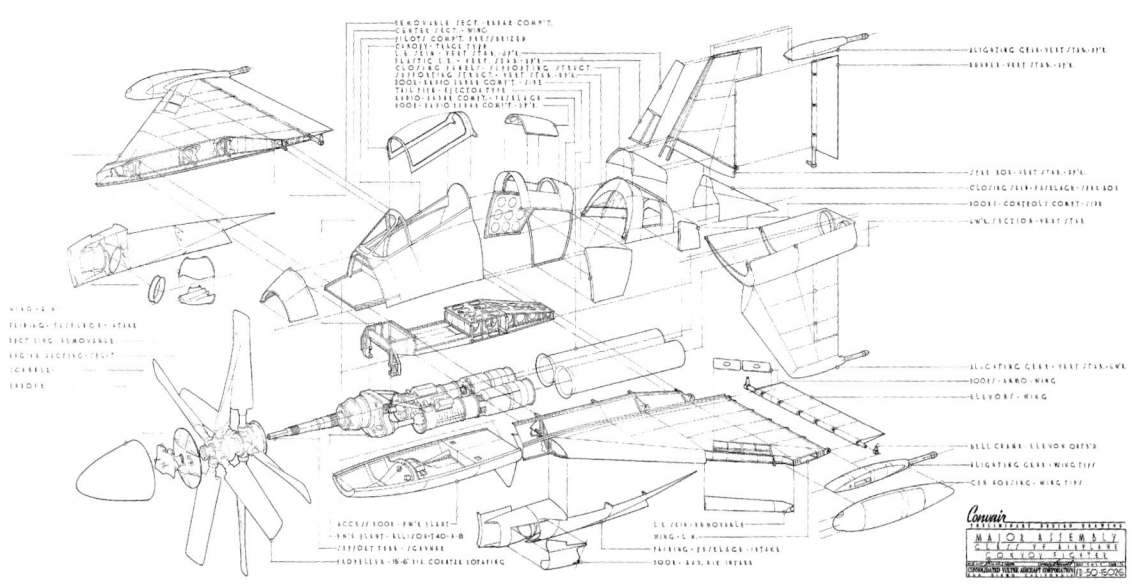

Assembly diagram of the Convoy Class VF Convoy Fighter proposal.

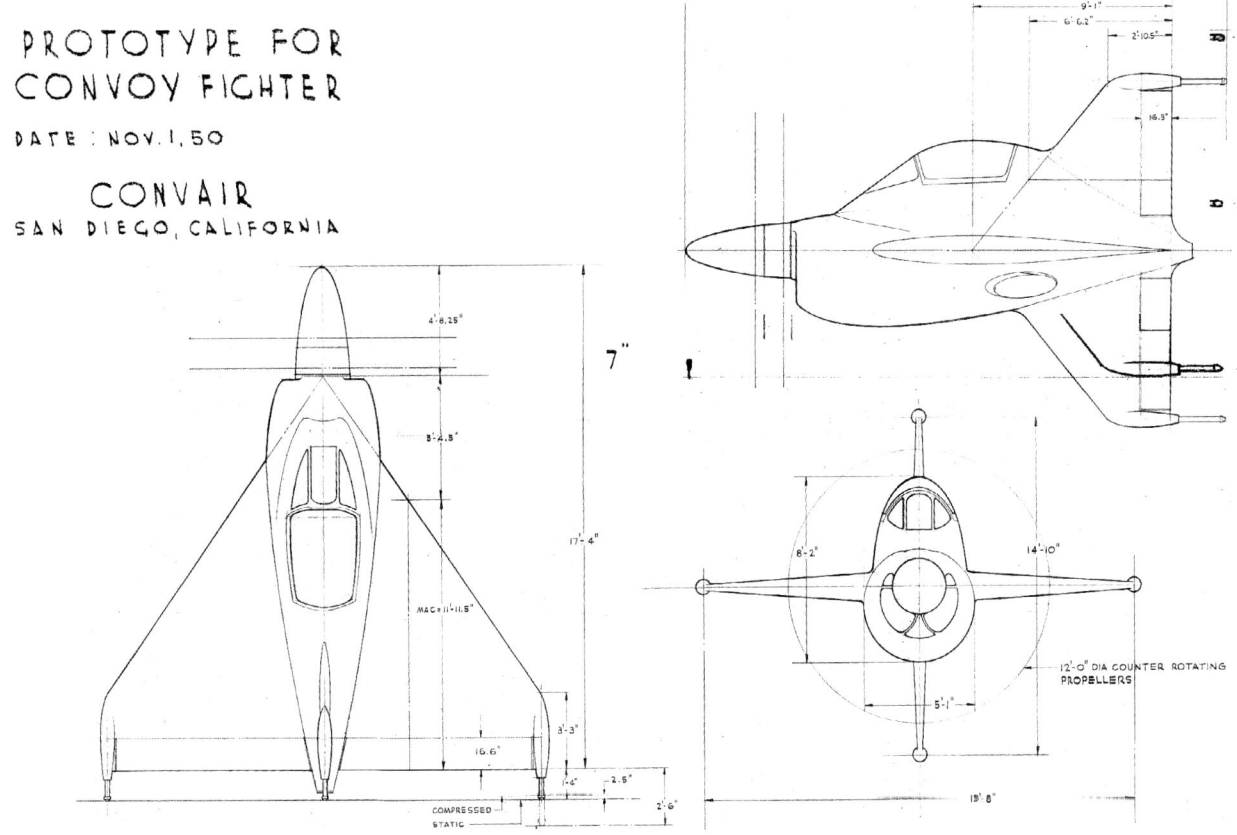

Another general arrangement of the prototype, this one with a sketch by a Convair engineer indicating that a smaller ventral fin was considered.

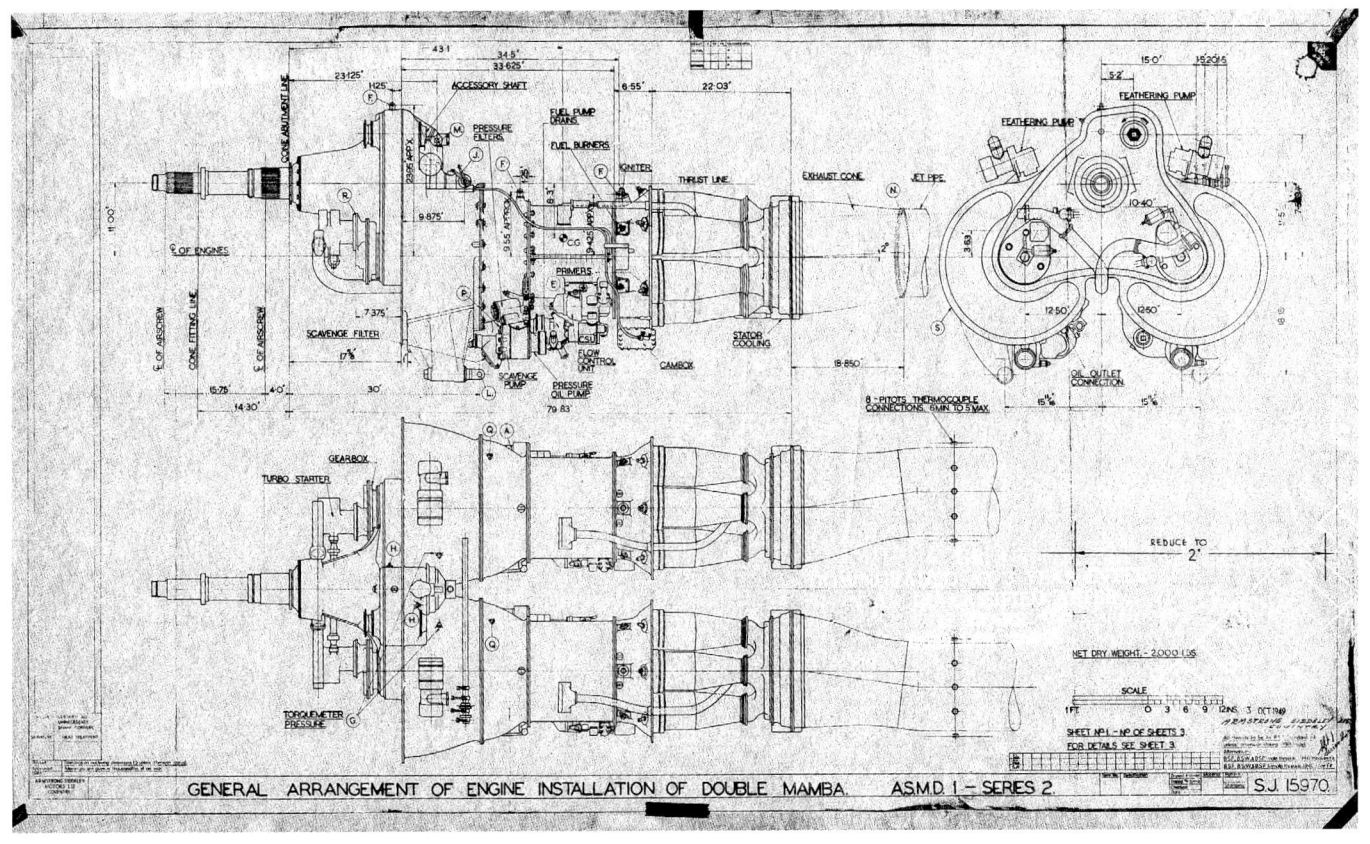

Armstrong Siddeley provided this blueprint of its Double Mamba III turboprop engine to the US Navy, which in turned shared it with Convair and other participants in the Convoy Fighter competition as reference in designing their reduced scale prototypes.

PROPOSAL

MISSION AND DESCRIPTION

MISSION:

THIS AIRPLANE IS INTENDED AS A PROTOTYPE FOR THE CONVOY FIGHTER PRESENTED IN REPORT ZP-50-15002 AND CORRESPONDING STANDARD AIRCRAFT CHARACTERISTICS CHARTS.

DESCRIPTION

THE PROTOTYPE AIRPLANE IS A ONE-PLACE, FLYABLE, DIMENSIONALLY AND DYNAMICALLY SIMILAR 0.766 SCALE PROTOTYPE OF THE CONVOY FIGHTER. THIS AIRPLANE IS CAPABLE OF VERTICAL, UNASSISTED TAKE-OFF FROM AND LANDING ON, SMALL PLATFORM AREAS.

THE AIRPLANE HAS A DELTA WING AND BOTH AN UPPER AND LOWER VERTICAL TAIL SURFACE, ALSO OF DELTA CONFIGURATION. CONTROL SURFACES ARE POWER OPERATED. THE AIRPLANE IS PROVIDED WITH A NAVY STANDARD EJECTION SEAT WHICH MAY BE ROTATED THROUGH 45° FOR VISION IN THE VERTICAL ATTITUDE.

ALIGHTING GEAR IS PROVIDED ON THE TIPS OF THE WING AND VERTICAL TAIL SURFACES. AN AUXILIARY, CONVENTIONAL LANDING GEAR IS PROVIDED FOR TEST PURPOSES. THE AIRPLANE IS POWERED BY A DUAL ROTATING PROPELLER DRIVEN BY A TURBO-PROP ENGINE.

WEIGHTS

	G.W.	L.F.
EMPTY	5982	–
BASIC	6070	–
DESIGN	7008	+7.5*
MAX. T.O.	7500	

(LIMITED BY MIN. FLIGHT ACCEL. DURING TAKE-OFF – SPEC. OS-121)

| MAX. LANDING | 6762 | +3.0** |

* LIMIT MANEUVER L.F.
** LIMITED BY ALIGHTING GEAR DEFLECTION

FUEL AND OIL

LOCATION	CAPACITY
WING	191 GAL.

TANKS ARE INTEGRAL – NOT SELF SEALING – FUEL CONFORMS TO SPEC MIL-F-5616 (JP-1)

OIL

LOCATION	CAPACITY
FUSELAGE	2 GAL.

(2 SEPAR. TANKS) – OIL CONFORMS TO SPEC. AN-O3 – GRADE M

POWER PLANT

ENGINE: 1 ARMSTRONG-SIDDELEY "DOUBLE MAMBA" TURBO-PROP MODEL ASMD-1, AS PER ENGINE SPEC., ISSUE #3, MARCH 1950; LENGTH 79.83", HEIGHT 43.85", WIDTH 52.8" – PROPELLER: SIX-BLADE-DUAL-12' DIA., ACTIVITY FACTOR = 150, DESIGN C_{ℓ_i} = 0.50 REDUCTION GEAR – 10.3:1

RATINGS

	RPM	PROP SHP	JET THRUST (LB)
T.O.	15000	2640	810
MAX. CONT.			
CRUISE	14500	2095	710

PER ENG. SPEC. ISSUE#3 MARCH 1950

ORDNANCE

NONE

DIMENSIONS

WING SPAN	19'8"
FUSELAGE LENGTH	22'2"
VERT. TAIL SPAN	14'10"
WING AREA	203 SQ. FT.
WING MAC	11'11.5"
WING A.R.	1.9
WING L.E. SWEEP	55°

ELECTRONICS

AN/ARC-27 UHF

AN/APN-1 RADIO ALTIMETER

NOVEMBER 1950 — CONVAIR CONVOY FIGHTER PROTOTYPE

PROPOSAL

PERFORMANCE SUMMARY

LOADING CONDITION		FLIGHT TEST
TAKE-OFF WEIGHT	lbs.	7500
Fuel	lbs.	1230
Bombs	lbs.	0
Wing/Power Loading (A) lbs/sq.ft; lbs/bhp.		36.9/2.53
Stall Speed—Power off	kn.	
Stall Speed—Power off – No Fuel	kn.	NOT APPLICABLE
Stall Speed—Power on	kn.	
Maximum Speed/Alt (B)	kn/ft.	518/S.L.
Take-off Distance, deck -- calm	ft.	0
Take-off Distance, deck	kn.	0
Take-off Distance, Airport	ft.	0
Rate of climb -- sea level (B)	ft/min.	10,800
Service Ceiling (RC=100 FPM)(B)	ft.	43,000
Time-to-climb 20000 ft. (B)(C)	min.	3.11
Time-to-climb 35000 ft. (B)(C)(D)	min.	7.81
Combat Range/V av	ft. n.mi/kn.	NOT APPLICABLE (SEE NOTE H)
Combat Radius/V av	ft. n.mi/kn.	NOT APPLICABLE (SEE NOTE H)

LOADING CONDITION		FLT. DES. G.W.
GROSS WEIGHT	lbs.	7008
Engine power		MAXIMUM
Fuel	lbs.	738
Bombs/Tanks		0
Max. speed at sea level	kn.	520
Max. speed/Alt	kn/ft.	520/S.L.
MAX. SPEED/35000 (D)	kn/ft.	495
Rate of climb SL	ft/min.	11,700
Ceiling for 500 fpm R/C	ft.	42,500
Time-to-climb/Alt.	min/ft.	7.0/35000

NOTES

(A) BHP AT MAXIMUM CRITICAL ALTITUDE (TAKE-OFF POWER @ S.L.)
(B) MILITARY POWER
(C) FROM STANDING START
(D) EXCEPTION FROM MIL-C-5011
(E) PERFORMANCE IS BASED ON CALCULATIONS USING WIND TUNNEL TESTS IN CORRELATION WITH FLIGHT TEST RESULTS AND NACA REPORTS ON AN AIRPLANE OF SIMILAR PLANFORM, NACA STANDARD ATMOSPHERE
(F) FUEL CONSUMPTION IS BASED ON ENGINE SPECIFICATION FUEL CONSUMPTION DATA USING FUEL OF 6.7 #/GAL DENSITY
(G) FUEL CONSUMPTION DATA ARE INCREASED 5%
(H) ENDURANCE AT SEA LEVEL FOR FLIGHT TEST IS AS FOLLOWS:
    AT REDUCED POWER (300 KN AIRSPEED)—68 MIN.
    AT TAKE-OFF POWER (520 KN AIRSPEED)—24 MIN.
    AT REDUCED POWER AFTER FUEL ALLOWED FOR 10 MIN WARM-UP, TAKE-OFF AND LANDING AT TAKE-OFF POWER (300 KNOTS AIRSPEED) — 45 MIN.

NOVEMBER 1950 — CONVAIR CONVOY FIGHTER PROTOTYPE

SAC charts summarizing the key physical and performance characteristics of Convair's 0.766 scale prototype of the Convoy Fighter.

CONVAIR MODEL 5 CONVOY FIGHTER

PERFORMANCE SUMMARY

WEIGHT SUMMARY

Performance and weight summary for Convair's Stripped Down Convoy Fighter.

Illustration showing Convair's proposal for tethering the Convoy Fighter during preliminary hovering flight tests; a similar setup was actually used during testing of the XFY-1 Pogo at Moffett Field in April 1954.

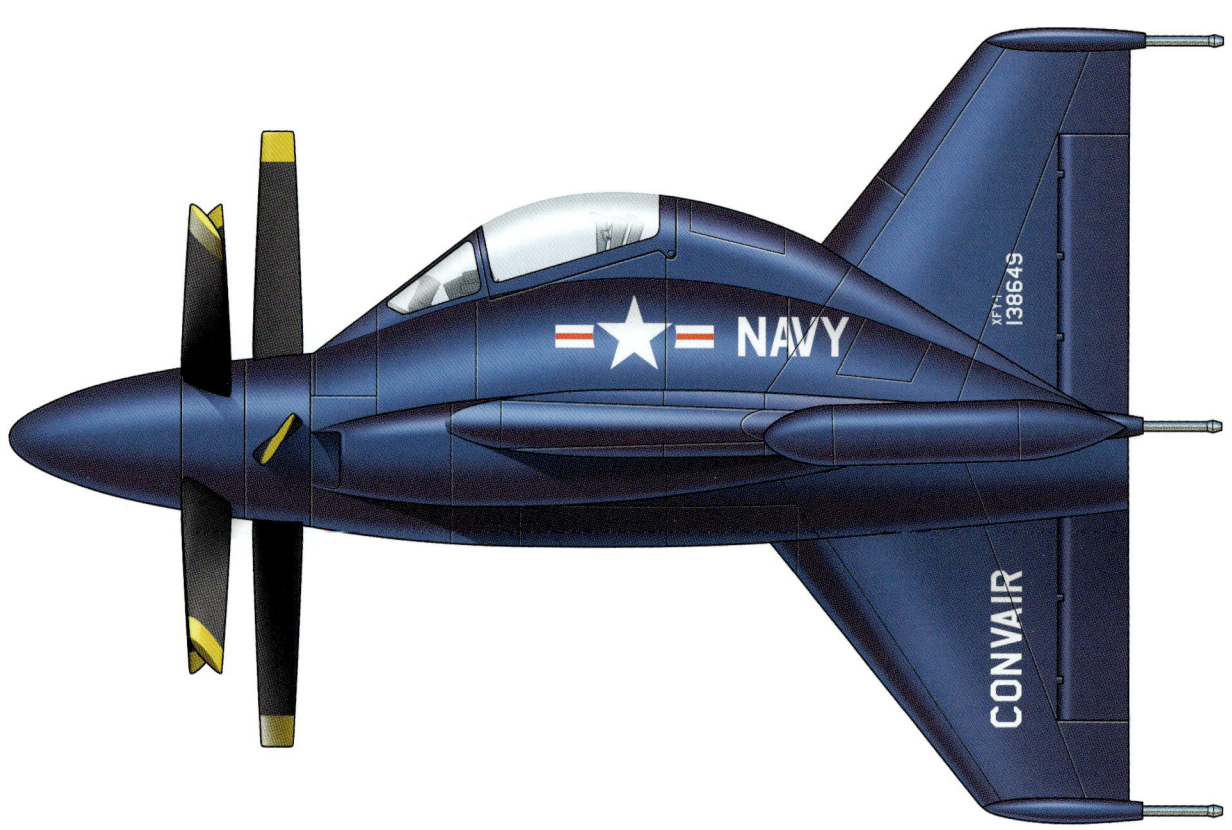

A speculative colour profile of the original Convair Model 5 Class VF Convoy Fighter proposal armed with rocket pods on the wing tips, each holding twenty-five 2.75in folding fin rockets. The overall Glossy Sea Blue scheme was typical of Navy aircraft of the early 1950s.

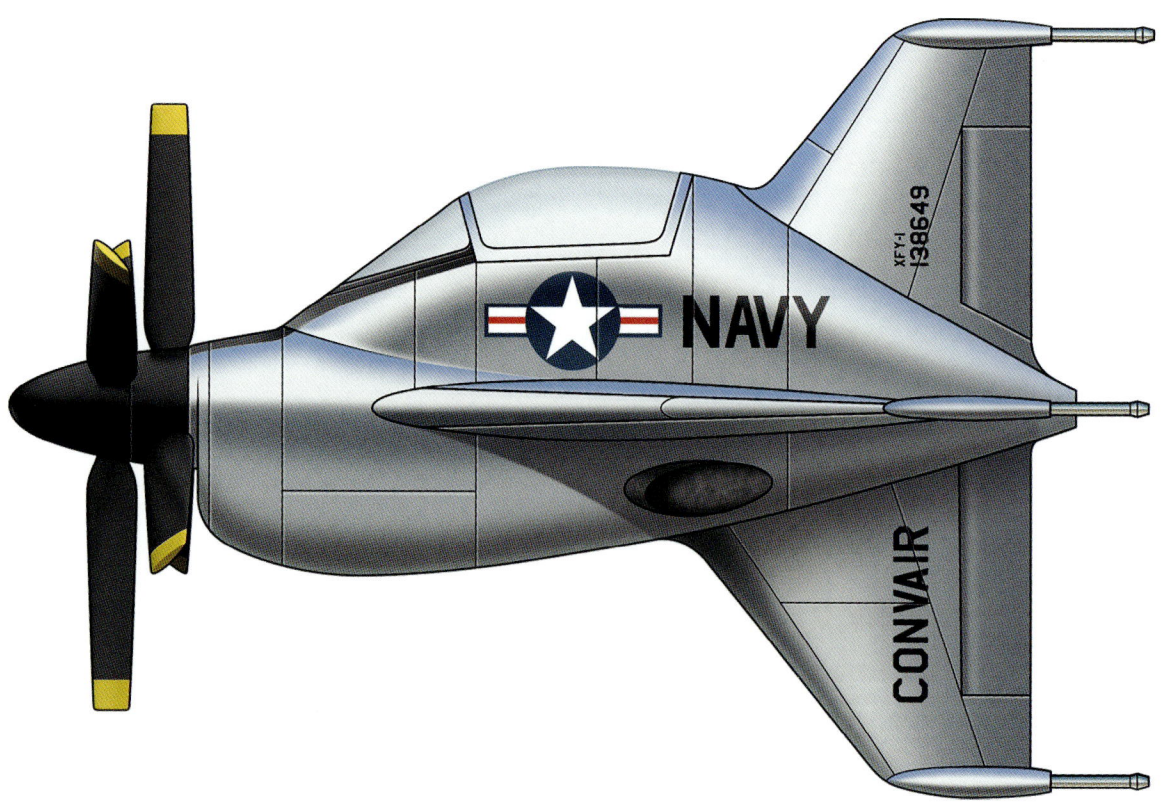

Colour profile of the 0.766 Scale Prototype of the Convair Class VF Convoy Fighter, a compact demonstrator powered by an Armstrong Siddeley Double Mamba III. The Navy ultimately decided not to build this aircraft in favour of a stripped version of the full-scale Convoy Fighter. The colours are based on the natural metal scheme worn by the actual XFY-1 Pogo experimental aircraft.

3

Goodyear GA-28A and GA-28B

WHILE THE Goodyear company of Akron, Ohio is primarily known these days for the production of tyres and blimps, it once had a division known as the Goodyear Aircraft Corporation which specialized in the design and production of military aircraft. Its most famous product was the FG-1 Corsair of the Second World War, a licence-produced version of Vought's famous gull-winged fighter.

Having developed a significant aircraft manufacturing capability and an experienced engineering staff, Goodyear was eager to land new military aircraft contracts in the postwar era and keep its new division going. While it didn't score a win in the Convoy Fighter competition, the company's unorthodox proposal shows that Goodyear was willing to take risks and think well outside the box to address the Navy's challenging requirements.

Goodyear's offering consisted of two aircraft types—a three-quarter scale demonstrator designated as the GA-28A and a full-scale fighter known as the GA-28B. Like Convair, Goodyear chose a tailless delta wing layout (modified with a modestly swept trailing edge) for their VTOL fighter, though it differed considerably in other aspects, particularly in the landing gear. The proposal documents for both the GA-28A and GA-28B were quite similar, with many sections being basically identical in content. For reasons of economy, they have been combined in the summary below, key differences being indicated where necessary. The text is rather technical as it was originally intended for BuAer engineers, but the majority of it is presented for those who appreciate such abstruse material.

FEATURES & INNOVATIONS

General design considerations. The Goodyear GA-28A was a three-quarter scale model prototype (proof of concept demonstrator) of the GA-28B Convoy Fighter, both of which were designed to fulfill the requirements laid out in OS-122. After careful consideration of the requirements and several different configurations, Goodyear determined that it was not only entirely feasible to eliminate auxiliary handling gear and incorporate the landing gear within the basic aircraft, but actually desirable to do so.

Before embarking on an all-out effort for such an "ultimate airplane," a careful analysis was made of the dynamic landing loads involved in the vertical landing configuration from both the standpoint of alighting on its own gear and alighting on auxiliary gear. Surprisingly enough, the weight penalty required to handle the loads imposed on the airplane structure by merely setting the airplane on its "tail feathers," together with anticipated gust loads, was of such magnitude that carrying a completely adequate alighting gear at all times was reasonable. Further

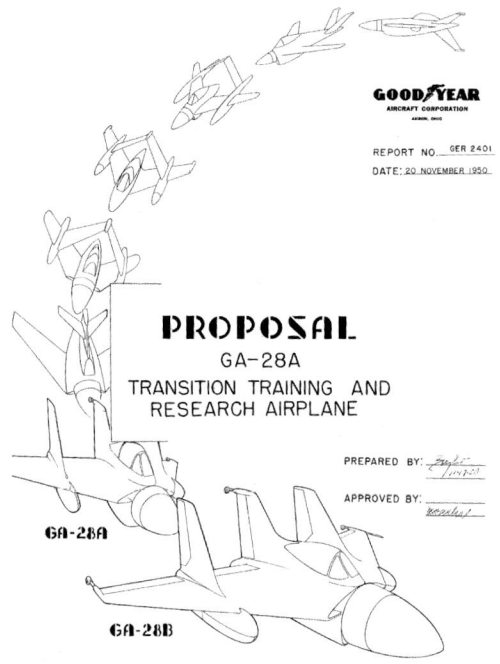

Cover to the proposal brochure of the Goodyear GA-28A, a three-quarter scale prototype of the GA-28B Convoy Fighter, which is shown at the very front of the procession of aircraft designs. Note the evolution of the configuration as the aircraft approaches the viewer.

Artist's impression of the GA-28A in operation, showing its remarkable ability to take off and land both vertically and horizontally. Goodyear designed the type to be not only a research airplane, but a trainer as well.

examination indicated that for very little additional weight penalty the airplane could be designed to land conventionally as well as vertically, giving the GA-28A/B a tremendous flexibility and safety above and beyond its initially conceived employment.

Ground stability. Once this trend was definitely established, considerable time and effort was spent in endeavouring to arrive at a configuration which would provide satisfactory static stability on a pitching and rolling deck without auxiliary stabilizing means in order to eliminate all specialized handling gear. Obviously, this meant that some means had to be devised to lower the centre of gravity of the airplane when it was in the vertical takeoff position. This resulted in shortening the airplane and eventually defined the GA-28A as a fundamentally tailless machine embodying 27.25 degrees of static ground stability. That is, the deck would have to be inclined 27.25 degrees before the stabilising moment would reduce to zero. Since 20 degrees of roll was generally believed to be the maximum likely to be encountered, the additional 7.25 degrees provided a margin for inertia forces generated by deck roll in power off and landing conditions. The GA-28B was basically identical to the smaller airplane, except for having a static ground stability of 27.5 degrees.

Takeoff presented a different and somewhat more complicated problem in that at high angles of deck roll, the upsetting thrust moment was of such magnitude as to overbalance the normal restoring moment. As a result, higher powers up to takeoff power could be applied and takeoff accomplished only within certain angular limits of deck roll, before overturning the airplane, unless means of restraint were employed.

General purpose of design. As the basic configuration progressed to the point where an operationally feasible airplane appeared possible, it occurred to Goodyear that the GA-28A, in addition to being useful as a research airplane, would be ideally suited to use as a transition training airplane. Therefore, this concept was pursued in the preparation of the design of the three-quarter scale model airplane.

General appearance of design. The artist's impressions depict the general appearance of the aircraft and define the basic configuration as a semi-delta, semi-midwing design with triple vertical tail surfaces and containing an alighting gear completely adequate for either conventional or vertical landing and takeoff. The inclusion of a rotatable ejection seat, as suggested by a report from the Johnsville Naval Air Development Center, was completely adequate

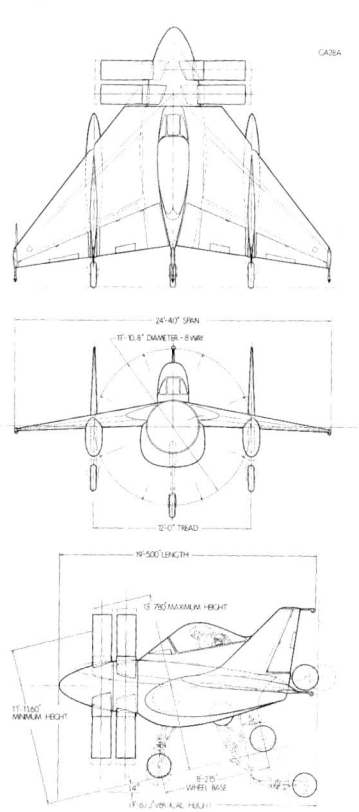

Basic three-view of the Goodyear GA-28A showing landing gear positions for both horizontal and vertical attitudes.

and logical, but obviously required a much larger cabin than was originally contemplated for the DR-72 and DR-72A airplanes (see Chapter 1).

Seat ejection. In keeping with the training airplane concept, clearance was provided in the GA-28A for ejecting the pilot with the seat in any position. For the GA-28B, clearance was provided for ejecting the pilot with the seat in the down position only. In the interest of space and weight saving, Goodyear noted that conventional and vertical adjustment of the seat was omitted and vertical adjustment was accomplished by rotating the seat up or down as desired.

General data. The normal gross weight of the GA-28A transition training airplane with 45 minutes' fuel aboard was 8,183lb; wing area 204ft²; wing span 24.3ft; overall height in vertical position 19ft 8.75in; overall height in conventional landing position 13ft 7.8in maximum and 11ft 11.6in minimum.

The normal gross weight of the GA-28B with 3,280lb fuel aboard was 16,994lb; wing area was 345ft²; wing span 31.3ft; overall height in vertical position 28.0625ft; overall height in conventional landing position 14.8125ft minimum and 16.9375ft maximum. In the conventional attitude the dimensions were such that the GA-28B could be stowed

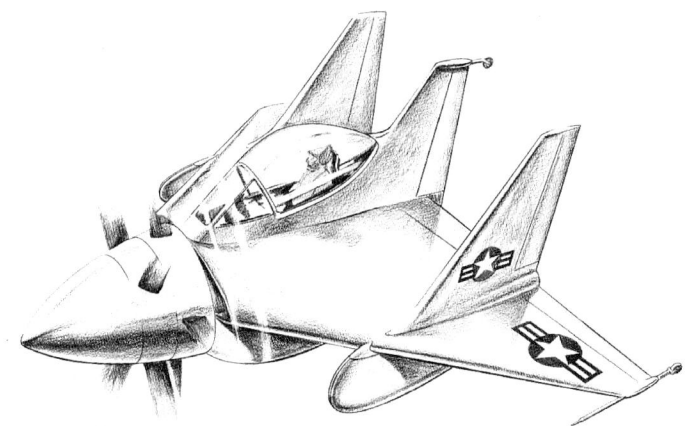

STANDARD AIRCRAFT CHARACTERISTICS

DATE: 20 NOVEMBER 1950

GOODYEAR AIRCRAFT
MODEL GA28A

Artist's impression of the GA-28A prototype/trainer. Goodyear's interpretation of the OS-122 requirements resulted in a design which looked more like a caricature of an aircraft than an actual flying machine, but the company believed that it was the best solution to the specification.

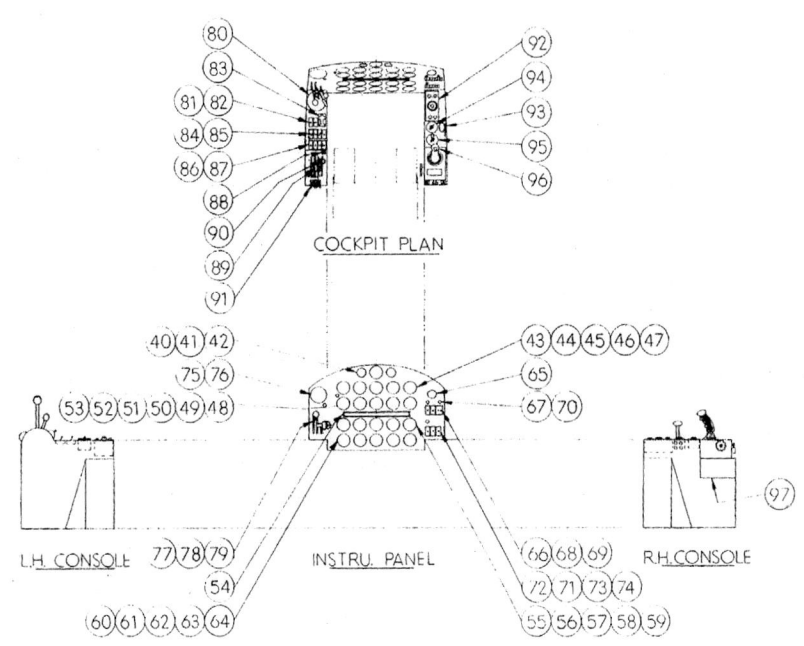

A detailed plan of the cockpit layout taken from the inboard profile blueprint.

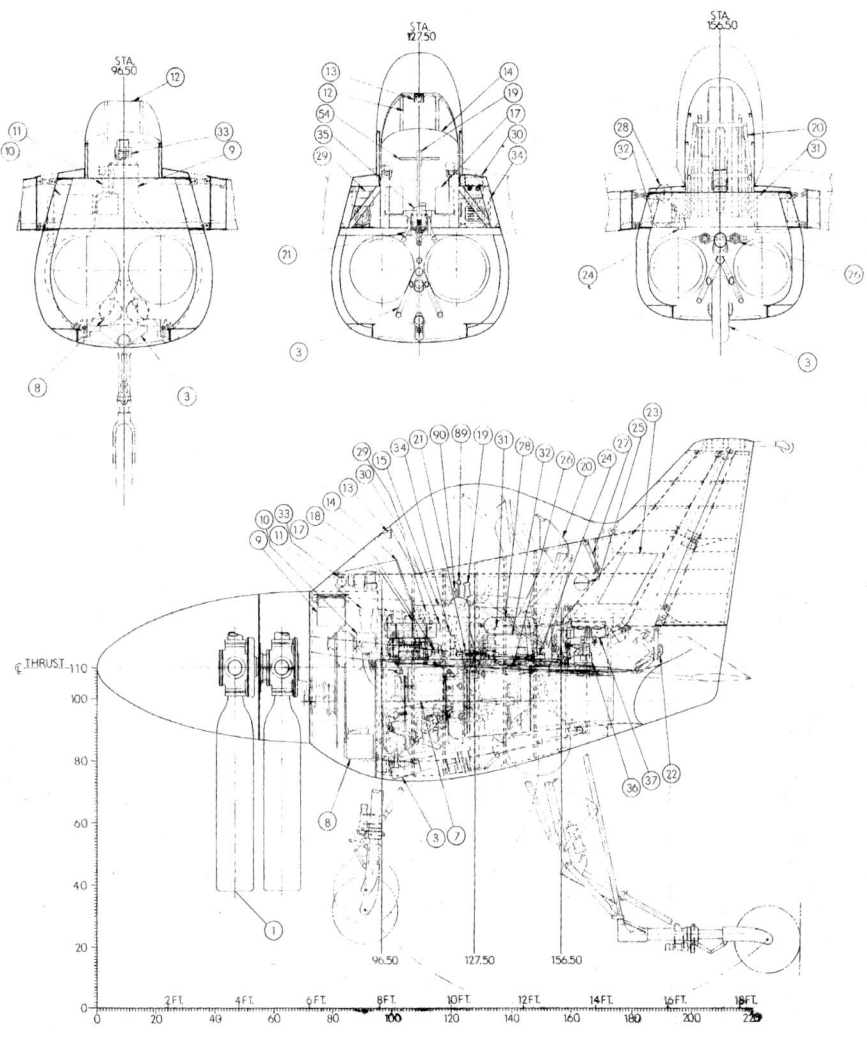

Inboard profile of the GA-28A showing the location of major equipment and structural members.

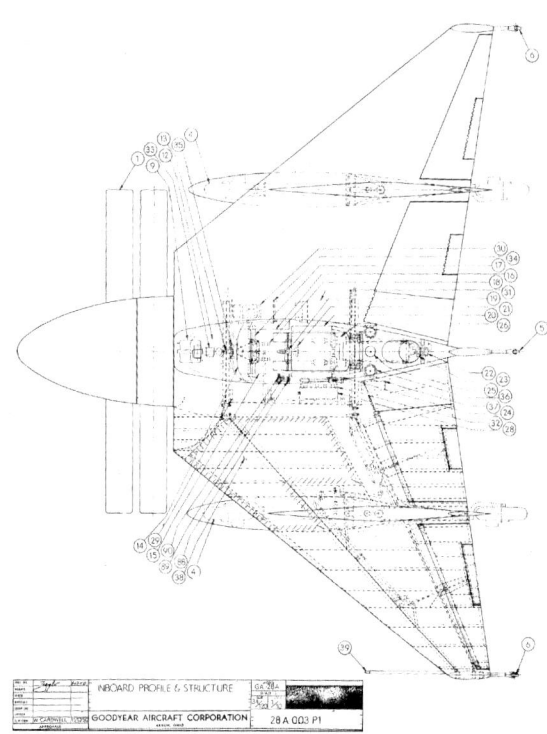

Top inboard profile of the Goodyear GA-28A.

Legend for the inboard profile blueprint reproduced previously.

in the hangar deck of a carrier.

Ground attitude. Goodyear believed that both the GA-28A and GA-28B should be self-contained airplanes completely capable of assuming either the conventional or vertical position on the ground without an external power source and able to land either conventionally or vertically by simply positioning the landing gear selector lever. A 3,000psi hydraulic system was provided as a source of power.

Single engine performance. Further, since the nature of the Double Mamba engine made single engine operation possible, propeller feathering was provided for the GA-28A. The airplane could be trimmed to fly satisfactorily on a single engine, one propeller feathered, and landed in the conventional fashion. With both propellers feathered, the airplane would be fully controllable in a 'power off' emergency conventional landing.

The GA-28B also featured propeller feathering and could be trimmed to fly on a single engine and land conventionally.

Producibility. To improve the producibility of both aircraft, Goodyear took some small weight penalties in their design. Examination of the inboard profile drawings shows that the aircraft were of conventional stringer and light skin construction with flush joints and riveting throughout. However, Goodyear noted that it was actively engaged in the production of a metal core sandwich material for industry-wide use and that as the actual design progressed, material weight saving could be realized by judicious use of sandwich construction. Further, provisions were made for feedershop installation of equipment in major structural subassemblies which would result in high production efficiency since the number of stations necessary on the final production line would be materially reduced.

Experimental adjustments. Provisions were made on both aircraft to permit varying the dihedral angle by means of different length rigid links between the fuselage and the upper wing root attachment fitting.

Control system. The basic mixing of the elevon and aileron signals was accomplished by a bevel gear "mixer" unit to which the control stick was attached. Stability calculations indicated that it would be desirable to have the outboard elevons operate as elevons continually, whereas it was desirable for the inboard flaps to function as elevons only during vertical flight, and as ailerons only throughout the high speed and conventional landing conditions.

To achieve this, a "sifter unit" was designed which enabled the pilot to select the control configuration desired. To reduce pilot burden, this selection was automatically made for the vertical landing gear configuration. However, the gear could be immediately retracted upon vertical takeoff without effecting the inboard elevon control, which could only be altered by direct selection of the pilot.

A 100% boost artificial feel mass balanced elevon control system is shown in the inboard profiles. While Goodyear did not feel this system to be necessary on the small airplane, it was felt to be expedient to include it in order to achieve an indication of the suitability of this type of elevon control system for larger airplanes of this type under consideration. Only conventional mechanical linkage without boost was provided for the rudder controls of the GA-28A. The rudder controls of the GA-28B employed a "hydro power" proportional feel booster unit of 3.1 ratio.

Autopilot. To provide for automatic stabilization in cruising, transition and hovering flight, and to enable programming of takeoff to cruise and cruise to landing flight paths, an automatic pilot was installed in both aircraft. This unit was basically a modification of the Sperry X-4 autopilot, with three integrating rate gyros and an accelerometer which served as sensing elements of yaw, pitch, roll, and thrust line velocity motions of the airplane respectively. These indications were fed into amplifiers and thence were transmitted into airplane movement through the control system power units, and the engine power-control linkages. Manual stick override was achieved by a simple button engage switch on the pilot control stick, while throttle override was direct, manually imposed motion of the throttle overcoming the small inertia of the light power control servo during all non-automatic operation.

The pilot could utilize the autopilot in one of four possible modes. Selection was accomplished by a Command Selector knob on the instrument panel with dial positions for:

1. Autopilot control of the plane in cruising or hovering flight;
2. Pilot control of the airplane in any flight configuration through the autopilot via a "formation stick";
3. Automatic control sequencing for cruise to hovering transition;
4. Automatic control sequencing for hovering or takeoff to cruise transition; and
5. Autopilot 'off' — the formation stick control was the conventional miniature control stick conveniently located forward and to the right of the right arm rest, and was operated in the same manner as the standard control stick to give pitch and roll, and to give yaw via a small heading knob at the top of the pistol grip shaped arm.

It was entirely feasible to also include two additional automatic control features. One was the "load-limit computer" or regulator to prevent inadvertent overstressing during manoeuvres. This instrument would operate as a g-restrictor for the airplane, using accelerometers as sensing means. The other device, governed by solenoid action and monitored by the airspeed indicator, would control the operation of the 'sifter' mechanism.

Quick power plant change unit. The power plant proposal drawings for both aircraft depict a power plant quick-change unit and transporting dolly that would permit a complete power plant change to be accomplished in 30 minutes or less. If the self-contained unit had been previously checked out by running on a test stand or another airplane, no additional time would be required to adjust the power plant controls.

To accomplish a power plant change for the GA-28A, the lower forward plenum chamber cowl and the lower rear fuselage panels were removed, standard Navy jacks placed under the front beam jack pads and the nose gear extended in the knuckled position by means of closing certain hydraulic valves and using the hydraulic hand pump. This permitted a crossbar to be inserted between the two rear engine hoist points.

Three men then positioned the three-legged dolly under the engine; one man adjusted the extendable engine nose case support by means of a hydraulic hand pump located on the rear vertical leg and each of the other men simply adjusted one of the two vertical leg extensions until they engaged and the crossbar attached to the two engine (rear) hoist points. To complete the removal of the quick change unit, it was only necessary to remove the safeties and rotate three small levers. One of these levers was located at each of the three engine mount fittings. The unit was then free to be towed away with the first inch of forward travel disconnecting all fluid and electrical connections.

The replacement unit, installed on another dolly, was simply pushed into the proper position, adjusted to the proper height by means of the three hand hydraulic pumps and the tapered male engine attachment fittings pushed into the tapered female engine mount fitting which aligned and connected all fluid and electrical connections. The installation was secured by rotating and safetying the three engine mount levers. The lower plenum chamber cowling was replaced and the power plant unit was completely installed since all power plant controls operated through common centre balance plates, one attached to the basic airplane structure and one attached to the engine, with movement of the controls being accomplished by wobbling of the plate with no through attachment being involved. It was only necessary to retract the nose gear by means of the hydraulic hand pump and the airplane was ready to fly. In addition to access to the engine compartment via the landing gear wells and by removal of the lower forward plenum chamber and lower rear fuselage cowls, other necessary small inspection doors were provided.

The power plant change for the GA-28B was very similar, except for some minor details unique to the construction of the larger aircraft.

Cockpit access and egress. With the airplane in the vertical attitude on the ground, access to and egress from the cockpit was easily achieved by flush type hand and toe holds located in the cabin side and top surface of the wing root skin. When the airplane was in the conventional attitude on the ground, access was by means of a small ladder over the leading edge of the wing and thence by conventional flush type hand holds to the cockpit. Since space was not available to open the canopy by sliding it aft in the conventional manner, it was hinged at the trailing edge and the front was elevated for entrance. Cabin jettisoning was provided.

Landing gear. Free swivelling of both the main and nose gear was provided in either the conventional or vertical landing configuration. Appropriate self-centring shimmy dampeners were provided on each of the gears as well as a selector lock so that if desired, the main wheels could be prevented from swivelling. When the GA-28A was to be landed cross wind conventionally, it was probably desirable to land the airplane with all three wheels unlocked in order to avoid the generation of marginal overturning moments. Ground stability of the GA-28B in the conventional landing configuration was normal, and the airplane would, in most cases, be landed with the main gear locked. It was obviously mandatory that both airplanes be landed in the vertical configuration with all wheels free to swivel. Centring devices were provided on all wheels so that the direction of rotation was parallel to the ship's centreline for retraction.

Goodyear noted that with the unusual disposition of weight, that is more weight on the nose wheel in the conventional landing attitude than on the main wheels, the brakes were not quite as effective as might normally be expected. However, including the desirable low speed drag effects of the contra-rotating dual propeller, the GA-28A could be stopped in 22.3 seconds with a landing roll of 1,850ft. The GA-28B could be stopped in 21 seconds with a landing roll of 1,950ft. Reversible propellers were provided for the larger aircraft which enabled it to stop in 500ft.

The landing gear was designed to fall free and lock down unassisted in the conventional landing gear position, but a hydraulic hand pump was provided as an additional safety measure.

Brake system. Power boosted "hydro power" brake cylinders installed horizontally on the rudder pedal paralleling arms were provided. These were dual purpose cylinders and when manually boosted by the airplane's hydraulic system, provided a powerful, nice-feeling brake at normal pedal loads but contained the added safety feature of functioning as a straight master cylinder in the event of hydraulic power failure.

Fuel tanks. Goodyear proposed integral wing fuel tanks in the interest of simplification, space and weight saving. These were chosen because self-sealing tanks were not required; a relatively smooth propeller turbine power plant was installed on a very rigid wing structure; the combination of features peculiar to this configuration—vertical dive, vertical hovering, transition to horizontal flight; and the reevaluation of progress made in the development of rubber based aromatic fuel proof sealing compounds. For the GA-28B, bladder type seals would have been installed in the two fuselage tanks as an additional safety feature since leakage of fuel in this area might find its way into the engine compartment immediately underneath the cockpit floor.

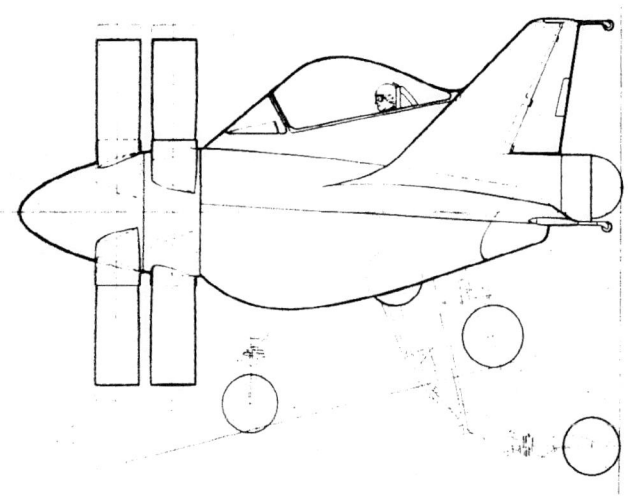

ALTERNATE MAIN GEAR FAIRING

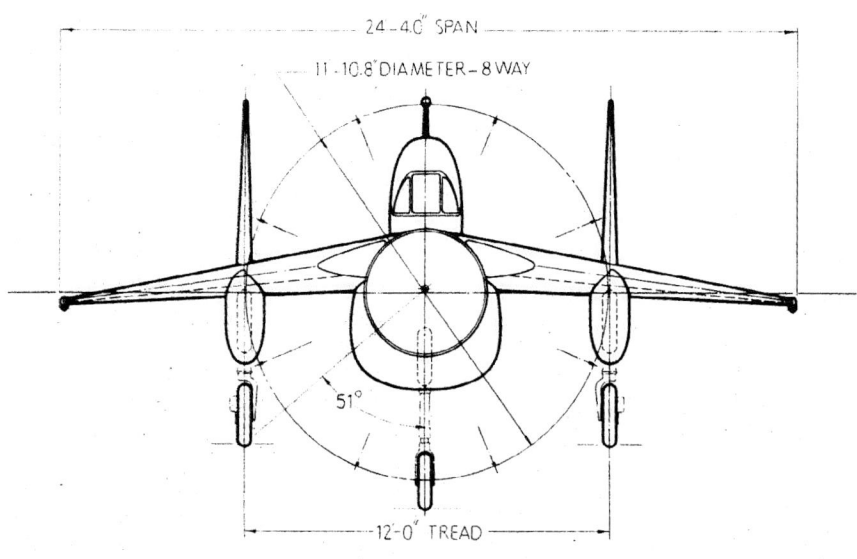

A more detailed three-view of the GA-28A with additional measurements and station numbers added. Note the scrap side view showing an alternate version with the wing pods deleted and main gear stowed at the base of the outer vertical tails; concerns over possible propeller vibration with the original configuration prompted the study.

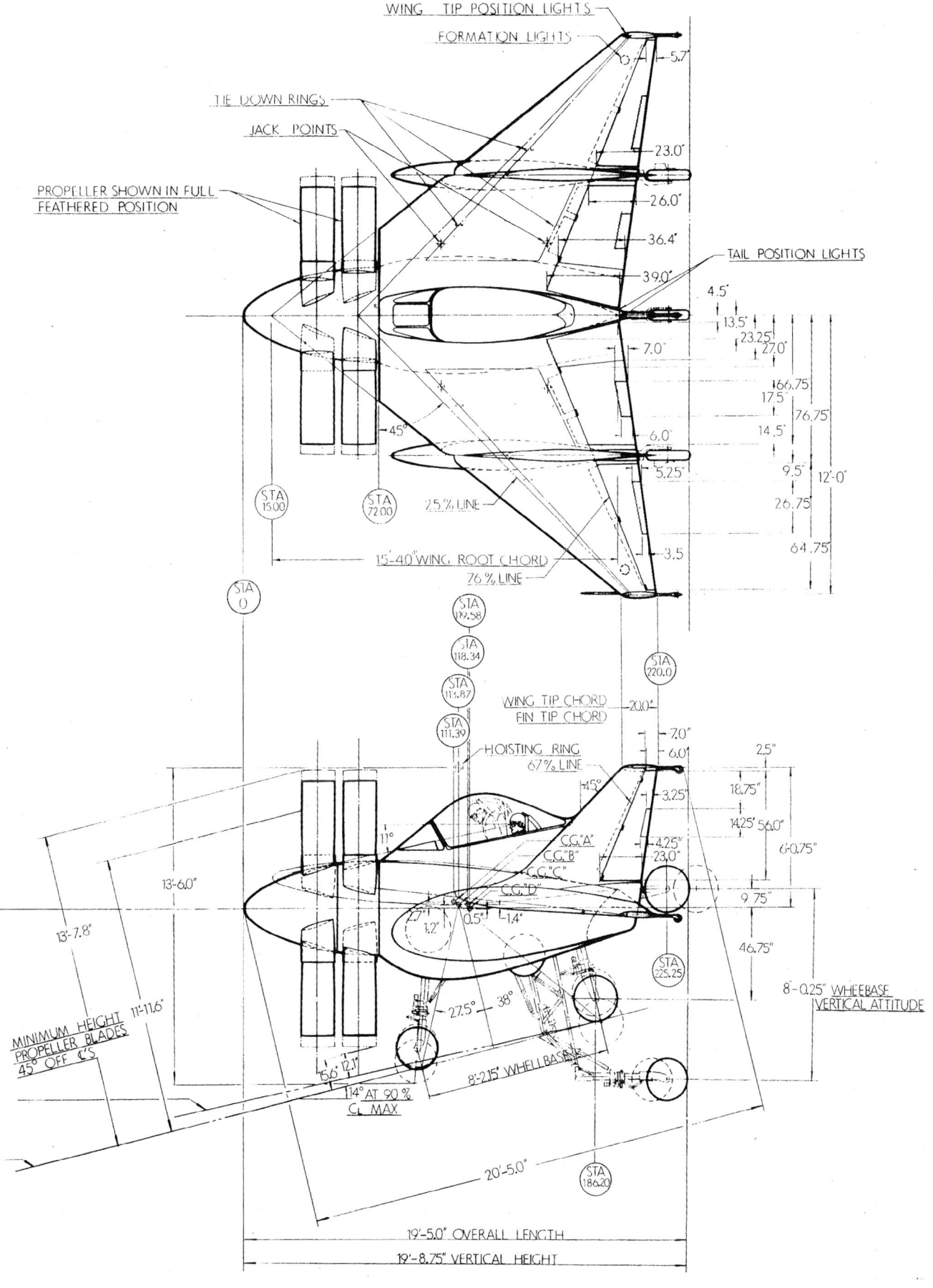

A comprehensive table of the GA-28A's major characteristics taken from the blueprint shown in the previous spread.

Deck protection. In view of the close proximity of the jet exhausts to the deck in the vertical takeoff and landing configuration, it was pointed out that it would probably be necessary to have the deck protected from the local concentration of heat.

Tyre protection. Since the deck would serve to deflect the heat sidewise across the deck, it might have been necessary to provide small vertical heat shields for the main and nose gear tyres as special equipment.

Goodyear also prepared an alternate main wheel landing gear fairing for the GA-28A. If the close proximity of the nose of the existing landing gear fairing proved troublesome from a propeller vibration point of view, resort to this type of fairing (approximately equivalent drag characteristics) was contemplated.

Landing load analysis. With the concept of the self-contained landing gear as being both feasible and highly practicable, the consideration of geometry and structural mounting became the paramount study. From the former standpoint, it was necessary to locate the wheels in conventional and tail first landing attitudes so as to achieve both stability in landing and in secured on-deck service conditions, as well as to restrict imposed landing loads to reasonable values. A byproduct of the short wheel base was the unconventional location of the centre of gravity closer to the nose wheel than to the main gear. This caused high loads at the nose gear. The latter condition defined the critical loads for the forward gear, and was analyzed in the same fashion as a bicycle arrangement.

In the tail first attitude, the airplane had to be capable of alighting on a rolling and pitching deck with extreme inclinations of 20 degrees and five degrees respectively, as well as maintaining position with a 15 knot ship way in a 20 knot wind from any quarter; heave of the ship was neglected.

Once the aircraft touched the deck with any extremity, it was effectively uncontrollable, with two extreme landing conditions being evident. The first considered forward inclination of the airplane thrust axis of 13.5 degrees for hovering into a 35 knot relative wind. This angle was additive to a bow-down ship pitch of five degrees for initial contact of the nose gear with subsequent rotation of the airplane to contact of the main gear. For the existing configuration, this mode was not critical.

The second, cartwheeling into a 20-degree rolled deck in the wing plane, yielded high design loads. In this mode the airplane, inclined in yaw 7.5 degrees against a 20 knot wind, struck first on the wing tip, necessitating the wing tip bumper wheel and shock strut, rotating down onto one main gear, and thence continued rotation to the other main gear. Goodyear noted that the 20-degree roll condition was considered extreme, and therefore, all loads calculated on the basis of that angle would be conservative.

Design loads were determined using landing design gross weight (i.e. takeoff gross weight minus 60% fuel) at 10ft per second sinking speed. Loads were also calculated for takeoff design gross weight at 10ft/second sinking speed, and for both landing and takeoff weights at 17ft/second, and the effects on the airplane for each of the latter three conditions established.

Armament installation (GA-28B). Once the configuration had progressed to the point that it was obvious that external landing gear fairings were mandatory, it became possible to make practically an ideal armament installation. Two 20mm guns were installed on the insides immediately under the wing and totally enclosed in the landing gear fairing. By locating the guns on their sides with the cradles facing each other immediately under the heavy landing gear rib structure, it was possible to provide a simple lightweight trunion mounting easily accessible through the landing gear doors for installing, removing and boresighting the guns. Together with the quick-opening small doors over the feed mechanism, it was possible to easily arm the weapon on the deck when under blacked-out conditions.

Built-in wing ammunition storage boxes (150 rounds per gun) were provided in the interest of weight saving (approximately 10lb per gun), easily accessible from the top side of the wing through a quick-opening door mechanism. Goodyear emphasized the compactness and straight-forwardness of this installation.

Armour plate in the form of bulletproof glass was provided ahead of the pilot and conventional armour placed immediately abaft of the seat. No floor armour was provided since it was felt that the installation of the power plant immediately below the cockpit floor provided adequate protection from below.

Further inspection of the armament drawing shows that installation of the MK6, Model 2 sight was not feasible and installation of an alternate sight would need to be considered.

Further configuration possibilities (GA-28B). Goodyear felt the performance figures contained in the performance summary of their report to be conservative. A careful review of the proposal suggested that a possible future configuration of improved vertical flight characteristics and capable of considerably higher speed could be achieved by a twin engine configuration embracing outboard located engines installed in an enlarged landing gear nacelle, single rotation supersonic propellers and a smaller, more conventionally shaped fuselage.

Alternate landing gear/armament installation (GA-28B). When it became apparent that the addition of the simple external landing gear fairing, also originally thought desirable from the armament installation point of view, was prohibitive from high speed considerations (-25 kts), Goodyear designed an alternate landing gear installation, retracting the wheels completely within the wing. This improved the overall appearance and performance.

Time constraints prevented Goodyear from showing the alternate 20mm cannon installation in their armament proposal drawing. However, a preliminary layout showed that the two cannon could be installed within the wing outboard of the landing gear position in a smoothly faired installation similar to the fairing shown on the DR-72 airplane.

HYDRAULIC SYSTEM

General. The main hydraulic system was a 3,000psi system and operated the landing gear, surface controls, sifter actuator and power boosted brakes. The landing gear section of the system was automatically depressurized in flight. Ground test connections were provided for maintenance purposes. A hand pump was installed for emergency operation. The fluid used was non-flammable in accordance with the BuAer specification.

In the original report, Goodyear went into great detail regarding the pumping system, surface control hydraulic system, and landing gear hydraulic system; this is omitted here due to the highly technical nature of the material, which is of minimal interest to the typical enthusiast.

Gear retraction time (GA-28B). The time required to fully retract the handling gear from the vertical takeoff position was estimated to be 11 to 14.95 seconds depending on the requirements of the boost system. This time was a result of the large fluid volumes and consequently large cylinders required to place the airplane in the vertical position without auxiliary power. However, if this retraction time was found to penalize unduly the time required to effect transition, the addition of two extra pumps would enable the gear to be retracted in 4.65 seconds.

PERFORMANCE SUMMARY

GA-28A. The GA-28A airplane was basically a ¾ scale version of the GA-28B Convoy Fighter design. Both versions were powered by turboprop engines driving dual-rotation propellers. The GA-28A was designed to be the prototype model which would furnish performance comparable to the GA-28B in vertical, hovering and transition phases of flight.

A unique design feature of the GA-28A/B was the ability to perform conventional-type landings and takeoffs, in addition to the specified vertical landing and takeoff.

Only moderate level-flight performance was obtained, mainly because there was no gearshift in the prototype specified engine-propeller combination. Maximum vertical acceleration, on the order of 8ft/second, could be expected from the prototype airplane. Transitions from vertical to horizontal

A speculative colour profile of the Goodyear GA-28A three-quarter scale Convoy Fighter prototype in the Glossy Sea Blue scheme that characterized early US Navy postwar aviation.

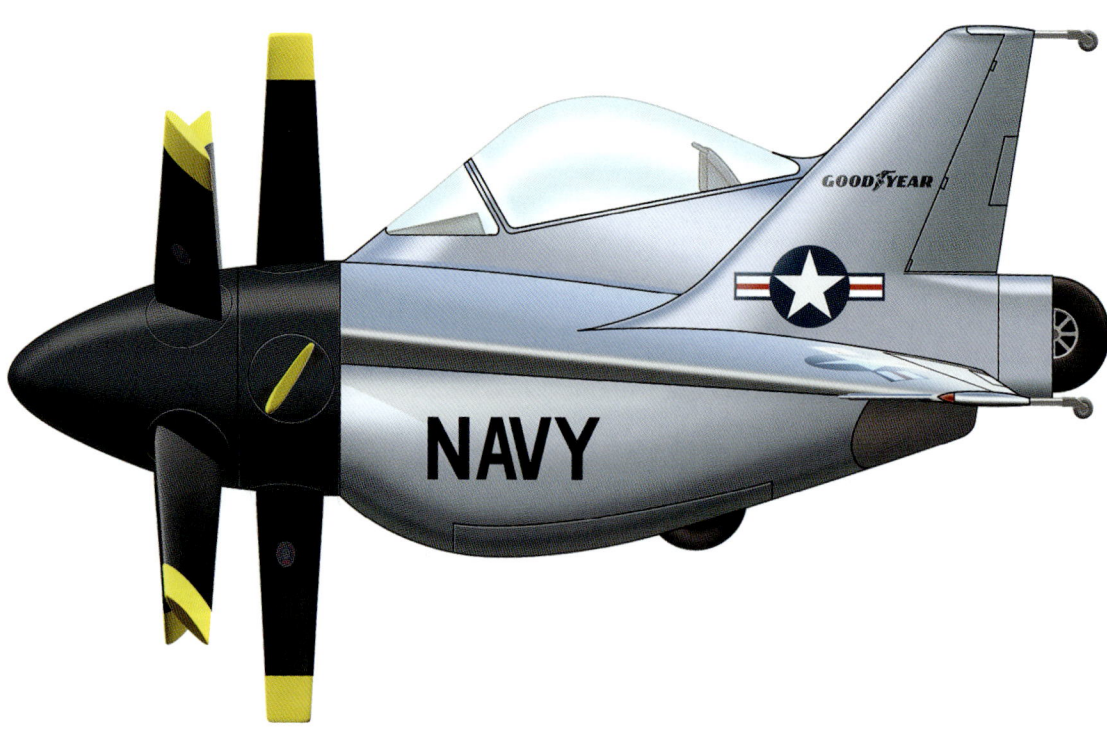

Goodyear proposed an alternate version of the GA-28A with the wing pods deleted and main gear retracting into the lower vertical tails; the natural metal scheme is inspired by the Convair XFY-1 Pogo.

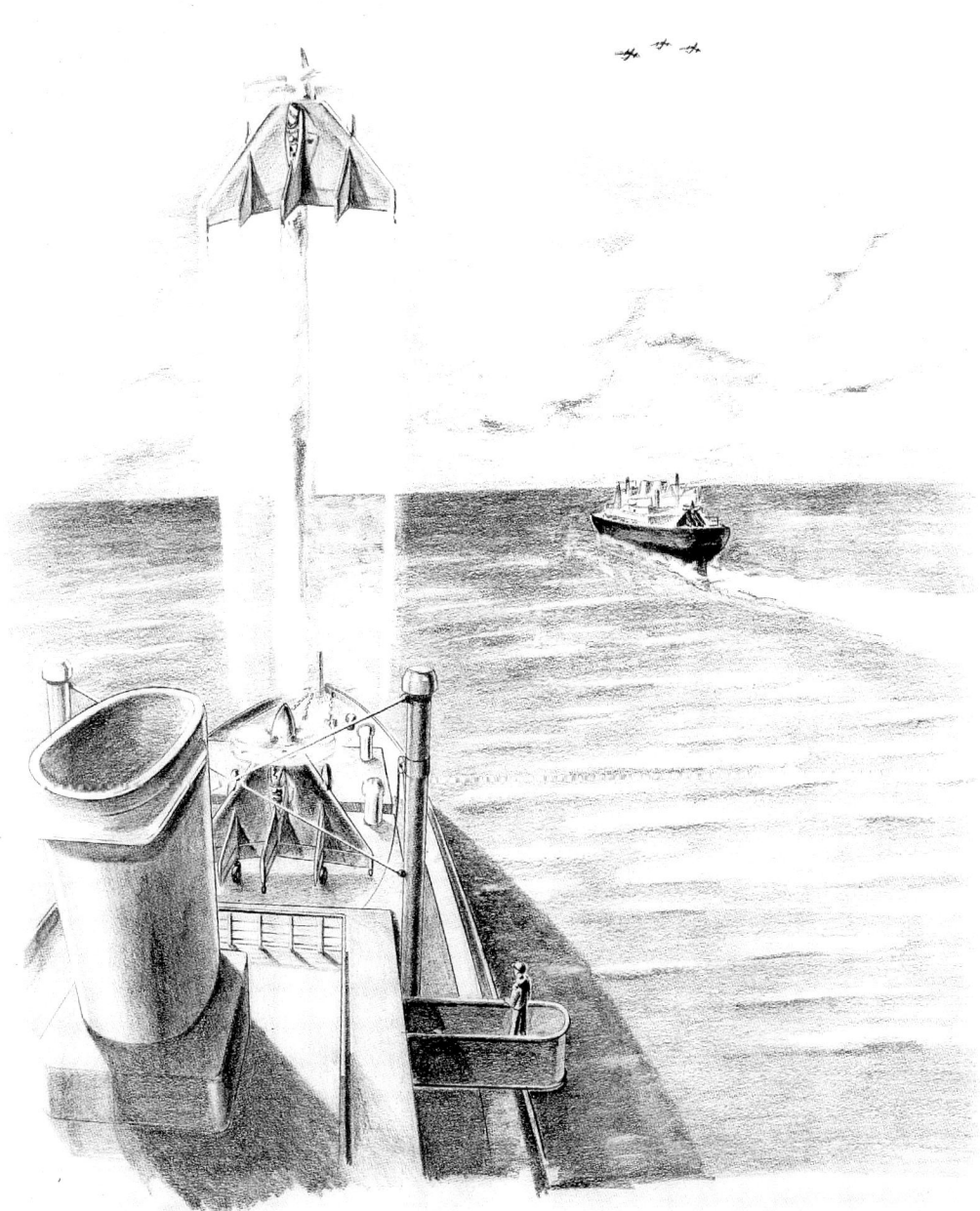

Artist's impression of the full scale Goodyear GA-28B Convoy Fighter taking off from the forward deck of a merchant vessel. Such ships would have required relatively minor modifications to operate the aircraft, as Goodyear's design featured retractable landing gear and a minimum of auxiliary handling and recovery equipment.

flight, and the reverse, could be made smoothly without abrupt manoeuvres. Maximum rate of climb at sea level was 6,250fpm. Service ceiling was 29,300ft, while time to climb to 20,000ft on military power was 6.2 minutes. Maximum level flight velocity was 320 knots at sea level and 315 knots at 20,000ft altitude.

GA-28B. The GA-28B airplane was a tailless type incorporating a modified triangular wing planform which utilised NACA low-drag airfoil sections. It was designed to meet the basic requirements of vertical takeoff from the deck of a ship, ability to hover in a vertical position and ability to make smooth transition into conventional level flight. In addition, it presented the high all-around performance required of a fighter type aircraft.

An additional feature was its ability to take off or land in a conventional manner with its specially designed landing gear. Initial transition could be accomplished as a gradual pushover with a minimum acceleration along the flight path of approximately 8ft/sec^2, and could be accomplished from zero velocity on a carrier deck to 150 knots level flight at 1,000ft of altitude.

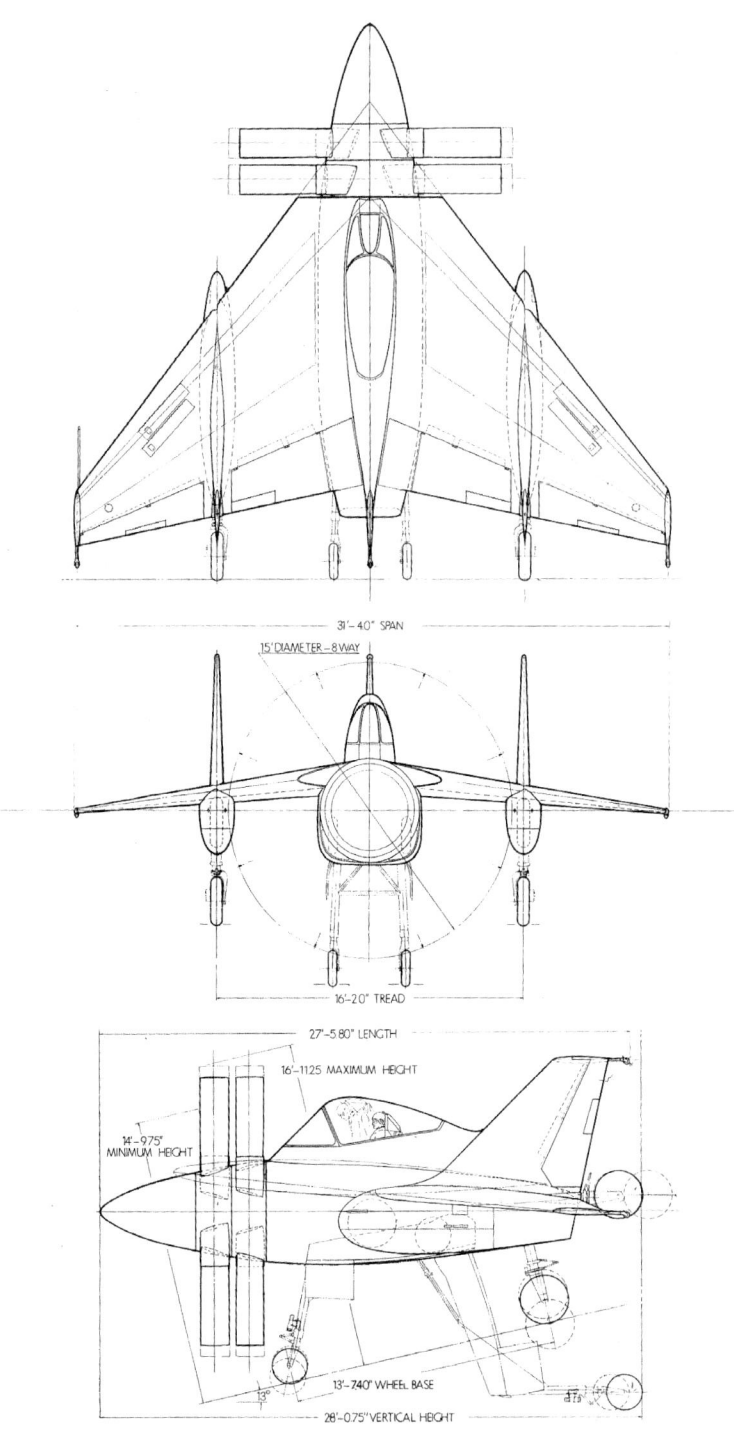

Basic three-view of the Goodyear GA-28B Convoy Fighter showing landing gear positions for both horizontal and vertical attitudes; note the dual wheel nose gear.

Final transition from level flight to vertical landing was accomplished as a gradual slowing of airplane velocity as the attitude angle was increased, along with variation in engine throttle settings, until an attitude of 90 degrees was reached and the forward velocity was zero. This transition could be made with a minimum total change in altitude consistent with OS-122. A conventional takeoff could be accomplished in zero wind with a ground run of 1,035ft. A 30 knot headwind reduced this distance to only 640ft. This indicated the possibility of conventional landings and takeoffs from aircraft carriers.

Maximum rate of climb at sea level on military power was 10,000ft/min. Time to climb from

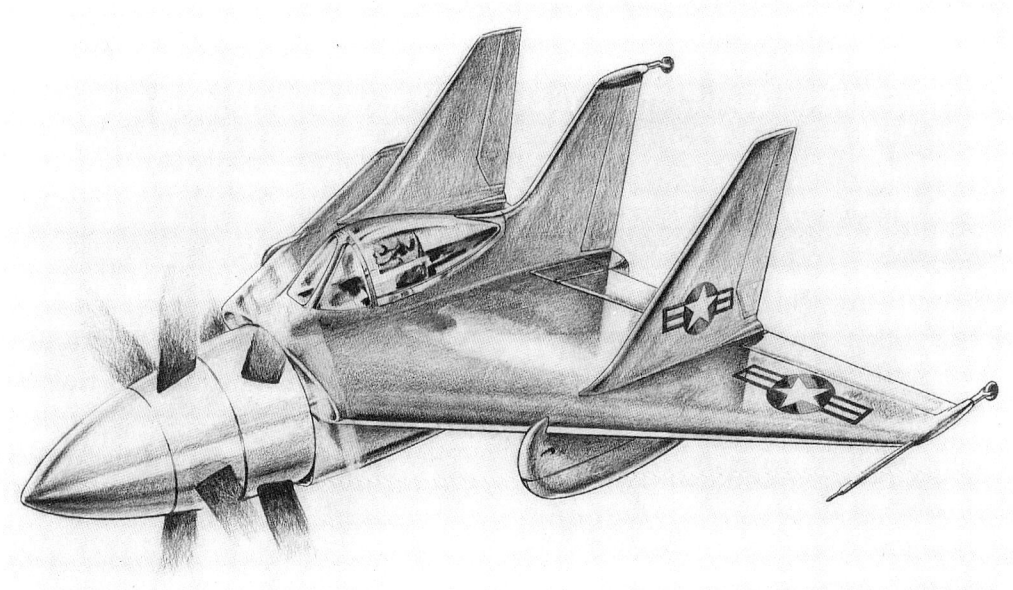

STANDARD AIRCRAFT CHARACTERISTICS

DATE: 20 NOVEMBER 1950

GOODYEAR AIRCRAFT
MODEL GA 28B

Contemporary artist's impression of the Goodyear GA-28B Convoy Fighter from November 20, 1950.

standstill to 35,000ft with vertical takeoff was approximately 5.7 minutes. Attainment of combat radius under specified conditions permitted one hour 20 minutes for loiter at 35,000ft before mission was started. Maximum velocity was 500 knots and minimum radius of turn was 18,700 ft at 420 knots.

The improvement in performance resulting from the elimination of the external landing gear and gun fairings would result in a 25 knot increase in maximum speed and a reduction in the time to climb to 35,000 ft to 4.7 minutes. In addition, the outboard gun installation permitted an increase in the diameter of the propeller, resulting in approximately a 2% increase in propeller efficiency which would allow the GA-28B to meet the specified performance requirements.

AUTOMATIC PILOT

A control system for both aircraft was proposed which met the control specifications. The design of the system made use of a commercially available autopilot with some modifications. The dynamic analysis of the control system-aircraft loop was carried out both by standard automatic control techniques and by simulation on the Goodyear electronic differential analyzer. From the preliminary analysis of the stability of the airplane without controls in the hovering case, the characteristic equation indicated that it would have been controllable without autopilot if the controls were given constant pilot attention.

CONTROL SYSTEM DESIGN

The control system had four major divisions:

1. pilot controls
2. autopilot
3. special features
4. airplane controls

The pilot controls consisted of the command selector, the formation stick, the pilot control stick and rudder pedals, the pilot throttle-control, and trim-tab adjustments. The command selector was a means by which the pilot could select the control system operation mode desired. It was a rotary switch having five different positions which could be selected in any sequence. This was done by declutching the selector knob and moving it

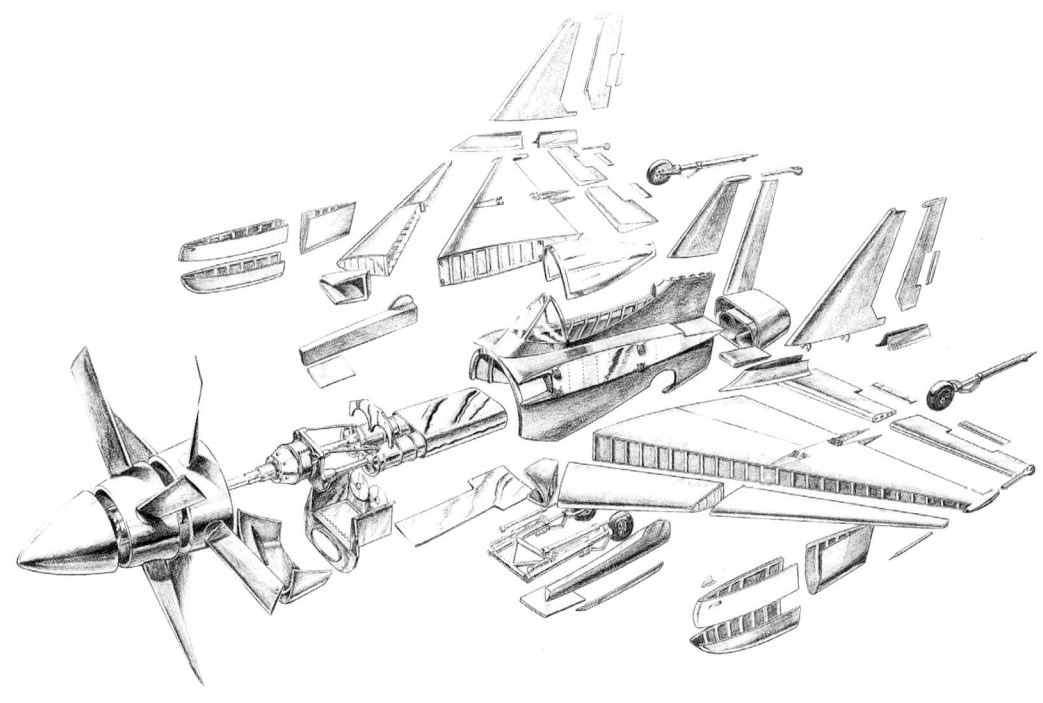

GENERAL PRODUCTION BREAKDOWN   GOODYEAR AIRCRAFT MODEL GA28B

DATE: 20 NOVEMBER 1950

Exploded view of the GA-28B showing the major components of this bizarre and interesting design. Note the eight-blade contra-rotating propeller, which set it apart from the Convair XFY-1 and Lockheed XFV-1, both equipped with six-blade units.

to the desired position. When the selector clutch was reengaged, the contacts were snapped to the new position. This feature was necessary since the sequence of modes of operation was not always the same. An interlocking safety device was necessary so that certain switching sequences were prevented. The description of these five modes of control system operation was as follows:

1. The OFF position made the entire autopilot inoperative by disengaging the servo clutches. In this position the pilot control-stick and rudder pedals were in operation through the booster.
2. With the command selector in the HOVERING TO CRUISE position, the servo clutches were engaged and pitch and throttle time programmes were initiated to automatically launch the plane and take it through the transition stage to the cruising flight condition. If the airplane was already hovering, the airplane pitch and throttle time programmes would not have changed.
3. With the command selector in CRUISE TO HOVERING position, a second pitch and throttle time programmer would have been used and was designed to reverse the procedure used in the HOVERING TO CRUISE mode.
4. When either transition phase was completed, the pilot could switch the command selector to the FORMATION STICK position which allowed the pilot formation stick to be a pitch-and-turn reference and the heading knob to be a rudder control, except when there was a turn called for by the formation stick—at which time the heading reference was locked out. This allowed the pilot to manoeuvre in either the vertical or the horizontal positions. It was also possible for the formation stick to have been used in the transition phase, although this was abnormal.
5. In the AUTOPILOT position, the command

selector tended to hold the airplane in the heading in which it was flying at the time the AUTOPILOT position was selected. Changes in this heading were made by switching the command selector back to FORMATION STICK position and controlling it to the heading desired with the formation stick.

DITCHING

Several aspects of the ditching characteristics of the aircraft were outlined:

1. If full power was available, a vertical descent was obviously not desirable due to the possibility of the airplane landing on its back once the control surfaces were immersed. The landing velocity for conventional type landings was high, and sinking speed relatively great for the low landing attitude desired to avoid heavy nose impact loads which could be incurred if the tail 'dug in' deeply in a high altitude approach.
2. It appeared that a power-on approach in the conventional manner was the optimum method of ditching the vehicles. Pilot technique should have been such as to prevent the propellers from immersing in the water until the airplane was slowed as much as possible. The high location of the cabin shielded by the large chord wing was a favorable feature in that the pilot had sufficient time to escape since the cabin was hinged at the aft end and was easily jettisonable.
3. If no power was available, the landing speed and approach sinking speed would have been in all probability far greater than the airplane structure could withstand.

COST PROPOSAL

In its Informal Cost Proposal dated December 1, 1950, Goodyear sought to negotiate a mutually acceptable cost plus fixed fee (CPFF) contract with BuAer. Under this arrangement, Goodyear estimated the cost of the GA-28A to be $4,554,300.10 and the GA-28B to be $7,639,287.08; the combined total was $12,193,587.18. Under a fixed price type contract, which Goodyear did not endorse, the cost of the GA-28A was estimated to be $5,692,875.13 and the GA-28B to be $9,549,108.88, for a total of $15,241,984.01. All figures are in 1950s dollars.

The GA-28A would have been ready for flight test approximately 17 months after authority was received to proceed. The second airplane would have been ready for flight test one month later. Goodyear stated that the automatic pilot was the most critical portion of their work in delivering the first prototype. The flight test programme for the airplane would have required about five months.

The static test article of the GA-28B would have been completed 14 months after the contract was signed. The second airplane would have been ready for flight test three months later and the third airplane three months after that. The flight test programme for the GA-28B would have required about five months.

Goodyear recommended that BuAer save time and money by skipping development of the GA-28A prototype and proceeding immediately with construction of the GA-28B. BuAer would soon agree with this advice.

ELIMINATION OF THE GA-28A

In a letter from Goodyear to BuAer dated December 22, 1950, the company indicated that a stripped-down version of the GA-28B airplane grossing 14,000lb at takeoff could be flown on a XT-40-6 power plant modified for vertical running and still satisfactorily demonstrate transition and adequate maximum speed with endurance approximately the same as for the proposed ¾ scale model GA-28A airplane. This was in accordance with an agreement reached between representatives from Goodyear and BuAer during informal conversations related to the Convoy Fighter project.

The minimum acceleration of the stripped GA-28B in vertical flight condition with takeoff gross weight of 14,000lb was 5ft/sec². The maximum speed at 35,000ft at military power (5,500 shaft horsepower) and combat weight of 13,000lb was 475 knots.

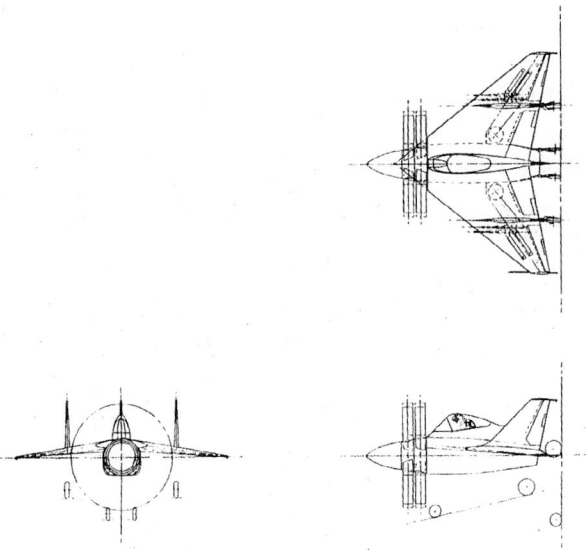
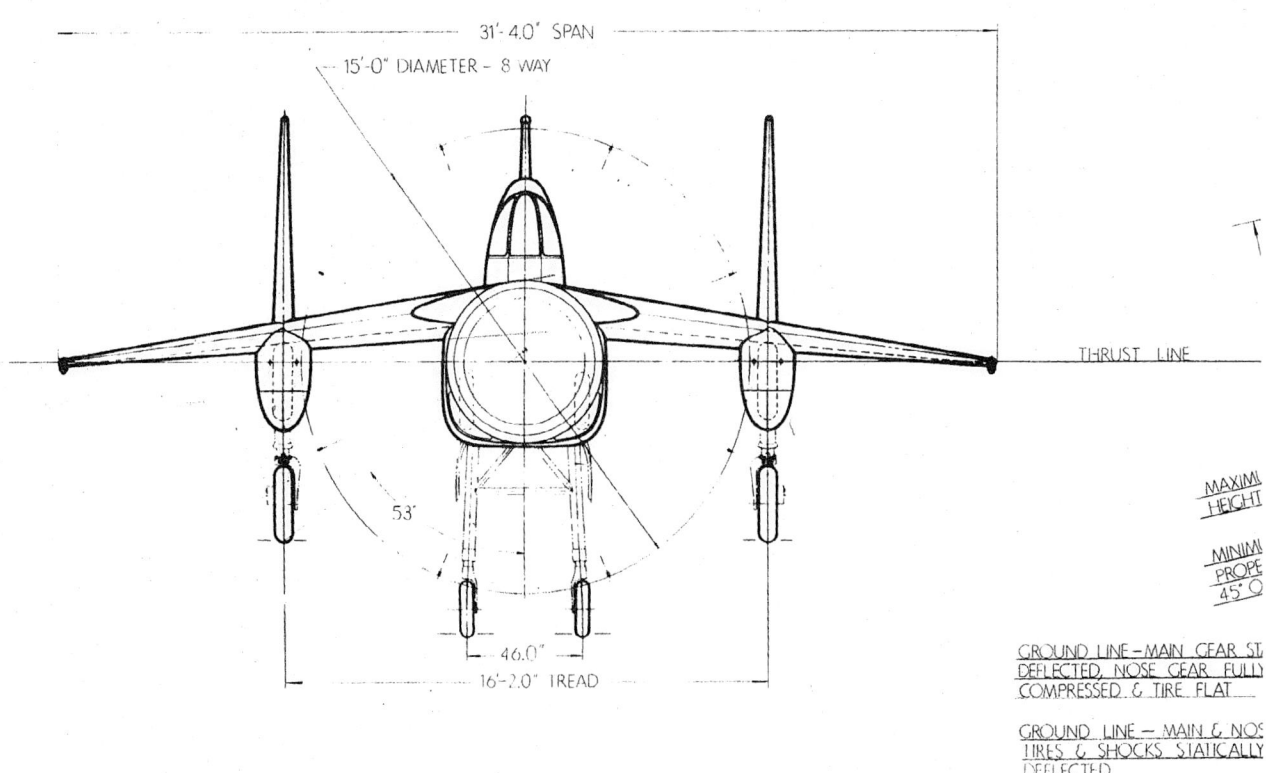

Detailed general arrangement of the GA-28B with additional measurements and other information added. Note the scrap three-view showing an alternate version with the wing pods deleted and inward folding main gear; this modification was estimated to increase the maximum speed by 25 kts and enabled an increase in propeller diameter, improving its efficiency.

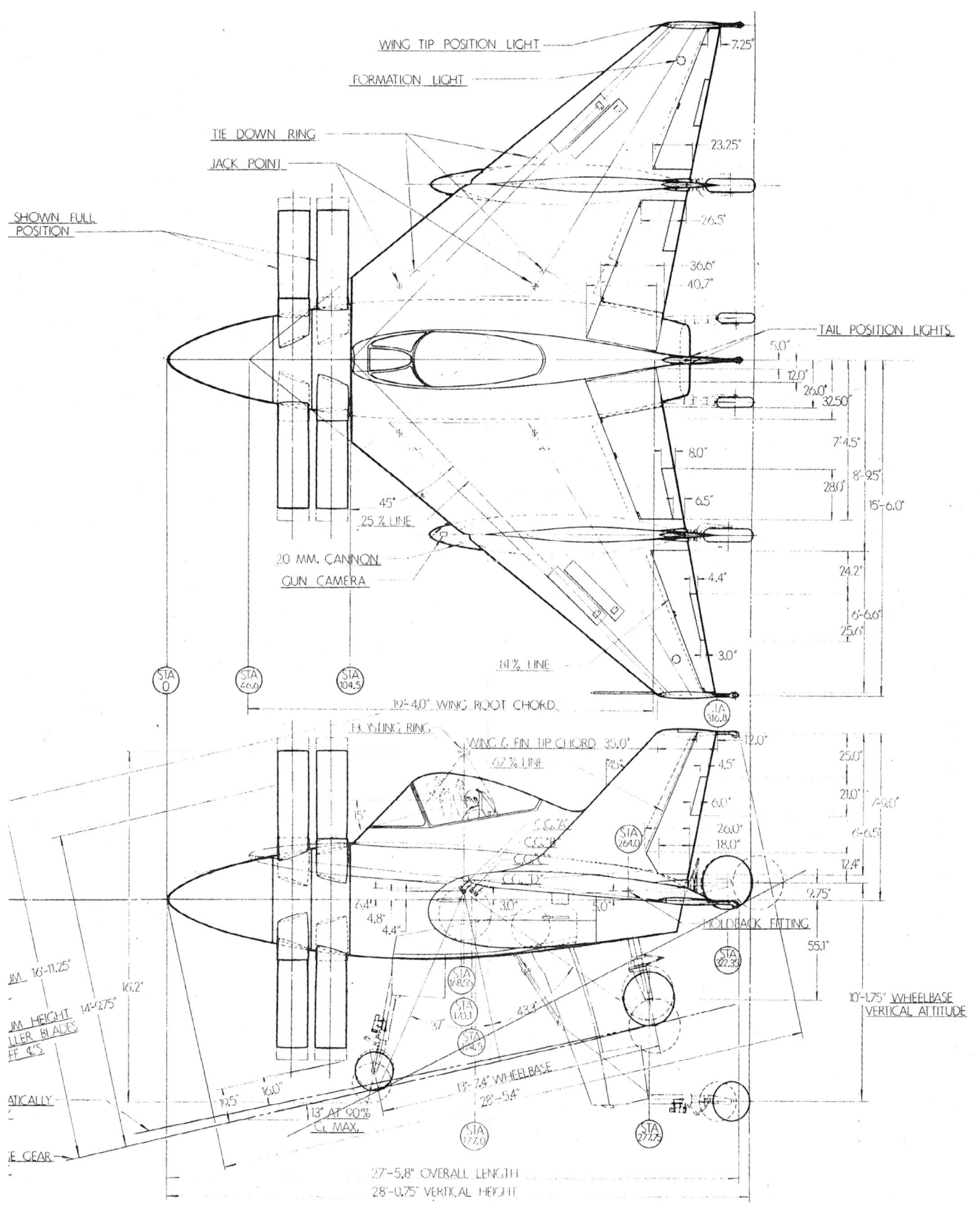

Detailed physical data for the GA-28B taken from the blueprint shown in the previous spread.

Legend for the inboard profile blueprint reproduced previously.

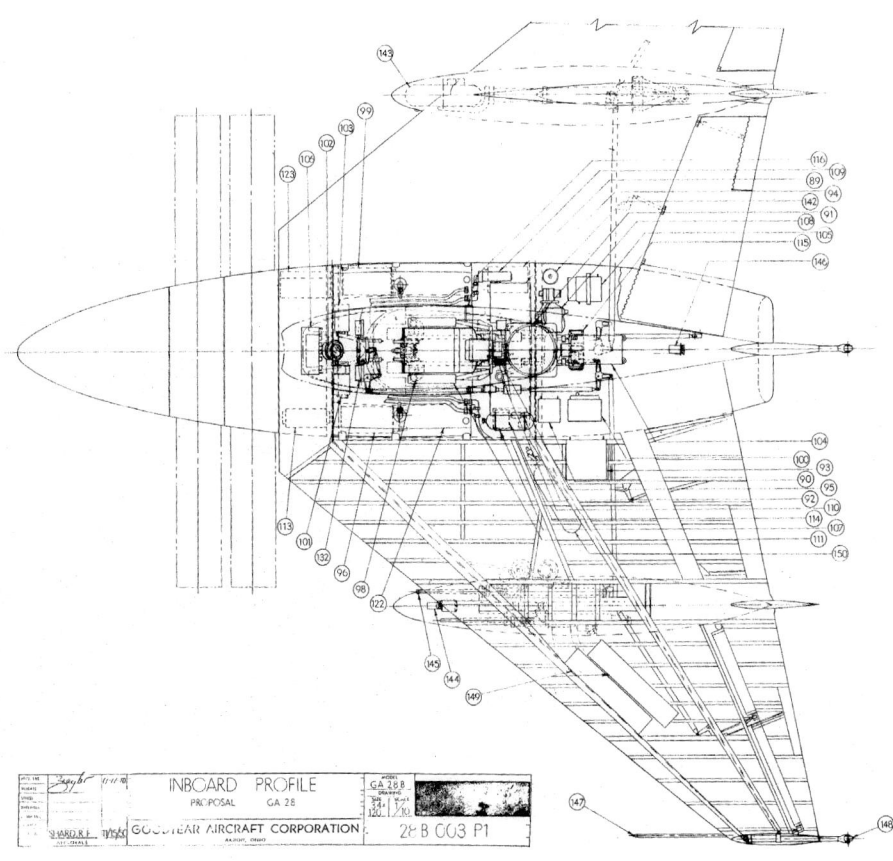

Inboard top view of the Goodyear GA-28B showing the disposition of major equipment and structural members.

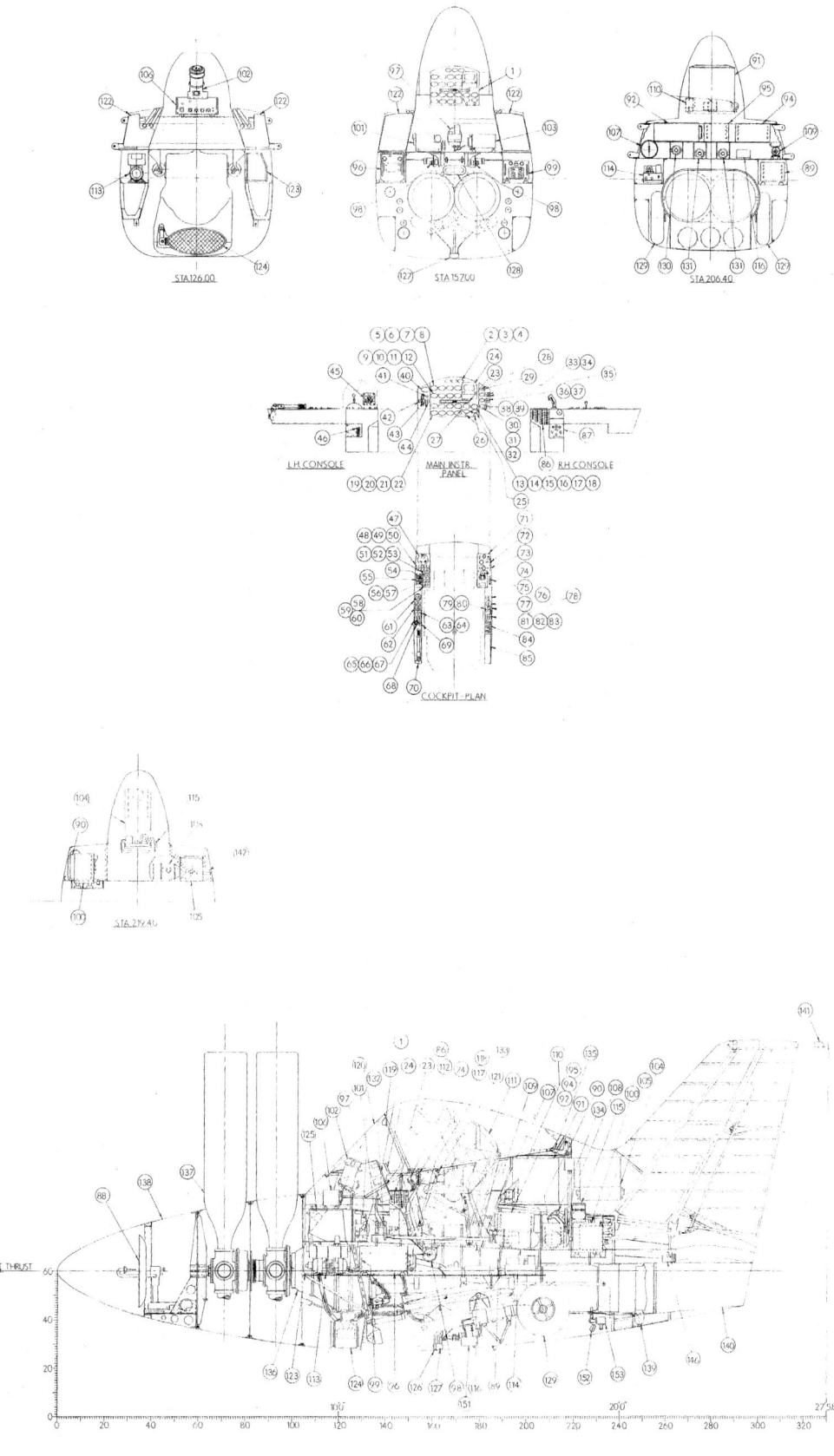

Side inboard profile, fuselage cross sections, and cockpit layout of the GA-28B Convoy Fighter; note the tilting seat to enable the pilot to land the aircraft in the vertical position, which in practice proved to be no easy feat with either the Convair or Lockheed vehicles.

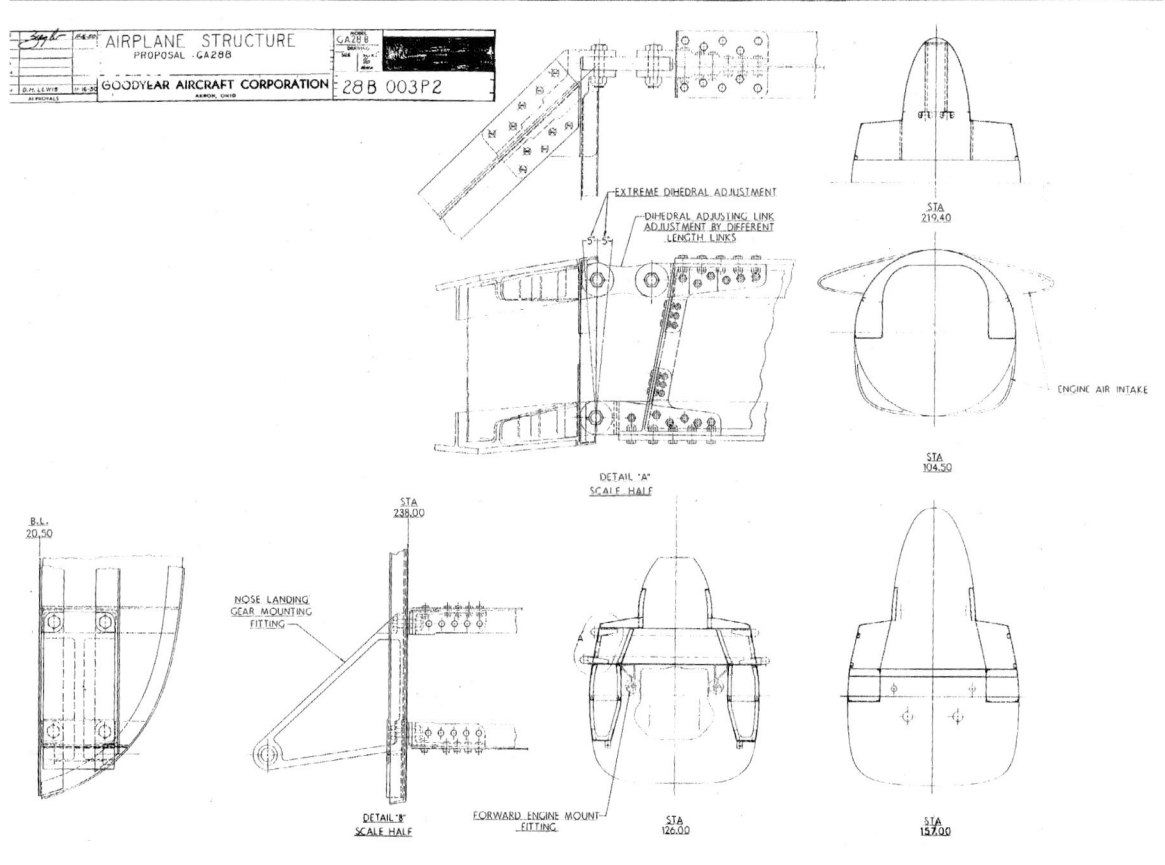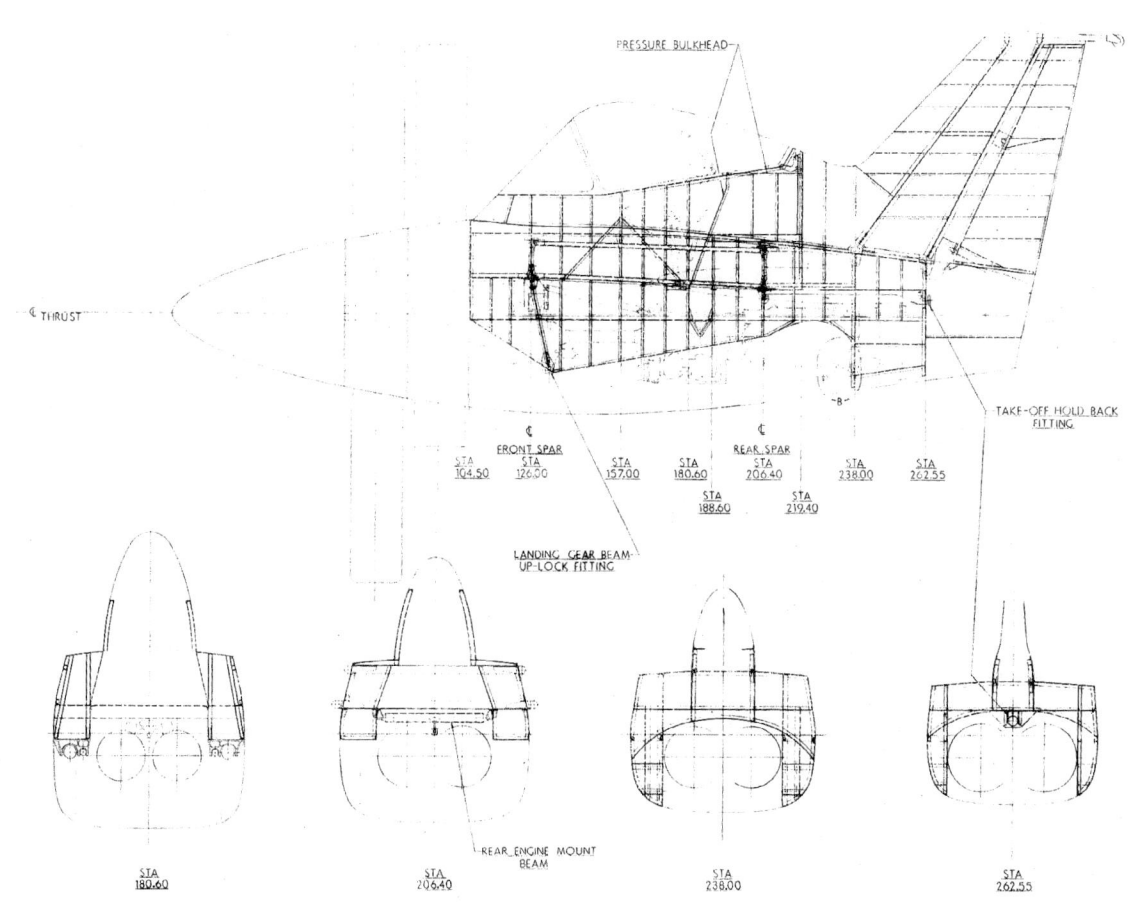

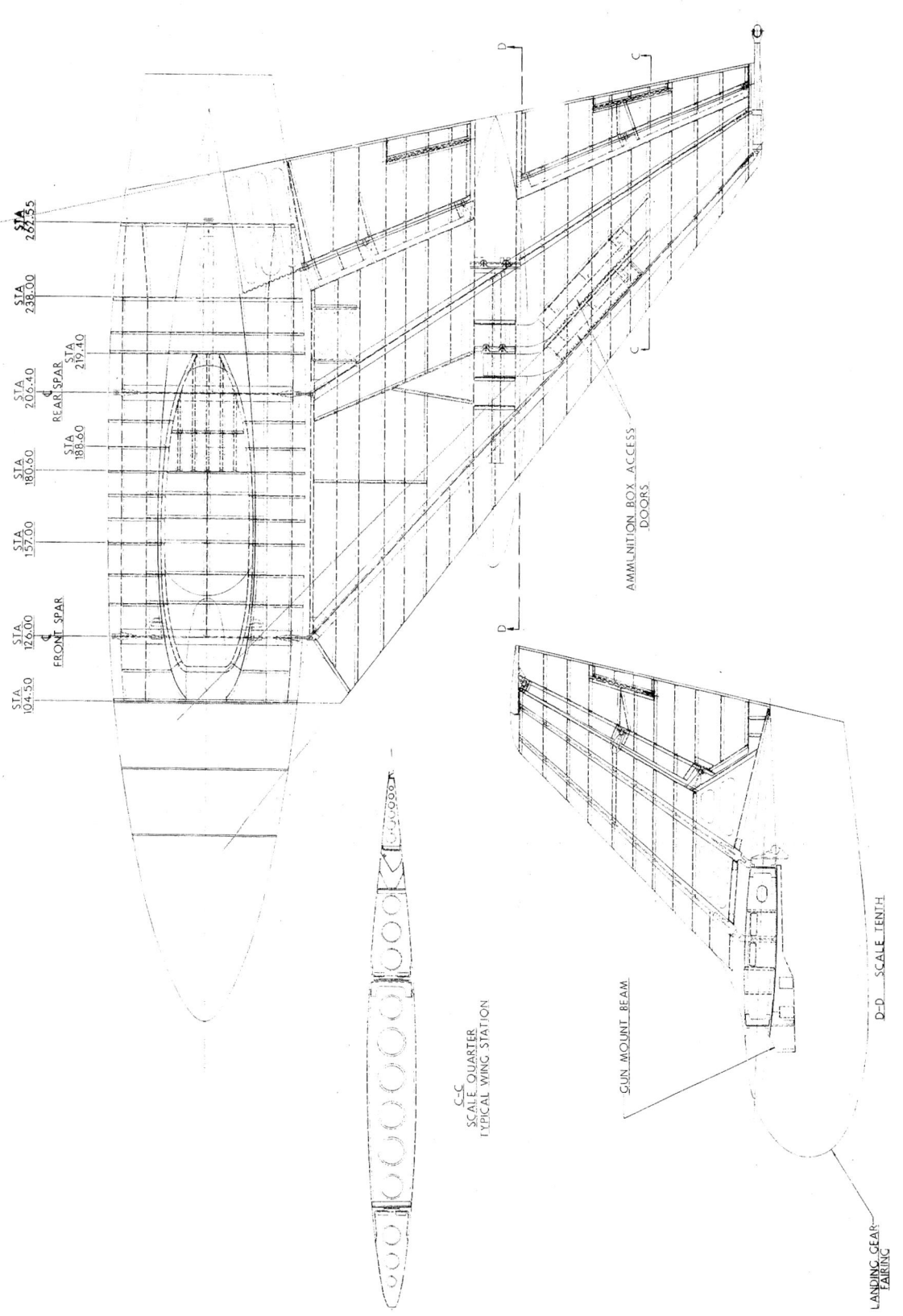

Structural blueprint of the GA-28B, which was of conventional stringer and light skin construction with flush joints and riveting throughout. Incorporation of a metal core sandwich material as a means of weight reduction was also considered. (opposite and above)

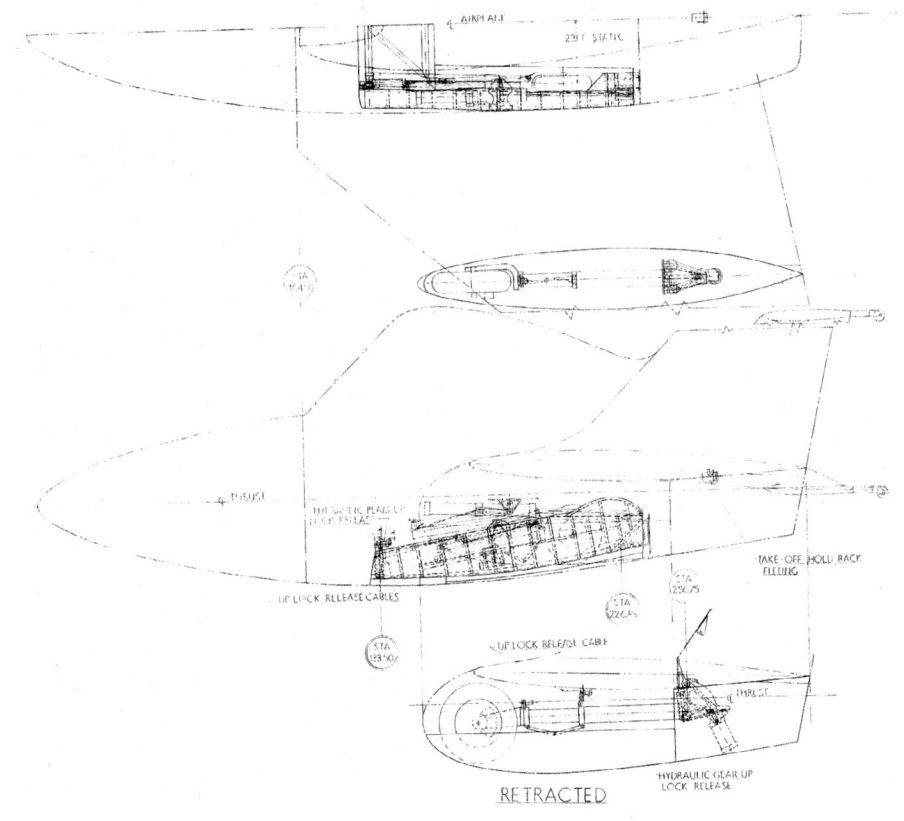

A blueprint of the Goodyear GA-28B with the landing gear retracted.

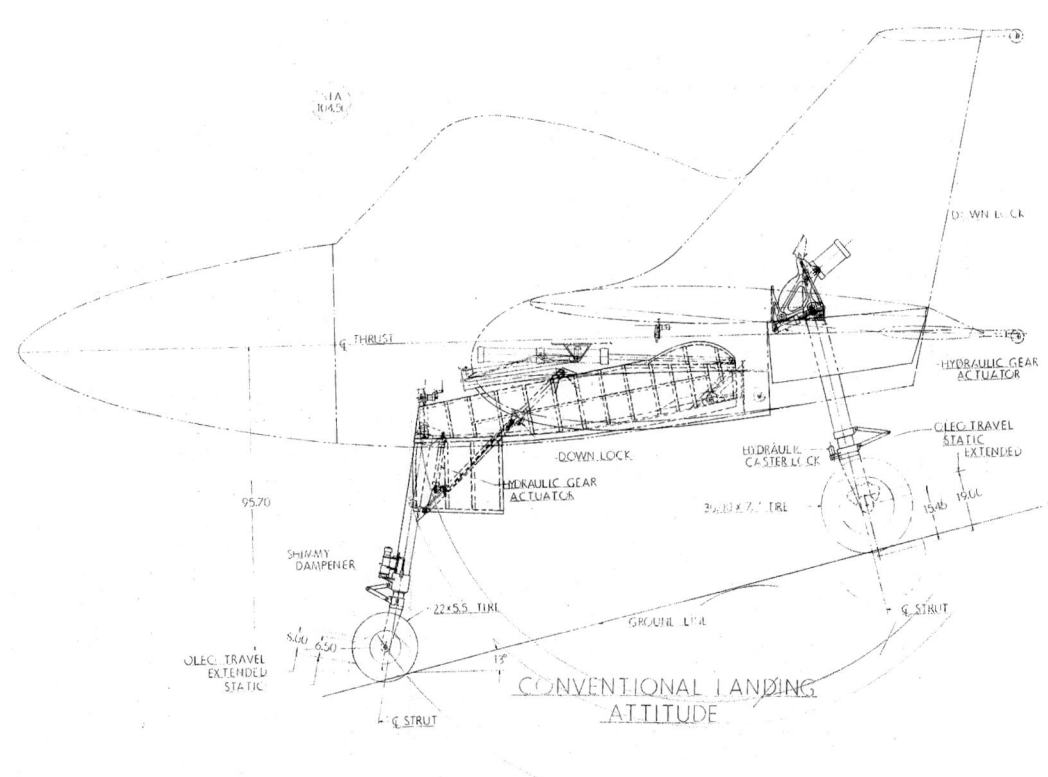

The landing gear extended in the conventional landing attitude.

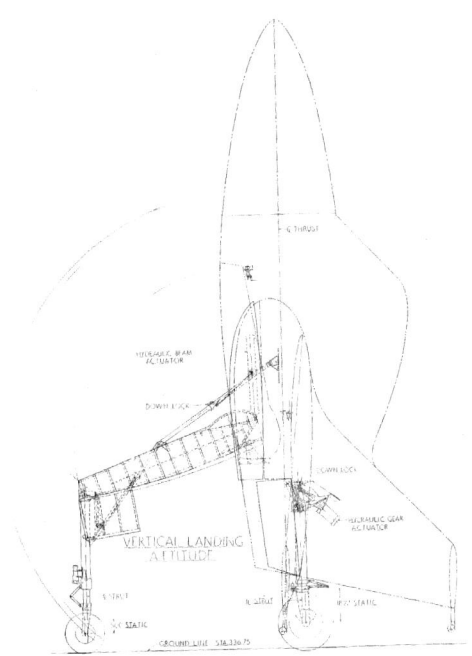

Drawing of the GA-28B with the landing gear extended in the vertical landing attitude. The complicated rotating nose gear assembly must be among the strangest ever contemplated for an aircraft. The 180-degree rotation of the main gear is also noteworthy.

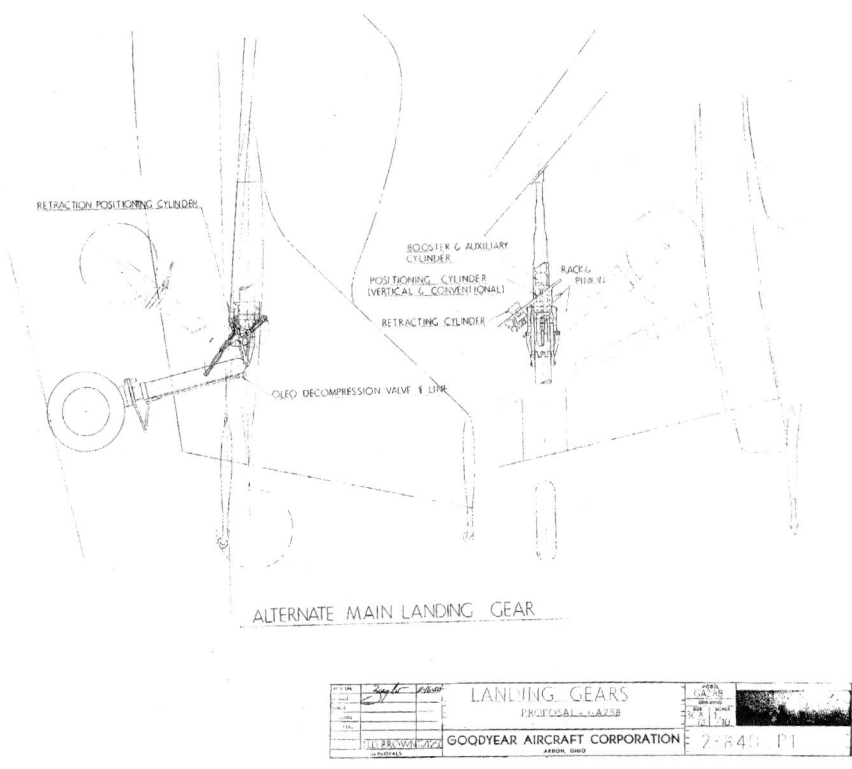

Late in the development of their proposal, Goodyear drew up an alternate main gear arrangement where the outboard pods were deleted and the gear retracted inwards towards the fuselage.

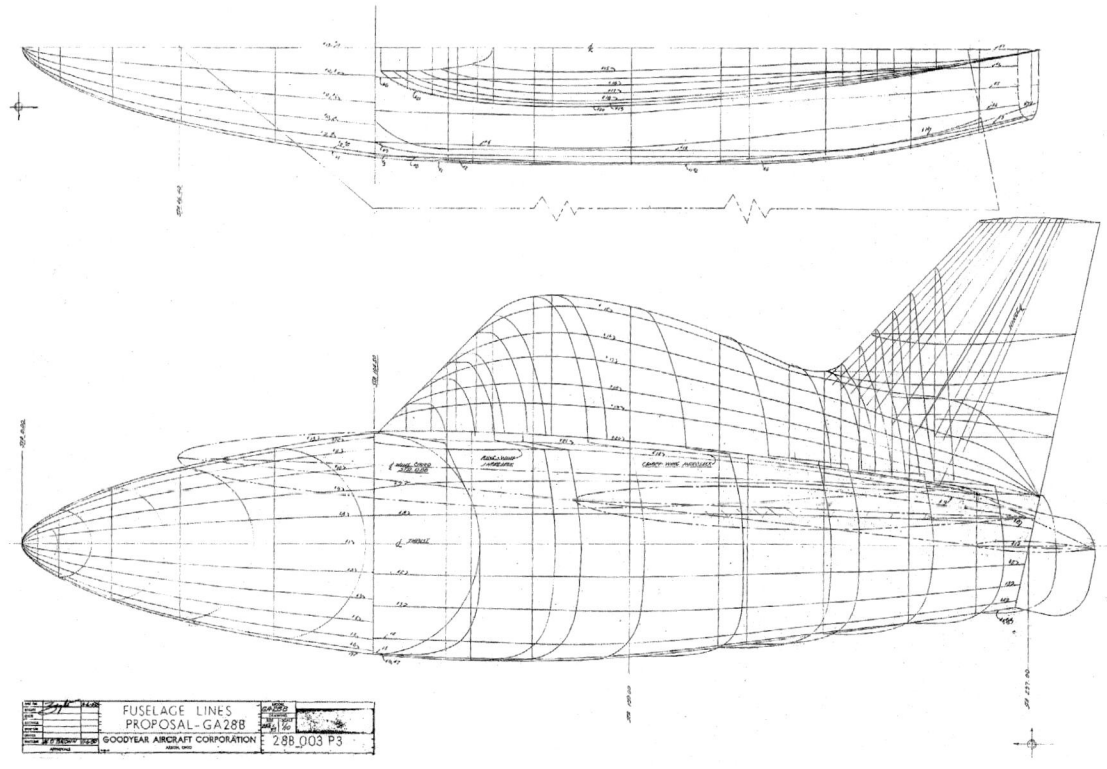

The fuselage loft lines of the GA-28B.

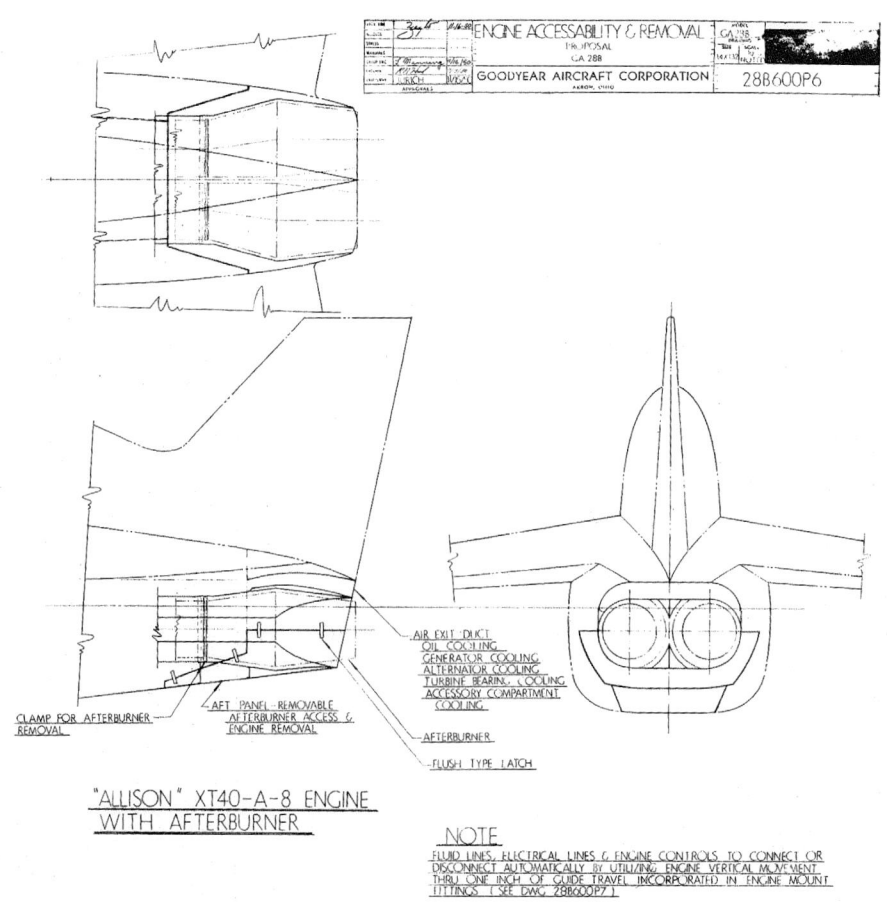

Engine accessibility and removal diagram for the afterburner-equipped version of the fighter.

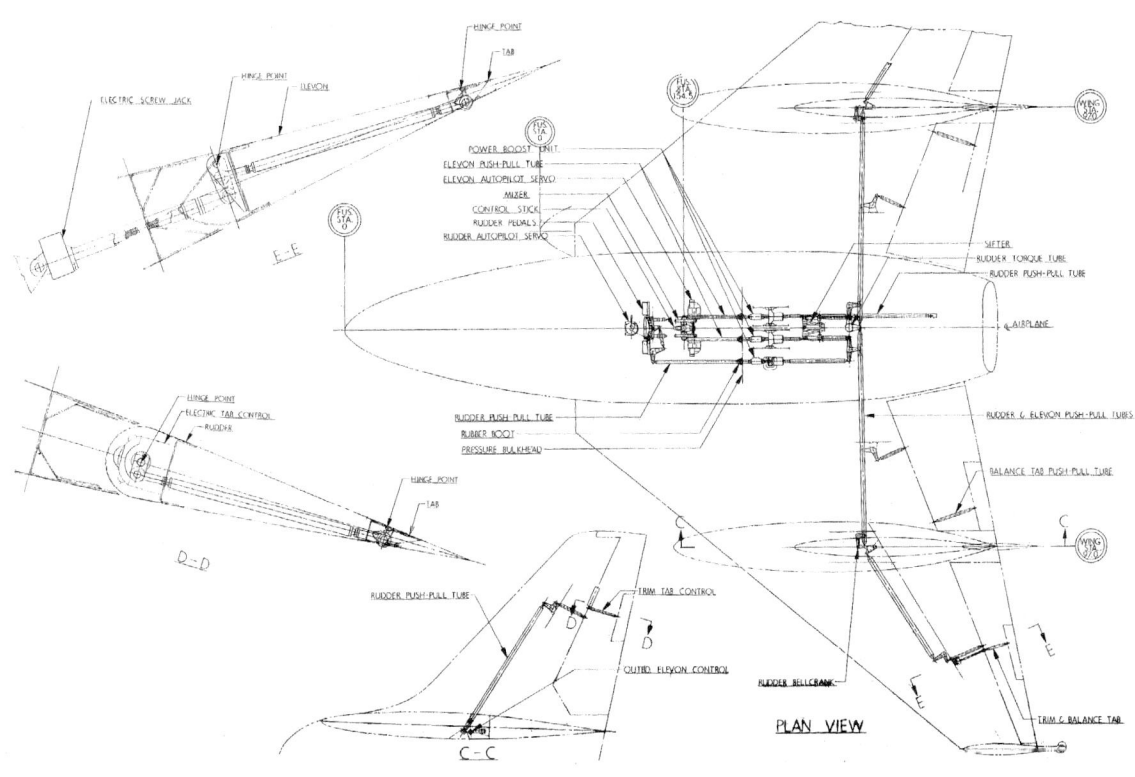
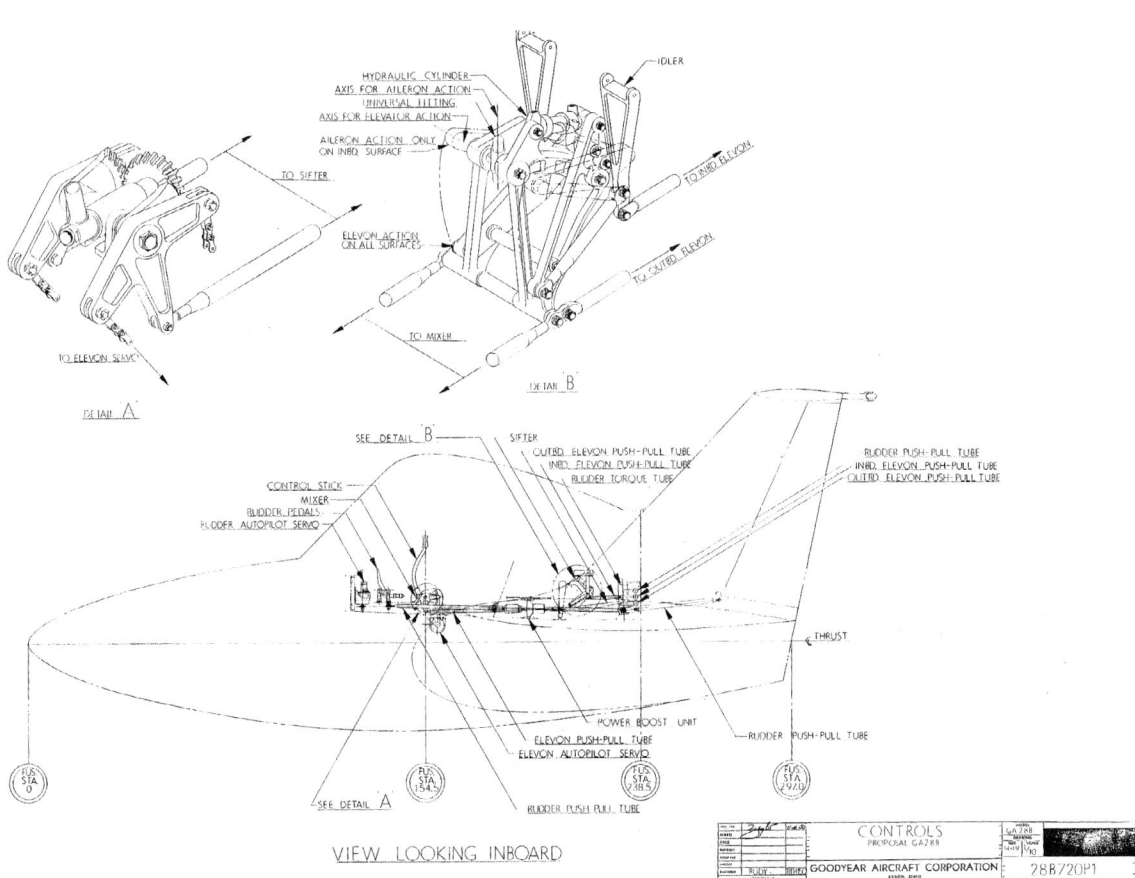

Diagram of the GA-28B flight control system.

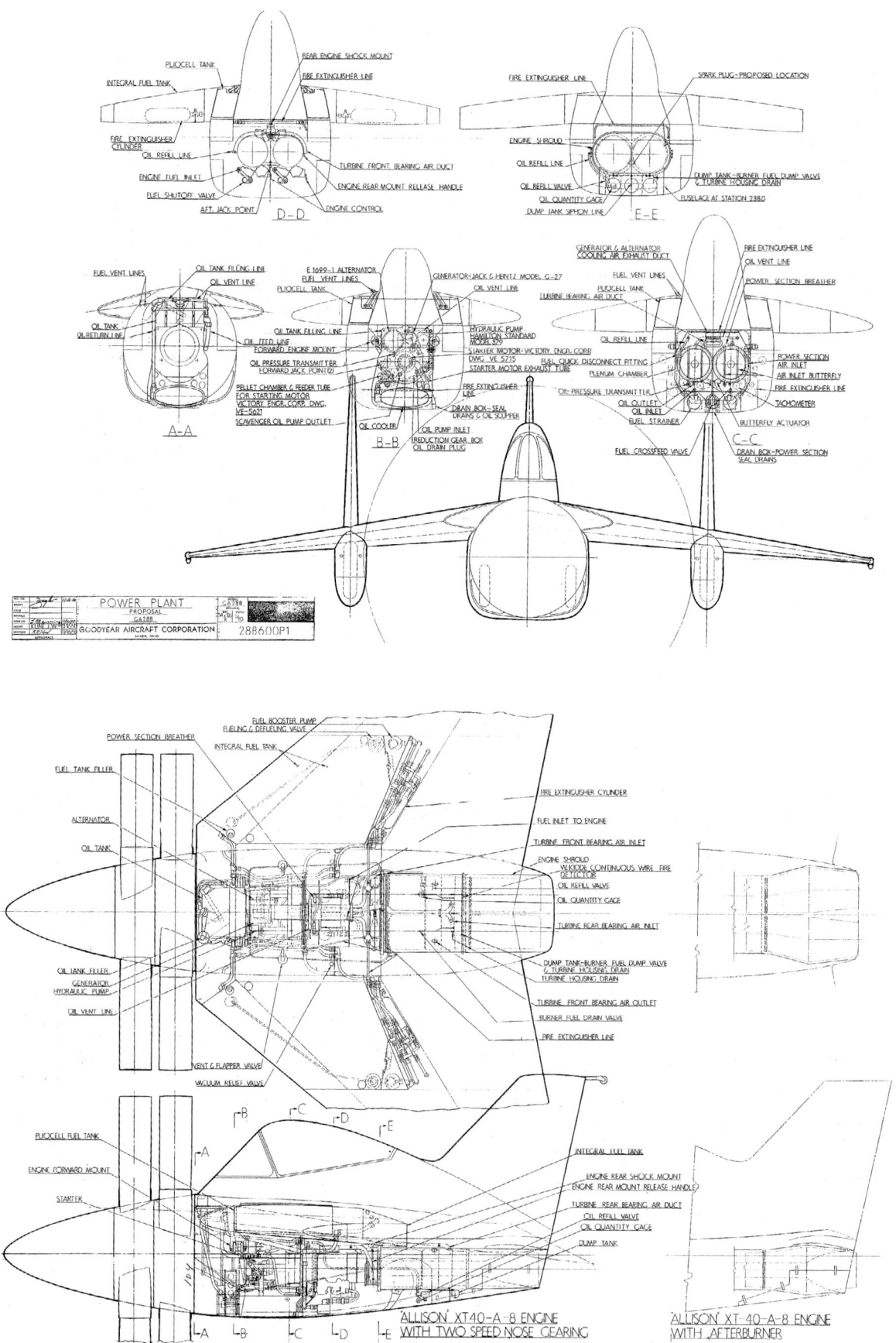

Blueprint of the power plant installation which includes an alternate version with an afterburner.

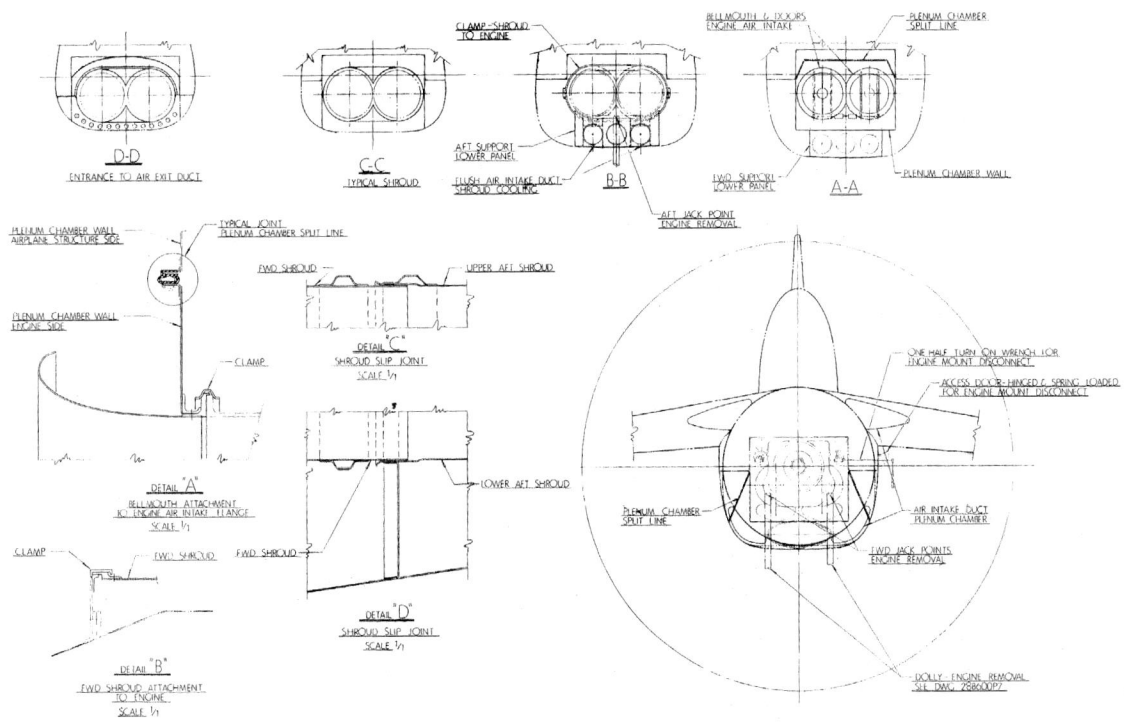
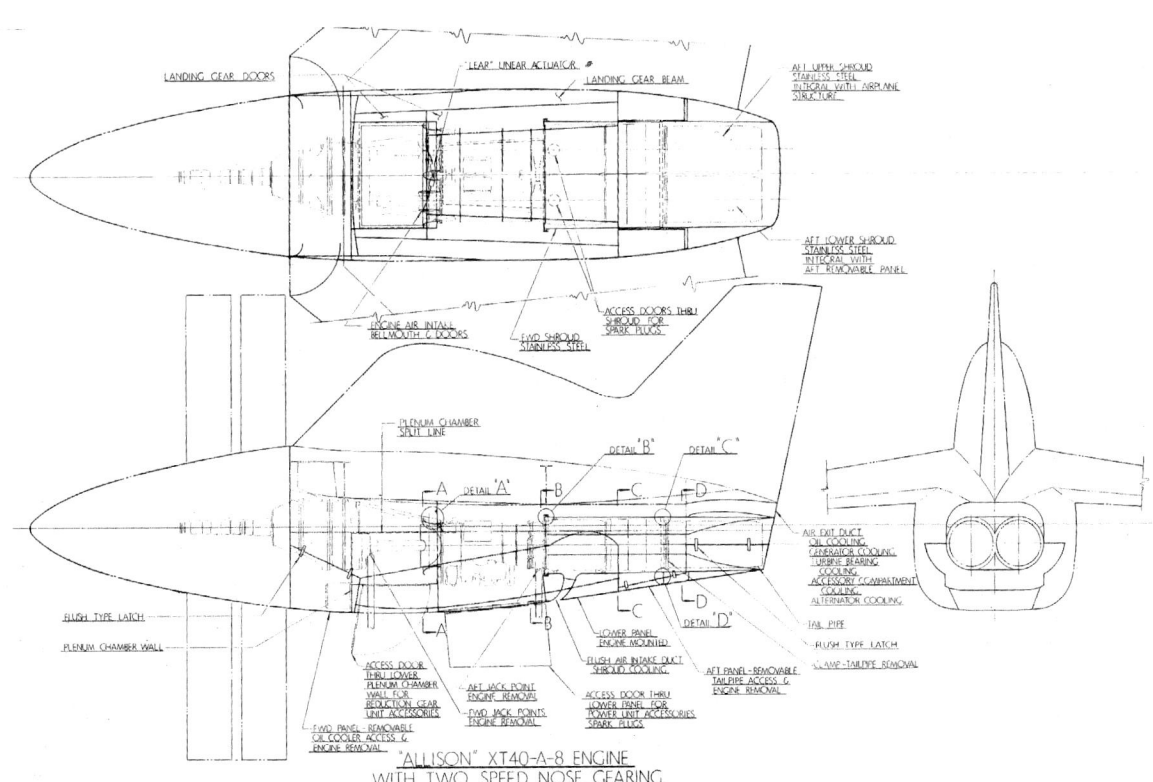

Engine accessibility and removal diagram for the Goodyear GA-28B.

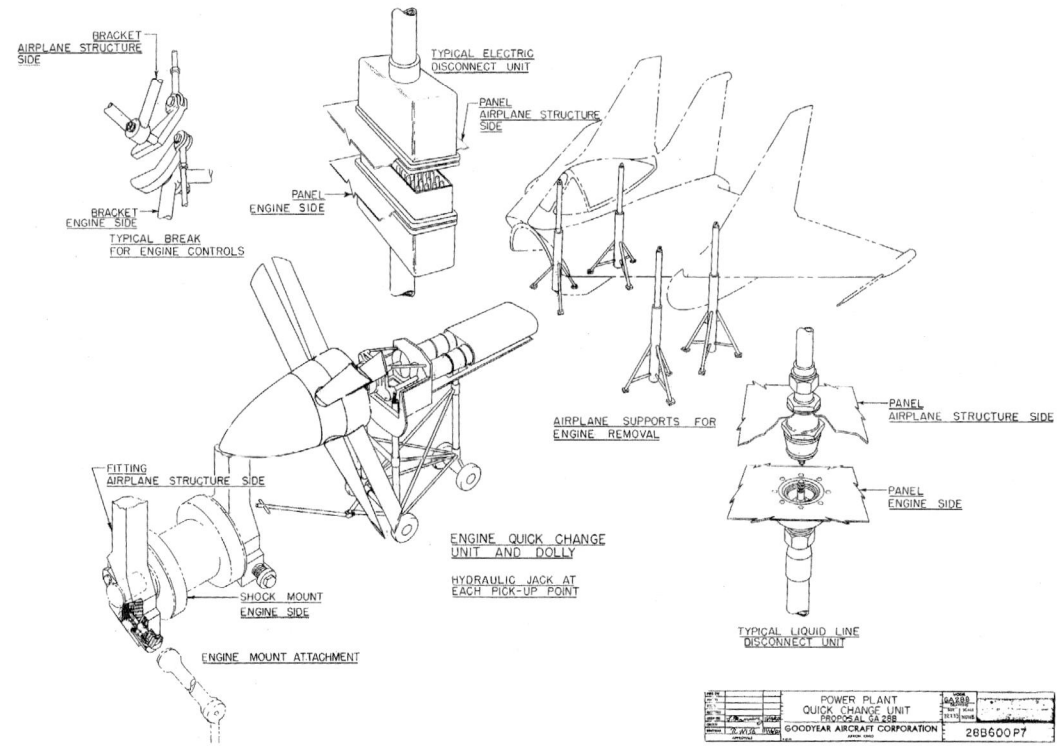

Perspective drawing of the power plant quick change unit.

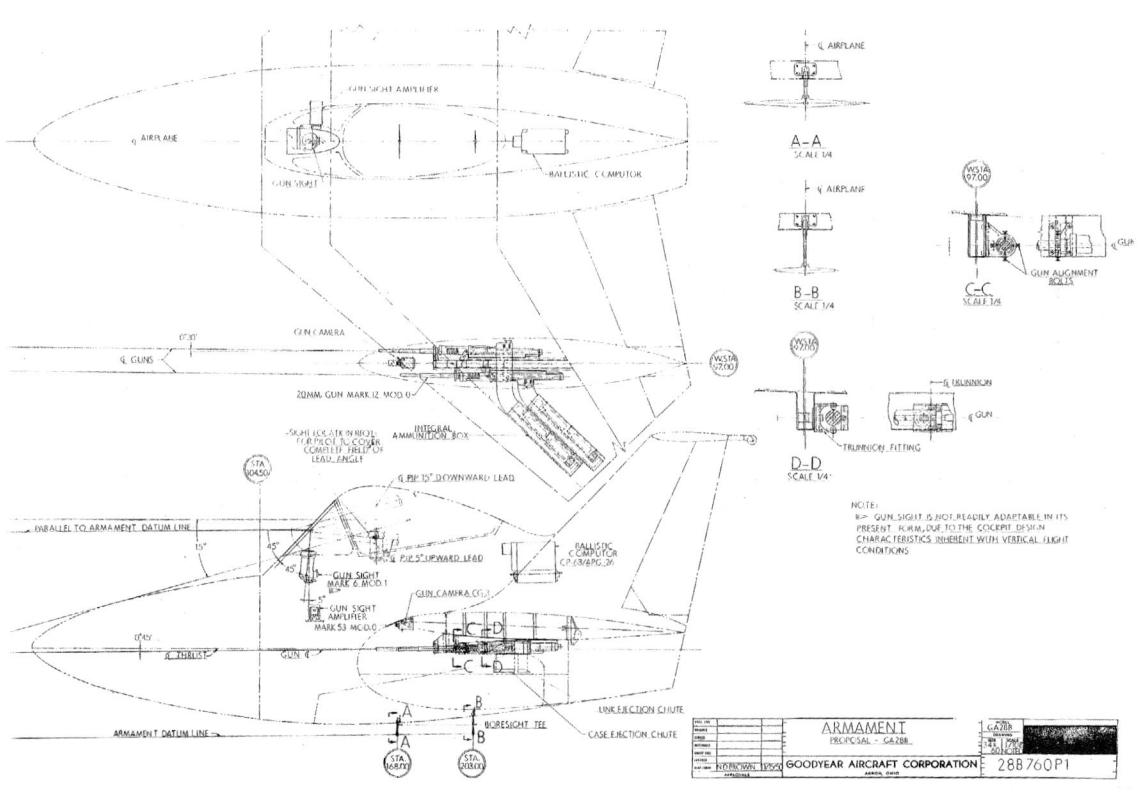

The GA-28B was armed with four 20mm Mark 12 Mod. 0 cannon, two in each wing pod.

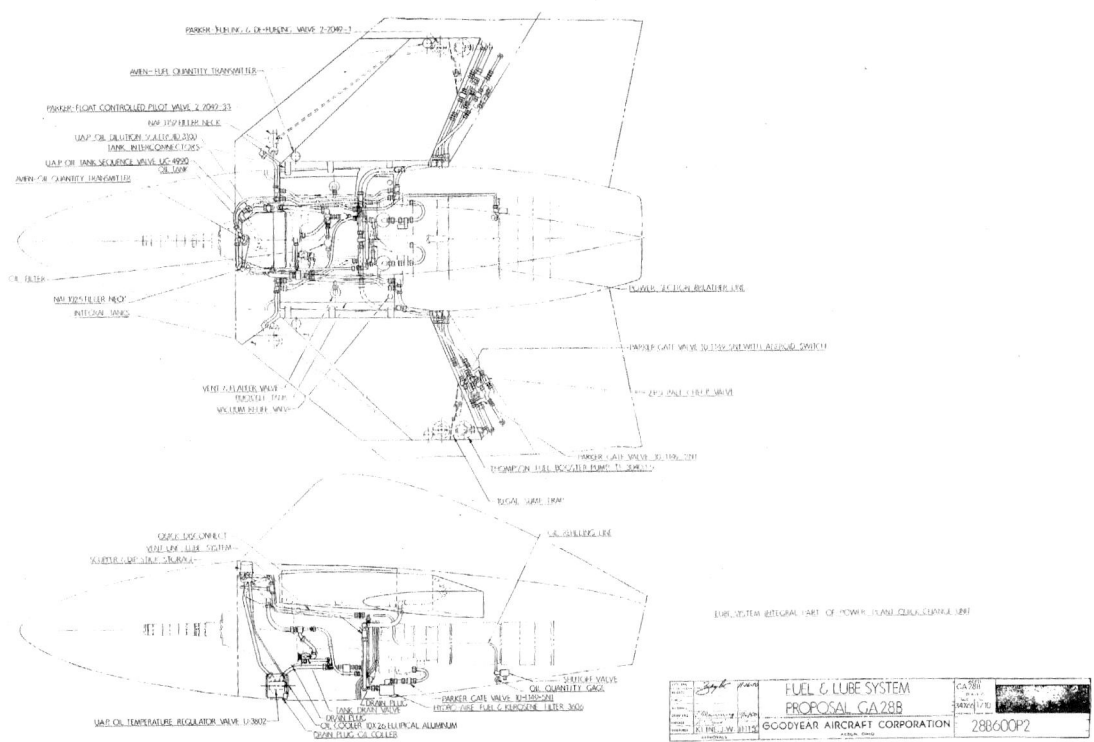

Blueprint of the GA-28B's fuel and lubrication system.

A speculative colour profile of the Goodyear GA-28B in the overall Glossy Sea Blue scheme which was standard at the time it was proposed.

Goodyear also considered removing the wing pods from the full scale GA-28B Convoy Fighter, primarily to decrease drag and increase performance; this natural metal scheme is inspired by the Lockheed XFV-1 Salmon.

Lockheed L-200

IN ADDITION to general design work to fulfill the requirements of BuAer Specification OS-122, Lockheed instigated and completed a research programme which included powered wind tunnel tests, a free flight powered model, and a movable cockpit mock-up. This work assisted in evaluating the magnitude of the problems inherent in making a vertical rising high performance airplane and suggested several possible solutions.

Since the task was unconventional and would involve considerable research before a successful airplane was developed, it was difficult to select any one complete design which fulfilled the specification in all categories. As a result, the Lockheed proposal, although showing one final design, presented alternates which had unique advantages. These, for the most part, had features which required further research to confirm their feasibility. Lockheed hoped to study these alternate design ideas and institute research programmes during the first phases of a further design investigation in order to achieve additional desirable features in the final airplane.

Lockheed's proposal, which appears to have been the most comprehensive of those presented by the companies participating in the OS-122 competition, would prove to be persuasive, resulting in the construction and testing of the XFV-1. Unfortunately, this aircraft would not live up to the expectations of Lockheed or the Navy, despite the optimistic analysis presented in the original proposal, which is summarized below.

OPERATIONAL FEATURES OF THE L-200
Flight Performance
The high speed performance of the airplane was equivalent to the specific requirements in that a Mach number of 0.94 was achieved at an altitude of 35,000ft. This performance, of course, did not indicate the complete capabilities of the airplane at other altitudes nor did it show the manoeuvrability or cruise performance of the airplane under other circumstances than those outlined in OS-122. Although the estimated performance equalled or exceeded existing carrier-based fighters, the primary and unique performance advantage of the airplane was its ability to land and take off in a vertical direction.

Pilot operating procedures
Takeoff and Landing Procedures
In the study of the vertical rising portions of the airplane operation, a mock-up was made of the proposed cockpit. A complete study was made of Navy information which accompanied Specification OS-122, and a trip was made to the Naval Aircraft Development Center in Johnsville to inspect the mock-up which had resulted from Navy studies of this problem. Although many alternate positions of

Cover to the Lockheed Convoy Fighter preliminary design summary report dated November 10, 1950.

the pilot were investigated, it was finally concluded that a vertical aircraft position where the seat only pivoted forwards somewhat, effectively requiring the pilot to bend his torso down towards his outstretched legs, was the most desirable. In this position, the pilot's head was pushed toward the instrument panel when the airplane reached its vertical attitude. No change in the controls of the airplane nor rearrangement of instruments appeared to be required and, as a result, an almost normal cockpit was achieved.

It was felt, however, that backward vision from this pilot's position was extremely uncomfortable and that somehow landing assistance had to be given to the pilot during his final touchdown manoeuvre so that backward vision was normally unnecessary. This led to the inclusion in the landing scheme of a landing signal officer (LSO) stationed on the landing platform on board the cargo vessel. Properly located, the LSO could adjust the vertical height of the airplane by signals to the pilot, accounting for the motion of the landing platform due to the

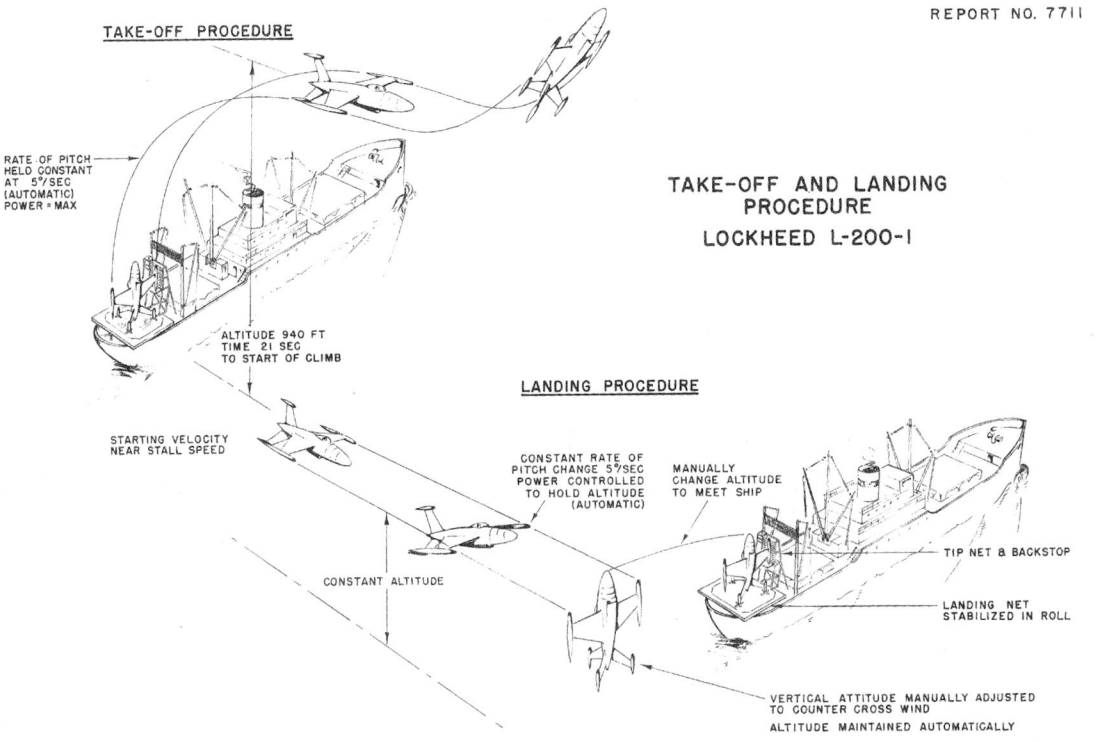

Illustration of the final takeoff and landing procedures which evolved from the study of the many problems encountered in this unconventional procedure. Although the ideal solution for some of these problems may not have been achieved in this system, Lockheed believed that the general solution almost completely fulfilled all the requirements and accounted for more contingencies than any other alternate design studied by the company.

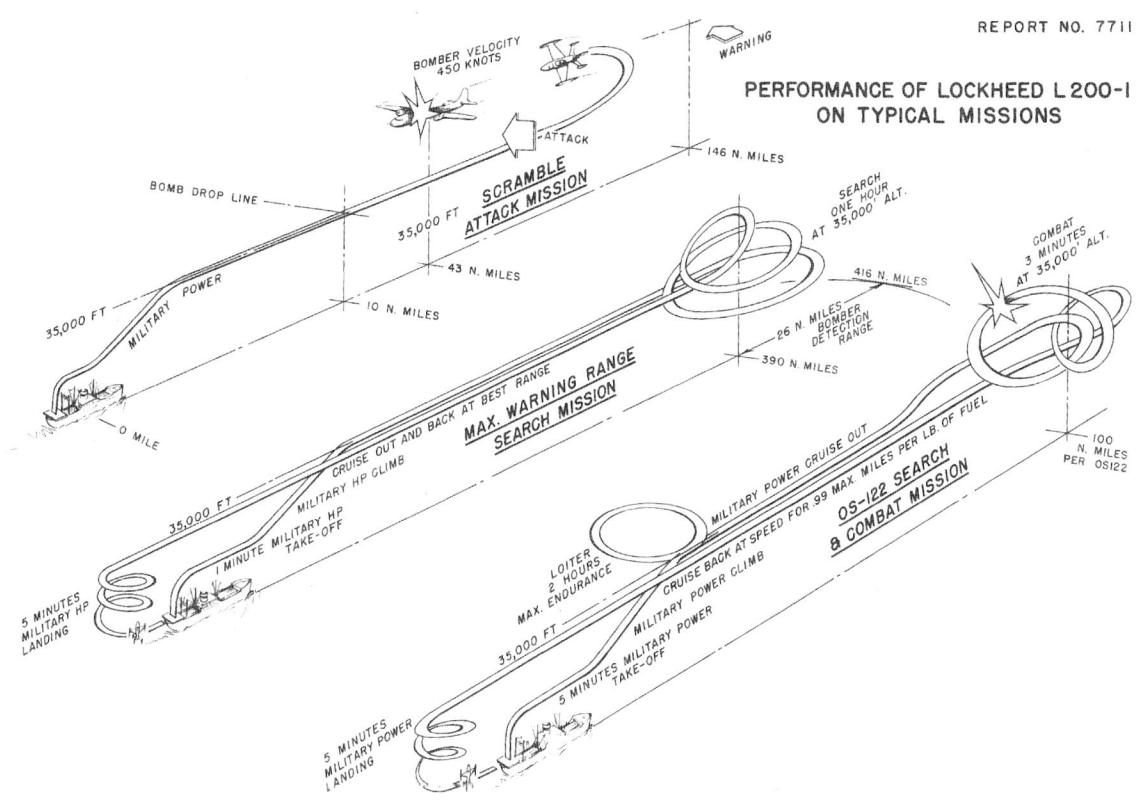

Graphic of the performance of the L-200 on interception and search missions using the deck of a tanker or cargo vessel as a landing and takeoff area.

COMPARISON OF REQUESTED PERFORMANCE
AND ESTIMATED L-200-1 PERFORMANCE

	Spec. OS-122	L-200-1 Estimated
Maximum Gross Weight	16,000	15,600
Engine	T40A-8	T40A-8
Armament	4-20mm cannon	4-20mm cannon
Ammunition	150 rds/gun	150 rds/gun
Radar	APQ/42	APQ/42
(Alternate Radar)	-	Westinghouse Advanced Beam Attack Fire Control System
Fuel Capacity	-	508 gal.
Combat Weight (~40% Fuel)	-	14,380 lbs.
Loiter Duration per OS-122	2 hours	2 hours
Combat Ceiling (500 ft./min.climb)	45,000 ft.	48,500 ft.
Acceleration in Transition	5 ft./sec.^{2}	7.3 ft./sec.^{2}
Time to Climb to 35,000 ft.	4.5 min.	4.9 min.
$V_{max.}$ at sea level, Military Horsepower	-	526 kn
$V_{max.}$ at 20,000 ft.,Military Horsepower	-	563 kn
$V_{max.}$ at 35,000 ft.(Combat Weight)	540 kn	548 kn
Horizontal to Vertical Transition Time	-	15 sec. approx.
Take-Off Transition Time	-	21 sec.
Altitude Change in Transition to Vertical	500 ft. max.	None

Table comparing the resulting performance of the Lockheed L-200-1 with BuAer requirements; according to the company, the airplane closely approached or exceeded all of the performance items required in the initial specification. The interception of a 450 knot bomber was possible at 35,000ft of altitude with an early warning distance as low as 146 nautical miles from the convoy. This warning distance was entirely possible using the L-200-1 airplane for early warning convoy patrol.

sea condition. Furthermore, if properly located, it only became necessary for the pilot of the airplane to look directly sideways out of the cockpit, a direction which afforded maximum visibility, maximum comfort, and, due to the location of the wing tip on the airplane, a visual reference as to the attitude and motion of the airplane itself.

With these major conclusions in mind, the deck facilities and procedures for landing and takeoff finally evolved. The major landing area consisted of a tautly drawn cable net with approximately a 6" mesh. A net was used so that many landing holes would be available for the tip spikes and no obstruction to the airflow past the tail would result from the ground plane effect on the propeller slipstream. This net was mounted on a platform and raised above the deck of the cargo vessel by means of support arms with bearings at each end. These arms were so arranged that the entire platform could be moved in a roll direction and stabilized to suit any roll of the vessel, achieving a consistently level platform no matter what the sea state.

Since the pitch angle of the ship was relatively small and could be accounted for in angular

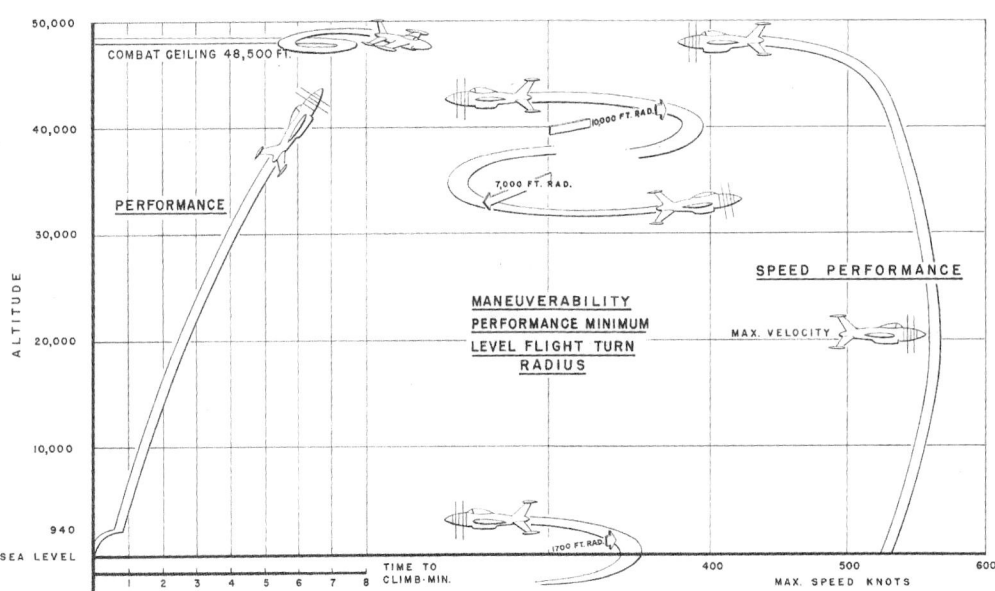

This summary of combat performance was created to illustrate the flexibility and utility of the L-200 Convoy Fighter. On paper, it was competitive with conventional carrier-based types while offering the unique capability of taking off and landing vertically.

tolerance between the airplane and net, no effort was made to achieve a level attitude in pitch. Lockheed noted, however, that the support arms for the landing platform were so arranged that the motion, which maintained the platform level, moved the platform in a direction to counteract the sideways translation of the deck due to the ship's roll. Thus, the platform itself remained level, rising up and down only about 18" due to roll and translated from side to side only +/-44" under the most severe roll conditions anticipated by the Navy.

The net itself, under ideal conditions, could be the entire landing facility for the airplane if accurate flight control could be assured and if it could be assumed that the LSO and the pilot together could accurately position the airplane in spite of a pitching and rolling deck without having the airplane drop into the net with any major translational relative velocity.

With further development of the landing scheme this could have been eventually achieved but it was felt that such an assumption was extremely dangerous at the time. As a result, additional features were added to the landing platform which provided further stability for the airplane once the landing had been effected, and permitted large tolerances in the translational velocity at the time of touchdown.

These facilities consisted of a tip net of similar construction to the main landing platform and a "tip backstop". With these additions attached to the stabilized platform, the landing facilities were complete. The procedure for the pilot during landing was as follows:

Landing procedure
1. The airplane approached perpendicular to the cargo vessel on the port side in a level flight attitude making a transition to vertical flight to the stern and above the landing vessel.
2. The pilot set the translational velocity required to hold against the crosswind into the autopilot control through the special vertical flight autopilot controller on the end of the control stick.
3. Flying parallel to the ship, the pilot closed distance in both altitude and position by decreasing altitude and translating the airplane sidewards toward the landing platform. This was done by reducing power and moving the vertical flight autopilot controller to the left.
4. With the airplane in close position in altitude, being slightly to stern of the vessel,

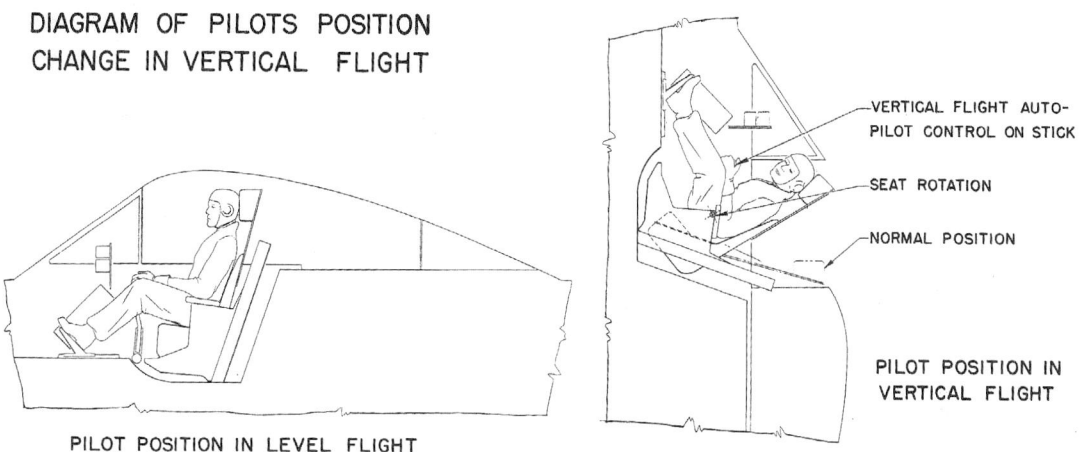

DIAGRAM OF PILOTS POSITION CHANGE IN VERTICAL FLIGHT

PILOT POSITION IN LEVEL FLIGHT

VERTICAL FLIGHT AUTO-PILOT CONTROL ON STICK
SEAT ROTATION
NORMAL POSITION

PILOT POSITION IN VERTICAL FLIGHT

Lockheed studied several options for the pilot position before settling on this design, where the seat rotated about an axis just below the pilot's elbow so that his head was raised toward the instrument panel when the airplane reached the vertical attitude.

flight direction responsibility was taken over by the LSO on the landing platform.

5. Under LSO direction, the airplane was then translated sideways directly over the landing platform at sufficient altitude to clear the highest plunge position of the ship stern. In this position, the pilot was ready for cut when the landing platform was at its proper position (slightly below crest and rising).

6. Just prior to cut, the pilot was given the signal to close in a sideways direction against the tip backstop which relieved him of the responsibility for accurate sideward positioning exactly at cut. The tip pod of the airplane struck the shock absorbing backstop which was equipped with rollers and, at the cut signal, the airplane dropped into the landing net standing on the three shock absorbing tail stands and pierced the tip net with the aft point of the tip nacelle. Thus anchored, the airplane could not tip even under the most severe conditions of ship pitch or roll, even though the platform stabilization devices were turned off.

7. In the event of difficulty, the pilot could elect a wave-off. Since the tip stop was located on the forward side of the platform where the normal ship's obstructions such as masts, booms, etc. were located, this structure did not constitute an additional hazard. Wave-off could be taken either up or to either side or aft.

Takeoff procedure

1. With the engines running and the airplane mounted in a vertical position on the landing and takeoff platform, engine checkout was performed, running to maximum thrust, with the airplane anchored through a tail cable to a catapult holdback.

2. After check-out by the pilot and setting in of roll, pitch and yaw stabilization of the autopilot about a vertical axis, the pilot signalled the LSO that all was in readiness for takeoff.

3. The LSO armed the breakaway feature of the hold-down cable (similar, but designed for more accurate loads, to break the ring of a catapult holdback) and the pilot opened the throttles to the maximum thrust condition. (A plain break ring could have been satisfactory if full power run-up was not required). When takeoff thrust was achieved, the break ring permitted the airplane to leave the deck (the tip net, which was retractable, was pulled away as soon as the tail anchor cable was secured).

These procedures for takeoff and landing may not have represented the ultimate in simplicity but it was felt that they satisfactorily accounted for the foreseeable contingencies and permitted sufficient inaccuracy in flying the airplane to prove the feasibility of the entire vertical rising scheme. Alternate landing methods were studied, each of which had unique features in permitting additional tolerances

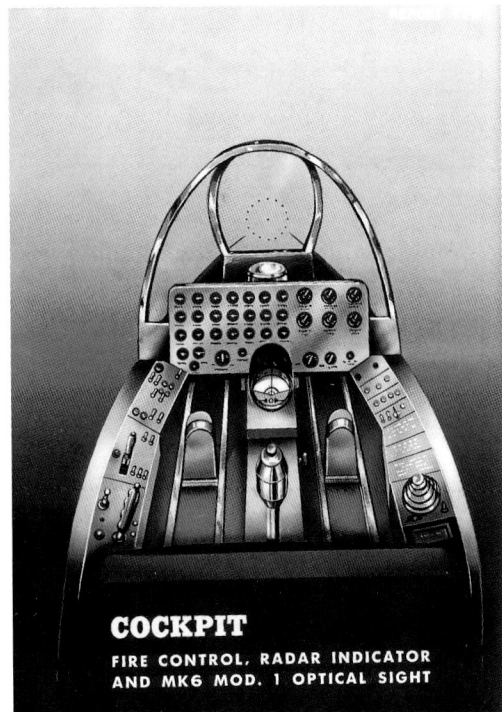

The cockpit of the L-200 had all flight instrumentation grouped in its standard arrangements, with all engine instrumentation carefully grouped on the left of the panel, and the APQ/42 radar screen located directly in the centre and immediately below the gunsight.

in flying the airplane at the expense of deck complications and which permitted considerable deck simplification if such tolerance was not required. Lockheed pointed out that one of the disadvantages of the plain net for landing was that the stability of a relatively normal airplane landing on its tail was small in a tip-over direction, permitting only very low translational velocities at touchdown. This was improved by the addition of a backstop and tip net in this proposal but could have been solved by the following schemes:

1. Extreme wide span or additional landing prongs on the airplane tail. This resulted in a large weight penalty which was not permitted on the airplane.
2. Wheels on the tip of the tail, which were fully swivelling so that no side reaction could be obtained on landing, required that side translation be stopped by a tail hook or dangling hooks from the centre of gravity (CG) of the airplane. This scheme eliminated the use of a net which made it necessary to have the tail close to the platform and it was doubtful whether proper control was achievable if a ground plane was this close to the tail controls. Furthermore, a large weight penalty resulted on the airplane.
3. A wide spread of the landing prongs could be achieved by designing a tail-first or all-wing airplane where the landing was effected on the wing tips rather than on the tail. This scheme had many possibilities but the airplane designs resulting from such a procedure had many unknown aerodynamic qualities that could not be investigated soon enough for a reasonable proposal to be made.

Transition phases of flight

Transition from horizontal to vertical flight was a relatively simple operation if the proper autopilot was installed. With this, the procedures for takeoff transition and landing transition were evolved.

For takeoff it was only necessary to set a given rate of pitch into the autopilot and open the throttles to their maximum takeoff power. This rate of pitch was determined by calculation to be such that the transition was made to best climb speed in the minimum time without exceeding any aerodynamic limitations on the airplane. The rate of pitch chosen for the presentation was 5° per second which resulted in a level flight speed slightly above power-off stall speed when level flight had been reached. Optimum methods for interception could be calculated when more was known of the airplane characteristics. Adjustment of this rate of pitch change could be made simply by overcontrolling the autopilot with manual control or by readjusting the initial setting with the standard autopilot controller.

The stick autopilot controller was not used during this portion of the flight transition but the hand could be held on the stick since the rate of pitch normally did not need to be readjusted and was automatically cut out when the airplane achieved a horizontal attitude.

The landing transition was simple since it could be achieved by setting a constant rate of pitch in the autopilot. This rate incidentally was approximately the same in a pitch-up as was the takeoff rate in a pitch-down direction. In this case, the maximum pitch rate was dictated by a desire not to have the airplane travelling too fast in a horizontal direction when the vertical attitude was reached. This seriously

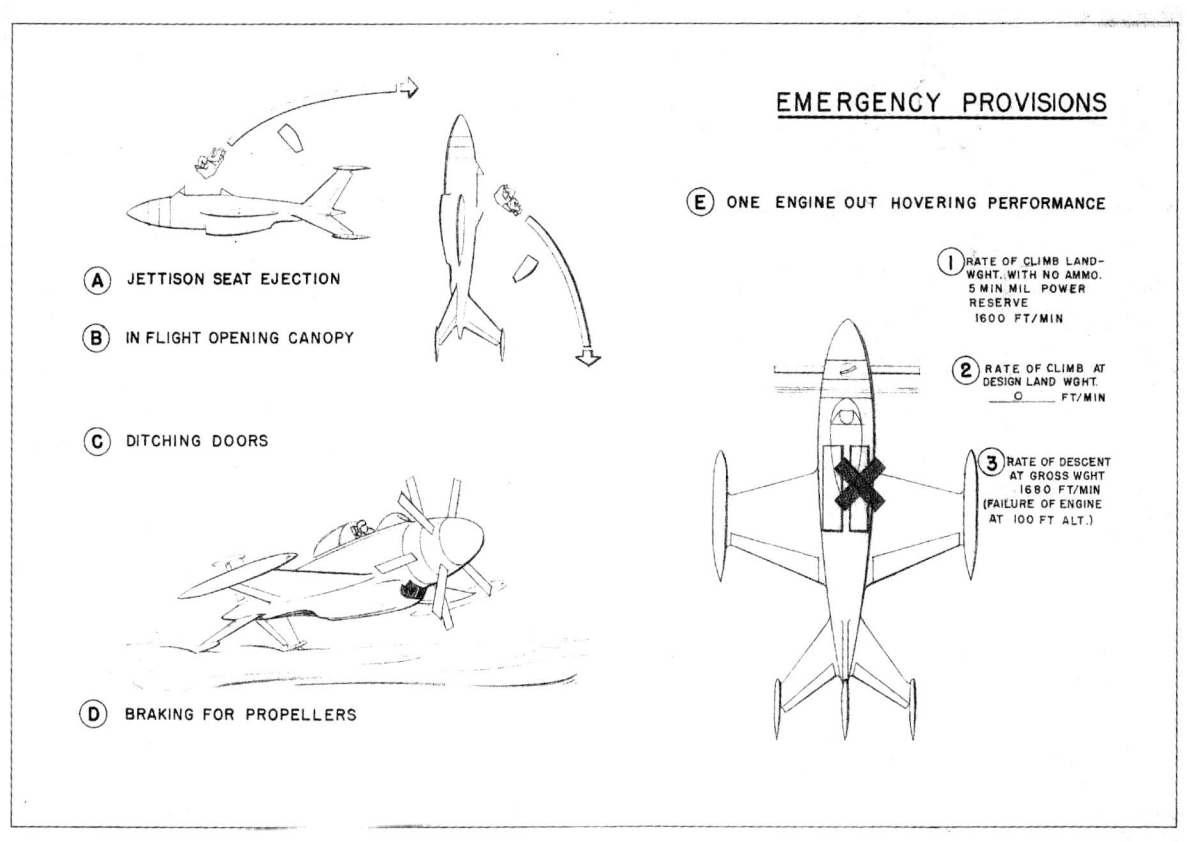

In the event of an emergency, the Convoy Fighter's ejection seat could be actuated whether the airplane was in the vertical or horizontal position.

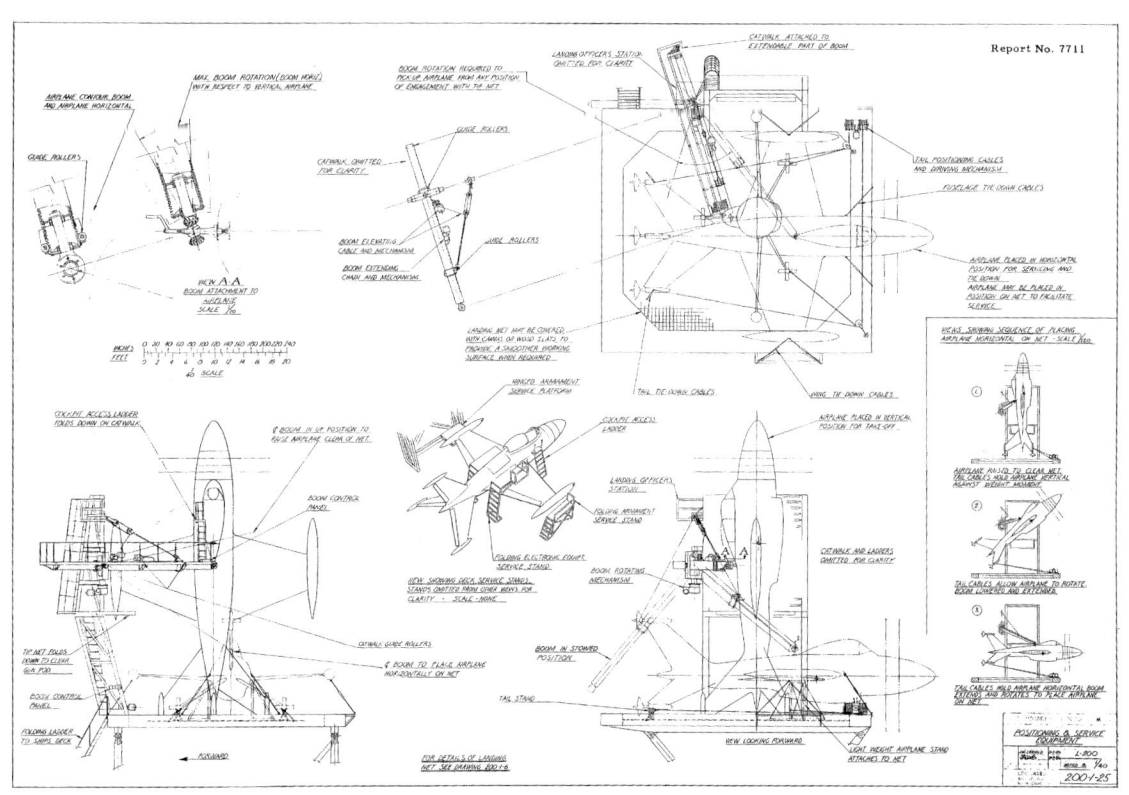

Detailed drawings of the major components of service equipment required for deck handling and servicing of the L-200.

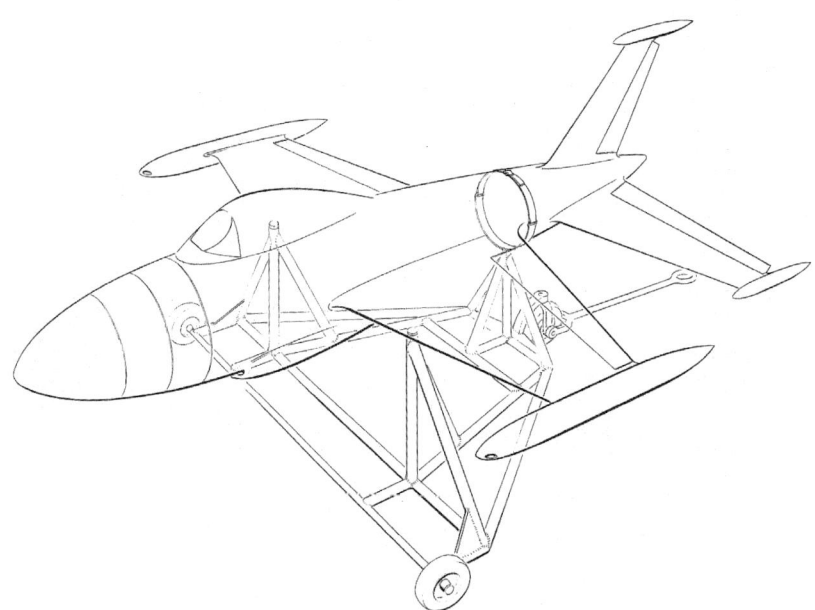

AIRPLANE ON SERVICE CART

Illustration of the cart which served as a maintenance or moving stand when the airplane was away from its landing platform.

impaired control.

In order to limit the altitude of the airplane during pitch-up, it was possible to set into the power controls a zero rate of change of altitude from an aneroid rate of change signal so that during the manoeuvre, the power was continuously adjusted to hold a constant altitude. Calculations showed that a transition to vertical flight could be made with no altitude change. This made possible an approach to the cargo vessel at relatively low altitudes under very poor visibility.

Once a vertical position was reached, an entirely different autopilot controller was used since the airplane became stabilized in a vertical attitude by the vertical gyro which was included in the autopilot system. At this point, roll stabilization was left in and rate stabilization was also utilised for both yaw and pitch, but the separate vertical flight controller on the end of the stick was utilised as described previously in the landing procedure.

Cruise and search procedures
The cockpit of the Lockheed L-200-1 was arranged to provide the best possible facilities for comfortable prolonged operations with the use of the APQ/42 radar and fire control system. (Similar facilities also applied to the advanced Westinghouse radar and fire control system proposed as an alternate by the company). All flight instrumentation was grouped in its standard arrangements, all engine instrumentation was carefully grouped on the left of the panel, and the radar screen was located directly in the centre and immediately below the gunsight. In this position, the radar scope was easily seen in a comfortable seating position by the pilot, either for search or combat.

Directly to the right of the right arm rest was the standard autopilot controller which was in a position within easy reach of the pilot's right hand and approximately in the opposite position in the cockpit from the power handle on the left. At cruise or search altitudes, the radar swept the sky within +/-75° from the centreline of the airplane. The search range was expected to be 26 nautical miles on a B-29 type of target. Lockheed determined that the L-200-1 could act as an aerial search airplane for convoy purposes and that a search radius of 406 nautical miles could be achieved with one hour loiter on patrol and with a 10% reserve after returning to the convoy.

This radius was actually not usable with the listed communicating equipment since maximum communicating range was approximately 120 nautical miles. This calculated early warning radius could have been used, however, if different communicating equipment were installed. This facilitated convoy protection since two airplanes (or groups) could be airborne simultaneously having taken off

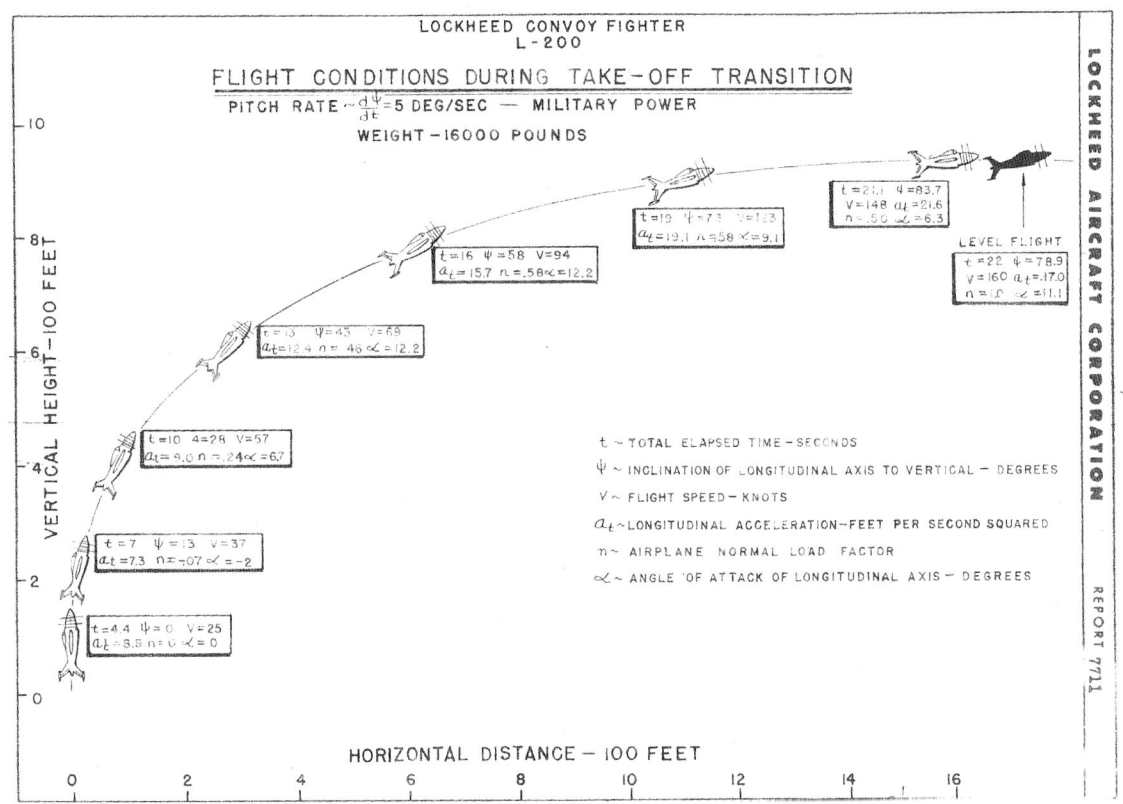

In its study of takeoff transition, Lockheed recommended a 5° per second pitch rate be used which resulted in transition in 21 seconds using 940ft of altitude.

at different times, one searching, the other being available to join combat in the event that the early warning airplane discovered a target. If no target was discovered within the search time of the first group, the second one could take over the search mission and another one could be made airborne to stand by for combat.

Combat procedure
As previously noted in the Flight Performance section, the manoeuvrability of the airplane was excellent at the altitudes where interception could be expected. The combat equipment comprised the APQ/42 radar combined with Mark VI Mod. I gunsight and the four 20mm cannon made the airplane an excellent interceptor for use against targets of the B-29 type. For using the radar for gunlaying, the radar controls were located within easy reach of the pilot on the left console.

If a target was located, the convoy was notified and an approach to a pursuit course interception was initiated. The run-in was made, keeping the target boresighted until lock-on range of the target was achieved, as noted on the radar scope range lines. At this point, the radar was locked on the target by means of the control handle on the left side of the cockpit at which point the radar stayed locked on the target until such time as the target went out of range or firing position was reached.

It was the conclusion of Lockheed's studies that the four 20mm cannon probably were not the ultimate armament of the airplane nor was the APQ/42 the ultimate gunlaying radar. Therefore, a study was made which showed an alternate radar installation which combined with alternate tip tanks containing 24 folding fin rockets each. With this armament arrangement, it was possible to achieve a beam attack rather than a pursuit attack course, thereby protecting the pilot of the fighter and improving the chance of kill on the bomber. These provisions were easily incorporated in the airplane and it was believed that any design which did not have the possibility of carrying rocket armament unnecessarily restricted the firepower of an airplane designed to be operational in 1954 or later.

Emergency procedures
One of the problems which became extremely acute in a vertical rising airplane was the manner in which the pilot could be safeguarded in the event

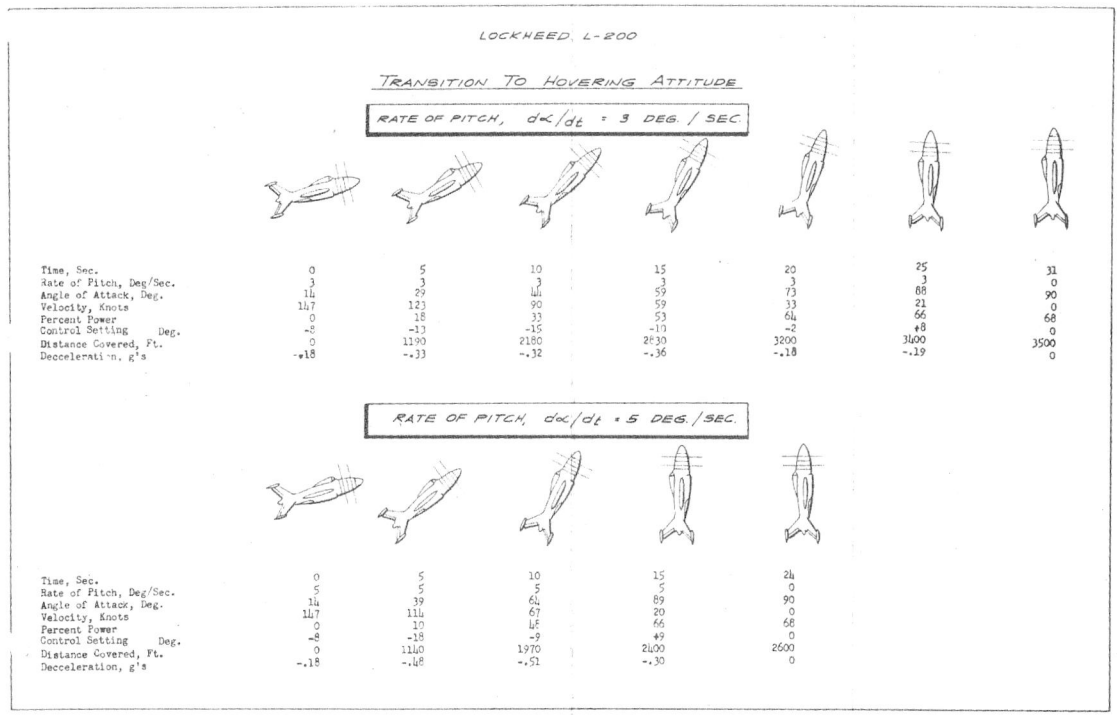

Lockheed considered a 5° per second pitch rate transition optimal in that it required 15 seconds and resulted in a translational velocity when the airplane attitude had reached the vertical of 31.4 kts.

of inadvertent power failure. An ejection seat was incorporated which could be actuated in either the vertical or horizontal position of the airplane, ditching doors were provided to ensure the airplane's stability in the ditching operation and to prevent water from entering the cockpit or destroying the airplane by ramming the inlet ducts of the engine.

An in-flight movable canopy was provided in order to ensure the pilot could exit in the event that the airplane dropped in the water and, finally, the airplane was kept at a light enough weight that hovering performance was feasible at the normal landing weight with one power section dead. Since single engine performance was possible in hovering, special care was expended in the systems which were common to both engines to ensure that no failure on one power section would be transmitted to the other power section. The two inlet ducts were separated, the oil cooling systems were separate with separate tanks, and the duct closure doors could be operated individually to close one power section and prevent destructive windmilling.

Facilities for maintenance and service

Lockheed studied many alternate schemes in achieving its final configuration of service and maintenance facilities, most of which consisted of utilizing the hoisting facilities already available on cargo vessels. The major disadvantage of this type of arrangement was that cables had to be attached to the airplane at a long distance from the deck which required stands for reaching the hoist fittings on the airplane, almost as complicated as the mechanism which permitted hoisting the airplane directly.

By handling the airplane in a stiff-legged apparatus, it was possible to keep it securely attached to its dock so that it did not sway with the roll of the ship. Furthermore, by thus handling the airplane, the entire service process could be done directly over the landing platform, taking up no additional space on the ship.

The alternative to this type of a maintenance rig was to leave the airplane in its vertical position for all servicing and maintenance. Disadvantages of the vertical positioning of the airplane which were obviated by the hoisting system were:

1. The airplane was more susceptible to ice collection and weather in its vertical attitude.

		90°	80°	70°	60°	50°
Angle of Attack						
Without Flaps:	Velocity, Knots	0	14	28	42	56
	Stabilizer Angle*	4°	2.7°	0.5°	-1.7°	-4.0°
	Percent Power	68%	67%	65%	59%	50%
With Flaps:	Velocity, Knots	16	27	36	46	58

*Wind Tunnel data in terms of stabilizer angle for convenience

	0°	10°	20°	30°	40°
Angle of Yaw					
Velocity, Knots	0	43	86	–	–

This illustration demonstrates the airplane attitude required to translate in the fore, aft and side directions as derived from wind tunnel test data.

2. Entrance of the pilot to the cockpit was complicated because of the lean-forward position of the seat while the airplane was vertical, leaving little space for the pilot to crawl into the seat; therefore, an additional seat adjustment had to be made.
3. Workmen were high up during maintenance operations and the motion of the ship would be aggravating unless the platform stabilization devices were operated during maintenance work.
4. Airplane could land in an oblique attitude which was not satisfactory from a maintenance or takeoff standpoint. This probably required a hoisting device for readjustment.

Some advantages existed, however, with the vertical maintenance scheme and these were worthy of further study:

1. It was simpler than the previously described rig if it did not have to hoist the airplane and did not require a mechanism for rotating the airplane.
2. The position of the airplane during storage was such that it was immediately available for takeoff in the event of a rapid scramble.
3. All sides of the airplane fuselage and both sides of the wing were equally accessible for maintenance operations since the airplane was supported by its natural landing gear on the tail.

It was suggested that one of the major contributions which a prototype programme could make was in the development of the servicing and maintenance equipment. The design of the fuel system, oil system, and ammunition-carrying system was such that loading and replenishing the airplane could be done either horizontally or vertically.

Lockheed designed a cart which served as a maintenance or moving stand when the airplane was away from its landing platform. This rig could be carried aboard ship and attached while the airplane was suspended above the landing platform, then hoisted overboard with it when unloading.

AERODYNAMICS SUMMARY

The discussion of performance, stability, and control of this airplane was divided into two parts following the two contrasting flight regimes. The first regime included the transition from zero velocity

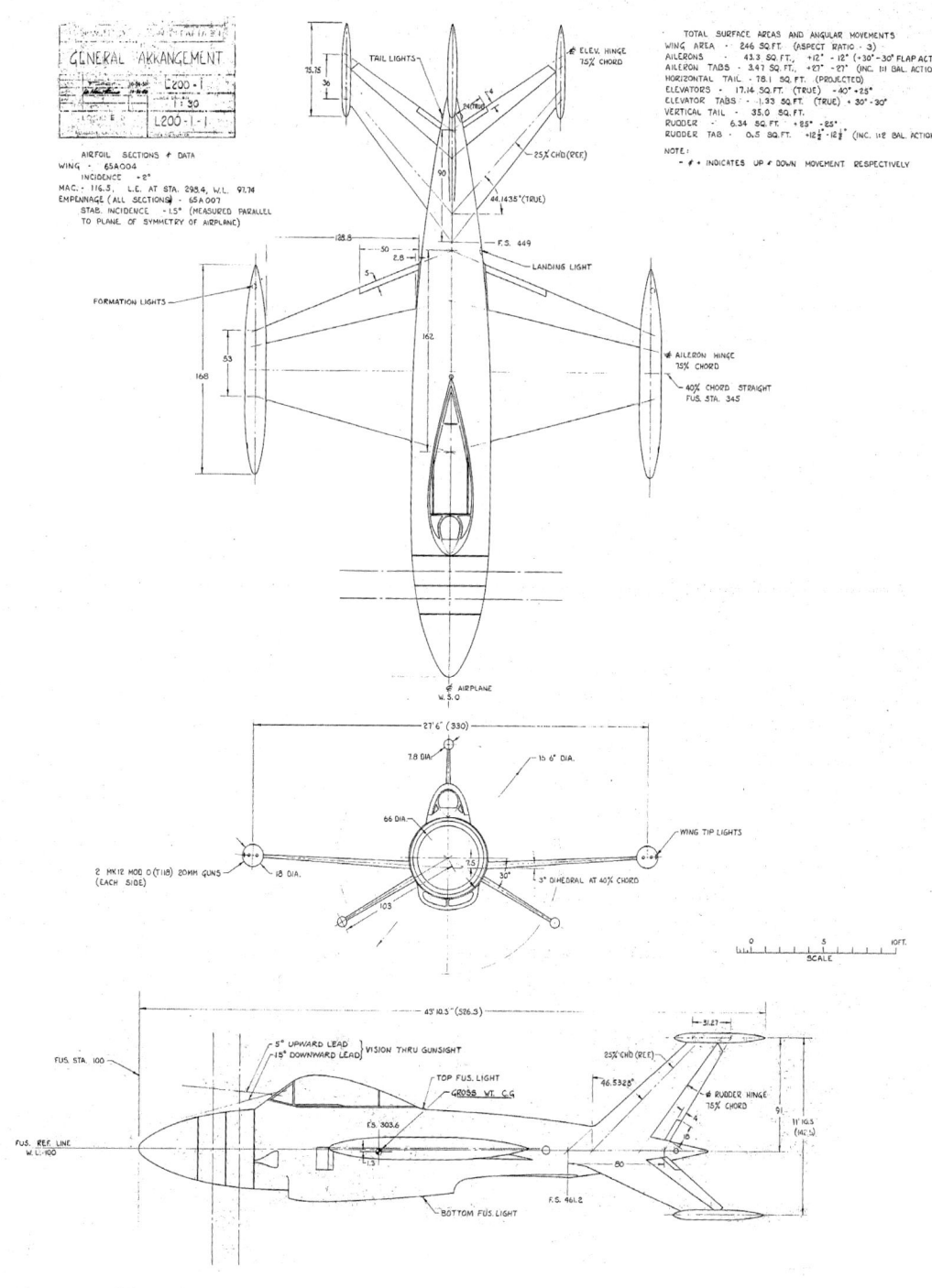

Three-view of the Lockheed L-200-1 Convoy Fighter proposal; the key difference between this configuration and the XFV-1 Salmon concerned the tail, with the former only having three surfaces and the latter having four.

to power-off stalling speed, and from power-off stalling speed to hovering flight, which constituted an aerodynamic study unique to this convoy fighter. The second regime, which included all the normal flight operations of the airplane above power-off stalling speed, followed conventional aerodynamic analysis for such high speed fighter aircraft. Three reports accompanied the proposal which carefully analyzed the transition performance, stability and control, and the performance and control characteristics in the normal flight regime. The following discussion presents a synopsis of these three aerodynamic studies.

Part I—takeoff and landing transitions
The transition range considered was between zero

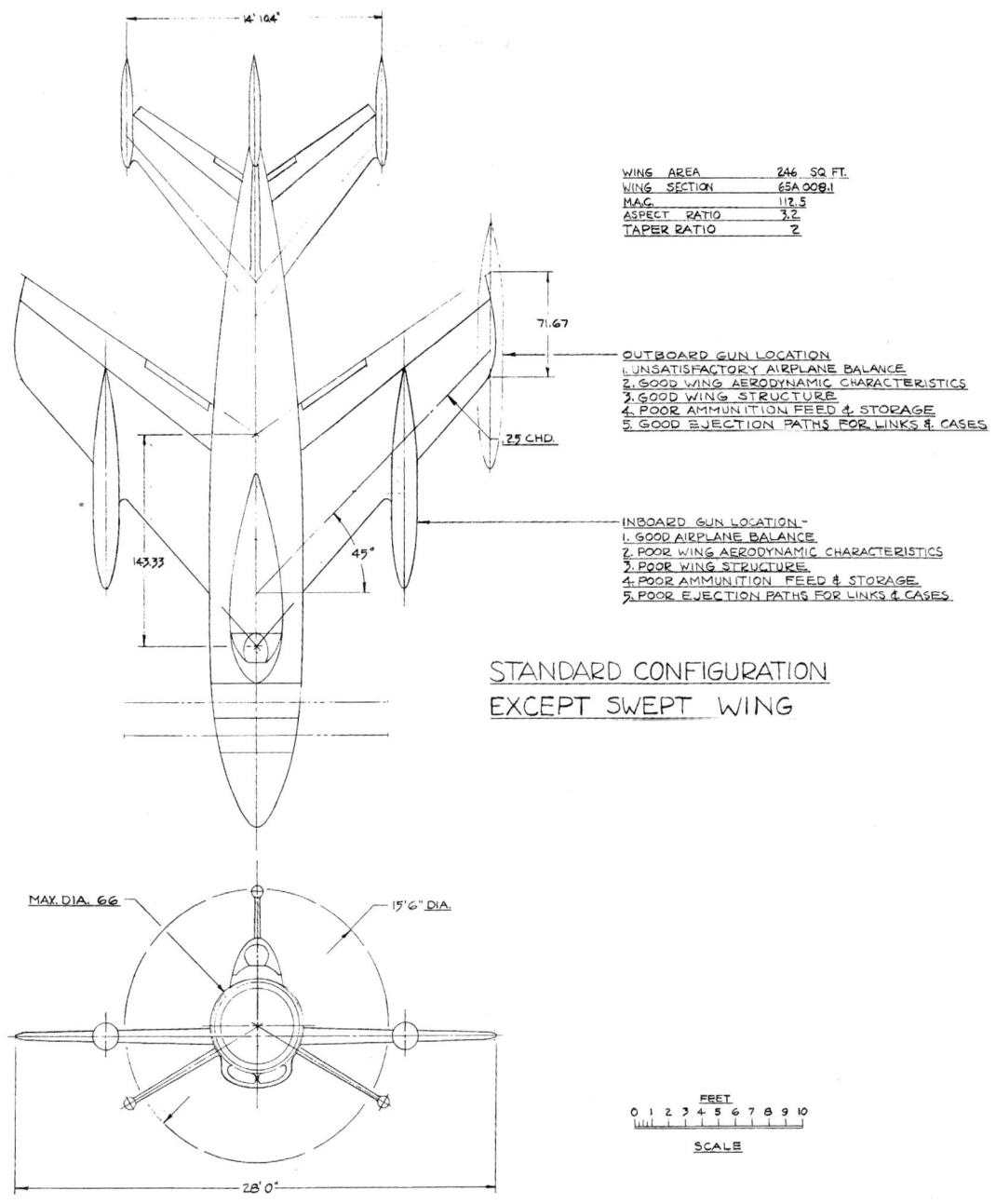

The L-200-6 was an alternate design with a swept wing; compared to the straight wing L-200-1, it was heavier and had inferior aileron effectiveness in hovering flight.

flight speed and approximately 150kts. Throughout this range, two aerodynamic effects were predominant which usually required very little consideration. These were:

1. The characteristic forces and moments on the components of the airplane under conditions where airflow separation was the governing influence.
2. The effects which resulted from an abnormally large relation between propeller slipstream speed and normal free stream speed.

Although it was possible to obtain crude estimates of these effects by calculation, Lockheed considered it necessary to utilize wind tunnel results in order to feel confident that the desired transition performance and control could be realized. Wind tunnel tests were made, therefore, in the Lockheed

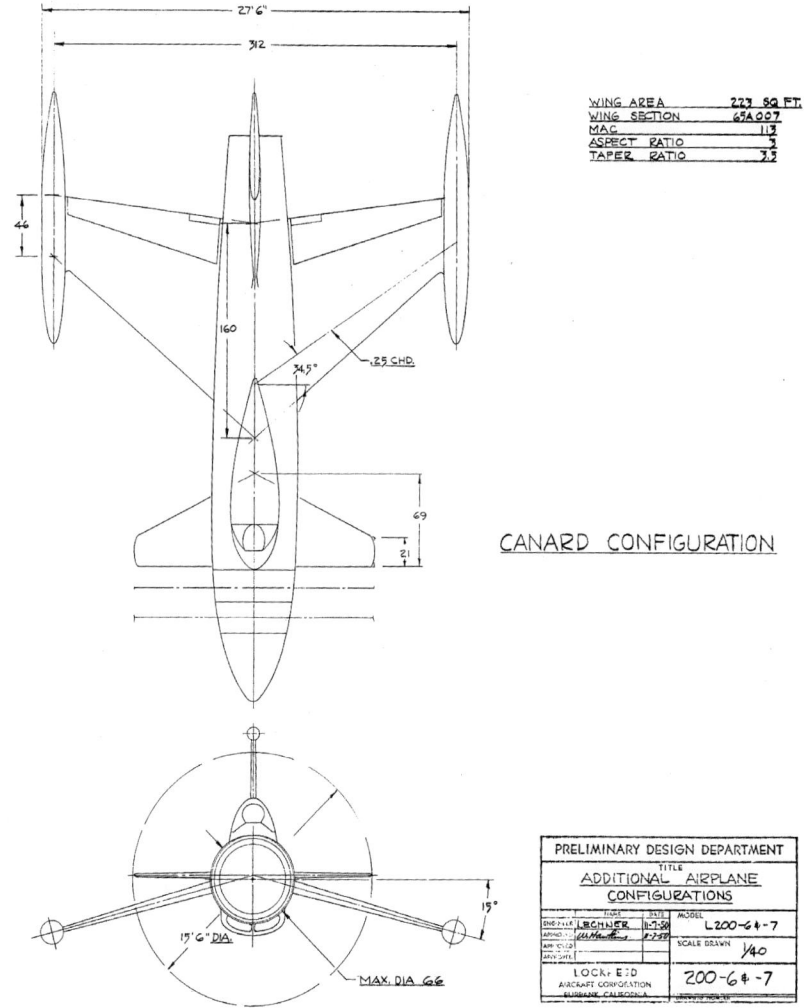

The L-200-7 canard configuration was also judged inferior to the L-200-1; in wind tunnel tests, it suffered from an extreme diving moment at a high angle of attack due to the aft position of the wing compared to the propeller.

Aerodynamics Laboratory.

The component aerodynamic forces which influenced the airplane at each particular relative angle of attack could be summed up as follows:

1. The propeller created both a thrust and crosswind force. The thrust gave rise to slipstream velocity. The crosswind force gave rise to a pitching moment, particularly at the extreme angles of attack approaching the hovering position, which was of major importance. These moments created the positive pitching effect required for stability in hovering flight which was opposite in nature to the conventional requirement for negative pitching moment in normal airplane flight.

2. The wing was utilized for its normal functions in normal airplane flight, but at high angles of attack, wing lift due to slipstream acted in the drag direction. This slipstream effect on wing forces and moments, particularly at air speeds approaching zero, was a large factor in determining the ability of the airplane to follow a desired transition procedure.

3. The tail surface operated in a conventional manner in producing diving moments. The important effect of slipstream on the tail surface was to create sufficient dynamic pressure such that pitch and yaw control would operate at extremely low translational velocities. The problem of tail surface control effectiveness became one of

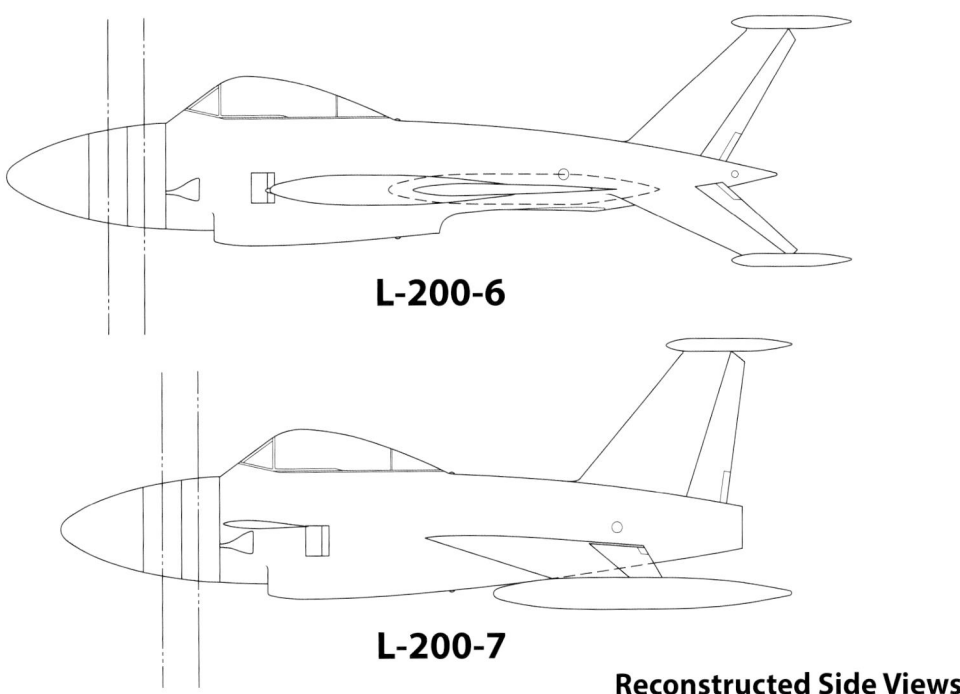

L-200-6

L-200-7

Reconstructed Side Views

Lockheed provided no side views of the L-200-6 and -7 in its report; to aid modellers in scratch-building these interesting designs, these side views have been reconstructed. In drawing these, it was assumed they had the same basic fuselage design as the L-200-1.

ensuring that the slipstream influence for the high angle of attack range was sufficiently strong such that control was maintained up to higher translational velocities than would be required in any transitional manoeuvre.

4. Roll control was required to allow for propeller torque inequalities and gust effects, particularly in the hovering regime, and for sufficient control in the normal sense so that the pilot could execute his desired manoeuvres. Although it appeared possible to use propeller controls for roll control, it was not deemed desirable, and normal aileron type controls were included in this design. Thus, the slipstream effect had to be sufficient to a relatively high translation speed to ensure no lack of roll control at the speeds required throughout the transitions.

Transition control devices

The airplane incorporated conventional types of aerodynamic surfaces with the possible exception of the use of full-span ailerons and flaps. For the normal flight regime, the pilot's controls were the standard stick and rudder type plus the modern interceptor automatic pilot system. For flight in the transition range, the pilot's controls included one with which he could request angular pitch rates, plus one with which he could request absolute pitch and yaw angles for the range close to the hovering attitude.

In addition, he had the ability to request a power variation which was automatically controlled to maintain constant altitude. Roll stabilization was automatic such that he could be assured of zero rate of roll. For fore and aft horizontal translation from the hovering position, which was required to allow the pilot to position and hold the airplane over the landing platform, the aileron could be deflected as flaps. This was desirable to maintain airplane angular attitudes within 8° of vertical for the highest steady state horizontal translational speeds required for landing.

Takeoff transition

After vertical takeoff, the capabilities of the airplane with full power were such that the pilot could choose a constant pitch rate on the order of 5° to 10° per second as he desired. It was recommended from this study that a 5° per second pitch rate be used which resulted

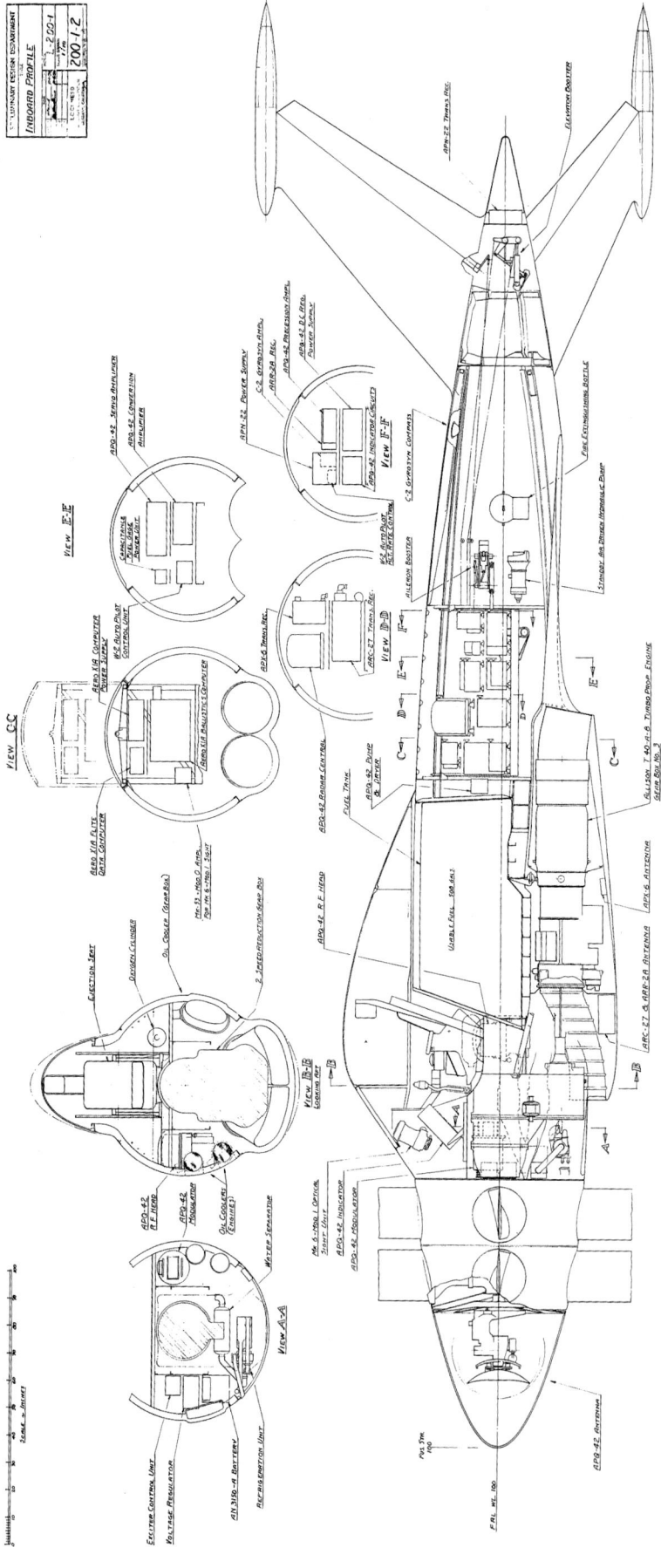

Inboard profile of the production Lockheed L-200-1 showing the general arrangement of the equipment, power plant, cockpit and internal structure. This version was powered by the Allison T-40-A-8 turboprop engine.

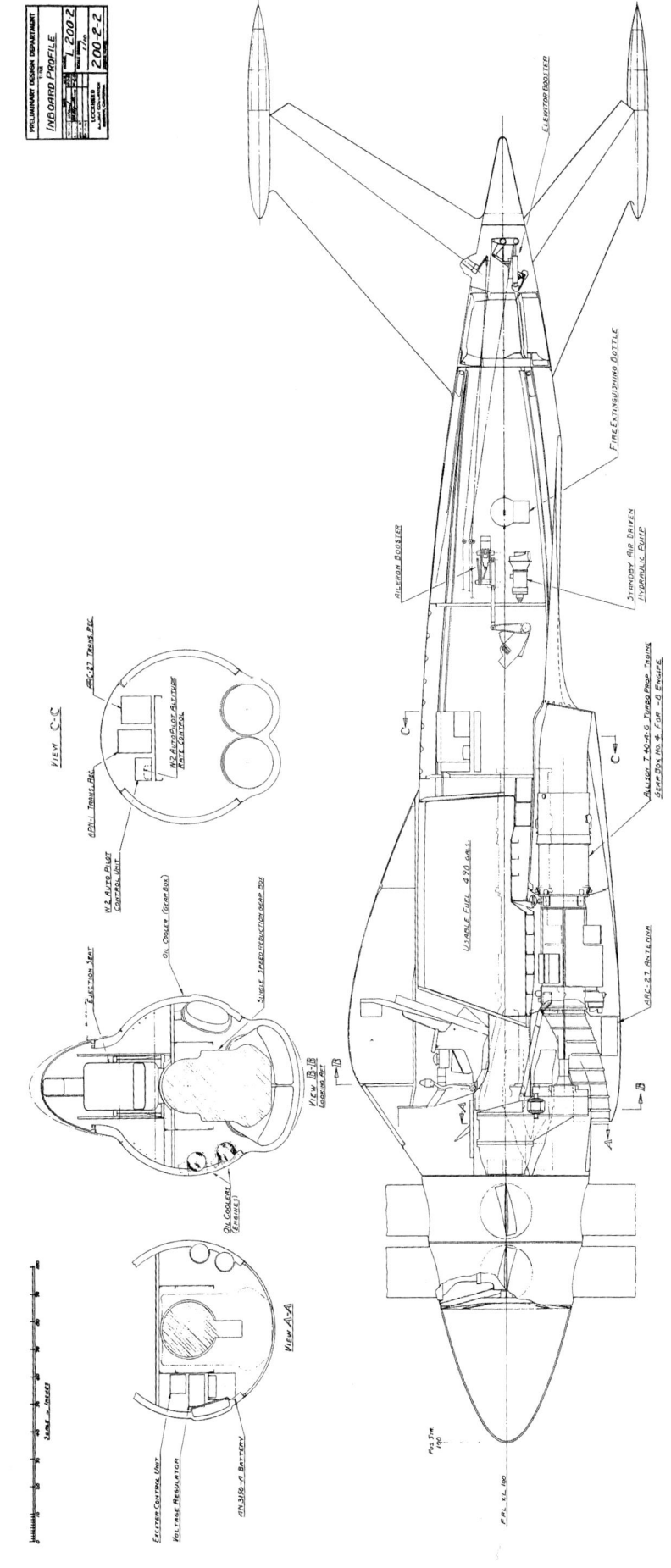

Inboard profile of the Lockheed L-200-2 full-scale prototype powered by the less powerful Allison T-40-A-6, an alternative to the 0.766 scale demonstrator discussed later in this chapter.

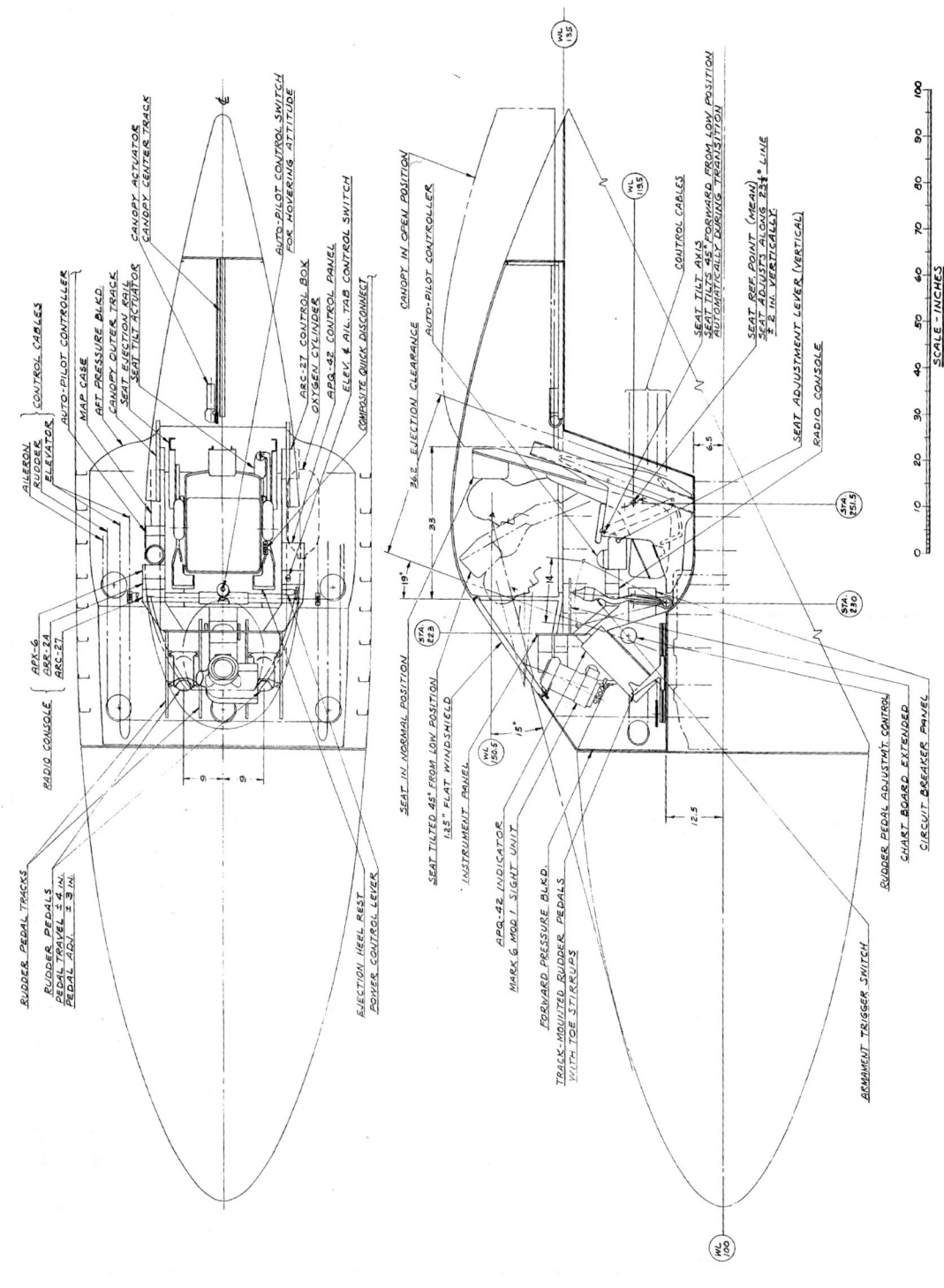

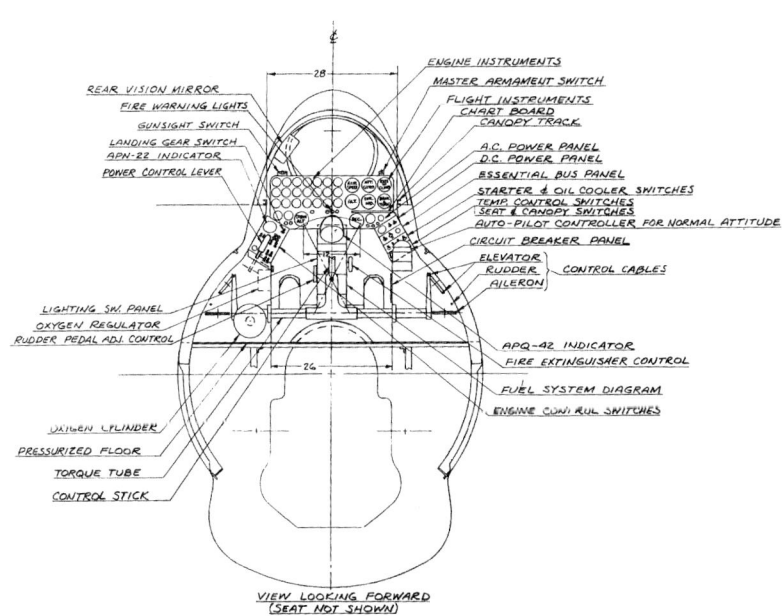

(Above and opposite). The cockpit arrangement of the L-200 was relatively conventional in spite of the unconventional requirements placed on the pilot. The most noteworthy feature was the seat, which was arranged to pivot about an axis parallel to the wing axis in order to raise the pilot's head as the airplane approached its vertical attitude.

in transition in 21 seconds using 940ft of altitude.

The choice of pitch rate was based on the optimum change of altitude and on the desire to obtain normal level flight airplane speeds. The constant pitch rate system was chosen both because it was the simplest system to provide in the airplane and because it resulted in a favourable speed, altitude, and acceleration combination. In this takeoff transition, the effective airplane angles of attack never exceeded those used in normal airplane flight, and controllability was not a difficult problem.

However, the nature of the changes in control position with speed was such that an initial deflection created the manoeuvre and, subsequently, a control position had to be maintained to keep from increasing above the desired angular pitch rate. It appeared possible, after practice, for the pilot to accomplish this manoeuvre without the automatic devices, but he had to be limited in the maximum possible pitch rate such that he did not arrive at the horizontal attitude with insufficient altitude and speed.

Slowdown transition

The problem of slowing down below power-off stalling speed and attaining hovering flight was solved by a method very similar to the takeoff transition in that it was based upon the use of pilot chosen pitch rates. However, it appeared that the critical and most desirable slowdown transition was one that was made at constant altitude, and calculations based on wind tunnel tests demonstrated that this was possible.

It was assumed in the calculations that the

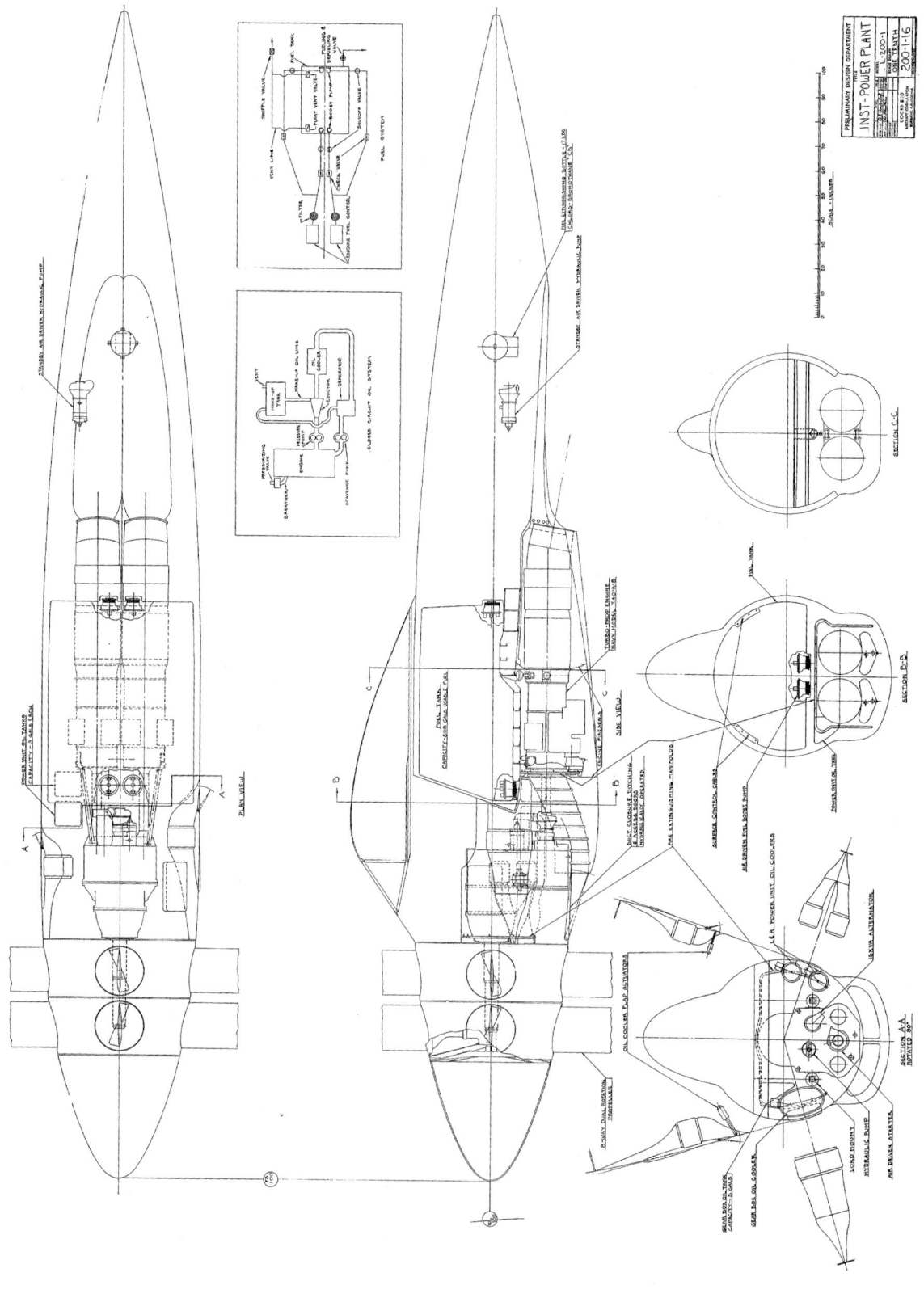

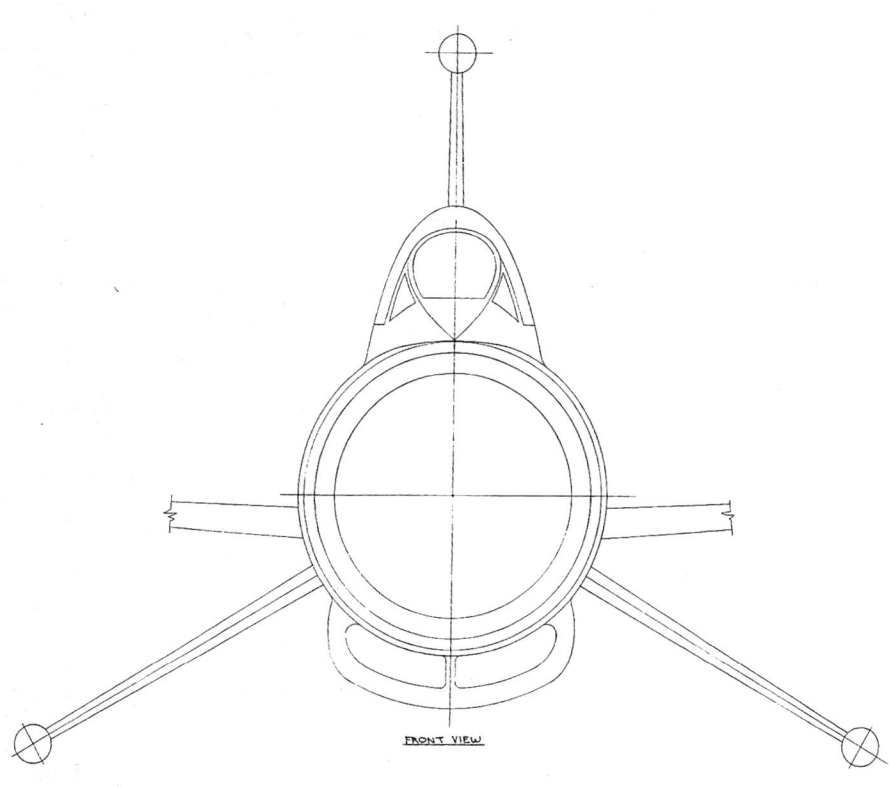

(Above and opposite). Blueprint of the L-200-1 airplane final power plant installation; the basic power sections were those of the Allison T-40A-8.

automatic power control maintained the thrust required to ensure constant altitude. This thrust variation had been shown to be one in which the transition was initiated from a speed of about 150kts at an angle of 15° and with relatively low engine power, and thereafter power was automatically increased to a maximum of around 70% as hovering altitude was approached. For the critical heavyweight conditions, the control movements were such that the elevator was moved to initiate the pitch rate, and sometime thereafter had to be reversed to ensure that the pitch rates did not become excessive.

A 5° per second pitch rate transition was considered optimum in that it required 15 seconds, and resulted in a translational velocity when the airplane attitude had reached the vertical of 31.4 kts. It appeared that the pilot could make this slowdown transition manually, although it was probable that he would make some gain in altitude due to his logical manipulation of the power control.

Care had to be taken to limit the maximum pitch rate so that the airplane did not reach the vertical attitude with too high a translational speed. The wind tunnel had demonstrated that there was a critical top speed for travelling horizontally in the vertical attitude at which the pitch control deteriorated. However, from the slowdown transition calculations there appeared to be adequate margin below this critical speed. Aileron roll control also had a critical maximum speed which appeared sufficiently high such that the airplane would not run out of control effectiveness in any normal transition manoeuvre.

Descent and landing

The airplane was capable of hovering at altitudes up to a maximum of approximately 19,600ft. Wind tunnel tests showed that rates of vertical descent up to values of 2,000ft per minute were entirely satisfactory from considerations of control. The constant altitude power control could also be used by the pilot to request rates of descent such that he could ensure not exceeding the critical descent speed. Wind tunnel tests had not been run to determine the maximum descent rate because a 2,000ft/minute value appeared adequately high.

When the airplane reached the landing altitude it was necessary to translate horizontally, maintaining

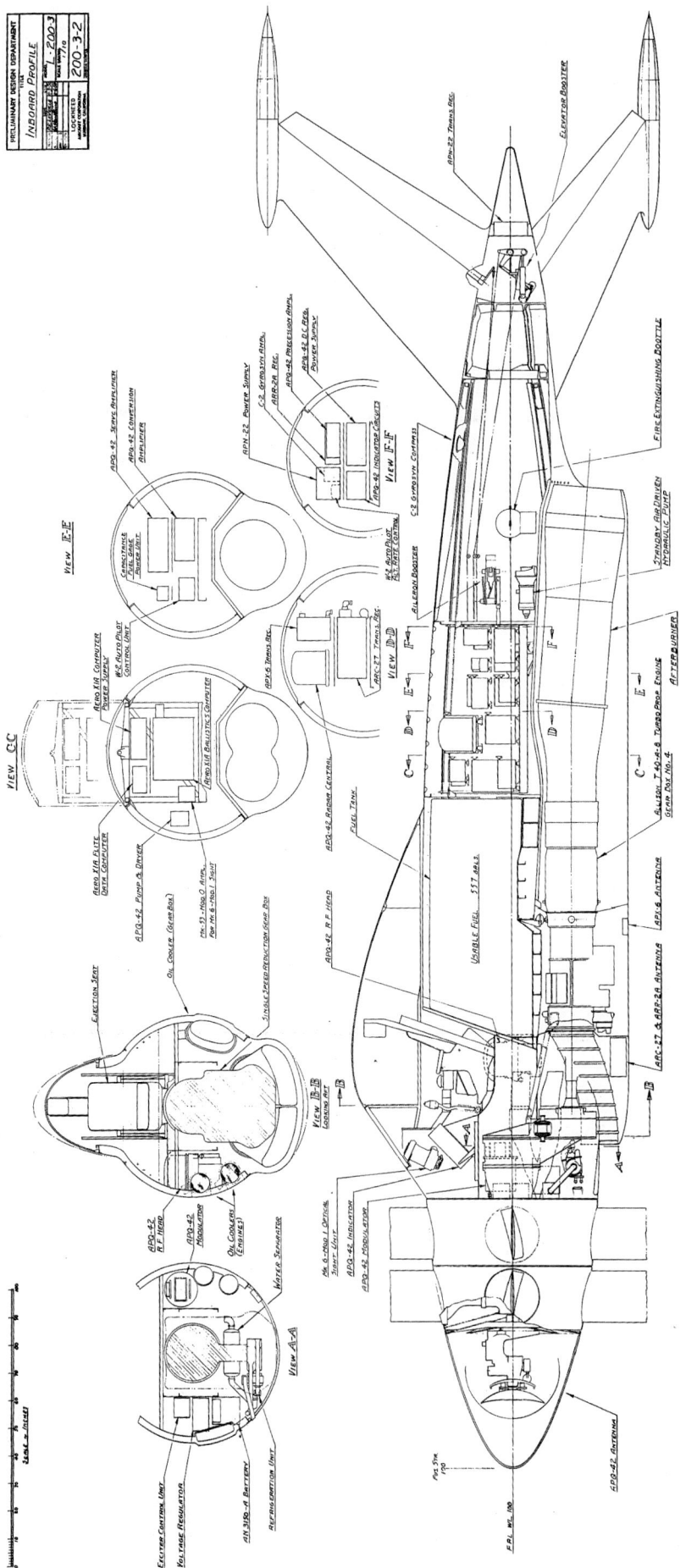

Lockheed studied the option of adding an afterburner to the Allison T-40A-8 to eliminate the necessity of a two-speed gearbox. While it increased the top speed of the aircraft, the afterburner also increased its overall weight. The company recommended additional study to determine whether the high speed advantage was worth this penalty.

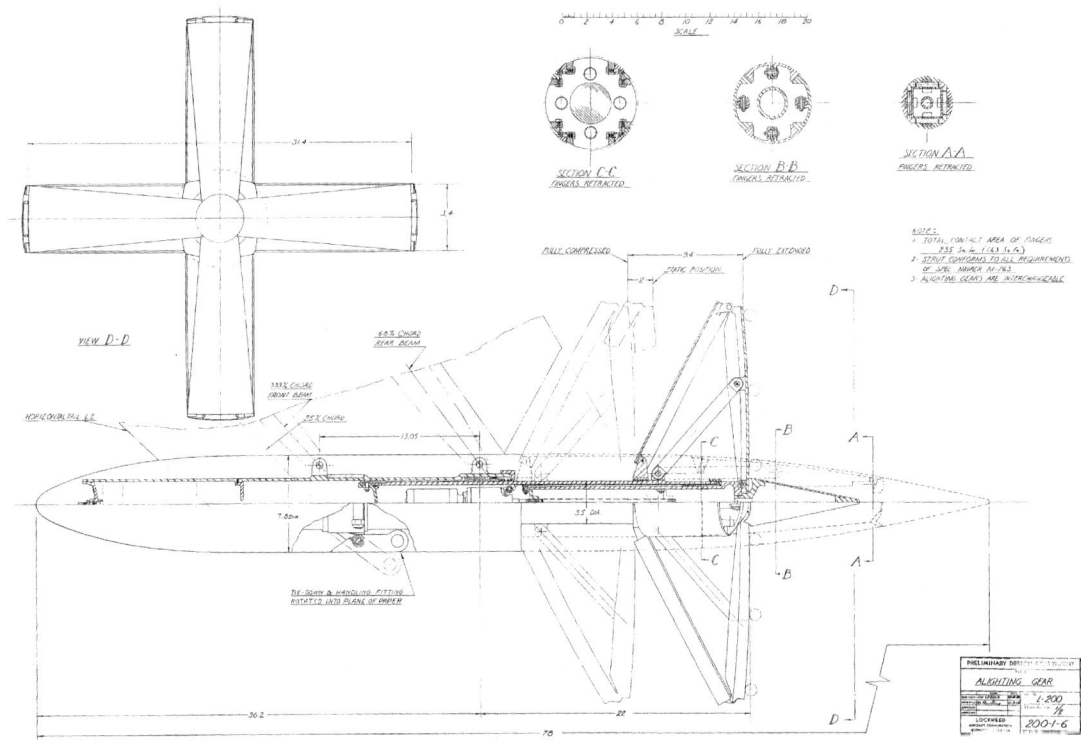

When landing on a ship, Lockheed's Convoy Fighter was designed to touch down on a landing net with its three-pronged empennage. This drawing of a typical tail tip pod shows the mechanism for retracting the landing pads and obtaining adequate shock absorption.

constant altitude to position the airplane above the landing net. The pilot executed his translation by setting the angle of the longitudinal axis at the desired amount from the vertical with the constant altitude control maintaining the power required. The airplane accelerated horizontally towards its steady velocity and it was easily checked by the pilot by a rapid change in airplane attitude.

Very little angular change in yaw from vertical was required to attain desired velocities in side translation. Fore and aft translation, however, normally required sizable pitch angles from vertical to translate at equivalent speeds, due to the influence of the forces acting on the wing. Therefore, the ailerons were deflected as flaps for rapid fore and aft horizontal translation. These flaps were very effective in reducing the required pitch angle from vertical in the desired horizontal speed range.

Hovering stability

Preliminary dynamic stability calculations for the hovering attitude indicated that at high weights, an airplane disturbance from vertical would be counteracted by the inherent static stability, but that the dynamic characteristics would be unstable. Conversely at low weights, which meant lower engine power, the results of a disturbance from hovering flight tended to be increased by an unstable static stability, but with dynamic subsidence about the static variation.

However, the importance of these natural stability characteristics of the airplane in hovering flight had to be judged relative to the dynamic oscillatory period and to the divergence time. Fortunately, calculations indicated relatively long periods and times to increase amplitude as compared to the normal pilot's reaction time. This meant that without the automatic stabilization devices, the pilot could still hover the airplane satisfactorily. These facts had been calculated theoretically, and had also been indicated by experience with a flight model of the airplane.

Part II—normal flight characteristics

The complete transition analyses were based upon L-200-1 airplane because, relative to the transition regime, there was little difference between the aircraft presented. However, in this section, the

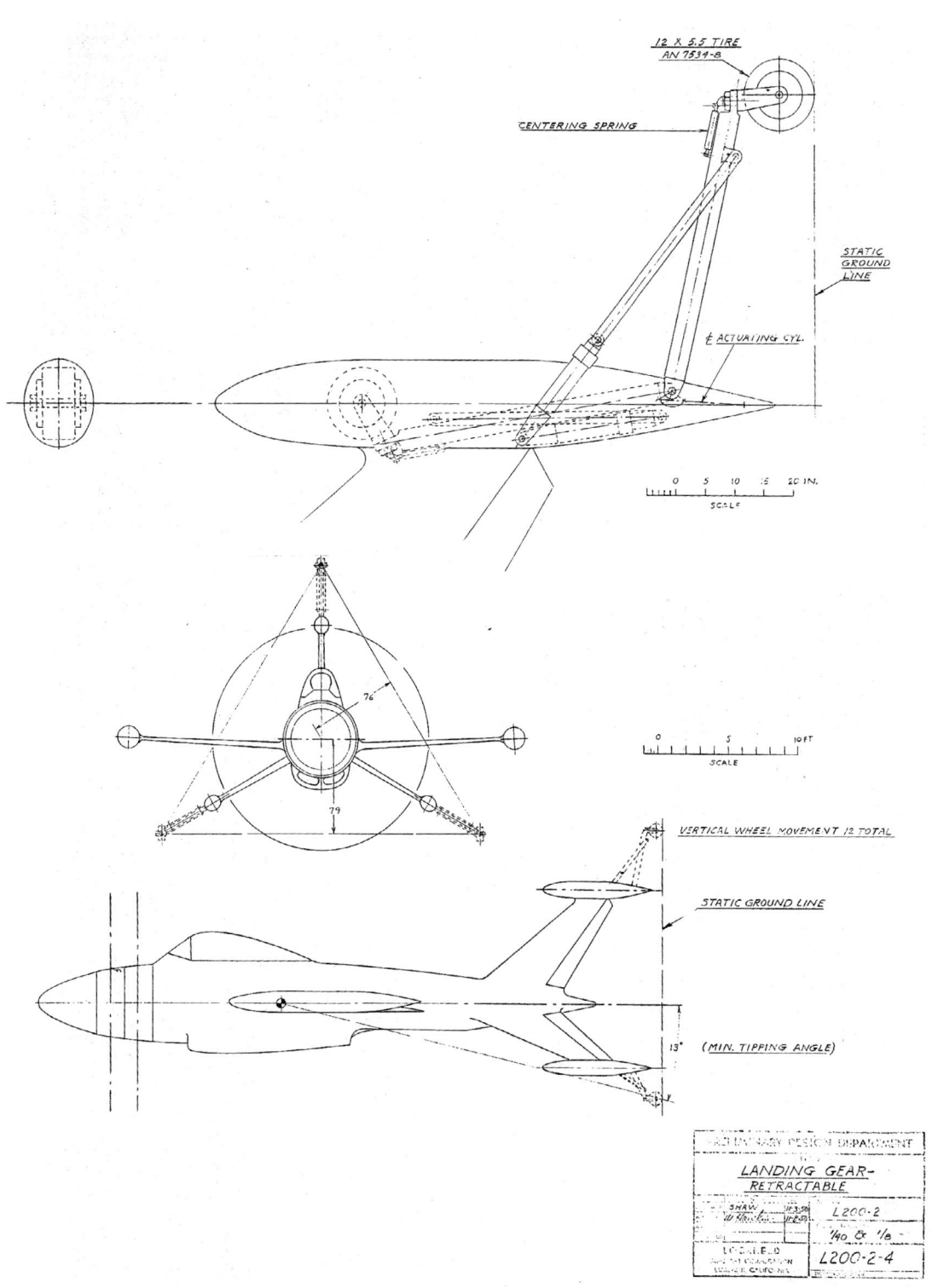

Lockheed studied this alternate retractable wheeled landing gear as an interim installation which could be used for the first vertical flights of the airplane or prototype while developing the simpler tactical system. The actual XFV-1 was equipped with a less complex fixed wheel gear on its empennage.

ARMAMENT INSTALLATION

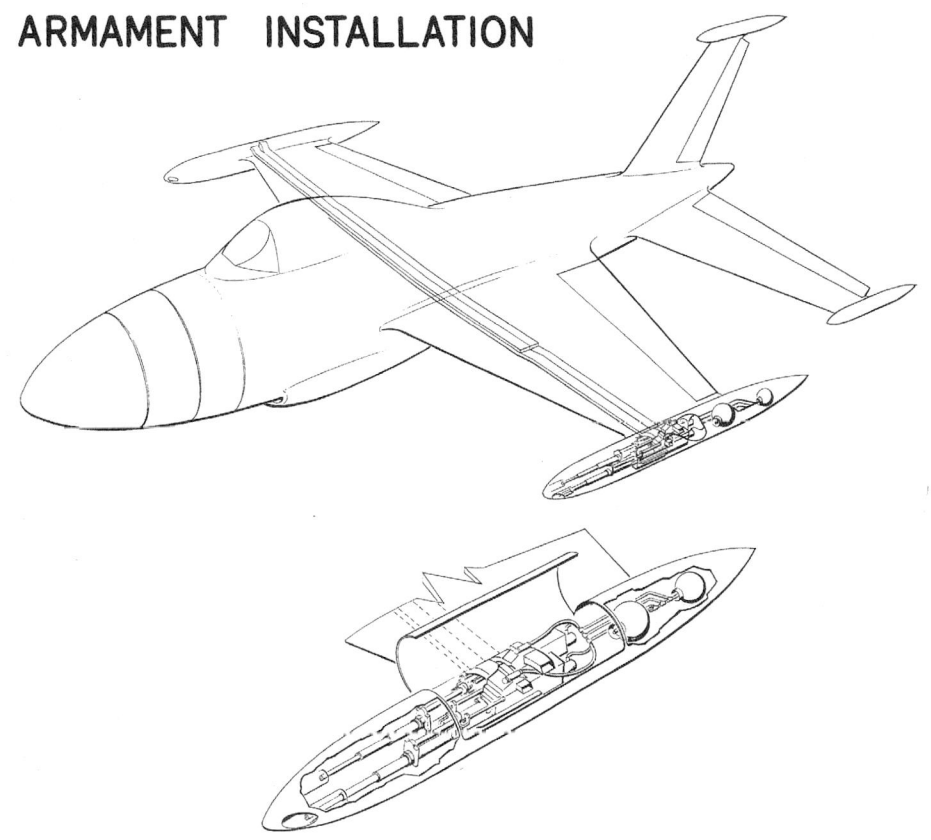

The standard armament of the operational L-200-1 would have consisted of two 20mm guns installed in each wing tip pod. Lockheed considered this installation the best from an aerodynamic standpoint.

normal flight characteristics of high speed, climb, and range were discussed and, in these respects, the various airplane proposals differed.

Propeller selection

Information available from the propeller manufacturers showed that the largest propeller was desirable in that increased propeller thrust more than compensated for increased propeller weight. Therefore, the objective of propeller selection was to choose the maximum thrust propeller which, for the basic airplane L-200-1, was the eight-bladed dual rotating 15½ft diameter type.

Transition analysis only required that the maximum possible thrust be available. The increased propeller cross wind forces which increased propeller solidity would not adversely affect the transition performance. The optimum choice envisaged the use of this propeller with a two-speed gear ratio to obtain the best thrust for transition and the best efficiency for climb and high speed flight.

The choice of a single versus dual gear ratio determined the added weight and complication in the airplane. The prototype L-200-2 and L-200-5 airplanes had single propeller gear ratios chosen as optimum for the transition regime. For the production airplane, there was a question which had to be considered further. The proposal requirements could be satisfied with either a two-speed gear ratio or a single speed system plus an engine afterburner. A two-speed gear ratio and an afterburner improved high speed and climb performance. The thrust available varied depending on gear ratio and afterburner installation maintaining the optimum thrust characteristics for transition with a propeller rpm of 750 and 1100.

An entirely different approach could be taken with a single speed propeller whereby one of the dual units could be put into high pitch with essentially zero thrust and the other propeller allowed to absorb all the power; however, tip speed corrections appeared to be excessive for this type of operation.

All configurations were able to meet the high speed requirement except possibly the last which could not

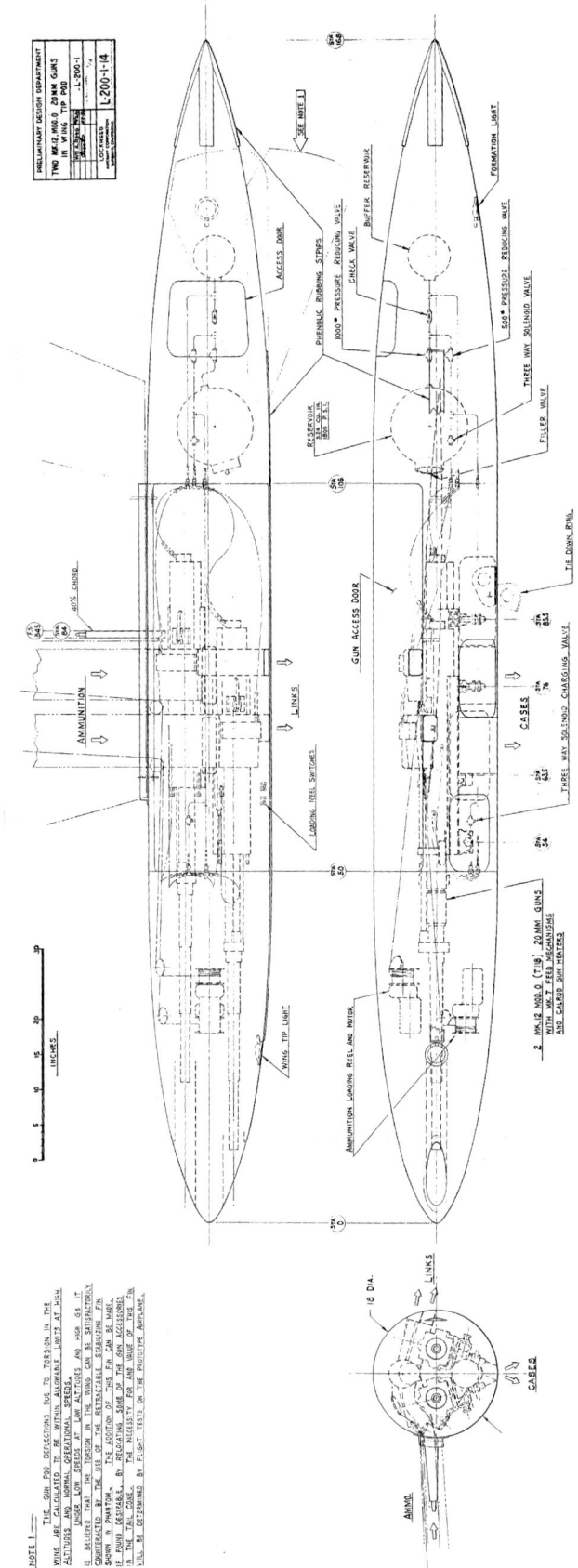

Detailed schematic of the Mk.12 Mod. 0 20mm gun installation on the Lockheed L-200-1 Convoy Fighter.

be truly evaluated with the test data available.

L-200 tactical airplane performance
The L-200-1 exceeded the Navy requirements for high speed and combat ceiling, fulfilled the complete desires for the basic radius mission and came close to obtaining the required time to climb to 35,000ft. Since the drag on this airplane was low and the calculated propulsive efficiency was as high as could reasonably be expected (88%), the possibility of improving this climb time was doubtful.

Optimistic assumptions may have been made on the DR-72 airplane which accompanied the Navy's proposal material and permitted the time shown in that report. Since NACA tests had proven a poorer 'e' for a higher swept wing as compared to an unswept wing, and the proposed airplane was lighter in weight, the time to climb of the L-200-1 should have been slightly superior to that of the DR-72. Neither airplane met the requirement unless propulsive efficiencies were considerably superior.

Prototype and alternate airplane performance
The performance of two prototype airplanes was studied, the first being a 0.766 scale model of the Convoy Fighter powered by a Double Mamba turboprop engine, and the second a full-scale reduced weight airplane with an earlier Allison XT40-A-6 engine. Both airplanes exceeded the performance requirements of NAVAER OS-121. However, the performance capabilities of the L-200-2 (full-scale prototype) were well above those for the 0.766 scale model. This model would also provide full-scale characteristics of the tactical airplane since no disproportionate cockpit size and incorrect fuselage interference effects existed, thus providing an answer to the full-scale flight characteristics at a much earlier date.

Stability and control
The stability and control of the L-200-1 in the normal airplane flight regime fulfilled most of the requirements of NAVAER Specification SR-119-B. In many ways that specification was not applicable to an airplane which could fly and maintain altitude to zero velocity. The requirements for normal flight stability and for stick force variation in high speed manoeuvres were satisfactory.

A particular condition of SR-119-B which was obviously difficult to fulfil concerned stalling characteristics. Since this airplane incorporated relatively greater thrust than any of its predecessors and, since it entered regimes in which the outer wing sections were completely stalled, it was obvious that control manipulation was required of the pilot when operating without automatic pilot devices. However, the reliance on automatic stabilization devices was becoming more acceptable by now because they appeared necessary even for relatively conventional aircraft. Since this design envisaged identical types of automatic stabilization, the use of these devices to some degree in the satisfaction of requirements of SR-119-B was acceptable.

The most marked difference relative to stability of this airplane was the usage of an extremely far forward centre of gravity. This was required to compensate for the large destabilization effect of the propeller, but fortunately was in the proper direction for the optimum configuration for transition performance. In addition, the deletion of any normal landing requirement, which always created difficulties of elevator control due to ground effects, made the usage of a far forward centre of gravity feasible.

Alternate design discussion
The study leading up to this proposal was started by a careful comparison of the design submitted with the Navy proposal material as compared to alternate designs. The use of a coaxial propeller in the conventional propeller position was largely dictated by the requirements of minimum all-up gross weight. With the existing limitations of engine thrust-weight ratios, it appeared undesirable to deviate from this engine-propeller-fuselage relationship. The most promising configurations were the normal swept and straight wing types, plus the tail-first canard configuration. The choice of the straight over the swept wing was largely a matter of weight, but there was one aerodynamic advantage which recommended straight wing configuration. Wind tunnel tests indicated that the aileron effectiveness on the swept wing was very poor and that extremely unconventional means would be required to obtain roll control with this swept configuration in the hovering range. Aileron effectiveness on the straight wing configuration appeared entirely satisfactory throughout all the speed and angle ranges which were required.

The canard configuration appeared to have advantages relative to the design provisions for takeoff

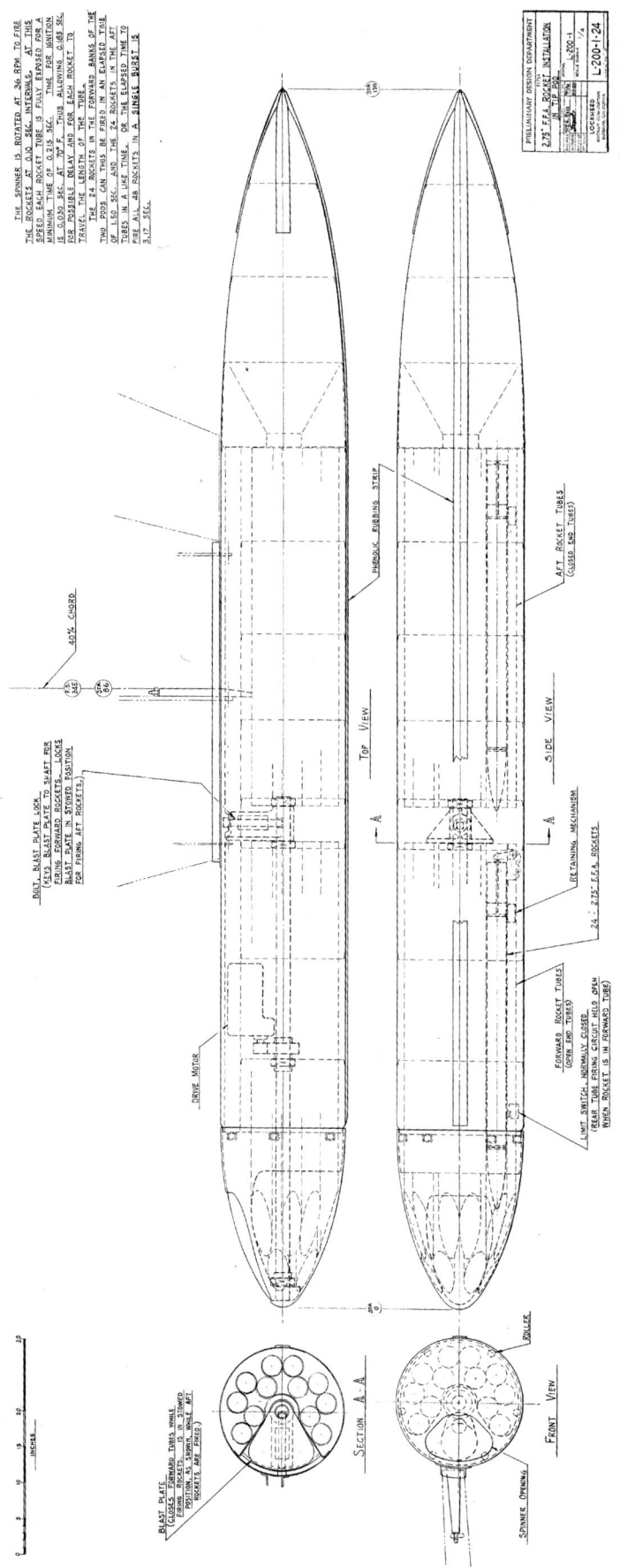

Lockheed also studied an alternate armament of forty-eight 2.75" folding fin aerial rockets, 24 in each wing tip pod.

and landing. However, aerodynamically the canard configuration appeared questionable and subsequent wind tunnel tests had shown that this configuration would be unsatisfactory. The greatest aerodynamic difficulty of the canard was the extreme diving moment at high angle of attack due to the aft position of the wing compared to the propeller. Without a considerable weight increase to lengthen the fuselage, it appeared that this problem was insurmountable. Concurrent with this, the effectiveness of the nose trimmer for longitudinal control appeared too small due to the reduced slipstream velocity in the position close to the propeller.

DEVELOPMENT PROGRAMME

Considerable study was made of the various methods of developing an airplane of the L-200 type. Obviously, a very complete research and mock-up programme was required to guarantee the ability to make vertical ascents and descents under complete control. Also, a great deal of work was required to obtain the optimum transition between vertical and horizontal flight. Lockheed concurred with the Navy in that it was extremely important to flight test a prototype airplane as soon as possible.

Having designed the Mamba-powered prototype, suggested by the Navy, and compared it to the full-scale L-200-1, it was immediately apparent to Lockheed that the sizes of these two airplanes were so nearly the same that there would be very little time saved in making the Mamba-powered airplane against constructing a stripped full-scale L-200-1. In comparing the development trend of the Allison T40 engine, studies seemed to indicate that it was entirely feasible to fly a prototype L-200-1 airplane in the same time that one could fly the Mamba-powered airplane. The advantages of this procedure are outlined below:

1. The method of gearing the power sections to propellers on the T40 was preferred because it permitted both propellers to operate from one power section, increasing the available thrust during one engine power-off operation which could save the prototype airplane. This was not possible with the Mamba engine without a complete gearbox development.
2. Having the prototype airplane complete to size and identical to the eventual tactical airplane permitted development of valid aerodynamic data beyond the development of the landing and takeoff procedures.
3. Having the prototype identical in shape and size to the tactical airplane meant that controls, cockpit arrangements, booster system, hydraulic system, and electrical system would all be essentially identical to the final airplane; any development problems that arose in prototype operation would essentially be solved and would not recur in the flight of the tactical airplane.
4. Ground handling equipment, including landing facilities and maintenance facilities, would be identical between the prototype and first tactical airplane, and the development would not have to be duplicated.
5. The major detail design problem aside from the landing and takeoff was the problem of maintaining sufficiently low weight so that a usable margin between thrust available and required would be realized on the final design. With this in mind, it was proposed that the full-scale prototypes be used for a) measuring loads in flight and b) that an identical article be used as the static test airplane for the tactical version in spite of the fact that its design gross weight would be considerably lower than the tactical airplane. This required more modifications during the static test programme but would result in a substantially lighter final aircraft.
6. Since it appeared that modifications were required on the Mamba power plant for vertical flight, and, since it appeared entirely feasible to fly a full-scale T40 prototype without the two-speed modification to the gearbox, the time for developing either prototype could be approximately the same. From a cost standpoint, it was estimated that developing full-scale prototypes in Part I of the development programme would be far cheaper in arriving at a tactical airplane than if a Mamba type prototype was developed. This was due to the fact that only three articles needed to be manufactured instead of five and the transition from prototype to tactical airplane would be much simpler. This did not imply that the prototype part

of the suggested programme itself would be any more expensive since the development of any airplane of this type appeared to involve just as many research problems, whether it weighed 7,500lb or 12,500lb.

A possible disadvantage of the suggested procedure concerned the reliability of the T40 engine compared to that of the Mamba. At the time, the Mamba was probably more reliable, but considered from a view of 18 months' additional time, and also the advantage of the gearing method of the T40 type engine (allowing flight on one engine), it seemed that satisfactory reliability should be available from the T40A-6 engine by the time it was required for a prototype airplane. It was necessary to ensure proper delivery of the propeller gearbox to meet the required time schedules. The Allison Corporation provided the following dates applied for the various configurations of their gearboxes:

No.	Type	Months
1	Standard gearbox revised for vertical flight	15
2	Standard gearbox plus two propeller speeds	24
3	Large offset gearbox plus two propeller speeds	24
4	Large offset gearbox single speed	21

Note: Front accessories, if desired, took an additional three months on Numbers 3 and 4.

The prototype L-200-1 airplane could be flown with the standard gearbox revised for vertical flight which would be available in 15 months. This procedure was not desirable, however, since the high position of the engine changed the fuselage structure, caused very poor duct inlets, broke the wing structure and lengthened the tail pipes. It was considerably more desirable to obtain Gear Box No. 3 or No. 4 in 15 months or delay the programme enough to obtain these units.

With the above development spans, the production airplane deliveries would be delayed six months for Gear Box No. 4 and nine months with Gear Box No. 3. Even with the maximum gearbox delay, a production airplane was available at the same time as with Programme A. In spite of this, every effort was made to get Gear Box No. 4, or preferably 3, developed in 15 months.

The reasons for proposing such a prototype programme could be best obtained from consideration of Programme A and Programme B. Programme A indicated the proposed Navy programme as regarded time for various phases of development. Lockheed believed it reasonable to make tunnel tests, preliminary design, and the required research within a six month period. In an additional 12 months, the first of the two Mamba prototypes could be available for flight test.

Under the BuAer proposal, 20 months after start of the project, the design and tunnel testing for the tactical airplanes would be undertaken, following about 2 months of flight testing. It was the Navy's desire to build the first two tactical airplanes and a static test unit on production tooling. This meant that pre-planning for production would start at the earliest possible date; in fact, almost simultaneously with the basic designing of the tactical airplane. Production tooling could not be done until tunnel tests and structural design were well along so, approximately six months from the start of Part II of the contract, the production tooling could start.

It required at least 15 months from the start of construction of the production tooling to build the first of the two airplanes. The flight test on the tactical airplane would then start 41 months from the very beginning of the programme at the very earliest.

A very considerable hazard existed in building production tooling for an airplane which had not undergone any of its experimental testing. Continued production would inevitably be faced with a flood of flight test changes. The expense involved in tooling changes would be prohibitive.

Programme B presented an alternative procedure based on the conception of maintaining a given aerodynamic configuration which would gradually develop into the tactical airplane as the T40 engine power increased to allow higher gross weights. The time required for initial tunnel tests, mock-up, and preliminary design, was the same for both Navy and Lockheed programmes. Likewise, the time to flight test the prototype was the same. The prototypes were built, however, on experimental tooling. This was also true for the static test article.

In this programme, it was advisable to have a static test article at the earliest possible date as the weight trend to the tactical airplanes was based upon obtaining the most efficient structure through continued static testing. After a period of eight months of flight testing of the prototypes, one

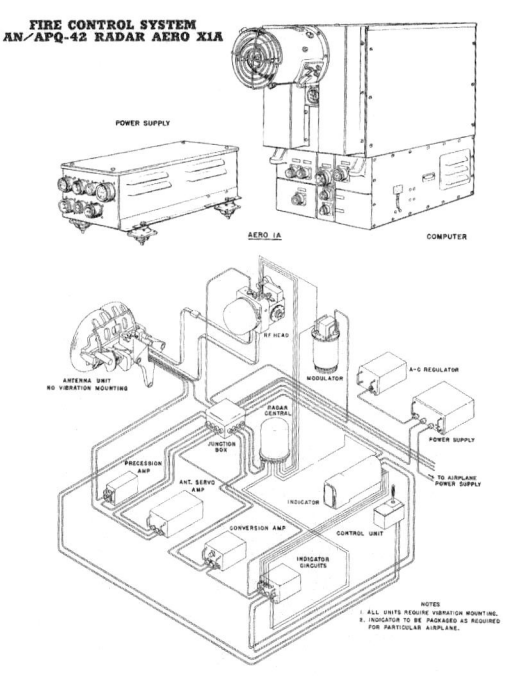

Diagram of the L-200 fire control system and AN/APQ-42 Radar Aero X1A.

airplane could be converted to the tactical airplane, subject to the availability of the high-powered Allison engine. During conversion of one airplane, the other prototype would continue flight testing. It would later be converted to a tactical airplane also in a manner which would best fit in with overall conditions existing at the time.

The tactical airplane could then be flown approximately 32 months from the start of the overall programme compared to 41 months in the case of the plan outlined in Programme A. Production pre-planning would start about 18 months from the beginning of the programme because the basic airplane configuration for the first prototypes and the tactical airplane would have a great deal in common, particularly structure, equipment, and arrangement. Production engineering could start 22 months from the beginning of the programme, or after having obtained four months of flight testing on the prototypes rather than two months as in the original plan.

In spite of these advantages, it appeared that Programme B would make available a tactical airplane for flight test nine months earlier than Programme A, and production airplanes would be

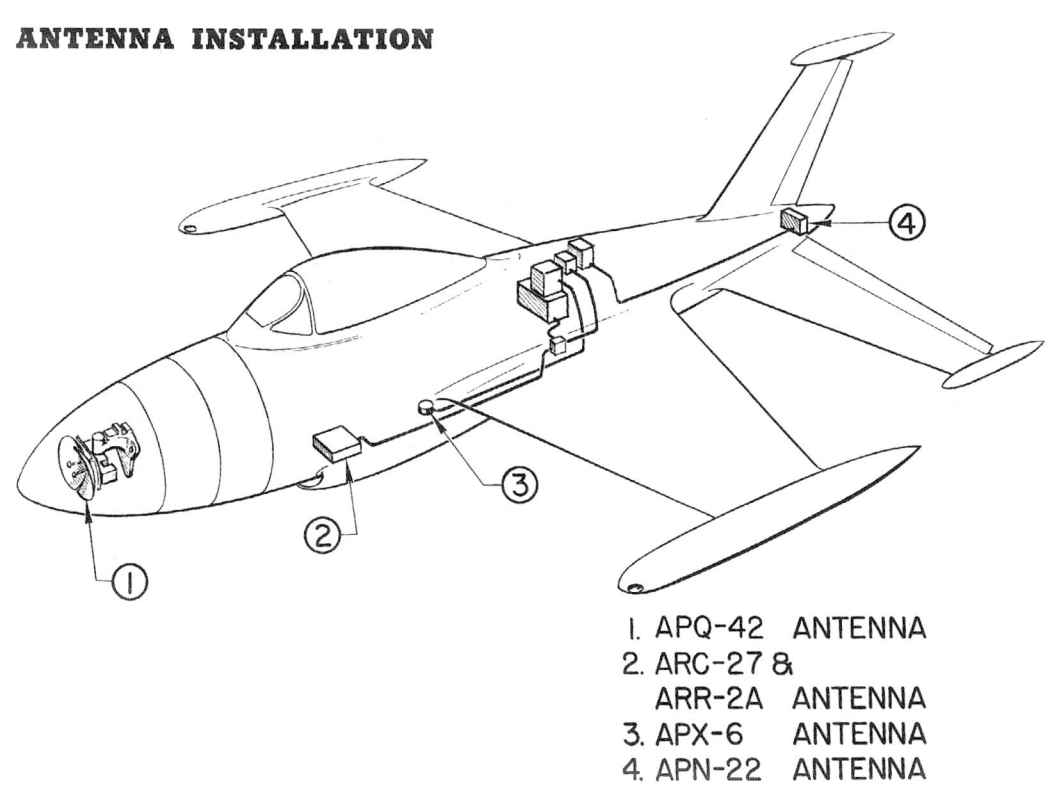

Drawing showing the layout of the communication system antennas.

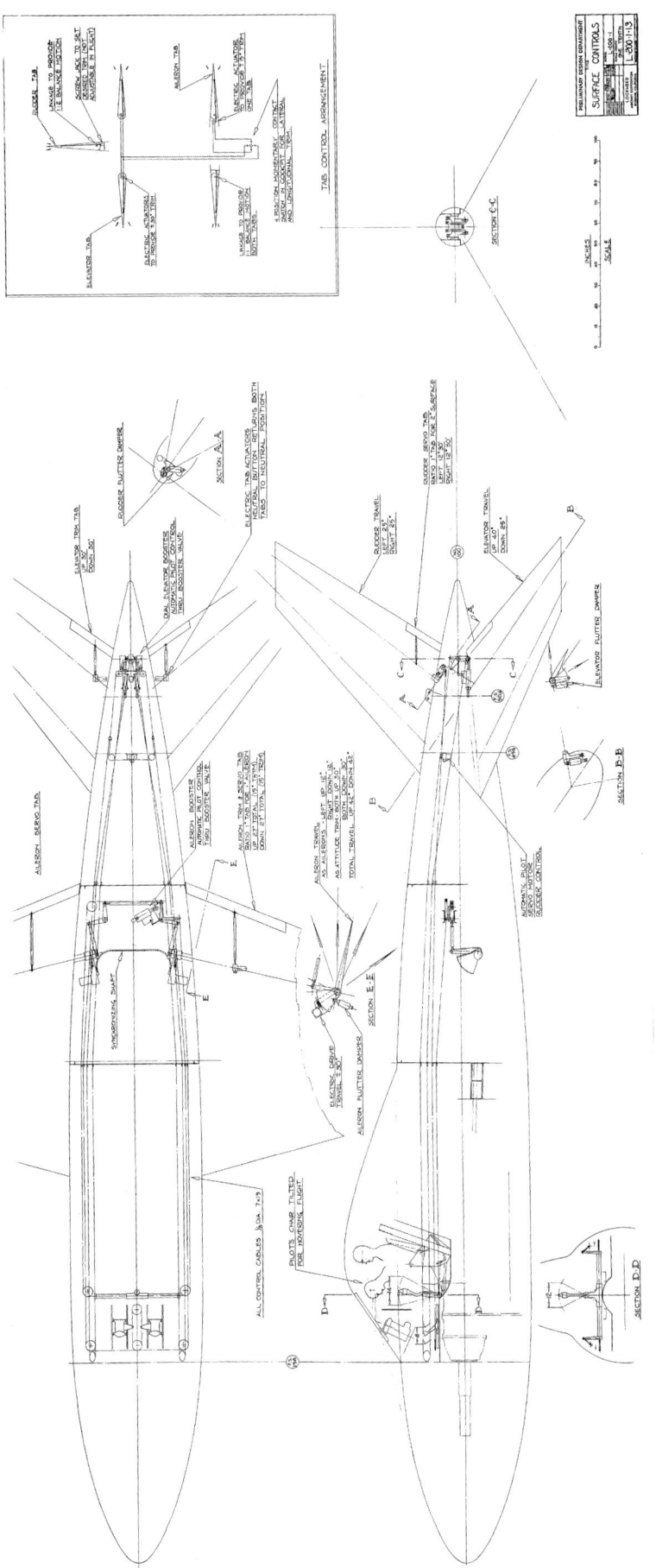

Diagram of the major surface control system of the Lockheed L-200 Convoy Fighter.

available nine months sooner. There would be much less risk and overall expense involved in the plan outlined in Programme B than in Programme A. More orderly planning and production engineering could be done with the engineering load considerably reduced by engineering only one configuration. Programme B was, therefore, recommended. Cost information, however, was provided for both methods of carrying out the development.

Lockheed emphasized that it would bid and pursue whichever programme the Navy considered most desirable.

DESIGN DETAILS
External arrangement
The general arrangement of the L-200-1 was relatively conventional in spite of the fact that it was designed to rise and land vertically. It incorporated a straight wing, a relatively normal cockpit, external tip fairings for armament installation, and a three-way arrangement of the horizontal and vertical tail surfaces. Alternate arrangements of the airplane were studied in considerable detail and two of the most promising of these alternates were included in Lockheed's proposal.

Alternate A, the L-200-6, incorporated a swept wing with 47.5° sweep of the quarter chord line and 8.1% thickness which compared to the 4% thickness of the straight wing with no other changes. The reason that this wing was not proposed as the primary configuration of the airplane was that weight studies showed the straight wing to be lighter by 210lb for a given drag rise Mach number than the swept wing configuration.

Furthermore, since the swept wing was identical in aspect ratio to the straight wing and since its effective aspect ratio was considerably lower, the straight wing had better manoeuvrability at altitude and higher ceilings. Finally, the cockpit installation, control system, and armament installation were considerably simplified and the balance improved since the major structure of the straight wing went directly through the fuselage, and the gun pods were close to the airplane CG which would have been impractical with the swept wing.

Lastly, it was determined that the straight wing had better aileron control in hovering than the configuration with sweep. Alternate B, the L-200-7, was an extremely attractive configuration because of its widespread footprint pattern since the wing itself was used as a landing gear support device. Lockheed recommended continued investigation of this configuration along with the swept wing configuration in the early phases of the design programme because it believed that having such a widespread landing support would simplify facilities on deck and could permit simpler operation of the airplane from land bases. The major reason that Alternate B was not proposed as the primary configuration was that the stability and control, both in horizontal and vertical flight, were open to considerable question and further research was required before the feasibility of obtaining desirable flight characteristics could be proven.

Internal arrangement
An inboard profile of the tactical airplane showing the general arrangement of equipment, power plant, cockpit and internal structure was provided by Lockheed. An inboard profile of the proposed T40-A-6 powered prototype, illustrating the similarity between the two airplanes, was also included in the proposal. In presenting the internal arrangement, Lockheed noted that a major revision was made to the standard gearbox supplied with the Allison T40-A-8 engine in order to provide a larger offset between the power section centreline and the propeller centreline. This offset resulted in a weight penalty to the power plant section of the airplane but provided such major simplification advantages for the rest of the configuration that this weight penalty was cancelled out in the overall design. Some of the advantages of the offset gearbox were as follows:

1. The duct inlet to the engines was shorter and had much less bend than with the standard gearbox. This resulted from the fact that the spinner diameter for proper propeller efficiency was fixed regardless of the engine location and essentially covered up what would be the normal engine inlet for the standard gearbox.
2. The lower position of the engine resulted in an extremely simple tail pipe installation since the exhaust was practically external on the airplane and would require almost no tail pipe.
3. Lowering the engine permitted the wing and its carry-through structure to be more nearly along the centreline of the airplane, making available a large unencumbered

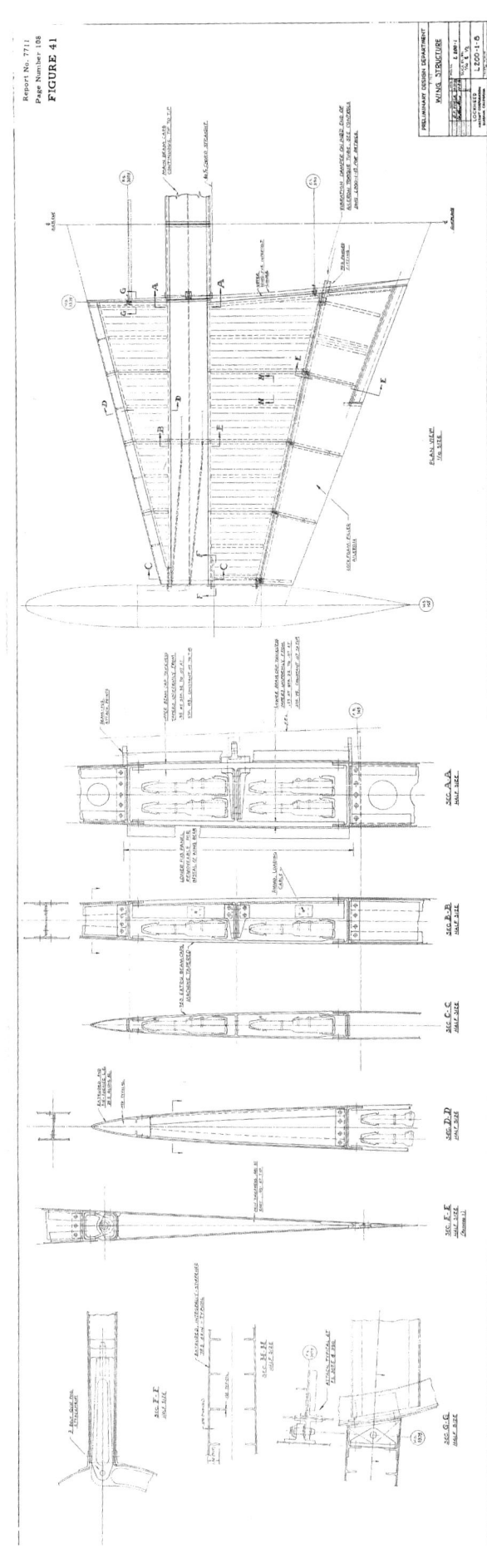

This structural diagram of the L-200-1 wing shows the major improvement in simplicity and structural integrity enabled by the elimination of the landing gear, armament, fuel, and flap provisions within the wing.

space above the wing for a single simple fuel tank.
4. The space available below the cockpit was considerably larger with the lower engine position permitting mechanisms for control, seat movement, etc. more space, thereby simplifying their design.
5. With the lower engine position, the main fuselage structure was deeper, thereby effecting a major saving in fuselage weight and rigidity.
6. Having the engines partially exposed outside of the basic fuselage structure simplified and lightened the fire protection provisions which were required above and between the power sections.
7. If it was later considered feasible to add an afterburner to the airplane to improve its high-speed performance, the installation of this device would be quite simple with a low engine position but would be practically impossible if the standard gearbox were used.

A more complete discussion of the alternate power plant possibilities in the airplane is found in the Power Plant section. The simplicity of the service and maintenance of this airplane is well illustrated in the inboard profile, since all of the electronic gear could be arranged in essentially one place and, since the accessories as proposed for the engine and the electronic gear which had to be located close to the radome, were all available through external doors without breaking through major structure or engine ducts.

Cockpit arrangement
The design of the cockpit arrangement followed conventional lines in spite of the unconventional requirements on the pilot. It was felt that this was very nearly the optimum approach since it corresponded to the conclusions reached by Navy investigators and did not impose upon the pilot unconventional positions and locations of controls and instruments simultaneously with the unconventional attitude of the airplane. Accordingly, the seat was arranged to pivot about an axis parallel to the wing axis in order to raise the pilot's head as the airplane approached its vertical attitude.

This was found to be absolutely necessary from the standpoint of comfort. It was proposed that this seat position be governed entirely by the attitude gyro in the autopilot and that the movement between the normal position to the lean-forward position be gradual, starting from a nose-up attitude of approximately 45°. As a further item of comfort, stirrups were added to the control pedals in order to support the weight of the feet and legs in the vertical attitude.

Additional items of note in the cockpit arrangement included:

1. A central location for the radar scope was found between the pilot's feet with the scope face aimed directly at the pilot's normal eye position. This location did not interfere with the Mark VI Mod. 1 gunsight and it required that the flight instruments be moved only slightly to the right of the centreline of the cockpit. All of the engine and miscellaneous instruments were located on the left side of the panel.
2. Autopilot and communicating controls were located in a convenient position on the right-hand side shelves and were easily reached by the pilot in either the vertical or horizontal seat position.
3. The left-hand console was reserved for the power and radar controls. These also were convenient to reach with the seat in either position.
4. A separate autopilot controller for vertical flight was installed directly on top of the control stick. This control consisted of a switch which would translate the airplane in a belly-forward or aft direction by moving the thumb away from or toward the pilot, respectively. The airplane would translate laterally by moving the switch to the left or right of the pilot, depending upon the direction desired. A considerable discussion was held with several experienced pilots and the autopilot manufacturers concerning the desirable location of the autopilot controller for vertical flight and the desirable sense in which these controls should operate. This stick location was finally selected for the following reasons:

   a. Since the controller actually selected attitudes of the airplane and, therefore,

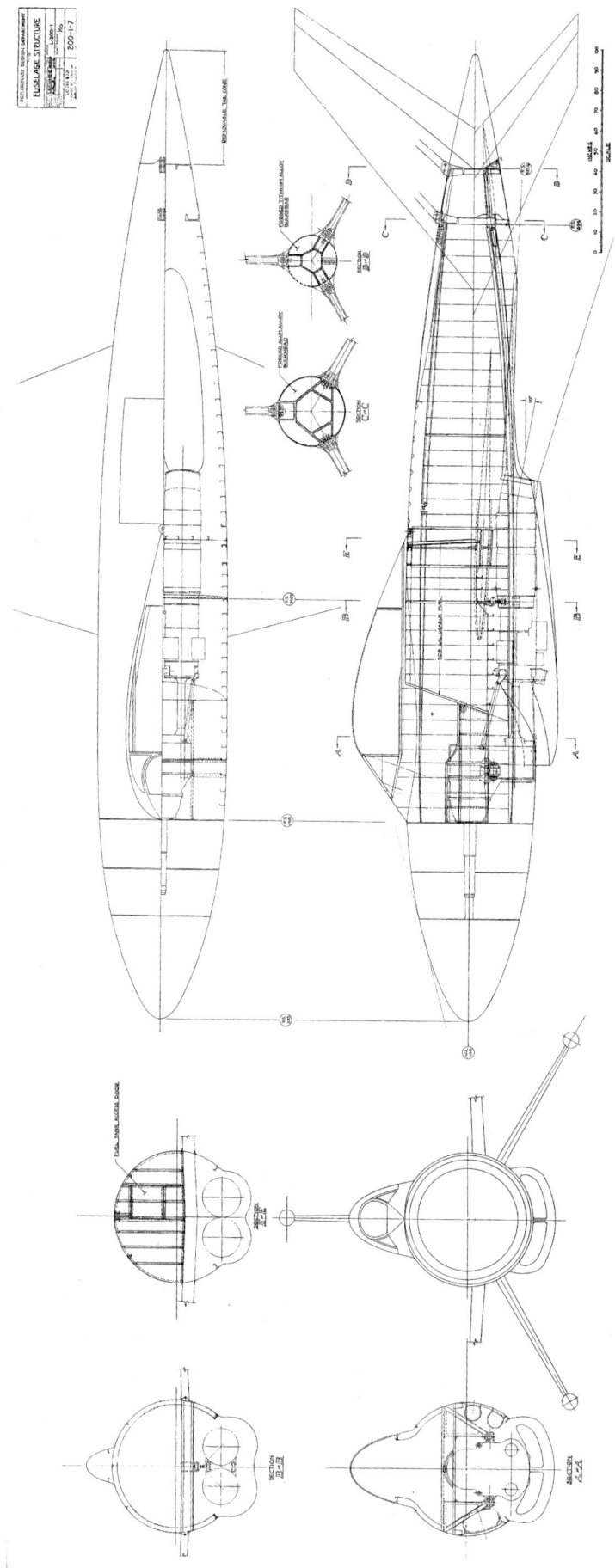

This structural diagram of the L-200-1 fuselage shows that a major simplification was achieved due to the airplane's VTOL tailsitter configuration and the incorporation of a gearbox with a large offset.

as a result, selected translational velocities. It was believed that a sidewise motion of the controller was more suitable for side movement of the airplane than the yawing type of control normally associated with an autopilot controller which turned the airplane to the left or right.

    b. Thumb movement away from the pilot for forward motion and back toward the pilot for backward motion was believed to be in the satisfactory sense since this corresponded to the motion of the stick to achieve fore and aft translation.

    c. Although similar motions could have been worked out in a very reasonable fashion on the standard autopilot controller by changing the control head, it was believed to be impractical to remove the pilot's hands from the control stick in order to achieve this simplification since a failure of the autopilot might occur at such a time that the pilot could not afford the delay of moving his hands from a separate autopilot controller to the stick in order to retain control of the airplane.

5. A radio altimeter was placed in the lower left-hand corner of the instrument panel with its face attitude adjusted for easy reading by the pilot when the seat was in the vertical position. It was found that this location of the altimeter was the most desirable when the pilot was flying the airplane in its vertical attitude since it required the minimum change in eye position from the normal outside vision line past the wing tip toward the ship.

Other seat positions, including 180° rotation positions, cocked axis positions, etc. were all investigated in an attempt to improve the comfort and usability of the cockpit. None of these seat ideas achieved any advantage except that of better and more comfortable direct rearward vision. It was concluded after these mock-up studies that this vision, when obtained, was not nearly as valuable as maintaining a normal cockpit with normal controls since the accurate vertical height adjustment of the airplane could better be handled through the inclusion of an LSO on board ship who could better judge the actual motion of the ship and more accurately determine the precise altitude for power cut. For operation from land bases where no ship motion need be accounted for, Lockheed believed that the pilot could adjust himself sidewise in the existing seat to obtain sufficient down-vision. Furthermore, if the landing platform as proposed for the ship was utilized for ground bases, an altitude target directly to the pilot's left and just beyond his wing tip was provided by the wing tip backstop.

It was believed that the pilot would have a more accurate determination of altitude at touchdown from such a target than he would from direct aft vision even under the most ideal location of his seat.

Power plant

The basic power sections were those of the Allison T40-A-8 engine as described in Allison Specification No. 272B, revised May 31, 1950. The power ratings of the engine used for the performance of the airplane were:

Military Power	6,955shp at 14,300rpm
Military Thrust	1,363lb
Normal Rated hp	5,790shp at 14,000rpm
Normal Rated Thrust	1,225lb

This engine was modified by changing the standard single speed gearbox to a two-speed gearbox with a larger offset between the power section centrelines and the propeller centreline. Since this gearbox was the subject of considerable discussion between BuAer, the Lockheed Company, and the Allison Company, a clarification of the gearbox development programme was obtained from Allison. As a result of this investigation, it was decided to use Gear Box No. 3 since its development time did not exceed that of the standard gearbox with the two-speed unit and since it had a sufficient offset to improve the airplane arrangement in a major fashion.

Study of the power plant and aircraft accessories indicated that the most desirable accessory location would be on the front side of the gearbox where all of the accessories for the airplane could be reached through external doors or through the duct closure doors. Since this location of accessories caused an additional development period on the gearbox, it was decided to forego these advantages in the proposal and accept Gear Box No. 3 with the normal

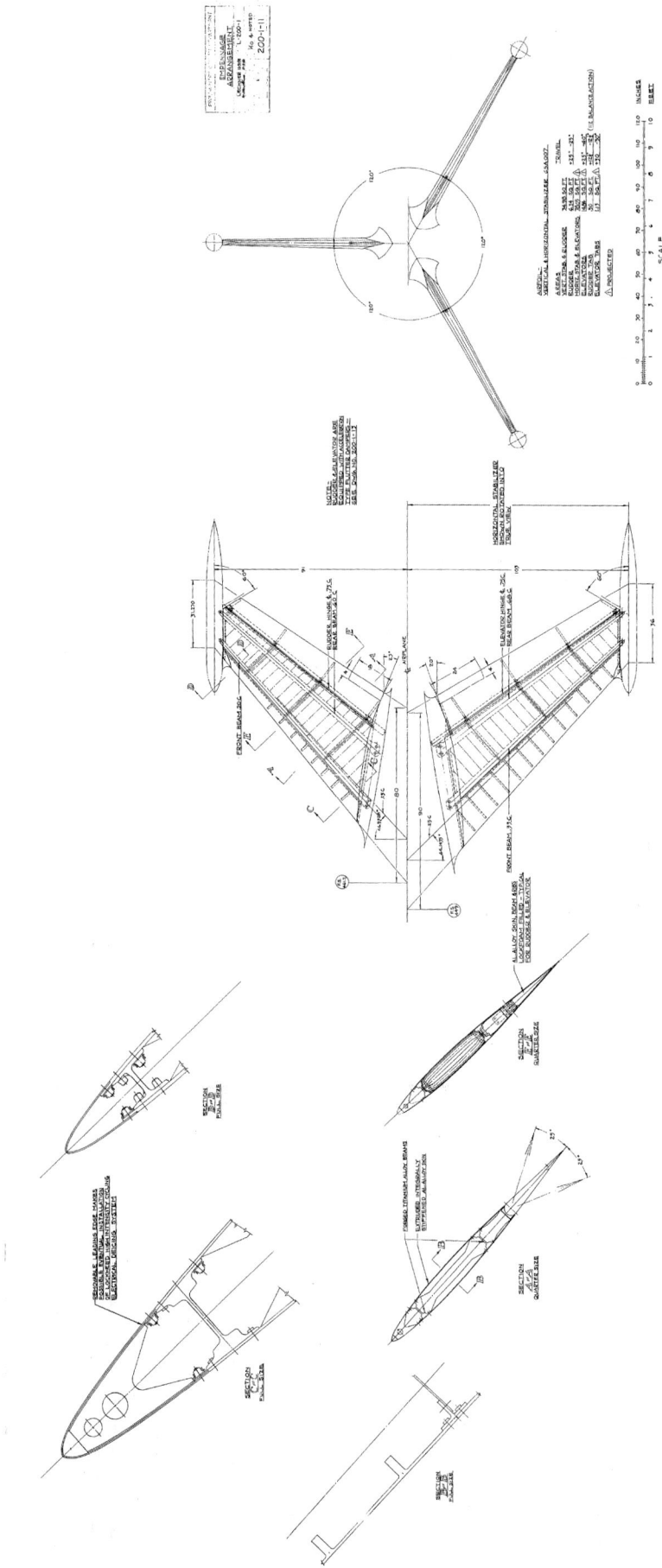

The L-200 empennage presented the greatest structural difficulty with the VTOL tailsitter configuration; not only did the tail have to withstand the air loads but also the landing loads. It had to be both strong and light; to achieve this, Lockheed proposed utilizing titanium forged empennage spars and an aft fuselage attachment bulkhead.

accessory positions on the aft side. Further study of this item was required during Phase I of any contract to determine whether the forward location of the accessories actually necessitated the additional gearbox development time or whether the programme of aircraft development could reasonably stand such a delay.

During the analysis of the airplane performance, a further discussion was held with the Allison Company and BuAer as to the feasibility and desirability of using an afterburner to eliminate the necessity of a two-speed gearbox in the airplane, or in addition to the use of the two-speed gearbox. It appeared that the use of an afterburner would improve the high-speed performance by approximately 13 kts without the two-speed gearbox and 38 kts if both the two-speed gearbox and afterburner were installed.

These high speed advantages could only be obtained, however, with a weight empty penalty over the proposed airplane of approximately 435lb with the single-speed gearbox or 635lb with the two-speed box. This included a weight penalty of about 498lb of fuel which was required if the afterburner was used for climb and combat. Only 282lb of this fuel could be conveniently carried. Lockheed recommended that this type of power plant also be considered in any Phase I study in order to determine whether or not the high speed advantage was worth this penalty and whether or not it was necessary to incorporate two speeds in the propeller gearbox where an afterburner was installed.

The ducts for the oil coolers and the ducts for the engine intake were separated in order to ensure that flow through the oil coolers would always occur in one direction and would not be disturbed by variation in flow into the power sections. Several alternate schemes were studied in which the oil coolers were fed by a large plenum which also fed the inlet to the engines.

This system had several advantages since it eliminated the fuselage cutouts for the oil coolers, permitted the oil coolers to be located further forward in the airplane and more equally distributed the air entering the engines. It was discarded, however, since a major thrust loss was calculated for this type of plenum chamber and it was impossible to determine what kind of pressure ratio would be available for oil cooling in the event of:

1. A single power section failure or cutoff for cruise purposes.
2. Plenum chamber evacuation due to suction from the power plants at zero forward velocity.

A final deterrent to the use of this type of duct was the inability to separate the two power sections to prevent backfires from causing a fire hazard in a plenum chamber containing accessories and inflammable materials.

Lockheed believed it desirable to have two power sections in a vertical rising airplane and, therefore, the T40 type power plant arrangement was preferred. Since the development of propeller turbine power plants, however, had been extremely complicated and difficult, it was believed to be important that an alternate engine be available to do the job since this airplane could depend neither on an alternate reciprocating engine nor on an alternate pure jet engine.

Lockheed provided a summary of power estimates with time for the Allison T40 engine, showing the availability of the power ratings used in Lockheed's proposal from a date standpoint. Plotted also on this curve was a power development proposal for the Pratt and Whitney PT2E engine. This indicated the feasibility of incorporating a Pratt and Whitney engine in an airplane of this type at approximately the same time. The airplane could be made entirely usable and have similar performance with the Pratt and Whitney engine. This required, of course, additional gearbox development to obtain a two-speed offset counter-rotating gearbox but such a development would probably not be as difficult as the similar development incorporating the two power sections.

Airplane landing provisions
As described previously, the landing with this airplane was made by flying sidewards into a shock absorbing wing tip stop under the direction of an LSO and, upon being given the cut, the power was reduced and the airplane dropped into a landing net on the three-pronged empennage with the spike from one of the wing tip pods penetrating a small wing tip net to provide stability against inadvertent tipping over of the airplane due to ship roll or excess side velocity at touchdown.

The provisions for landing on the tip pod consisted only of a rub strip of either hard wood or titanium

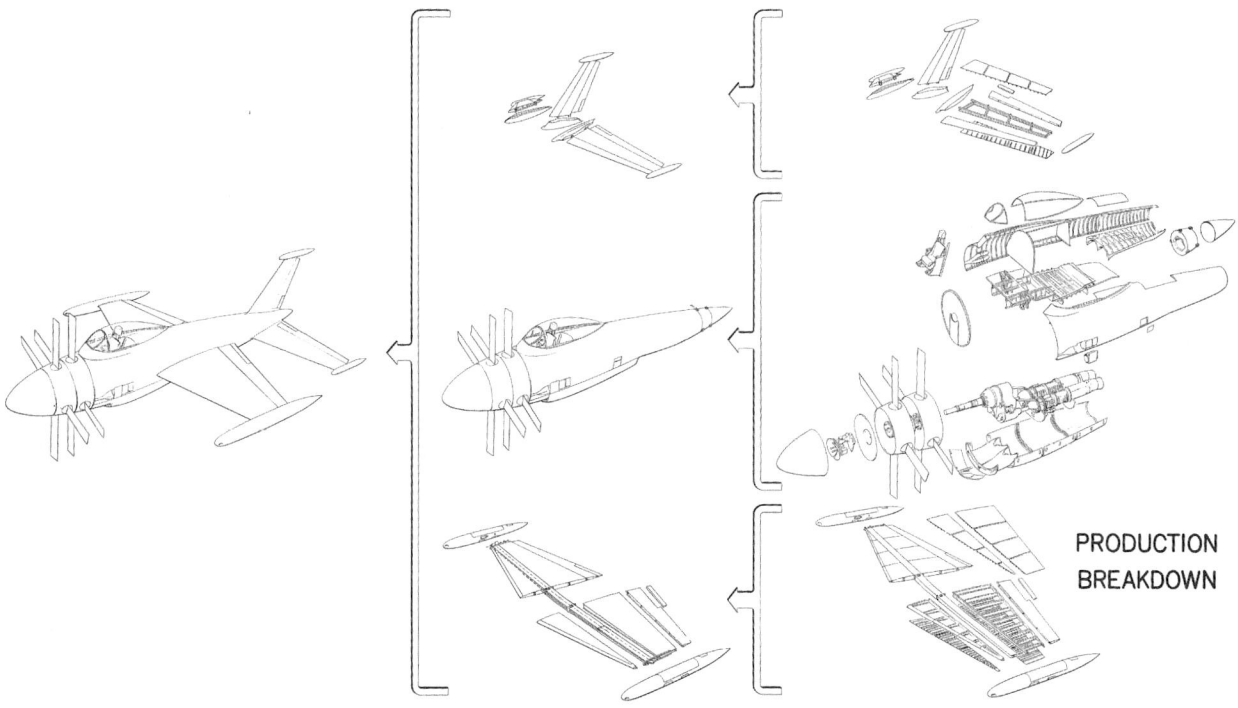

The L-200 production breakdown was identical for both the full-size prototype and production airplanes, and was based on service, spare part requirements and producibility.

attached to the outside of the normal tip pod and rub ribs near the tail of the pod to prevent scraping by the tip net cables. Landing provisions on the tips of the vertical and horizontal tail, however, were more exacting and required not only pads to prevent the tail from penetrating the landing net but also required shock absorption means to prevent damage to the airplane and reduce the load factor which needed to be carried by the empennage.

The typical tail tip pod had a mechanism for retracting the landing pads and obtaining adequate shock absorption. The facilities for landing the airplane, though heavy, were not excessive when compared to a normal landing gear for an airplane of this weight. Alternate landing procedures for prototype or land-based flying were studied, and Lockheed believed that none of these procedures could easily be incorporated into the airplane as a permanent type of landing system, since they involved complication and weight which would seriously impede its tactical utility.

Lockheed developed alternate wheeled gear to show what could be done with the airplane during the early development stages of flying. A very simple gear was designed for operating the airplane in a normal fashion without reverting to vertical flight for landing or take-off.

Armament

The installation of the four 20mm guns required in OS-122 appeared to be optimum when a wing tip pod was installed on the airplane. Installing this pod on the extreme tip of the wing was obviously best from an aerodynamic standpoint and it had many additional advantages when compared to any other arrangement for the guns. These advantages included:

1. Permanent location of the guns at the extreme tip of the wing saved approximately 440lb in the wing bending material.
2. The gun location in these pods made access for removal and loading as nearly perfect as could be obtained.
3. By locating the guns at the extremities of the wing, the entire wing span became available for simple integral ammunition chutes.
4. The existence of the wing tip pod provided an external stabilizer prong for the landing

system and an external buffer for the wing tip backstop.
5. When combining the wing tip installation with a straight wing, the CG advantage of having the ammunition practically on the centre of gravity of the airplane was very apparent.
6. Locating the guns and pods near the airplane centre of gravity made it simple to change the type of aircraft armament at will without seriously affecting the airplane design. Forty-eight folding fin rockets could be installed in lieu of the four 20mm guns. This installation affected none of the basic airplane design and, in fact, it appeared feasible to have alternate wing tip pods for installation on the same airplane.

Loading of the ammunition for the tip guns was done by means of a built-in electrically powered ammunition loading cable. This cable was attached to the end of the belt and the belt was pulled into the wing by the activation of the control button on the tip through which the ammunition was being led. Since the electric winch for this ammunition loading was installed in the opposite tip pod, no difficulty was encountered if it had to be replaced. Furthermore, the cable, if frayed, could easily be replaced by threading through the straight ammunition chutes. The cable remained attached to the ammunition belt and was immediately available at the wing tip for reloading when an ammunition belt had been expended.

One problem introduced by the tip pods was the ability to maintain boresight accuracy during manoeuvres at high speeds and high load factors. For extreme altitude operation where load factors were low and the Mach number was high, Lockheed found that tip deflections were within the desired accuracy. However, low speed high load factor manoeuvres at sea level were believed to cause some divergence. A solution to this problem was suggested by contemporary high load factor manoeuvres with tip tanks equipped with tail fins. These indicated that tip torsional deflections could essentially be eliminated and, as a result, a tail fin option was proposed by Lockheed.

Electronic and electrical provisions

The autopilot installation for the Lockheed L-200 was probably the most important item of electrical equipment since its proper functioning and installation ensured the ease of making the flight transitions in an ideal fashion and in effecting the landing and takeoff satisfactorily. Discussions were held with several autopilot manufacturers and it appeared, for proposal purposes, that the Westinghouse W-3 autopilot, suitably modified, met all of the requirements.

The major component of the autopilot was a rate gyro system which was used under normal flight operations. The added feature for use in vertical flight was the vertical gyro. As described previously, this gyro was used to ascertain the vertical axis about which the airplane was manoeuvred when in a vertical position. A signal from the autopilot controller on the stick asked for a zero rate of motion about a fixed airplane attitude in space, either laterally or in a pitch direction. This attitude was held through the rate gyro system whose signals were implemented by a teeter valve on the elevator boost control and by servos on the rudder. Zero roll rate was maintained for all attitudes through an aileron servo. The system which eliminated a servo on the elevator control had been checked out on flight tests of the Lockheed F-94 all-weather fighter and performed in an entirely satisfactory manner.

Aside from having all of the components which were required for vertical flight, this autopilot also provided a controller which could be used by the pilot in gunlaying and tracking manoeuvres for combat purposes. Under these circumstances, the normal autopilot controller on the right-hand console was used.

The Antenna Installation (see p111) drawing illustrates the remaining electronic components of the airplane, showing the relative locations of the radar scanner and the various components which had to be located throughout the L-200-1 in order that the radar could be satisfactorily operated. As noted previously, all of the components of the radar were available for simple servicing and maintenance procedures.

The system for operating the electrical and electronic equipment had as its main power source a 15 kva alternator driven from the standard alternator drive pad on the rear of the propeller gearbox. Other alternate electrical systems were studied but it appeared that this was the lightest and most usable, considering the limitations on the power drives which could be available with reasonable development time.

The Antenna Installation drawing of the communication system antennas shows the following required communication and navigation sets:

AN/ARC-27	UHF Command Set
AN/ARR-2A	Navigation Receiver
AN/APX-6	IFF Transponder

These sets were those listed in OS-122, and the ARC-27 essentially limited the operating range of the airplane to approximately 120 nautical miles at the operating altitude of 35,000ft. As discussed under 'Operational Procedures,' it appeared that the communication range of the airplane was the actual limiting factor if it was used for search purposes. Therefore, a Phase I study of the actual operational use of the airplane required a detailed study of its utility as a search airplane in order to determine whether or not recommended changes in communication equipment were desirable. An alternate gunlaying and radar system had also been studied.

The utility of this radar and gunlaying system was that it permitted the airplane a larger flexibility in the final attack manoeuvres since it permitted collision course flying as well as tail chase operations. With this equipment it was possible to carry and effectively use 2.75" folding fin rockets. Lockheed believed that this also required study in considerable detail during the early phases of the contract development of the Convoy Fighter since the margin in time by which a 450 knot bomber was intercepted was extremely small and collision course attacks would materially improve the pilot attack procedures and shorten the attack time.

Control system

The Surface Controls blueprint (see page 112) shows the major surface control system of the L-200. The control system featured the following basic ideas:

1. The elevator control system comprised two completely independent power boost units so arranged that battle damage could not reasonably eliminate both units. This removed the necessity of a complicated and undesirable aerodynamic balance system which would have to be devised for boost-off flight and which could not reasonably be made to control the airplane over its entire speed range.

2. The aileron control system utilized a single boost cylinder for normal flight operation and incorporated a servo tab located in the propeller slip stream for boost-off operation. This servo was not completely satisfactory for combat flying and, therefore, the boost could not be eliminated but it did serve to give adequate aileron control during the flight transition between horizontal and vertical position and did give adequate aileron control during vertical flight. An aileron droop or flap movement had been incorporated to act as a vertical flight attitude trimmer. It was found from wind tunnel tests that cross winds perpendicular to the wing plane required high tilt angles to position the airplane. A full-span flap alleviated this condition and permitted the tilt of the airplane to be approximately the same in pitch as in yaw for a given translational velocity.

3. The rudder system incorporated a servo tab and required no boost since satisfactory pedal forces could be achieved for both combat and transition flying.

The trim control system was conventional in every respect. It was notable that the vertical flight operation of the airplane eliminated some of the most severe trim conditions since no major moment change existed which was comparable to the operation of the landing flap for a normal airplane.

Lockheed emphasized that the lowered position of the engine made possible by the special gearbox simplified moving the controls out of the cockpit and arranging them within the airplane. This resulted in major weight savings and major improvements in simplicity which were directly responsible for easy servicing and infrequent maintenance attention.

Studies of the flutter characteristics of the control surfaces indicated the need of closely controlled dynamic and static balances since the flight speed range of this airplane caused the flutter problem to be relatively severe. The solution of this problem in the normal fashion by the addition of weights was hampered by the thinness of the control surfaces and the desire to maintain the maximum possible control requiring sealed hinge surfaces. This led to the incorporation in the control system of dynamic dampers in place of the normal weights mounting

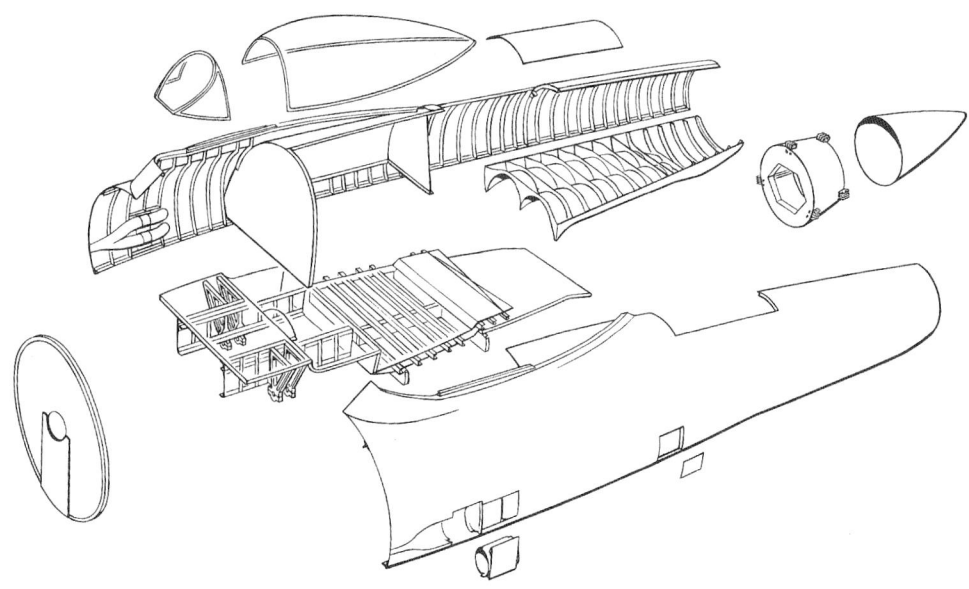

FUSELAGE BREAKDOWN

The L-200 fuselage broke down into five major assemblies: left and right side panels, floor and engine mount sections, lower aft section, and empennage attaching section.

ahead of the hinge line.

The dynamic damper was developed by Lockheed in its study of the phenomena of wing aileron buzz and was a simple self-contained unit which permitted normal rates of movement of the control surface but did not permit rates of motion of the surface above those amounts. An actual flutter damper was put through many exhaustive tests in the Lockheed Research Laboratory, including tests at low temperature, at high temperature, and for prolonged periods of operation. The basic principle of the unit involved a frictionless lead screw (ball bearing race) driving a rotating shaft. On this shaft was a rotating collar driven through another frictionless lead screw.

With this arrangement the collar rotated at the same speed as the shaft if no acceleration or deceleration took place. If acceleration was present, the shaft rotated faster than the collar and the outside lead screw then drove the collar against a thrust bearing which applied a brake to the entire system. Since the brake had a double surface, the reverse took place during deceleration. This braking system was found to be entirely effective in eliminating buzz and flutter.

An alternate scheme to solve this problem would have been the incorporation of irreversible boost components in which two boosters per control surface were installed, either one of which would be satisfactory for flutter control and surface control in the event of boost failure. Since this system required the installation of false feel or the elimination of any pilot feel, it was felt that it was the least desirable of the two because the airplane needed to be as conventional as possible to fly – at least until the ease with which vertical flight could be accomplished was determined.

Hydraulic system

As noted previously in the 'Control System' section, hydraulic boost was installed on the aileron and elevator and a dual boost was installed on the elevator. Two completely independent sources of pressure were needed. This was achieved by driving the main utility system from the one hydraulic pad which was available on the engine and driving the other system through an air driven hydraulic pump connected with air bleed holes on both power sections of the Allison engine. This power system was made as simple as possible to eliminate any possibility of failure since it was the relief system upon which the elevator control depended in the event of main system failure.

As pointed out previously, no connection whatsoever existed between the two hydraulic systems, thus ensuring that no failure would occur which would

GROSS WEIGHT SUMMARY - L-200-1

WEIGHT EMPTY		11,315
Pilot	200	
Oil	69	
System Fuel and Oil	57	
(4) 20 mm Guns	450	
600 rds. Ammunition	408	
Oxygen and Equipment	53	
EQUIPPED WEIGHT		12,552
40% Fuel (203 gals.)		1,218
LANDING WEIGHT		13,770
20% Fuel (102 gals.)		612
COMBAT WEIGHT (Design Gross)		14,382
40% Fuel (203 gals.)		1,218
TAKE-OFF WEIGHT (508 gals. fuel)		15,600

Table showing the breakdown of all the required weights for performance calculations based on the recommendations and definitions of 0S-122.

WEIGHT EMPTY SUMMARY - L-200-1

WEIGHT EMPTY		11,315
Wing Group		967
Panels	836	
Ailerons	131	
Tail Group		792
Stabilizers	498	
Elevators	65	
Fin	191	
Rudder	38	
Fuselage		1,250
Alighting Provisions		300
Power Plant		5,777
Engine (XT40-A8)	3,310	
Accessories	34	
Power Plant Controls	37	
Propellers	1,918	
Starting System	44	
Lubricating System	177	
Fuel System	257	
Fixed Equipment		2,229
Instruments	128	
Surface Controls	311	
Hydraulic System	163	
Electrical System	303	
Communicating	216	
Armament Provisions	803	
Furnishings	278	
Anti-Icing	27	

A complete summary of the L-200 weight empty breakdown estimate based on the suggested prototype programme.

eliminate elevator control. Lockheed recommended an engine driven pump for the reserve boost system if additional pads became available on the gearbox.

The basic system used 3,000psi hydraulic pressure and was of the constant pressure type. The lack of landing gear and flap operating devices made the system extremely simple and no unconventional features existed which could cause service trouble.

Air conditioning system

No new components were required for the performance characteristics and altitudes expected of this airplane and, therefore, no development programme was envisioned other than a test checkout of the air conditioning system and its mechanical components. Power sources from both power sections were utilized with a crossover valve which permitted cabin pressure and anti-fogging even though one power section was inoperative. The cabin pressurization was set for a differential pressure of 3.3lb per square inch under standard operating conditions, and 1.3lb per square inch for combat conditions as per Navy Specification SR-163a.

BASIC STRUCTURE

Wing design

A structural diagram of the L-200-1 wing is shown in the Wing Structure blueprint (see p114). This structure graphically illustrates the major improvement in simplicity and structural integrity which was made possible by the elimination of any landing gear, armament, fuel, and flap provisions within the wing. Further augmenting this simplicity was the decision not to use sweep to obtain low drag at high speeds but rather to attempt the use of extreme thinness.

This effort was amply rewarded with the straightforward structure which was light and simple to produce. Since the problem of wing design was primarily structural and structural alone, many alternate possibilities suggested themselves to improve weight or to improve cost of construction. As previously mentioned, the location of the wing tip gun pods was extremely important since it eliminated approximately 440lb of the normal wing weight whereas, if the guns had been installed within the wing, not only would this weight penalty have been paid but additional weight cost would have been incurred due to the complication of the structure.

Fuselage design

The structural diagram of the L-200-1 fuselage is shown in the Fuselage Structure blueprint (see p116). Here again a major simplification was achieved because of the nature of the airplane and because of the incorporation of a gearbox with a large offset. The major structure consisted of four main longerons which were located well to the top and well to the bottom of the fuselage, thereby achieving maximum structural strength and stiffness for a minimum weight. Attached to these longerons were several main bulkhead rings, all of which had

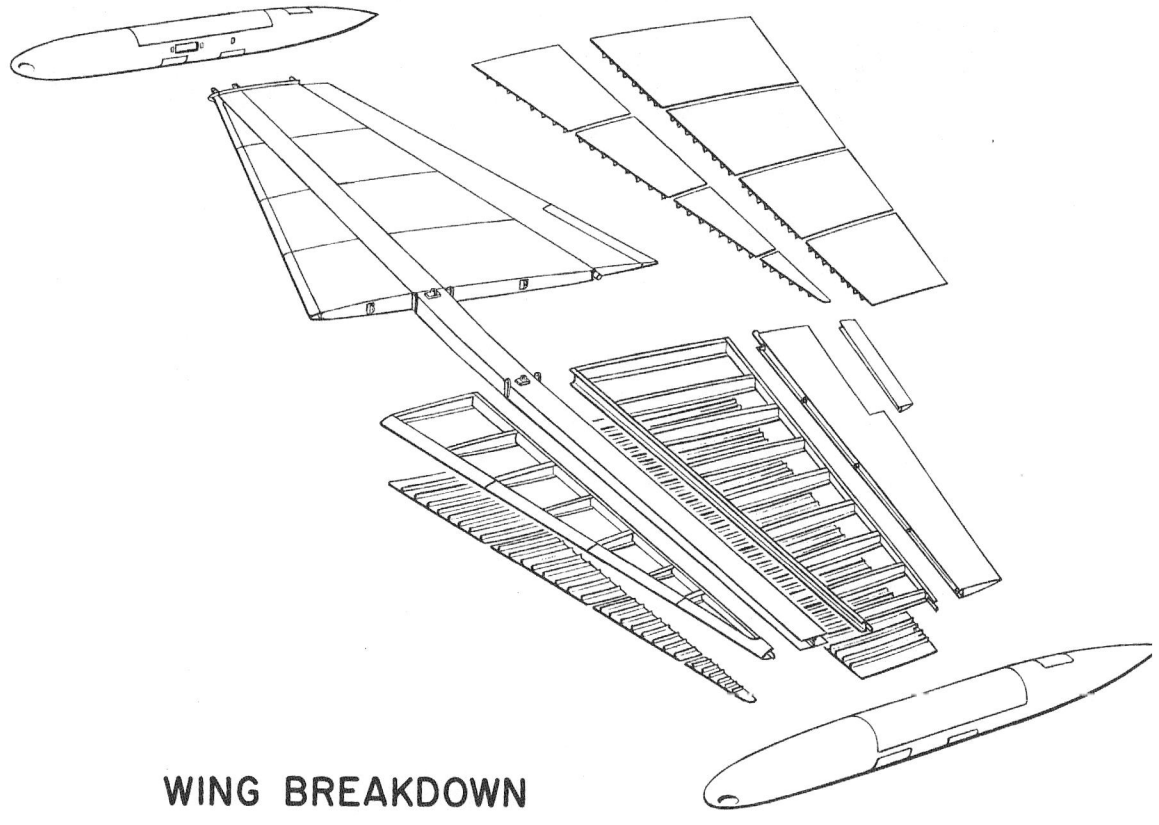

The L-200 wing had a much simpler structure than was typical on high performance airplanes. Only one control surface was used—the aileron-flap combination—and it was hinged at five points.

a circular cross section. This further simplified the production of the airplane.

Provisions for the power plant were made by providing engine mounts on two of the major bulkheads. Since the power plant was located outside of the major fuselage structure, the provision for tail pipe heat and for the installation of the tail pipes themselves were extremely simple and consisted only of external shrouds attached to the normal fuselage structure. By separating the oil coolers, the ducts required were quite small and did not cause any serious structural difficulty. For the main engine inlet, the duct was directly forward of the engine and occurred between the two lower longerons where it could be easily removed and where no additional penalty was paid since a cutout was already made for the engine installation itself.

Titanium was used in the fuselage structure which was also used for fire protection. This included a floor over the engine and splitters between engines.

Empennage design

The Empennage Arrangement drawing illustrates the one major structural difficulty which was incurred in an airplane of this type since in the empennage not only did the air loads have to be withstood but also the landing loads. This required a concentrated effort to reduce weight and Lockheed proposed that this be done by the utilization of titanium forged empennage spars and one of the aft fuselage attachment bulkheads.

These parts appeared to be reasonable to build in production and promised a major saving in structural weight. For prototype work it was proposed that an effort be made to obtain these titanium forgings if the programme accepted by the Navy included the full-scale prototype airplanes. If the small-scale prototype was utilized, it was very probable that machined dural spars or built-up steel spars could be incorporated with a weight penalty.

The use of extruded integrally stiffened skin

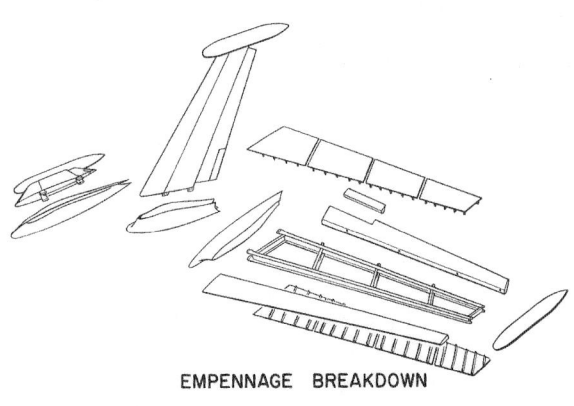

EMPENNAGE BREAKDOWN

The L-200 empennage consisted of a fin and two stabilizers. Each section was interchangeable and the identical stabilizer assembly was used on each side.

panels was proposed for the fixed surface structure. Experiments and production use of this technique had shown that a considerable weight saving could be made along with a reduction in cost. A further structural innovation was shown in the elevator construction where a simple shell was filled with Lockfoam plastic. Production parts installed on F-94 all-weather fighters had demonstrated the great value of this type of structure in strengthening and stiffening control surfaces.

L-200 PRODUCTION FEATURES

While the L-200 was presented as a prototype airplane, considerable effort was expended to ensure that it could be converted into a production model in a minimum amount of time. The same assembly tooling used for the full-scale prototype could in some cases be used for the production version. And, as shown in the production breakdown drawing, the airplane was well adapted to a very high production rate. Nearly all the prototype layouts would be used unchanged on the production airplane; the only additions, other than changes resulting from flight test, would be the procurement of forgings and extrusions that would be uneconomical to use in small quantities.

Lockheed pointed out that its plan for a lightweight full-sized prototype would further reduce the span between the time when the prototype order was placed and the first production airplane was delivered. This was due not only to savings that could be made in design and flight test time, but to the fact that manufacturing pre-planning and tool design could be started early in the prototype design stage. This procedure, which was first started at Lockheed with the R6O Constitution, had been made standard to reduce the gap between the prototype and production airplanes.

Production breakdown

The production breakdown was identical for both the full-sized prototype and production airplanes, and was based on service, spare part requirements and producibility. For the first consideration, the major airplane components that were subject to damage or wear were provided with simple interchangeable joints requiring a minimum number of attachments. These components were:

1. Complete wing
2. Complete empennage
3. Fin
4. Stabilizer
5. Rudder
6. Elevator
7. Landing pods
8. Armament pods
9. Ailerons
10. Engine fairing
11. Canopy
12. Power plant
13. Electronic equipment package
14. Radome

Airplane components that were readily replaceable by simple drilling or trimming operations were:

1. Wing leading edge
2. Wing trailing edge

Shipping

As shipping was considered an important factor, the airplane was designed so any component would fit in the 30' x 9' x 7' crate noted in the crating specification AN-C-118a. In order to accomplish this, the wing was separated from the fuselage by removing the attaching bolts and the wing leading and trailing edge assemblies which were attached with a series of screws. The fin, stabilizer, propellers, and radome were removed and the fuselage was then ready for crating. This crating arrangement allowed transport by either rail or ship, and no special routing or handling equipment was necessary.

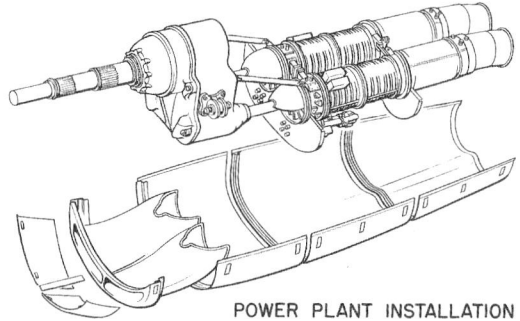

POWER PLANT INSTALLATION

The engines, engine accessories, and gearbox of the L-200 were accessible for repair or removal through the bottom of the fuselage. The cowling was removed in three sections, and the intake ducts were removed as a unit.

Producibility

The consideration of producibility which further influenced the design was made up of three main factors: minimum man-hours, economy in tooling and floor space, and subcontracting and expandability.

Minimum man-hours

The number of parts in the airplane would be kept to a minimum. An example was the use of integrally stiffened skin for the wing surface structure. This type of structure not only reduced the number of parts (by eliminating separate stiffening members), but also eliminated a large number of attachments; it was proven in actual practice at Lockheed to be a means of saving a large number of man-hours. Man-hours were also saved by designing for assembly access. The fuselage, for example, was made in half-shells which not only allowed unrestricted access for workmen during fabrication, but also permitted installation of part of the equipment before the shells were joined.

Economy in tooling and floor space

The use of simple joints, such as that of the empennage forged bulkhead to the fuselage, reduced the complexity of tooling and, in more complicated joints such as the four-bolt fin attachment, adjustment was designed into the forward beam fitting so it could be readily aligned with the rear beam fitting. The fairing between the fuselage and fin was attached to the fuselage, which further simplified the mating problem.

Tooling and floor space were both saved by multiple use of assemblies. This was done by designing the part so it could be used on both right and left-hand sides of the airplane. Major examples of this were:

1. Wing leading edge assemblies
2. Wing trailing edge assemblies
3. Ailerons
4. Horizontal stabilizers
5. Elevators
6. Landing pods

Subcontracting and expandability

The airplane breakdown was well adapted to subcontracting and expanding production at a mobilization rate. The main assemblies were small and, therefore, easily handled and transported. They were complete units with simple joints. This arrangement allowed short final assembly lines which could be established at several different airports with very low tooling duplication.

The manufacturing techniques used in the production of the L-200 were standard methods of manufacture and were all used on contemporary Lockheed production airplanes.

Critical materials

The use of the more critical materials was avoided as far as possible in the L-200 design. Titanium was used instead of stainless steel as it was expected that this material would be in better supply than stainless when the airplane went into production. In the event that titanium was not available, aluminium-coated low carbon steel would be used in place of titanium sheet and 8630 steel forgings used for the empennage bulkheads and beams.

Production research

Various research projects were underway at Lockheed to develop new types of structure which would result in either weight or cost savings. Among those that were possibly applicable to this airplane were: large thin-walled castings that could be used for leading edges and Lockfoam sandwich panels that could be used for wing surface structure. Lockfoam was a Lockheed developed, low density, poured-in-place structural foam that was used as a filler to eliminate ribs and stringers in structures such as control surfaces. In this application it had been thoroughly tested and was used in contemporary production airplanes. In addition, experimental designs utilizing titanium were in the process of construction. Large thin web forgings for wing surface structure were under development.

Fuselage

The fuselage broke down into five major assemblies: left and right side panels, floor and engine mount sections, lower aft section, and empennage attaching section.

This type of breakdown was advantageous as it produced sections that were easily handled, had good access for assembly, broke along natural lines (requiring no extra joints), and allowed installation of equipment in the subassemblies. The mating operation consisted of joining the first four assemblies and drilling the fittings for attachment of the empennage section. One of the primary aims in determining the breakdown was to keep as much work as possible in the subassemblies, thus relieving the congestion around the complete fuselage.

The fuel tank and cockpit floor assembly contained the engine mounts which were completely installed in this assembly. The lower aft section was titanium and was a spotwelded assembly.

The empennage attaching section connected to the forward section of the fuselage with six tension bolts. A joint of this type could be made interchangeable with very simple tooling; this was a desirable feature since, in the event of a subcontracting programme, the empennage attaching section could be built by the same manufacturer that constructed the fin and stabilizer. The more difficult problem of coordinating the fin and stabilizer joints with the empennage attaching section was thus confined to one manufacturer.

Wing

The basic design features of the L-200 wing produced a much simpler structure than was usual on high performance airplanes. Only one control surface was used—the aileron-flap combination—and it was hinged at five points. As all the fuel was carried in the fuselage, the wing was relieved of all complexities resulting from its use as a fuel carrier. The ammunition was carried on either side of the main beam web and was accessible from the wing tips or by removing the fore and aft surface structure.

The wing as a complete unit was interchangeable and was attached to the fuselage by five bolts on each side. The nose section and the aft section were attached to the main beam with screws and

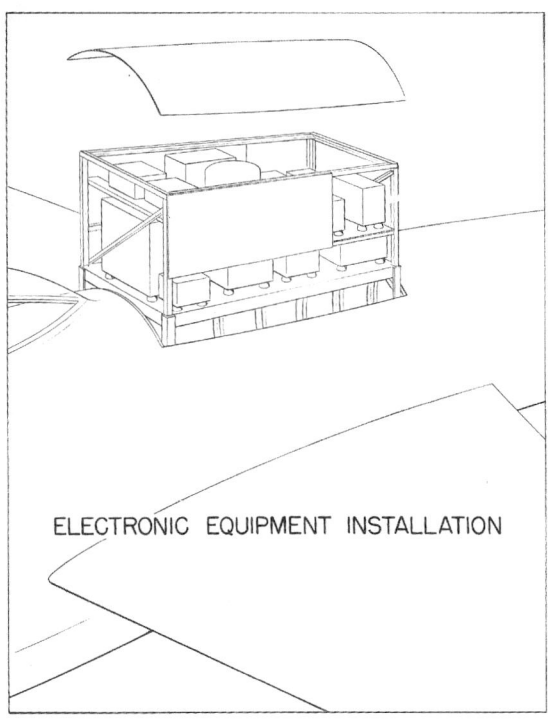

To reduce the maintenance-to-flight time ratio, most of the electronic equipment was installed in a compact rack in the aft section of the fuselage. The rack could be raised for checkout and alignment of the individual units, or could be removed entirely.

plate nuts and were removable or replaceable for quick repair of battle damage. The wing was assembled to the fuselage by removing the nose and aft sections and slid in the main beam through the slot in the fuselage. The fore and aft assemblies were then joined to the beam with screws, the aileron was installed, equipment connections made, and the wing was ready for service.

The main beam consisted of a web and two extrusions that were continuous from tip to tip. The extrusions were well within the limits of standard extrusion equipment. The fore and aft sections were beam and rib structures with extruded integrally stiffened skin which not only reduced the number of parts and attachments in the wing, but also allowed access in subassembly. The aileron was a Lockfoam-filled type. The armament pod was joined to the wing with three bolts accessible through the door in the pod. The joint was interchangeable to allow rapid replacement of either pod.

Empennage

The empennage consisted of a fin and two stabilizers. Each section was interchangeable and the identical stabilizer assembly was used on each side. The

Photo of the plywood L-200 cockpit mock-up, constructed on Lockheed's initiative to investigate the ideal position for the pilot.

The mock-up in horizontal flight attitude with normal seat position.

empennage sections attached to the fuselage with four bolts each, and the fairing, which covered the joints, attached to the fuselage only.

The structure of the stabilizer, which was typical for the fin, consisted of two forged titanium or steel beams separated by five aluminium alloy ribs. The skin was extruded integrally stiffened aluminium alloy which substantially reduced the number of ribs and attachments. The nose section was a separate assembly built of ribs and skin. It was attached to the front beam with blind fastenings or screws.

The elevators and rudder were Lockfoam-filled which eliminated a large number of parts. The hinges had adjustable brackets to ensure alignment, and the inner bracket took all the end load, allowing clearance between the remaining fittings.

The landing pods attached to the ends of the beams with two bolts; the identical pod assembly was used in all three positions.

Power plant installation
The engines, engine accessories, and gearbox were accessible for repair or removal through the bottom of the fuselage. The cowling was removed in three sections, and the intake ducts were removed as a unit. The duct closure doors could be either opened on hinges or removed for better access. In production, it was expected that the engine and gearbox would be installed as a unit.

The cowling and ducts were aluminium alloy and were attached with quick-acting cowl fasteners. The gearbox and power units were isolated from the upper fuselage by means of titanium fire barriers. The engine compressor section between the compressor inlet and burners was also compartmentalized in this manner. All potential fire compartments were adequately vented.

Electronic equipment
Experience gained on the F-94, the only contemporary automatic tracking radar-equipped interceptor that was delivered in production quantities, had demonstrated the importance of design in reducing the maintenance-to-flight time ratio.

To accomplish this, as much of the electronic equipment as possible was installed in a compact rack in the aft section of the fuselage. The rack could be raised for checkout and alignment of the individual units, or could be removed entirely. In production this entire unit would be a bench assembly

The mock-up could rotate from the normal position to a vertical attitude by means of an electric winch, simulating to some extent the sensation of flight transition.

and, to a large extent, could be checked out before installation in the airplane.

The radome was attached by two quick-acting fasteners that were accessible from the outside.

WEIGHT AND BALANCE ANALYSIS

Since the success of the vertical rising airplane depended not only on the aerodynamic problems of flight but also on the engineering ability to build a satisfactorily strong airplane for minimum weight, it was decided that drastic procedures to hold weight down were justified. Accordingly, Lockheed suggested the following procedure to ensure that minimum structural weight was achieved:

1. Design and construct two T40 powered full-scale prototypes in place of the small Mamba powered airplanes. These prototypes were to be designed for a low weight corresponding to their use as prototypes (11,578lb).

2. Construct one static test article identical to

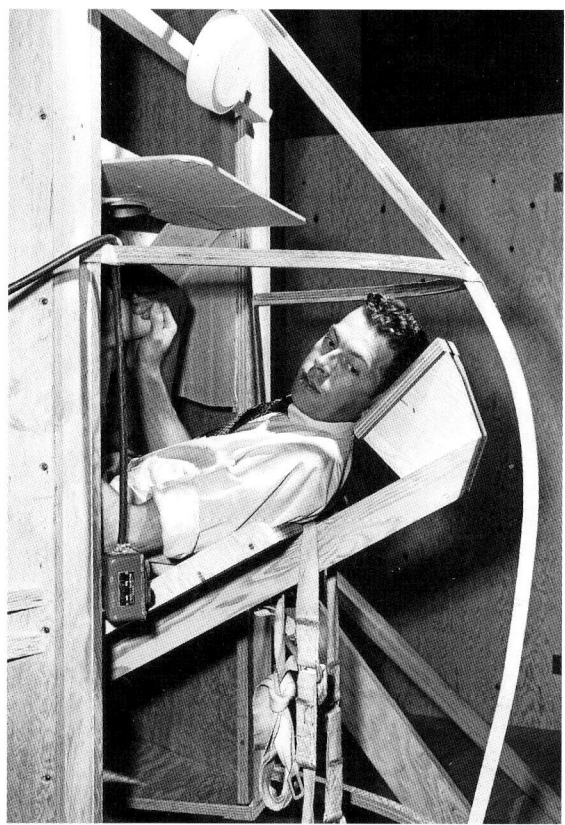 

These photos show the mock-up in hovering or takeoff attitude with the seat tilted 45°; the pilot stand-in is demonstrating the proposed method of over-shoulder sighting during landing operations.

prototypes, including design weight.
3. By means of flight test load determination on prototypes and application of these loads to the static test article, a structural beef-up programme was instigated to increase the design gross weight to the tactical requirements.

Following this procedure, the first tactical airplane prototype (a converted and beefed-up original prototype) would represent the lightest achievable vehicle, an essentially 'stretched' airplane which would compete in weight efficiency with airplanes which had years of operational stretch to achieve their actual operational gross weights.

It was the aim of the above programme outline to have the prototype airplane be the zero point for the L-200 airplane and have the weight of the tactical airplane determined by the weight trend of previous Lockheed aircraft. This, of course, applied only to the structural components of the airplane, the added equipment, changed power plant, armament, etc. having been accounted for by direct addition. Thus, the weight empty of the tactical airplane could be determined as follows:

9435	Weight empty of prototype with T40-A-6 engine.
757	Weight of added equipment: radar, extra fuel capacity, etc.
1022	Weight of changing T40-A-6 to T40-A-8 engine, including propeller change.
101	Increase in structural weight from weight history curve.
11315	Total weight empty of tactical airplane (lb)

To illustrate the possible weight saved by this programme, the weight empty of the airplane was estimated by normal procedures similar to those used in estimating the original prototype weights. These calculations showed:

12416	Weight based on conventional estimating procedures.
11315	Weight empty developed through growth programme.
1101	Savings through growth programme (lb)

The Gross Weight Summary table provided a breakdown of the required weights for performance calculations based on the recommendations and definitions of oS-122. Lockheed also provided

The hooded radar scope was located below the instrument panel with the sight head face forward of the panel.

a complete summary of the weight empty breakdown estimate based on the suggested prototype programme.

Of major interest to those who expected a swept wing on airplanes with performance approaching the transonic, Lockheed emphasized that an extremely thin wing was determined to be lighter than a swept wing for a given drag and drag rise Mach number. A direct comparison between the straight wing and the swept wing showed the straight wing saved 210lb or over 21% of the total wing weight.

CONCLUSIONS

It was the conclusion of the Lockheed Corporation from these preliminary design studies that a feasible and useful tactical airplane could be developed to fulfill Navy Specification 0S-122. Lockheed believed its design to have the best chance of fulfilling these requirements satisfactorily. However, the company emphasized that problems of this nature required a major research effort. Fulfillment of all of the requirements which could be uncovered in attempting vertical flight, and the transition between vertical and horizontal flight, could result in major design changes.

It was concluded from this study that the best type of research programme to obtain the most desirable end result would include a prototype airplane that

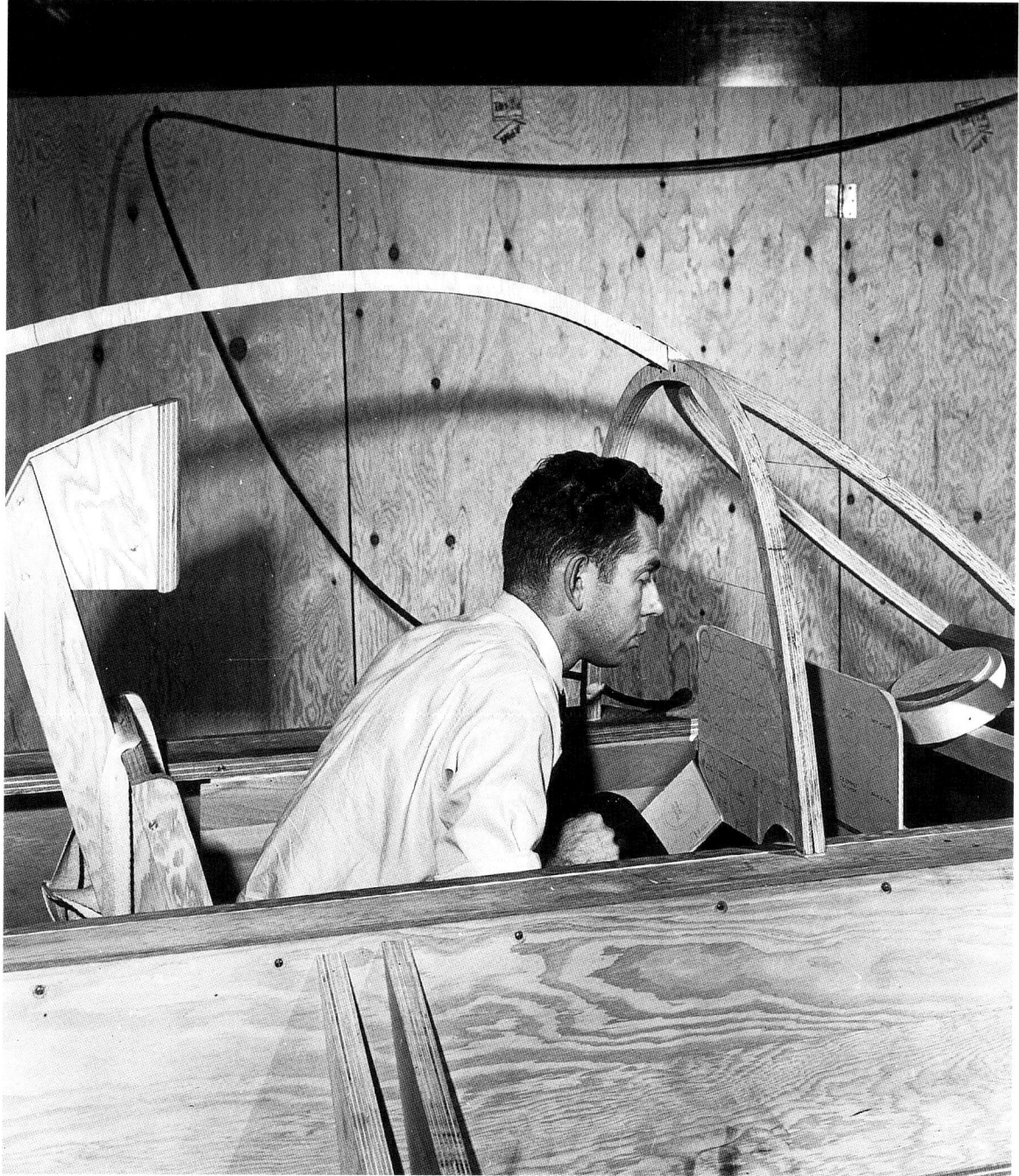

The pilot had to lean forward in order to cover the full range of gunsight lead angles without head movement.

was full-scale rather than small-scale as proposed in the original Navy request for proposal. Reasons were given for this suggested change in the programme, but it was concluded that the airplane could be developed either with a small-scale prototype or with a large-scale prototype, depending upon the desires of the Navy after studying the problem as stated in the proposal.

With respect to the performance requirements of the airplane, it appeared reasonable to achieve the combat altitude and speeds required; it appeared entirely feasible to make deck facilities which would ensure reasonably normal operations for takeoff and landing on a commercial vessel; and it appeared entirely possible to develop an automatic control from an existing autopilot which would relieve the pilot of many responsibilities that were of an unconventional nature. It was the major conclusion of this study that the unconventional performance requirements could be achieved with a conventional airplane, thereby relieving the research worker and the eventual pilot of the airplane of many incidental problems which were always a major stumbling block to progress.

The left-hand console, included the APQ-42 control panel with its tracking lever, the trim tab control switch, the power control lever and switches for various engine control functions.

Lockheed believed that any research programmes which would lead to the development of an airplane of this nature were extremely valuable to the progress of Naval aviation since a successful vertical rising airplane would open up many fields in which the airplane would become even more useful from a tactical standpoint than it had been in the past. These fields included ASW work, mining operations, reconnaissance, attack, and even advanced utility transports.

The dimensions of the L-200 were limited in such a manner that the airplane on its service cart would fit on any of the carrier elevators previously listed by the Navy for the carrier-based ASW airplane. The included the CVB-41, CV-9, CV-34, CVL-48, CVL-22 and CVE-105 carriers. Further efforts would be made in the first phases of the L-200 study to reduce the size of the airplane further by concentrated wind tunnel research programmes to determine minimum possible tail lengths. It was hoped that this would make possible a total airplane length below 41ft so that small carriers (then inactivated) could also be used to transport these airplanes or serve as service ships.

COCKPIT MOCK-UP

In addition to the L-200 design summary covered in the previous section, Lockheed also submitted a report to BuAer concerning its study of a cockpit mock-up for the aircraft. According to the document, although significant research had been carried out on the problem of pilot position in vertical flight at NADC in Johnsville, it was considered desirable to construct a simple cockpit mock-up for the

Photo of the right-hand console with the W-2 autopilot controller located aft of the console control panels.

Lockheed design proposal. In order to gain a better understanding of the movable seat problem with relation to visibility requirements, the Mark VI gun sight installation, ejectable seat clearances, and other equipment items, a full-scale wooden mock-up of the L-200-1 cockpit was constructed. It was designed to pivot in a supporting frame from a normal to a vertical position, power actuated to simulate transition in flight, and equipped with a tiltable, rotatable seat.

The tilting seat was decided upon for use in the airplane after mock-up inspection verified the advantages of this method over other proposed methods of positioning the pilot for vertical flight.

The mock-up permitted three-dimensional study of the various cockpit arrangement problems presented by the limited space envelope and the necessity of maintaining satisfactory control motions and visibility in all seat positions. The resultant arrangement appeared to be satisfactory in all respects, although compromises were necessary in resolving the problems imposed by the unconventional features with the requirements of applicable specifications.

Introduction

The construction of a full-scale wooden mock-up of the L-200-1 cockpit was prompted by the necessity of evaluating the relative merits of the various methods of moving the pilot from his normal position to a position suitable for the vertical flight attitude. Details of the mock-up design are described in Section I.

As a result of the mock-up programme, it was

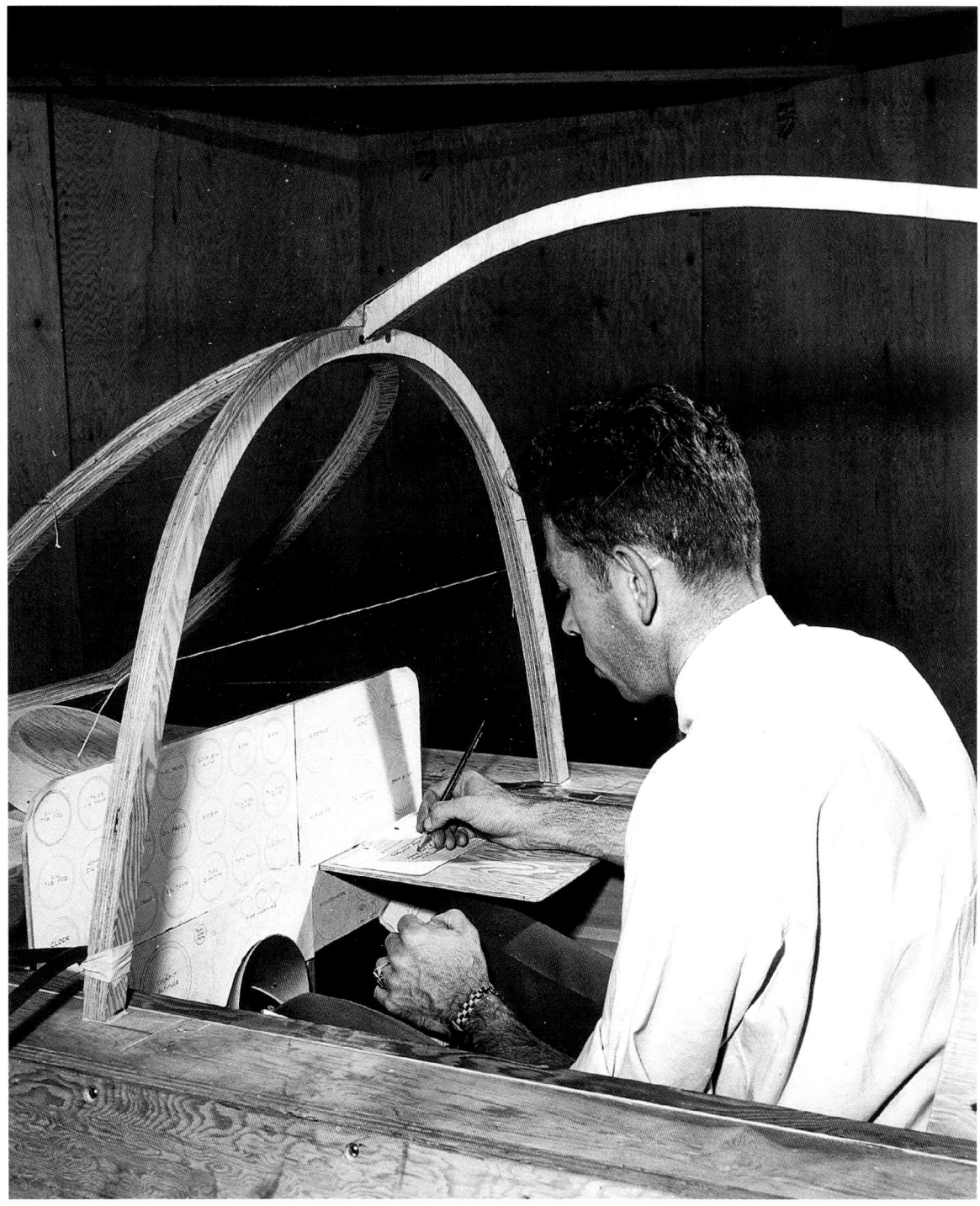

The chart board/writing pad holder in the extended position.

decided that the tilting of the pilot's seat forward about a horizontal axis was the most practical method. Section II is a discussion of the factors which led to that decision.

A cockpit arrangement based on the tilting seat design was studied in detail, and the mock-up proved to be very helpful in arriving at practical solutions to the problems presented by the unconventional features of the L-200. A description of the cockpit arrangement and a discussion of some of the major problems appear in Section III.

The mock-up was also used to demonstrate the practicability of entrance to and egress from the cockpit in the vertical attitude with the seat tilted back to its normal position.

Section I — mock-up design

The L-200 cockpit mock-up, constructed mainly of 1" plywood, was designed to approximate the airplane cross section above the floor line, using

Multiple views of the 1/16 scale free flight model of the L-200 Convoy Fighter. This represents an earlier configuration of the design, with the empennage rotated 180° and no wing dihedral.

A 1/16 scale takeoff and landing platform was also constructed and subjected to testing. Both it and the L-200 model were tipped to extreme angles to demonstrate the static stability of the airplane in place on the platform.

A good view of the L-200 model and platform; the latter consisted of netted areas for takeoff and landing. The large lower net was designed to mesh with the pads on the tips of the empennage and the smaller net to one side was provided to catch the aft tip of the wing tip gun pod.

straight sloping sides and a constant width to facilitate construction. The dimensions of the box were 76" long by 66" wide at the floor line. A 1 x 1 canopy and windshield frame was included, and the complete cockpit was supported in a heavy wooden A-frame on a horizontal steel-tube pivot axis located approximately at the centre of gravity of the occupied cockpit. Stops were provided at both horizontal and vertical positions.

The pilot's seat was designed to rotate on a turntable about a vertical axis as well as to tilt about a horizontal axis. The turntable was pinned in a fixed position when it was decided to abandon the rotational ideas in favour of the tilting seat.

The seat, canopy, floor line, cowl line, instrument panel, control stick, rudder pedals, and other items were corrected frequently as the airplane design progressed, and the final mock-up configuration differed from the proposal drawings only in minor details.

In order to simulate to some extent the sensation of transition, an electric winch was rigged up to hoist the cockpit from its normal position to a vertical attitude. The winch was controlled by a switch box attached to the cockpit, permitting a person sitting in the pilot's seat to fully control the transition. The tilting of the seat was automatically synchronized with the cockpit motion by means of a cable attached to the seat at one end and to the frame at the other, passing over pulleys mounted on the cockpit box.

Section II — pilot's tilting seat

Prior to construction of the mock-up, many different methods of changing the pilot's position were considered. The use of a prone position in normal flight was discarded because it was felt that the unconventional features of this airplane should not be augmented by the addition of an unconventional normal flight position.

After preliminary evaluation of remaining methods with regard to comfort, visibility, safety, and space requirements, all others were discarded in favour of a choice between rotation about a vertical or inclined axis and tilting about a horizontal axis, as tested in the Johnsville mock-up. Inasmuch as

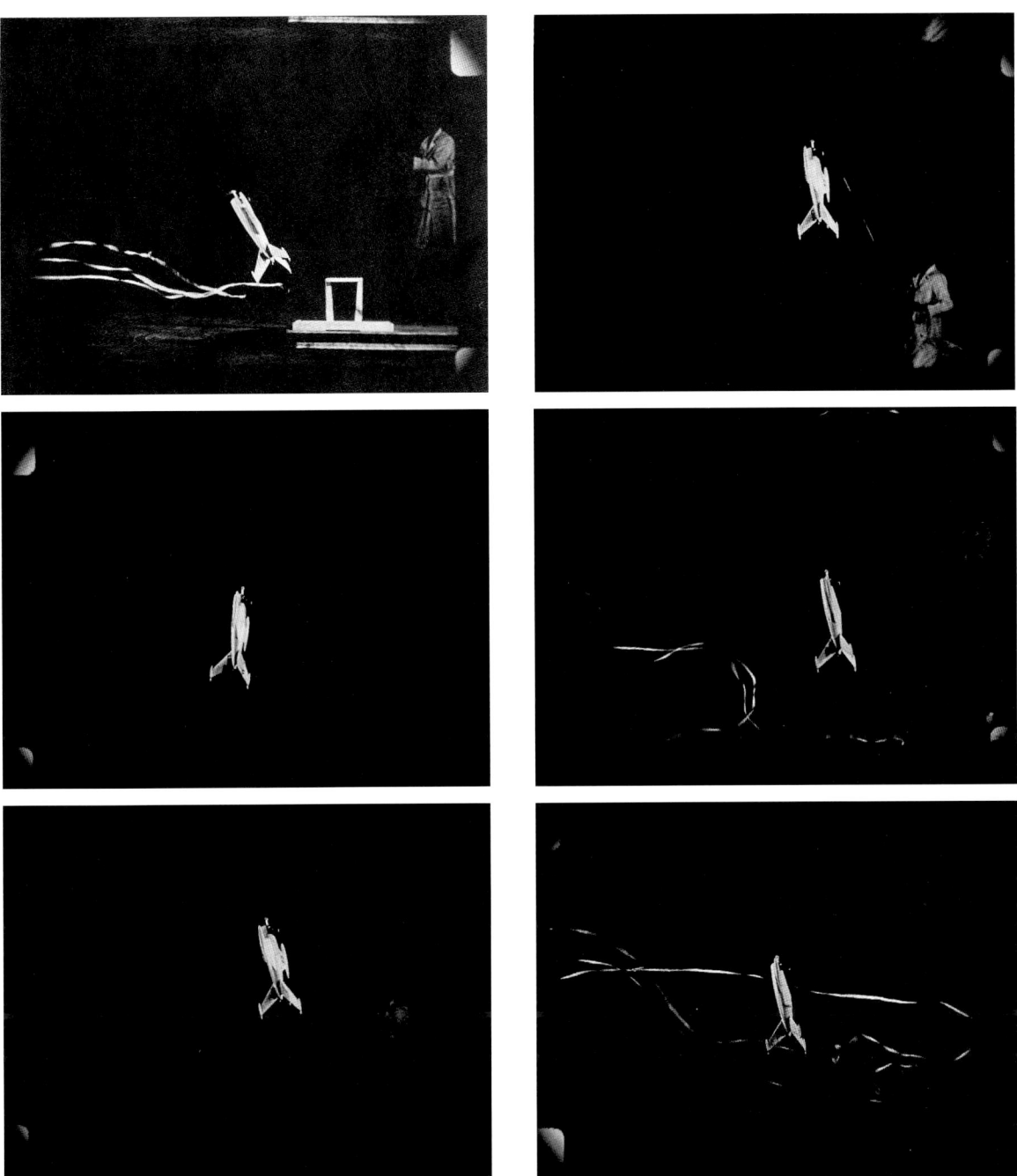

Stills from Lockheed's 16mm film of the L-200 free flight scale model undergoing testing. It was determined from the tests that hovering control of the model was very good and that the full-scale airplane would be even better.

the lateral space requirements of an inclined axis of rotation were incompatible with an aerodynamically efficient cross section, the mock-up was designed to include only provisions for seat movement about horizontal and vertical axes.

From inspection of the completed mock-up, it was readily apparent that rotation about a vertical axis, while possible within the chosen cross section, eliminated the use of considerable valuable space surrounding the seat. Furthermore, it would still be necessary to tilt the seat after rotation in order to achieve the desired position. Since the desired position was approximately one in which the seat and back locations were interchanged, consideration was given to the possibility of having the pilot change position without moving the seat. This idea was also abandoned as impractical and unsafe in such cramped quarters.

The tilting seat idea was adopted as the most practical because of the following advantages:

1. Hands and feet could remain on controls

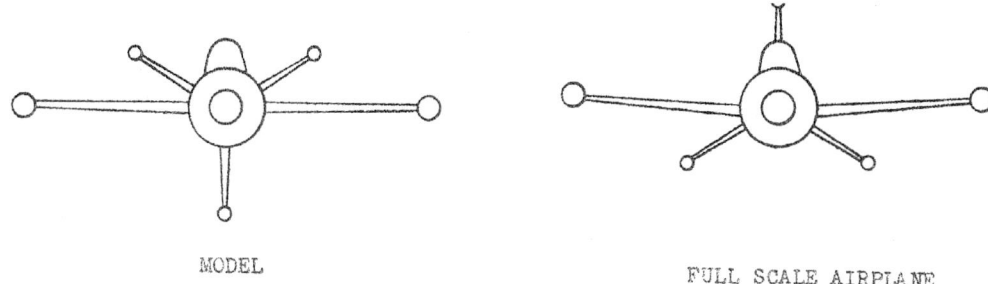

Drawing showing the configuration differences between the free flight 1/16 scale model on the left and the final proposal on the right.

during entire transition.
2. Duplication of controls and instruments was unnecessary.
3. Minimum space requirements.
4. Satisfactory visibility and comfort in hovering attitude.

The location of the tilt axis was dictated by the necessity of retaining satisfactory control stick and rudder pedal movement and instrument visibility, plus the clearance requirements of the canopy, which was already limited by the gunsight requirements on one hand and aerodynamic efficiency on the other. The proposed location of the tilt axis was 6.5" above and 6.5" forward of the seat reference point in mean position. A more forward location of the seat pivot, as used in the Johnsville mock-up, necessitated either a greater distance to the 35° windshield or a lower normal position in order to provide head clearance in the tilted position, resulting in sacrifice of specified visibility.

A tilt angle of 45° from the normal position was adopted as a reasonable compromise between comfort and visibility. In order to minimize the tilted position clearance problem, it was decided to provide for tilting only from the low position of the seat. This required the pilot to lower the seat prior to tilting during transition, a motion which could be utilized to arm the electrical circuit controlling the proposed automatic tilting procedure during transition. Since the low position was also a prerequisite for seat ejection, the seat could be readily positioned for ejection from either attitude.

Because the cockpit was sandwiched between a floor level fixed by the entire gearbox and a canopy whose height was limited by aerodynamic requirements, it was decided to limit the vertical adjustment of the seat to 2" up or down from the mean position, along a line sloping 23½° forward. This adjustment moved the seat within a frame supported at the tilt axis by a carriage which in turn rode on rollers in the ejection rails. The heel rests for ejection were attached to the carriage instead of to the tiltable portion of the seat in order to minimize floor clearance problems.

The mock-up was photographed in horizontal flight attitude with normal seat position. The hovering or takeoff attitude, with seat tilted 45°, was also depicted. These photographs indicate the proposed method of over-shoulder sighting during landing operations. A mirror attached to the canopy frame was also considered and could be used to relieve fatigue induced by prolonged hovering.

Section III — cockpit arrangement

The mock-up was used to good advantage in establishing the proposed cockpit arrangement. The limitations imposed by the visibility requirements over the large diameter spinner, the presence of the engine gearbox below the pressurized floor, the windshield slope as dictated by the optical sight head, and the ejection seat clearance requirements all had to be correlated with the tilting seat and the attendant control motion problems.

Because of the gearbox below the floor, the control stick was designed to pivot about a horizontal lateral torque tube just above the floor, with a pivot for lateral stick movement located 7" above the torque tube. By limiting the stick height to 20" above the torque tube, it was possible to use 14" fore-and-aft stick travel and 12" total lateral travel in either the normal or tilted position of the seat.

After consideration of normally hung rudder pedals and a mock-up tryout of a rudder bar, the use of sliding pedals was determined to be the most adaptable to the change of seat position. Pedals were inclined at a 45° angle and provided with toe stirrups to help retain the feet in position during hovering.

One of the necessary compromises worked out on the mock-up was the centreline location of both the Mark 6 Mod 1 sight head and the APQ-42 radar

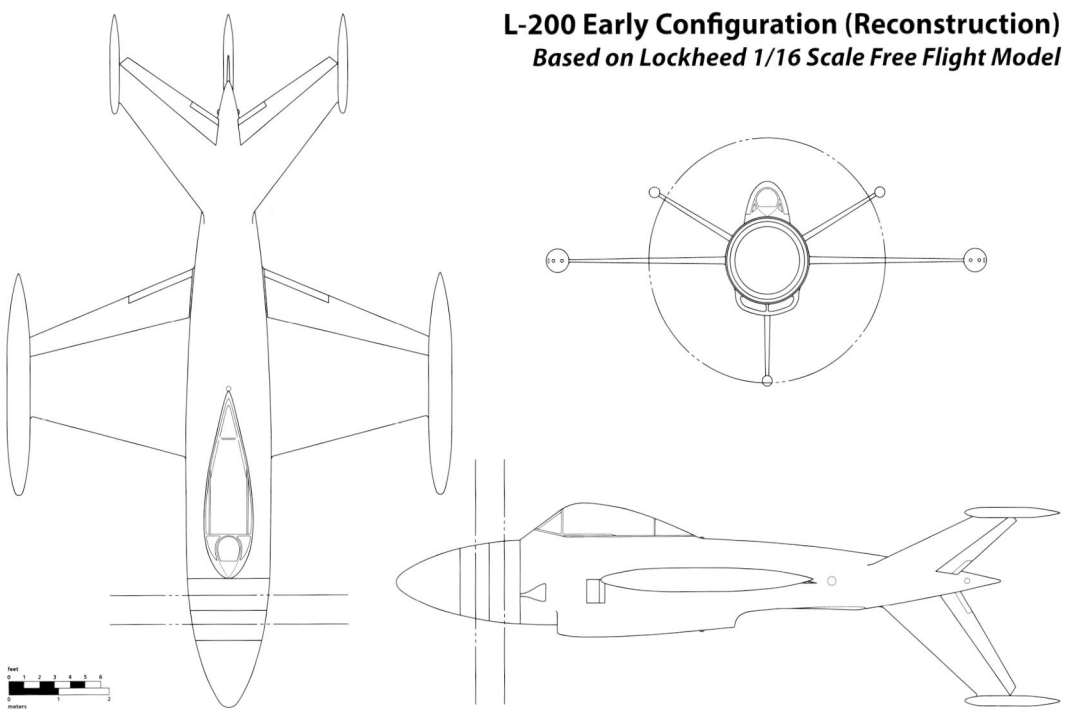

Reconstruction of an earlier configuration of the L-200 based on the free flight model; the aircraft has an inverted tail and the wing has 0° dihedral.

indicator, necessitating the offset of the primary flight instrument group to the right because of insufficient space between the sight head and the instrument panel, the location of which was limited by ejection clearance requirements and visibility specifications. The hooded radar scope was located below the instrument panel and the sight head face forward of the panel. In order to cover the full range of gunsight lead angles without head movement, it was necessary for the pilot to lean forward.

Maximum utilization of available space was obtained by sloping the side consoles for optimum visibility and continuing the consoles forward on an upward slope to the corners of the instrument panel. The left-hand console, included the APQ-42 control panel with its tracking lever, the trim tab control switch, the power control lever and switches for various engine control functions. The landing gear switch was located at the top of the sloping panel, flanked by the APN-22 radar altimeter indicator located for ready reference in the hovering attitude. The engine instruments were placed in a compact group to the left of the primary flight instruments. Also visible was the central switch panel for interior and exterior lights and the fuel system diagram, which included switches for the four fuel shut-off valves.

The oxygen regulator was located on the vertical panel below the sloping side panel. The right-hand console is shown with the W-2 autopilot controller located aft of the console control panels, which included the ARC-27, ARR-2A and APX-6 controls. The upward sloping panel included seat and canopy switches, cabin temperature controls, oil cooler switches, starter switch, DC power controls and "essential bus" switches. AC power controls were on the lower portion of the vertical panel below the flight instrument group.

The autopilot controller mentioned above was intended to be used only during normal flight operations; during hovering, the psychological advantage of manual stick operation could be retained by the use of a four-way momentary contact switch on the top of the control stick to effect required changes in the hovering attitude while stabilized by the autopilot. This allowed the pilot to change from automatic flight to manual flight without removing his hands from the stick. The switch button and

Original display model of the Lockheed L-200 Convoy Fighter, posed here as if in flight. Note the representation of the retractable 'fingers' on the tail pods, which were to be deployed when the aircraft was recovered on the shipboard landing net. (John Aldaz Collection)

the switch housing are shown on the stick in the mock-up photos.

A cylindrical knob for simultaneous adjustment of both rudder pedals was located on the left-hand side of the control console, and is visible in the mock-up photos.

The slot below the flight instrument group is the stowed position of a 9" x 12" chart board or writing pad holder, which is shown in the extended position in the mock-up photos.

FREE FLIGHT MODEL TESTS

Tests on a free flight model (unrestrained in pitch) were made as a part of the Lockheed design proposal for the vertical-rising Convoy Fighter. The purpose was to study model and airplane behaviour in hovering flight, ground or deck effects on control, airplane attitude with translational or side-wind velocity, and transition flight paths. Also included were in-place stability on the landing-takeoff platform and actual takeoffs and landings from the floor and the platform.

Two models were used in this programme; the first a conventional commercial kit and the second was a 1/16 scale model of the actual configuration proposed for the design competition. Both were powered with internal combustion model engines. The sixteenth scale factor was determined primarily from scale propeller diameter and model engine availability.

It was determined as a result of the tests that hovering control of the model was very good and that the full scale airplane would be even better. Manual flight of the airplane in the event of autopilot failure would be readily achieved. Good control of the airplane and model was possible with ground or deck close to the tail. Transitions from vertical to horizontal were made with minimum attainment of altitude.

Results of tests to determine pitch attitude versus translational speed of the model were checked very closely by wind tunnel measurements.

Introduction

At the inception of the design of the L-200 vertical-rising Convoy Fighter, the stability and control of the aircraft in hovering flight was a major problem. Free flight tests conducted by NACA with radio-controlled tail surfaces indicated instability or divergence with long periods which appeared to be easily controlled by the operator. Moving pictures of these tests were encouraging with respect to pilot control when hovering in what appeared to be calm air and at an appreciable distance from the ground. However, ground plane effects and various types of disturbances from retrieving and launching devices were still an unknown factor.

The construction and testing of a free flight model including the evaluation of side wind velocities, retrieving and launching gear effects, in addition to hovering, and transitional flight appeared to be very desirable. Dual rotation propellers and radio-operated control surfaces were considered for a scale model but were abandoned because the cost and

Additional views of the original Lockheed L-200 display model. (John Aldaz Collection)

time required for development were prohibitive for a design proposal.

Models powered with small internal combustion engines and controlled through fine diameter flying wires offered a solution to the time and expense problem.

In addition, it was known that the low disc loading of the propellers of model airplanes were favourable for hovering and vertical climbing flight. Satisfactory thrust-weight ratios could be achieved to study some of the problems of the full scale airplane. Further, this type of model flown by an operator with a "U-control" handle which operated the elevators, was completely unrestrained about the pitch axis and offered opportunities for studying many of the problems of the full scale airplane. With a slight modification, data could have been obtained in the yaw direction although this was not done in the tests described in the report due to lack of time. The model was necessarily restrained about the rolling axis because of the single rotation propeller and its resultant torque.

Having decided on a model incorporating an internal combustion engine, construction was started on a standard kit model which was available commercially. The problems of fuel feed in vertical flight and thrust control by the operator were solved with commercially available items. Meanwhile, the design of the full scale airplane in the Preliminary Design group progressed to a point where the general external configuration was firm. A 1/16 scale model of this configuration was constructed using the same type of fuel feed and power control. Preliminary tests were conducted on the conventional model, followed by a comprehensive test programme on the 1/16 scale model of the tactical airplane.

It was the purpose of the report to describe the model and its characteristics and to show briefly how some of the flying characteristics of the full-scale airplane were simulated. A few simple measurements of the significant parameters were made for comparative calculations with the full-scale airplane.

Comprehensive coverage of the model testing was made on 16mm motion picture film. Approximately 600ft of film was submitted with the proposal to the Navy. It was estimated that a showing time of 30 minutes was required with the projector operating at 16 frames per second. The test sequences in the film included the following scenes:

1. Operation of control handle and elevator.
2. Offset rudder and engine thrust and acute angle of flying wires to airplane centreline ensured tension in flying wires.

A beautifully preserved vintage display model of the original Lockheed L-200 Convoy Fighter proposal; note the clever placement of the metal supports, which doubled as a representation of the exhaust from the Allison T-40 turboprop engine. (John Aldaz Collection)

3. Thrust control achieved by use of two sets of ignition points (advanced and retarded) and an electrical relay actuated by operator through control handle and flying wires.

4. Scenes showing various methods of preventing models from crashing. A bamboo pole with two fishing lines to the wing tip pods was the first method used. The second method was to attach one fishing line to the outer wing lined up with the direction of flying wires.

5. With the bamboo pole, very small rubber bands were attached to fishing lines to prevent fouling in the propeller. Scenes showed that the tension in the rubber bands had a negligible effect on normal forces supporting the model.

6. Initial tests shown of hovering flight and controlled translation of the first conventional model prototype.

7. Extreme sensitivity due to the large elevator and its travel on the conventional model caused difficulty for the operator to make controlled hovering flights. After several modifications to reduce angular travel and the elevator area satisfactory control for hovering was achieved.

8. After preliminary testing of the conventional model a 1/16 scale model of the Lockheed L-200-1 was constructed.

9. Scale models of the landing, and takeoff platform proposed were made and static tipping tests were made to check in-place stability.

10. Some difficulty was encountered in oscillatory motions due to the bamboo pole method of safety and the coordination required between the pilot and the safety man.

11. The second method of safety devices consisting of an outboard fishing line with its line of action through the centre of gravity of the model caused fewer difficulties with non-inherent airplane oscillations.

12. Scenes showing takeoff and landings from the floor and from the proposed platform.

Lockheed promotional artwork depicting a trio of L-200 Convoy Fighters taking off vertically from a naval air station. (David Stern Collection)

13. Scenes showing the model hovering near the ground demonstrating good control effectiveness. Note that occasional contacts of the tail pods and the ground occurred.

14. Tests showing controlled translation back and forth in the flying model.

15. Scenes showing the model hovering in a side-wind velocity (normal to wing area) of approximately 6 to 7 kts. This corresponded to a 35kt translation velocity of the full scale airplane. A blower was set up on the floor which created a jet of air

Artist's impressions of the L-200 Convoy Fighter in action; these are from a Lockheed report on the aircraft's electronic equipage, which is not summarized here due to its rather technical nature.

approximately twice as large as the model planform at the distance from the blower corresponding to the 5 or 7 knot wind speed. Velocities of the jet were measured with a standard duct velometer.

16. Several transitions from vertical flight to horizontal were made after takeoff from the platform. Due to the small space available for testing and the limitations of the safety method, transitions from level to vertical flight were not attempted in this series of tests.

Description of the model and test equipment
Conventional model
This model was constructed from a standard commercial kit and represented a conventional wing and tail arrangement. It was also equipped with a conventional landing gear although this was not used in this series of tests.

The weight of the model was approximately 34 ounces and the maximum thrust obtained was 38 ounces. Hence the thrust-weight ratio for hovering was slightly more than 1:1 in the majority of the tests.

A complete description of the model was not provided since it was not pertinent to the project except for check-out of the fuel system and power control.

L-200-1 model
The 1/16 scale model was constructed in the early stages of design of the full-scale airplane, hence it did not represent exactly the final configuration. The wing configuration and the fuselage and body shapes were essentially correct and the empennage areas were accurately represented on the model. However, the empennage position was changed, moving the vertical surfaces on the full scale airplane from the bottom to the top.

This difference had little or no effect on the flying characteristics in the regimes which were tested since only pitch moments were of interest and the control surface effectiveness of the elevators was the same in either case in vertical flight and hovering.

General views of the model configuration were shown in the photographs accompanying the proposal. These views show the side view and planforms of the complete model as constructed. It was found to be necessary in testing to remove the spinner and canopy section in order to attain sufficient engine cooling and thrust-weight ratios for hovering.

A 1/16 scale platform consisting of netted areas for takeoff and landing was also constructed for the model tests. A large lower net was designed to mesh with the pads on the tips of the empennage and a smaller net to one side was provided to catch the aft tip of the wing tip gun pod. Photos of the

Additional artwork of the L-200-1 from the electronic equipage report.

model on the platform depict the platform tipped to extreme angles to demonstrate the static stability of the airplane in place on the landing and takeoff platform. The design of the full-scale platform on the ship was actually gyro-stabilized such that the platform was level at all times. The reason that the airplane and model were so stable on the platform was that three widely separated axes of rotation existed between the tip contact point and the three empennage contact points. In order to rotate about any one of the three axes required side motion of the other two points which was effectively provided by the pre-load tension in the net cables. Lockheed noted that the model still held in place even after one empennage contact point came out clear of the lower net.

TAKEOFF AND LANDING

According to Lockheed, the best compromise solution of the landing and takeoff problems of the Convoy Fighter was the Tip-Net Method, in which a roll-stabilized, taut cable net provided a landing platform for the aft end of the fuselage that offered minimum slip stream resistance to maintain tail control effectiveness up to the point of net contact. Tip pods on the tail surfaces contained a shock absorber strut, extending load distribution feet, and a net piercing point. The net resisted vertical and horizontal loads and restricted horizontal displacement of the piercing points.

The aft end of the left-hand wing tip pod pierced the tip net located above and forward of the main landing net. This tip net resisted horizontal loads only. At the forward edge of the net and extending vertically above it was a cable barrier with rollers threaded on the cables which could be contacted by the tip pod at the time of power cut and functioned as a position limiter and pilot target. The landing signal officer (LSO) was located alongside and near the top of the barrier.

The proposed landing procedure was a direct stern chase approach at low altitude with the aircraft approaching spanwise with left wing tip on, hovering and positioning over the nets, and cutting power for let down.

The only auxiliary equipment required for takeoff was a hold down at the aft end of the fuselage during engine run up. Release was automatically controlled by a break ring in the hold down. The aircraft was initially positioned for takeoff with the positioning and hoisting gear.

This landing method ensured positive stability of the aircraft under all anticipated conditions. Tests conducted with 1/16 scale nets and a flying model verified the theoretical data, indicating the stability of the model with an angular displacement of approximately 45°. Very high horizontal accelerations were required to dislodge the model at this

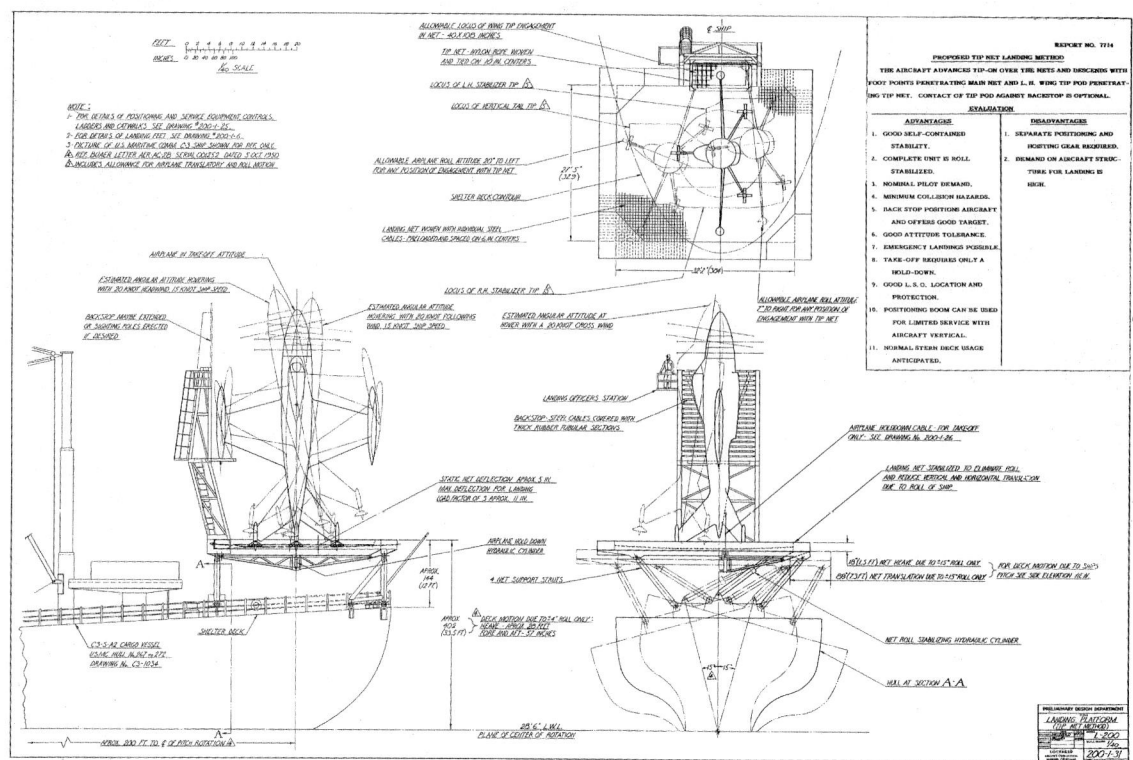

This drawing illustrates the Tip-Net method for the L-200, which was judged by Lockheed to be the most effective way of taking off and landing on a ship.

attitude eliminating the possibility of scale frictional effects.

Introduction

The purpose of this study was to investigate the problems of tail-first recovery and vertical takeoff of aircraft designed for operation from various types of ships and under the worst weather conditions anticipated for flight operations, to determine several methods of recovery and takeoff, and to evaluate these methods on the basis of the best compromise solution.

Assumptions were made to cover existing specification and data gaps and were presented as possible trends to be followed in establishing the design criteria for tail-first recovery and vertical takeoff of aircraft. Anticipated physical tests would establish the practicability of various methods under the limitations of conditions specified.

Description of problems and approach to solution

It was required that the aircraft land tail-first on a ship proceeding along a fixed heading at a constant velocity up to 15 knots maximum with winds up to a maximum of 20 knots at any heading relative to the ship heading. Hover over, or advance into, or takeoff from, the landing mechanism had to be made under conditions up to the maximum specified.

With the airplane thrust axis vertical and the pilot in a near backward prone position, his horizontal visibility was optimum spanwise. Ship approach and basic positioning therefore appeared to be desirable in the direction of the spanwise optimum visibility. Sighting the landing target relative to the wing tip provides approach-path and attitude perception.

Landings and takeoffs were to be effected on different types of ships, such as cargo, tanker, etc., normally comprising a convoy. The aircraft and associated landing and takeoff mechanisms were considered as a defensive weapon for the protection of the ship and were not to interfere with its normal operation or function.

The optimum location of the landing and takeoff mechanism on the deck was considered to be the stern. The clearance quadrant of the stern anticipated a basic chase, approach, retreat wave-off, and takeoff with little danger of colliding with ship superstructure. Likewise, study of ship plans indicated the stern as relatively clear of ship accessories vital to its normal operation.

The stern deck, due to its distance from the centre

LOCKHEED L-200

The recommended L-200 landing sequence was a direct stern chase approach at low altitude with the aircraft approaching spanwise with left wing tip on, hovering and positioning over the nets, and cutting power for let down.

Lockheed conducted tests with a 1/16 scale flying model and nets which indicated the stability of the model with an angular displacement of approximately 45°. Very high horizontal accelerations were required to dislodge the model at this attitude, eliminating the possibility of scale frictional effects.

of flotation, which was assumed as the centre of pitch rotation, would have considerable heave amplitude, and due to its general higher deck location, would have considerable roll translational amplitude. On the basis of a 200ft distance from the coefficient of lift (CL) of landing to the CL of rotation and an angular pitch amplitude of 8°, the heave amplitude was 28ft, and the deck attained a maximum vertical velocity of 10.8ft/sec for a cycle time of 8 sec. On the basis of 25ft height from the CL of a landing platform to the CL of rotation and an angular roll amplitude of 30°, the deck translational amplitude was 13ft, and the platform attained a maximum translational velocity of 3.5ft/sec. The roll amplitude and maximum translational velocity increased as landing mechanisms were placed higher above deck than the hypothetical platform.

Lockheed assumed that no desirable mechanism could be evolved that would compensate for the deck heave due to pitch and that landing techniques had to minimize its effect. Pitch angle was not considered of sufficient magnitude to justify stabilizing mechanisms, and translation due to pitch was small.

Roll stabilization of the landing mechanism was considered very desirable, both from the standpoint of eliminating roll angles and for reduction of deck translational amplitudes and maximum deck translational velocities.

While the deck of the ship was gyrating in pitch or roll, or a combination of both, the aircraft, whether hovering or advancing into a landing, would have an angular deviation of the thrust axis from space vertical. Preliminary tunnel data for horizontal velocities up to 60ft/sec and a gross weight of 16,000lb were plotted. These approximate angular deviations had to be considered additive to deck angles in considering stability of the airplane at landing contact and at takeoff, especially for platform landings. A reference curve of pitch deviation only with unidirectional aileron deflection to reduce wing lift was provided to indicate feasibility of angular reduction with surface controls.

Hover attitude of the aircraft at or over the landing mechanism immediately after completion of approach had to be considered. Using resultant velocities of wind and convoy and their headings as plotted by Lockheed, hover attitudes for a 180° stern approach quadrant with thrust axis deviated on a single axis only were also plotted.

It was apparent that the considerations of pilot visibility and aircraft attitude favoured the spanwise approach since the yaw deviation of the thrust axis was considerably less than the pitch deviation.

Since approach requirements of various landing mechanisms differed, data for hover attitude in either pitch or yaw for approach paths along the resultant velocity headings were supplemented by two plots of hover attitude in both pitch and yaw for either aircraft spanwise or belly-on approach. A plot indicated a hover attitude for aircraft spanwise approach 90° to the longitudinal axis of the ship or a belly-on approach parallel to longitudinal axis of ship. The hover attitude for spanwise approach parallel to the longitudinal axis of the ship or a belly-on approach 90° to the longitudinal axis of the ship was also plotted, both plots being relative to wind heading.

Landing mechanisms were simplified if aircraft approach was made along a single fixed path with sufficient approach quadrant allowance to compensate approach error. However, unless the approach to the landing problem, as discussed in the latter part of the report, was that the aircraft would turn over on landing, an orientated landing mechanism appeared necessary if landing contacts were made appreciably below the centre of gravity of the airplane and no turnover restrictive barriers were used.

The design and evaluation of landing mechanisms, and related takeoff mechanisms, was based primarily on the stability of the airplane on touchdown and in 'static' or 'power-on' position. In all conditions, dynamic loads were applied by the roll and pitch gyrations of the ship, and both the aircraft and the deck could be anticipated at attitudes other than respective vertical and horizontal.

Three design approaches could be made:

1. Assume that the aircraft would not turn over due to combined attitude of the airplane and the landing mechanism and/or anticipated loads, and design accordingly.
2. Assume that stabilization of the landing mechanism would reduce the combined attitude of the aircraft and the landing mechanism, and that the stabilization would reduce anticipated loads. Design accordingly.
3. Assume that the aircraft would turn over on landing and design accordingly. Stability for takeoff had to be provided.

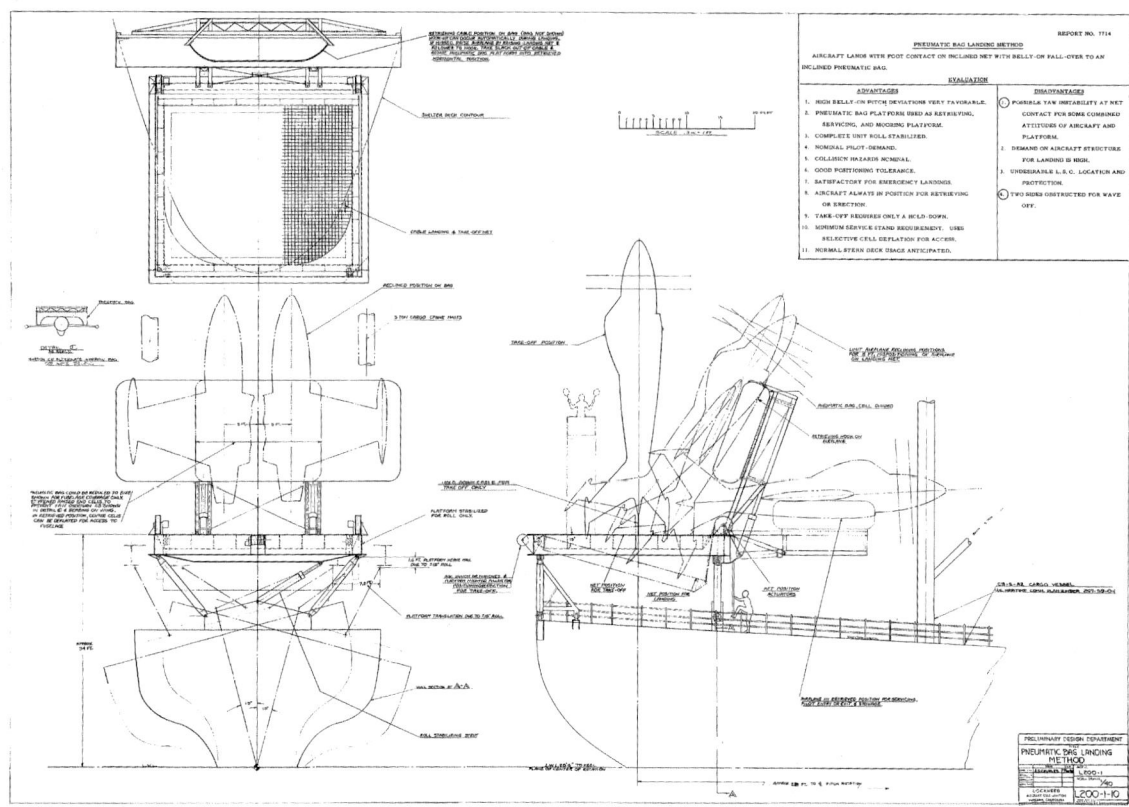

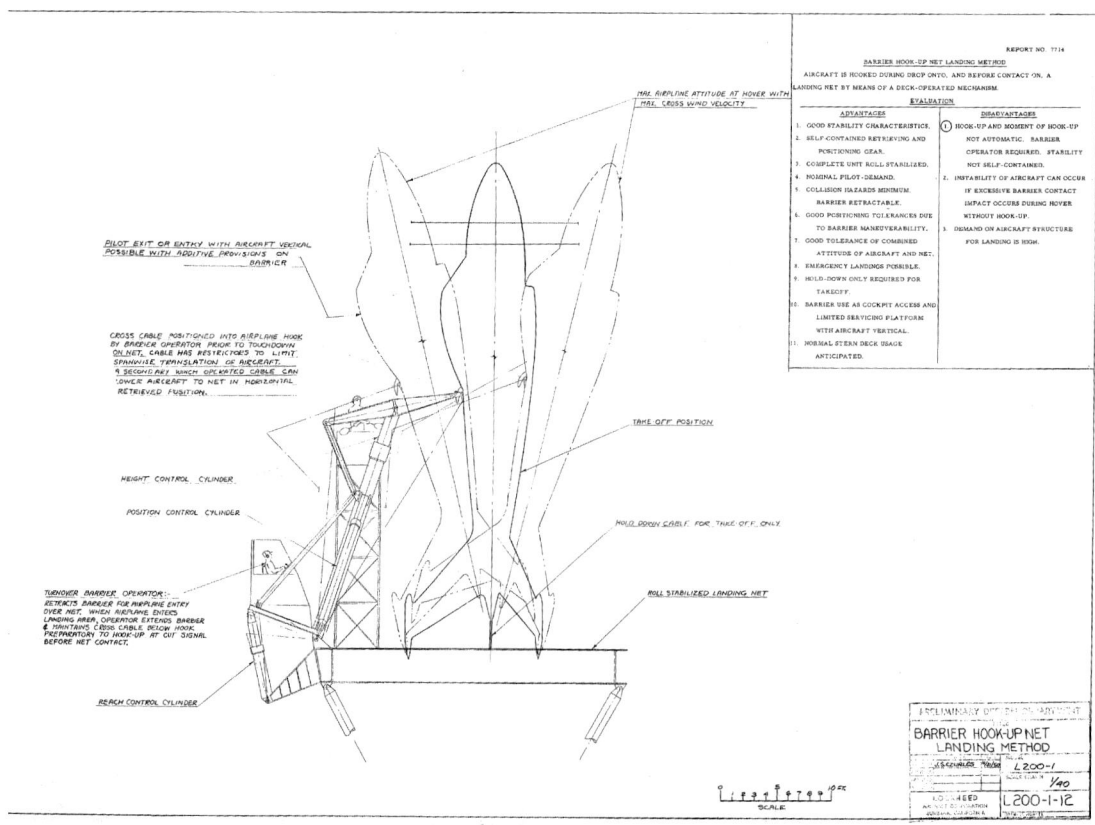

Detailed drawings of additional landing methods and equipment investigated and evaluated by Lockheed for the L-200 design.

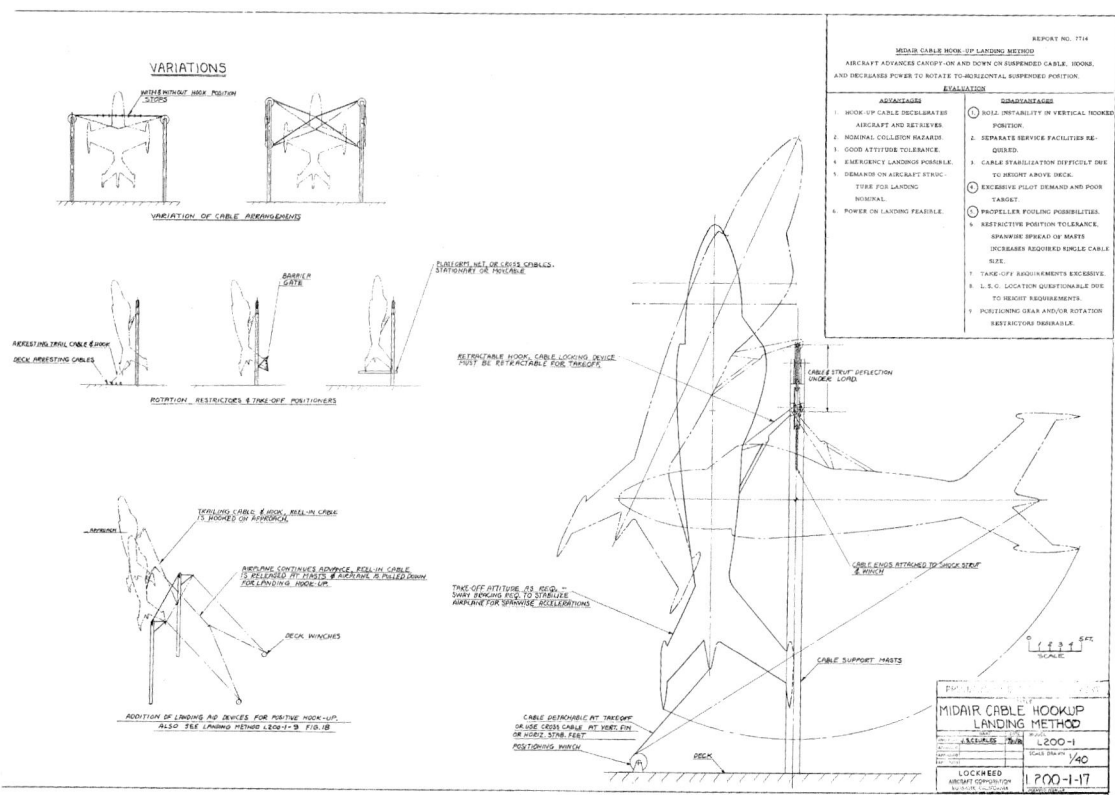
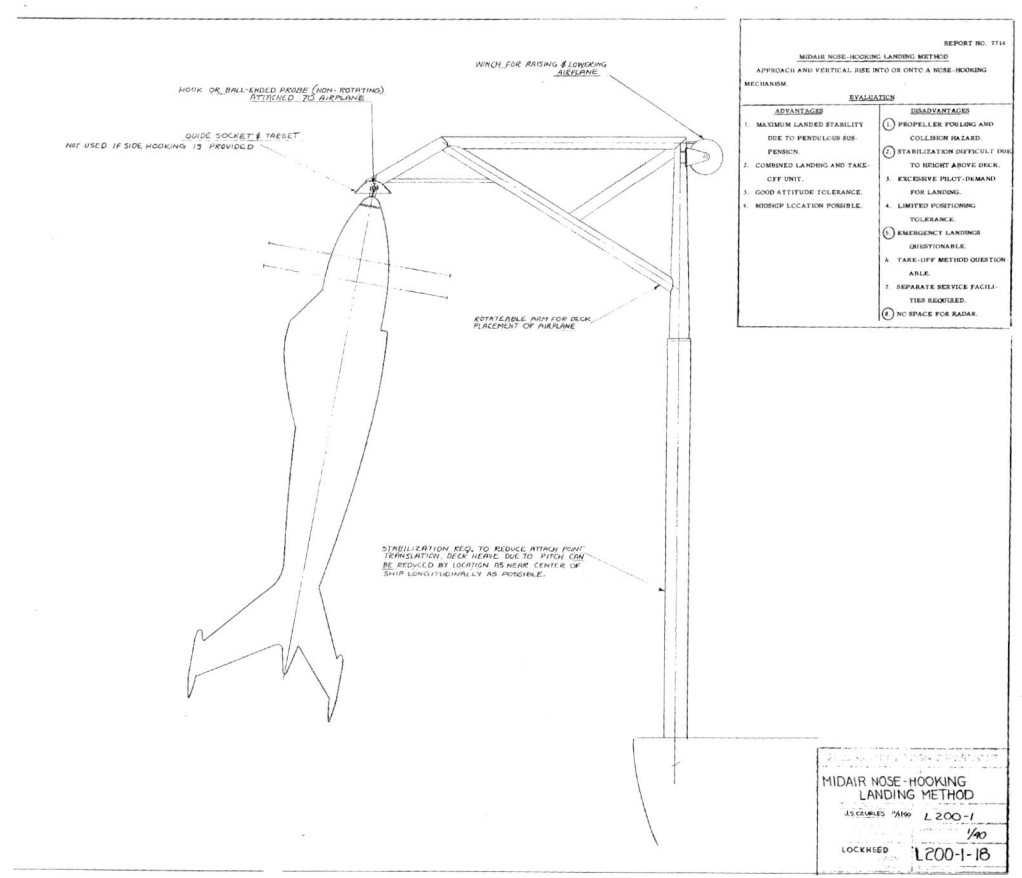

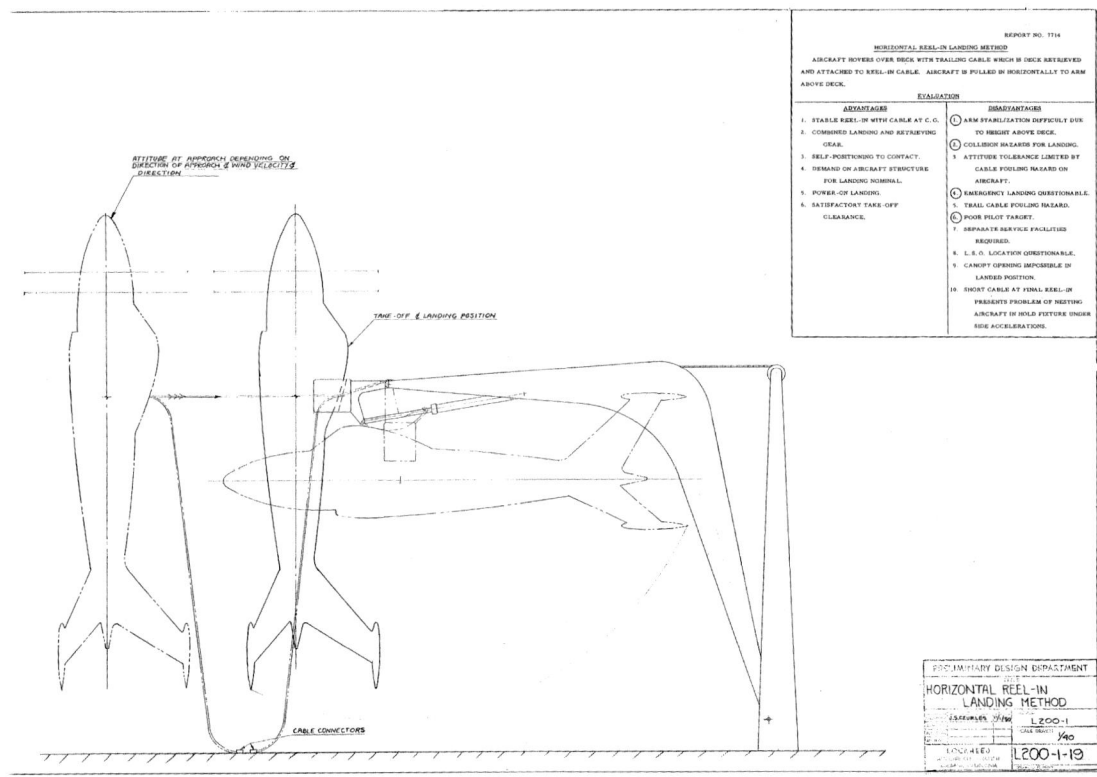

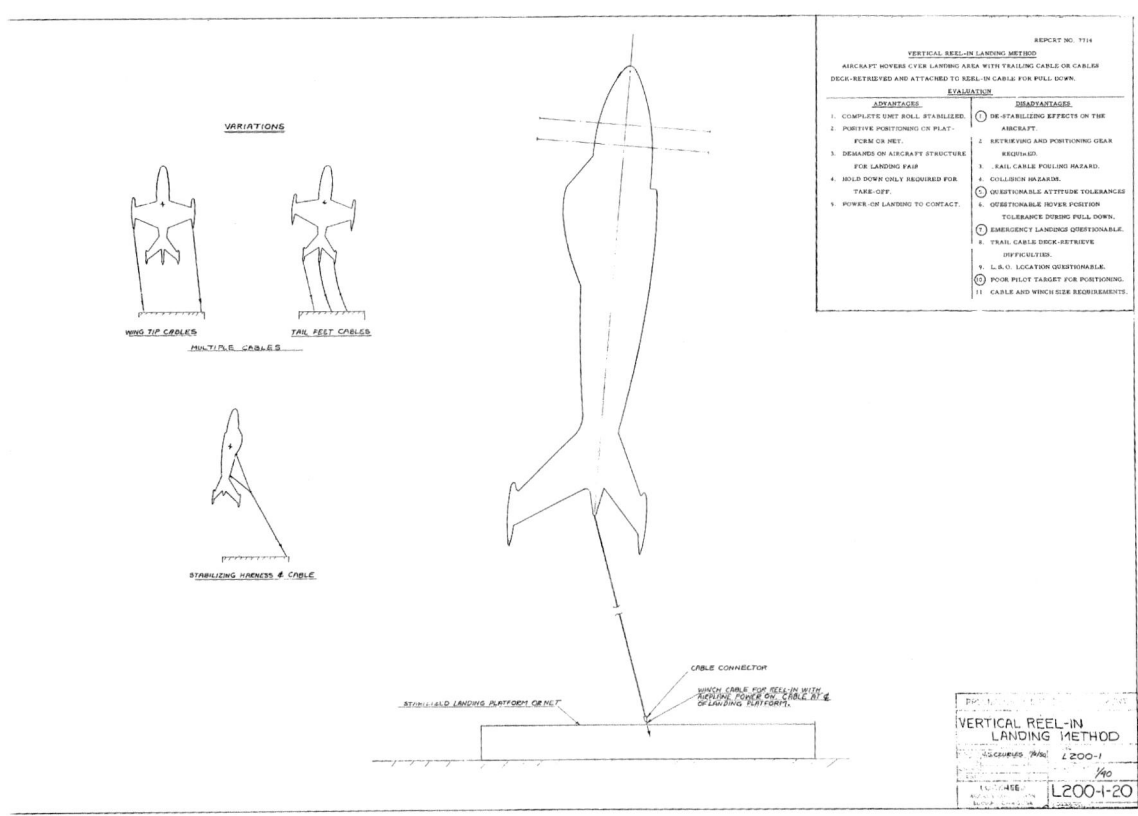

LOCKHEED L-200

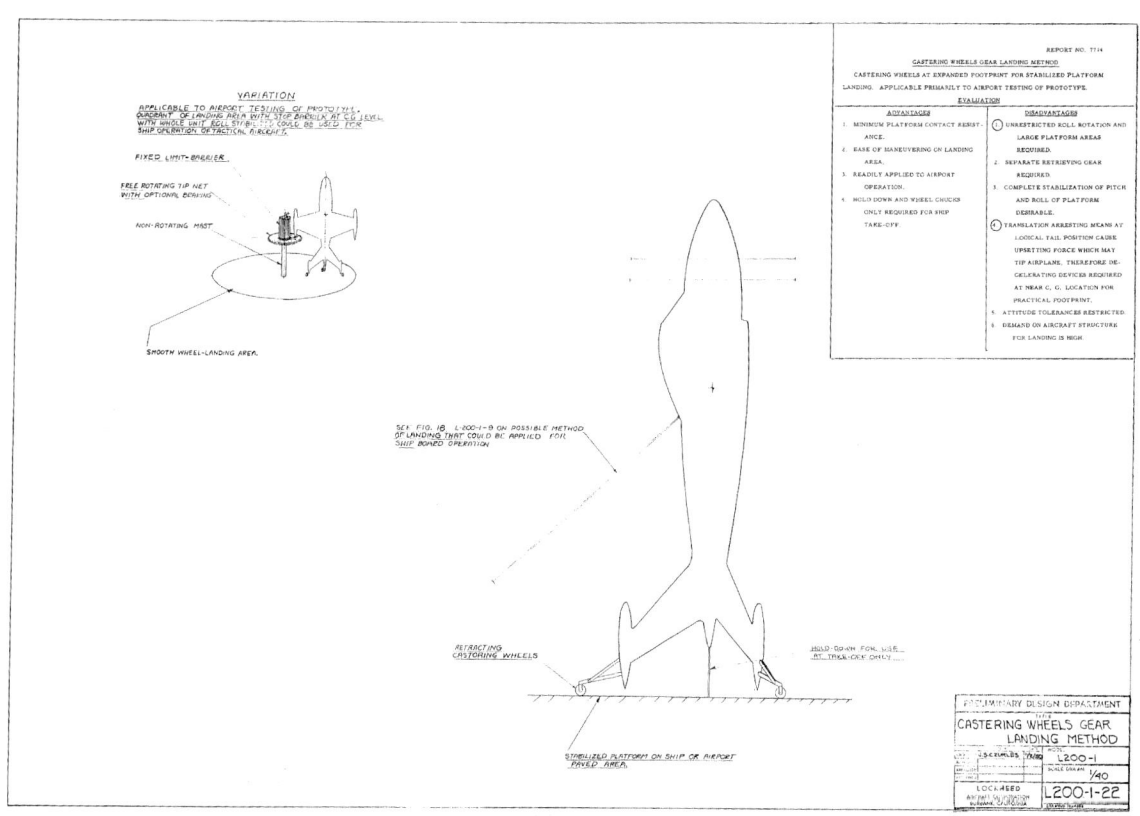

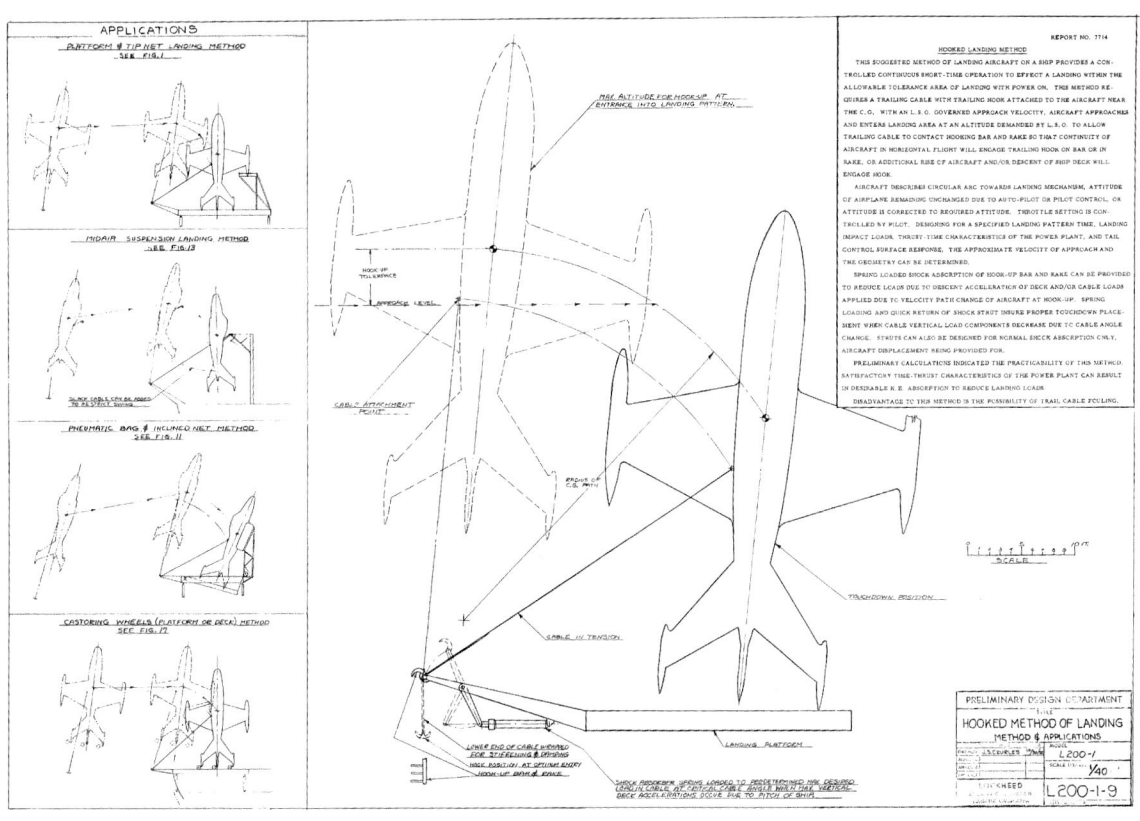

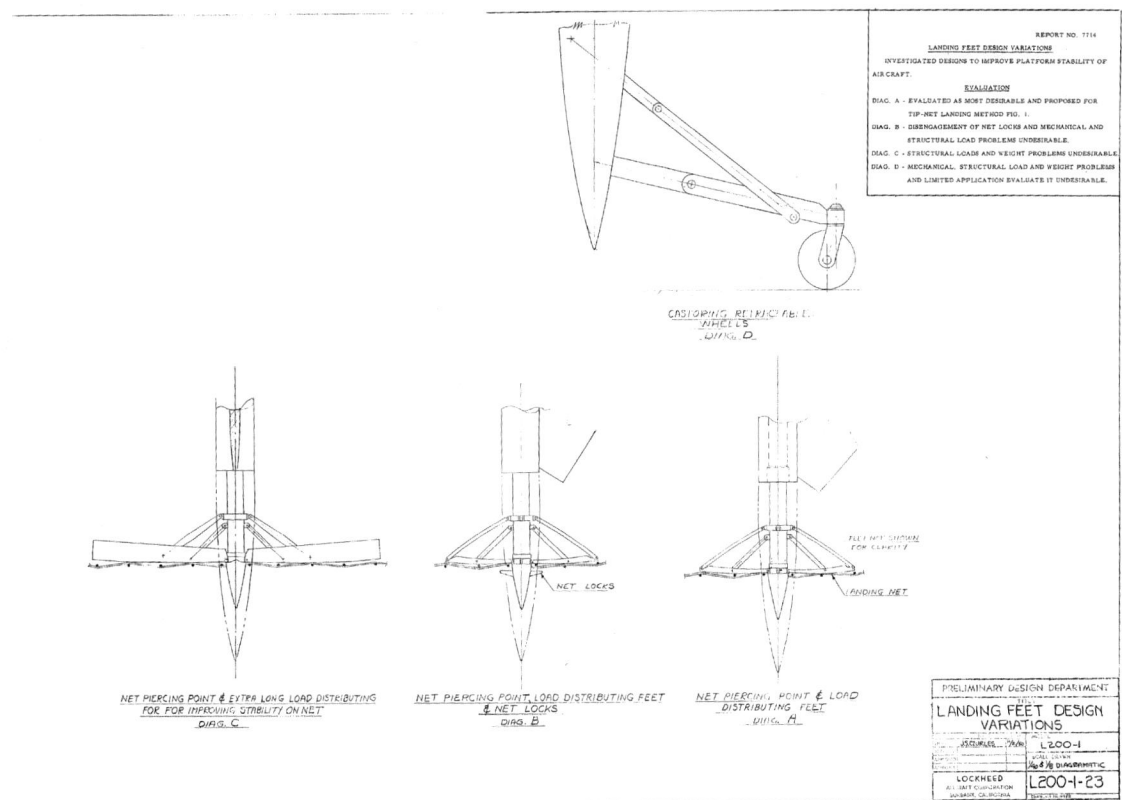

Different designs for the landing feet.

The term "turn over" applied to an undesirable attitude change of the aircraft resulting in aircraft damage or loss, landing mechanism fouling by propellers or other parts of the airplane, and difficult positions for retrieving.

Landing mechanisms could be divided into three categories:

1. Midair landing—involving cross cables, bars, hooks, or arms suspended above deck and an engaging hook or equivalent on the aircraft.
2. Platform landing—involving feet, geometrically arranged to prevent turnover either in pitch or yaw or both due to static and dynamic loads.
3. Combination landing—the final platform landing with aircraft stabilized by external agents.

In midair landings, when no ship contact was made at any point very far removed from the c.g. of the aircraft, tail surface controls would maintain effectiveness until the landing procedure was completed. For the platform landings, a perforated platform was considered to be required to maintain tail surface control effectiveness until touchdown and for penetration of the platform with the alighting gear to maintain contact position on a restricted area.

The three approaches to the problems and the three categories of landing mechanisms were proposed and evaluated in the following sections. It was noted that retrieving, servicing, and mooring requirements of the aircraft would reduce the initial anticipated simplicity of any landing method.

Anticipated hazards were taken into account, such as propeller fouling for midair landings, destabilizing effects of platform contact with horizontal velocities present, or destabilizing effects of pull down or pull in cables trailed by the aircraft, the possible fouling of these trailing cables, and time and crew requirements for effecting landing.

Landing and takeoff techniques were the final major considerations and were closely coupled with the gyrations of the ship in pitch and roll. Time requirements for retrieving, servicing, and takeoff positioning of the aircraft were considered as a part of this evaluation.

Final proposed landing method

The final proposed landing method required that

the pilot approach the ship spanwise (tip-on) with a pursuit approach path astern and parallel to the longitudinal axis of the ship with the aircraft in vertical attitude and with a horizontal velocity in direction of the convoy heading, approximating convoy velocity. Pilot approach to the landing area was at an altitude approximating maximum observed heave position of landing area. Final approach to the landing area was controlled entirely by the LSO.

The aircraft was approached over and into the landing mechanism prior to the top rise of the deck in pitch heave. The aircraft completed landing at cut signal with deck at peak rise or descent. Maximum hovering time required for proper phasing of aircraft descent and deck rise was five seconds if the aircraft was properly positioned.

The landing technique assumed satisfactory hover of the aircraft and no required time for effecting landing. Since the aircraft was vulnerable to attack during landing, it was probably desirable to complete a landing in a continuous operation. A hook-up landing aid was presented for such a continuous operation and its possible applications were indicated on the drawing.

No assumption was made in the basic evaluation that the angular deviations of thrust axis plotted in accompanying graphs were reducible by additive aircraft controls or landing techniques other than those discussed.

The problems of takeoff paralleled those of landing in that stability prior to takeoff was a prime consideration, and takeoff phasing with deck heave, aircraft rise, and retreat path, reversing the landing procedure. One fundamental difference was that power applications could be anticipated with aircraft in takeoff position and hold-downs were to be provided for this condition.

Problems of retrieving, servicing, mooring, and positioning were characteristic of the geometry of each individual landing mechanism and, as stated before, complicated the simplicity anticipated for every method considered.

Additional methods of landing and takeoff
The evaluation was based on the following:

1. Stability—at landing, on net, during hover. Self-contained or external.
2. Double usage—combined use of mechanisms.
3. Roll stabilization—was completely unit roll stabilized and difficulties of stabilization.
4. Pilot demand—requirement of pilot activity, judgment, perception, and vision.
5. Collision hazard—fouling and collision possibilities.
6. Position tolerance—position deviation possible for good landing.
7. Attitude tolerance—attitude deviation possible for good landing.
8. Emergency landings—landing possibilities with damaged aircraft or wounded pilot.
9. Aircraft structural demands—did structure loading for landing exceed flight loads? Structural complexities.
10. Crew hazard and time factors—danger to deck crew and simplicity of landing procedure.
11. Take-off requirements—special gear requirements.
12. Deck demands—interference with normal deck usage.
13. LSO location and protection—pilot visibility of LSO with landing area perception. LSO platform stabilized and collision protected.
14. Service requirements—crew, stand, and rigging requirements.

Conclusion
On the basis of the evolution of all the landing and takeoff methods examined by Lockheed, it was determined that the best compromise solution was the Tip-Net Landing Method. The proposed landing procedure established the direct stern low-altitude approach with the aircraft advancing spanwise in chase pursuit with the left-hand wing tip on.

PROTOTYPE PROPOSAL
As requested in Navy Specification oS-121, Lockheed studied the design of a prototype airplane for use in developing the flight characteristics and vertical landing and takeoff procedures leading up to the design for a tactical vertical rising convoy fighter. In the BuAer specification, it was requested that a prototype be designed about the dual Mamba engine and that, following a brief flight test period with this airplane, a further programme be instigated building two tactical and one static test article on production type tooling powered with the T40-A-8

engine. It was Lockheed's conclusion that time and flight test value would be lost if the prototype was designed for the Mamba engine and, as a result, two prototypes were presented to BuAer; one as requested with the dual Mamba, and the other as suggested by Lockheed, powered with an early version of the T40 engine. Lockheed produced a table comparing of the performance of the T40 and Mamba-powered prototypes including the design weights, the proposed power ratings to be used, and the proposed fuel capacities required to give the desired prototype duration as requested in OS-121. Satisfactory performance for vertical flight could be obtained with either prototype but a greater margin of performance was available if the T40 full-scale prototype programme was followed.

Introduction
The Lockheed Corporation had maintained a development staff for many years whose responsibility was primarily the design and construction of prototype airplanes. Several systems were followed with this group of specialists in the design and development of prototype airplanes which included the F-80, the F-80 production conversion for American engines, the F-90 penetration Fighter, the Saturn small transport, and many other prototype conversions for advanced models of the P2V and Constellation airplanes.

A similar programme was followed in the development of the Constitution transport which, because of manpower problems, utilized the project group of engineers rather than the specialist prototype group for development. Specific conclusions as to the proper type of prototype procedure had resulted from these operations. These conclusions were as follows:

1. Prototype development of an airplane with advanced and unknown flight problems should be done with consideration for production but without any major effort toward building the airplane on production tools. This imperiled early solutions for normal structural design problems and substantially delayed the early flight of the airplanes which was the primary reason for any type of prototype programme.
2. Consideration had to be given to the tactical utility of the airplane in the original design concept but constant parallel design effort to incorporate all tactical details during a prototype programme also imperiled early flight of the airplane, thereby delaying the entire programme.
3. If continual attention to production and tactical details followed a prototype development, major engineering effort was wasted since the flight changes usually required major changes in both production and tactical features, thereby wasting any prior engineering effort in this direction.
4. The proper time for production considerations and tactical conversion of the airplanes to take place was after a period of flight test with the prototype airplane.

Lockheed recognized that the Navy programme as outlined in letter Aer-AC-21 answered essentially all of the above desires. It was felt, however, that the production airplane was sufficiently different than any design which involved the use of a Mamba power plant that the prototype programme would have to be repeated with the T40-powered airplane before the start of full production tooling or planning.

Lockheed emphasized that even though prototype airplanes were built on essentially prototype tools, the very fact that the airplanes were full size and had construction which was similar and, in many places, identical with that desired for a production airplane, created a background of production experience and some stockpile of assembly tools which could be used for small production numbers. Thus, the prototype programme was of considerable value to final production in spite of the fact that no specific effort was aimed in this direction.

The following sections are devoted to a description of the comparative design details between a Mamba and T40-powered prototype, an outline of the weight analysis of the two airplanes, and a general description of the flight research programmes which could and should be undertaken with the prototype airplanes. Finally, a comparison in the form of a bar chart time history of two programmes was presented from which, if the basic design was chosen, the Navy could select the programme which most nearly coincided with their conclusions as a result of this study.

External arrangement
The major external differences between the two prototype airplanes involved the location of the

ducts and their size, and the location of the cockpit canopy. In the case of the T40 prototype, the external configuration of the airplane was identical to that of the final tactical proposal as long as the gearbox with a large degree of offset could be obtained for the prototype. The three-views did not show any alternate or auxiliary landing devices but showed only the configuration of the airplane when it was finally determined by sufficient flight experience that normal landings and take-offs in a vertical direction could be accomplished.

Internal arrangement

In the inboard profile of the full-scale T40 prototype, the gearbox depicted was a single speed offset type. Lockheed believed that a single speed gearbox was satisfactory for prototype development of the T40, but it was emphasized that the large offset should be obtained (if possible) so that the general engine location in the prototype was identical with that of the tactical airplane. An additional drawing shows the engine location if it was necessary to use the standard Allison gearbox. It was entirely possible to make the T40 prototype in this fashion in order to obtain an early flight date but delays would occur later in the programme which would make the production airplane available in approximately the same time. These delays would be caused by the following factors:

1. Duct data from the prototype would be invalid for the final airplane and any duct development would have to take place on a later version of the airplane.
2. Wing structure and fuselage structure would have to be modified considerably in order to account for the higher engine position in the fuselage, thereby reducing the advantage of having a light structure on the prototype airplane which could be static tested and converted for a minimum weight tactical airplane.
3. The tail pipe outlet would be considerably more difficult and would add a weight penalty to the prototype and sufficiently modify the exterior of the airplane that some differences between the prototype and tactical airplane might result.

Because of these problems, it was concluded that every effort should be made to obtain a gear box with a high offset for the first prototype airplanes.

The inboard profile of the Mamba-powered prototype shows that the arrangement of the airplane interior was entirely different than the final tactical airplane and no experience could be obtained with respect to duct characteristics, airplane drag, basic air loads, or systems performance. Recent experience with the high speed effect of external protuberances such as the cockpit enclosure led to the conclusion that some difficulty could be experienced on the prototype airplane due to location of the cockpit adjacent to the tail which would not be experienced on the final tactical airplane. This could be a major stumbling block in the development of the prototype flight characteristics.

Landing provisions

Lockheed provided a perspective diagram of a normal landing gear which could be installed on the prototype airplane. It was possible to install this type of gear on either the T40 or Mamba prototype although it was shown above only on the T40 airplane. Several variations of this spring shock absorbing type landing gear were possible and it could be satisfactory not to install the tail skid which was shown but rather to retain skids only on the tips of the horizontal tail.

The weight of the gear was not such, however, to prohibit vertical flight with the airplane. It was recommended that some such external gear be considered which did not have any major effect on the basic airplane design. Lockheed believed that since vertical rising and hovering were the major problems to be solved on the airplane that flight testing for high speed stability, etc. could more profitably be undertaken after vertical rising problems were solved and that a retractable gear was, therefore, not necessary for prototype use.

Some consideration, however, was given to the use of a ground gear for vertical landing and takeoff which was entirely divorced from special ground handling equipment. One proposal for this type of landing gear featured small castoring wheels, which were stowed into tip pods on the horizontal and vertical tail, replacing the landing spikes of the standard airplane. These wheels permitted the airplane to be placed vertically and operated from this position with greater ground tip-over margins to withstand landing translation velocity than with the tip spikes since the castoring wheels prevented any side load immediately upon landing.

It was recommended that the wheel landing gear be utilized for most of the prototype programme since with it, practical vertical rising and landing could be demonstrated, and additional flight tests could take place while landing nets similar to the shipboard rig were constructed and landing spikes developed. Furthermore, the use of the wheel landing gear probably eliminated any requirements for an LSO to land the prototype airplane. Lockheed recommended, however, that a tip backstop be constructed in the landing area for use as a vertical height target to assist the pilot in properly positioning the airplane on the airport.

The use of a tip net also ought to be considered during this stage of flight testing.

Power plant installation
The basic power plant installation for both the Mamba and T40 prototypes was shown in their respective inboard profiles. No additional diagrams were prepared since these were satisfactorily self-explanatory. Lockheed provided a drawing to illustrate the effect of the low or high position on the basic fuselage structure of the T40 prototype. This drawing served to emphasize the desirability of having a high offset gearbox not only in the prototype but in the tactical airplane as well.

If it became necessary to design the prototype for the high position of the engine, one of the structural changes which occurred was in the wing centre section and fuselage attachment, and structural development of the prototype would not be applicable to the final airplane. Since the major problem in incorporating the Allison engine into a prototype airplane appeared to be the availability of a suitable gearbox, a drawing was included which outlined the Allison estimates of the types of gearboxes which could be made available along with the weights and times which would be required.

Gear Box No. 3 was the gearbox desired for the tactical airplane and it was the gearbox which would be used in any kind of prototype programme assuming that it could be obtained in a reasonable time. The gearbox development appeared to be a major factor in delaying an early flight date of any prototype and, as a result, Gear Box No. 4 was selected for the prototype because it would be available three months sooner than the ultimate gearbox design and would still have a large offset which would properly position the engine within the airplane. It was urged that if the advantages of the T40 programme appeared to be valid, an effort be made to accelerate development of a high offset gearbox which was suitable for use in the prototype in around 15 months from the starting date.

Control system
The control system for the T40 prototype would closely correspond to that which was required on the tactical airplane, the major characteristics of which were:

1. A dual boost system was incorporated in the elevator, each boost cylinder being connected to an entirely independent hydraulic system.
2. Aileron control consisted of a single boost cylinder coupled with a servo tab located in the propeller slipstream. Satisfactory control in the transition, landing and take-off phases could be obtained with manual controls and the servo tab so that safe flight could be made with boost out.
3. The rudder control system also incorporated a servo tab which permitted satisfactory pedal forces in normal and transition phases of flight. No boost was required on the rudder.

For the Mamba prototype it appeared that some simplification of the control system could be countenanced if the flight speeds of the prototype were limited and if control forces could be obtained through use of external aerodynamic balances. This, of course, defeated some of the purpose of a prototype airplane in that controllability could be checked out completely without an identical control system or aerodynamic configuration of controls. Therefore, Lockheed suggested that a Mamba prototype incorporate an almost identical system in spite of the fact that control forces were sufficiently low that some of the boost components could be eliminated. This permitted high speed flights with the Mamba prototype and might serve as a partial flight test laboratory for the control system although identical conditions would not be reached when compared to the final tactical airplane.

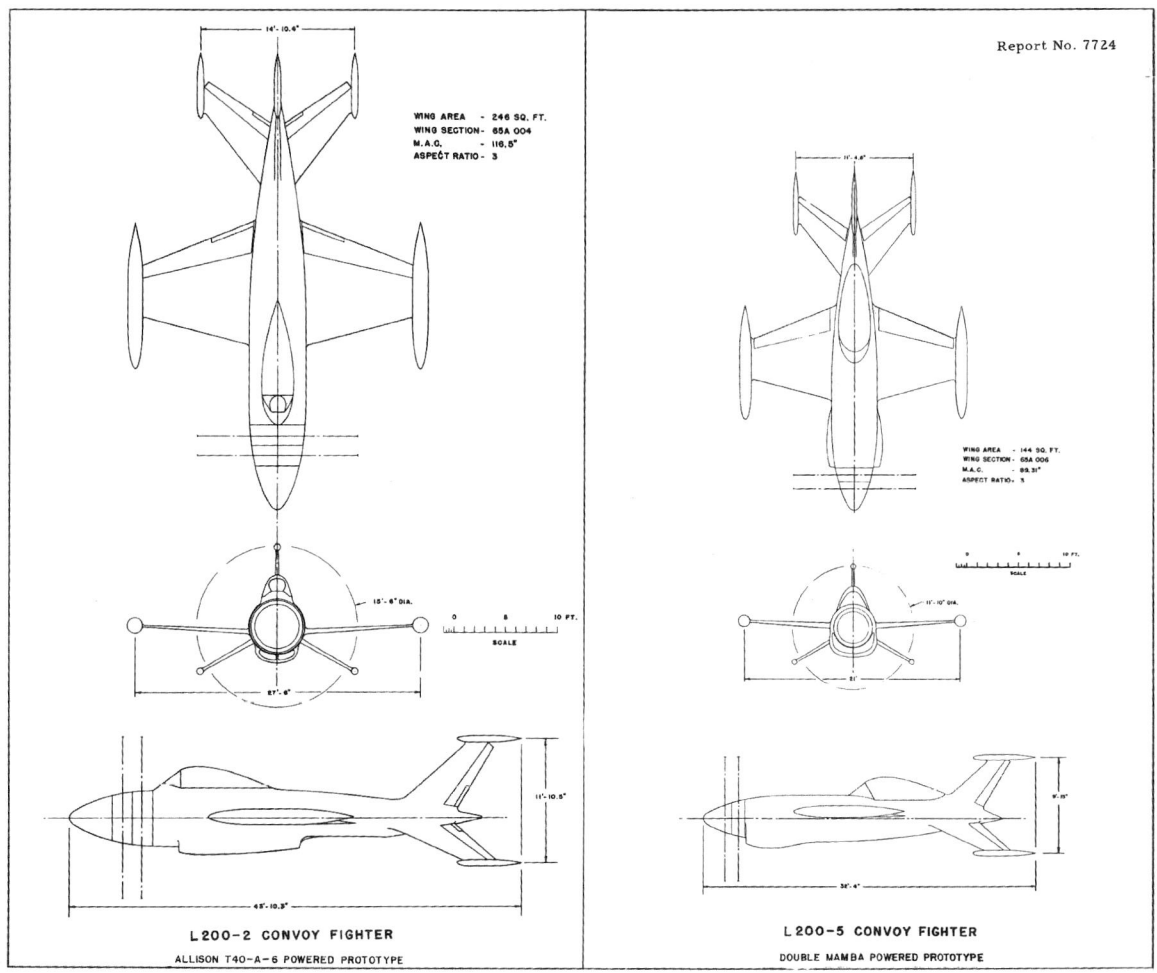

Comparative general arrangements of the two proposed prototype airplanes. The L-200-2 was essentially a stripped version of the full-scale Convoy Fighter while the L-200-5 was a 0.766 scale demonstrator powered by the Armstrong Siddeley Double Mamba turboprop engine.

Electronics

For the prototype airplane with either a T40 or Mamba, only sufficient communication equipment was installed for proper coordination of flights with a flight test base. No radar system was contemplated in either prototype. It was noted, however, that the prototype airplanes with the T40, if not converted entirely to tactical airplanes in one conversion process, could be used for prototype installations of all the electronic equipment for radar and radio checkout.

Of primary interest on the prototypes was the installation of the autopilot since the development of completely automatic flight transitions and development of aids to the pilot for vertical landings and takeoffs was the primary reason for construction of the prototype airplanes. Thus, it was proposed that either prototype have a complete autopilot system as a primary portion of its electronic equipment. In this matter the advantage of the T40 full-scale prototype again became apparent since the aerodynamic reactions of the airplane would be identical or very close to those of the final tactical airplane and the development of the autopilot responses would be directly applicable.

Developing the autopilot for use on the Mamba prototype would, of course, have solved a major portion of the autopilot problems but an additional programme would still be required when a similar type of autopilot was applied to the final tactical airplane.

Basic structure

The construction of either the T40 or the Mamba prototype was expected to be along relatively conventional lines except for those portions of the airplane which required special structural techniques in order to achieve adequate stiffness or strength.

The structural drawings of the standard tactical airplane constituted a thorough description of the type of structure which would be used in the T40 prototype. The use of forgings would, of course, have to be eliminated in any kind of prototype

LOCKHEED AIRCRAFT CORPORATION	REPORT 7724	
SUMMARY OF CHARACTERISTICS AND PERFORMANCE OF L-200 PROTOTYPE AIRPLANES		
	Model L-200-5	Model L-200-2
Engine	Double Mamba	Allison XT40A-6
Specification for Engine	A.S.M.D.1 Issue No.3	Allison 264-A
Military Equivalent Shaft HP	2390	5225
Design Gross Weight, lb.	6800	11578
Maximum Take-Off Gross Weight, lb.	7500	12754
Weight Empty, lb.	6750	9814
Fuel Capacity, lb.	1750	2940
Wing Area, sq.ft.	144	246
Wing Span, ft.	21	27.5
Length, ft.	32.33	39.8
Propeller Diameter, ft.	11.9	15.5
Description	8 blade, dual rotating	6 blade, dual rotating
Maximum Estimated Take-Off Thrust, lb.	10060	17370
Maximum Take-Off Thrust/Take-Off Weight, lb./lb.	1.34	1.36
*Maximum Level Speed, Sea Level, knots	448	456
*Maximum Level Speed, 20,000', knots	452	502
*Maximum Level Speed, 35,000', knots	455	516
*Service Ceiling, ft.	38900	46700
Flight Duration at Take-Off Power, min.	45	45
*Stall Speed in Normal Flight Range, knots	124	124
*At Combat Weight		

Table comparing the performance of the Mamba and T40-powered prototypes, with the latter having a greater margin of performance than the former.

programme and machined dural parts would be substituted wherever necessary. It was particularly noteworthy that the stabilizer and rudder spars and one attachment bulkhead in the fuselage were proposed to be titanium forgings in the final tactical airplane. For prototype purposes, these spars would have to be produced either as welded steel parts, machined dural parts, or machined titanium parts if titanium billets of sufficient size could be obtained.

For the Mamba prototype airplane, the structural problems were believed to be similar with the possible exception that the airplane may not need to be designed for the same high speed as would be necessary with the T40 prototype. Therefore, for the wing structure, it was probably possible to substitute normal ribs and skin in place of integrally stiffened wing panels proposed for the T40 tactical and prototype airplanes.

In either case, prototype airplanes had been made at Lockheed utilizing all of the techniques proposed for the basic structure of the L-200 and no difficulty was expected due to lack of facilities or proven methods for building an airplane of this nature.

Weight analysis

As noted previously, one of the major reasons for constructing a full-scale T40 prototype was to obtain experience both in the design of the airplane and in the flight testing of the full-scale configuration to enable the development of the tactical airplane from a basic structure which was designed for a lighter gross weight. Thus, the first tactical airplane could approach the weight ratio normally associated with a tactical airplane which had been developed over a period of years, the gross weight of which had increased without major structural changes. Since the ability to obtain more power to account for gross weight growth was seriously limited for this airplane, the weight empty control had to be extremely successful and every effort was required to ensure that an absolute minimum weight empty was obtained. The process of designing a prototype for light weight and stretching this structure into a tactical gross weight was believed to be a necessary adjunct to this programme.

The weight history of several Lockheed airplanes indicated how successive modifications to production airplanes to increase gross weight had a very minor effect on the weight empty. The company hoped that increasing the gross weight of the airplane to the tactical gross weight would have a resulting weight empty increase which corresponded to the historical trend of Lockheed aircraft rather than the more normal estimated increase which was based solely on standard weight analysis methods.

Aerodynamic analysis

Performance, stability, control, and duct analysis estimates for the prototype airplanes were provided by Lockheed in separate reports to BuAer. Adequate performance could be obtained for flight test purposes with either prototype design. Both airplanes had sufficient excess thrust for satisfactory transition performance, based on the comparative engine ratings shown in the following table:

Engine ratings for prototype airplane

	Mamba engine	T40-A-6 engine
Takeoff horsepower, max.	2640	5035
Takeoff thrust, max.	810	1225
Normal horsepower, max.	2095	4470
Normal thrust, max.	710	1115

Concerning high speed and rate of climb for either airplane, it appeared that the T40-powered

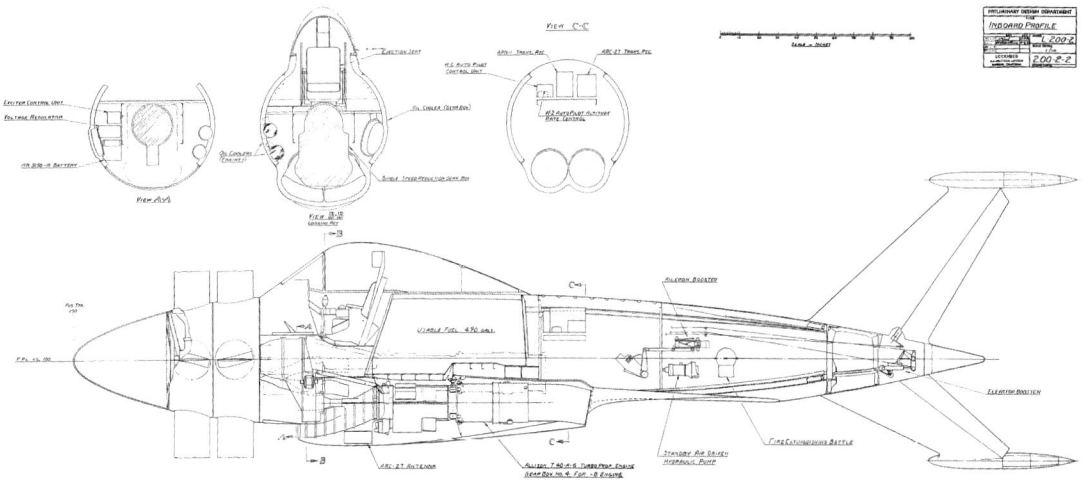

Inboard profile of the stripped full-scale prototype powered by the Allison T40A-6 turboprop engine.

airplane had a major performance advantage in all manoeuvres and in normal flight. Since the margin of success in this design development was probably a direct function of the margin of excess thrust for the initial prototype airplane, it was obvious that the T40 prototype was the more desirable of the two.

Summary of suggested research programme

In studying how the prototype airplane would be employed, an outline of the various research programmes required was made. Although some of these items did not actually come under the heading of flight test research, they were nevertheless necessary before the first flight was attempted and were directly applicable to any prototype airplane.

All of the programmes would be directly applicable to the final tactical airplane if the prototype and tactical airplane were identical in many respects. Many would have to be repeated for the tactical airplane if the Mamba-powered prototype was used.

1. Autopilot development testing—pre-flight programmes on the autopilot required check-out on a flight simulator such as the one available at the Massachusetts Institute of Technology.
2. Wind tunnel test programmes—powered wind tunnel tests had to be run under controlled conditions on special wind tunnel balances in order to determine the available stability and control under full power conditions during normal and transition flight.
3. Transonic wind tunnel testing—if flight was expected on the prototype airplane into the transonic region (and this could be done with a full-size T40 airplane), wind tunnel tests had to be run in this region on a powered model.
4. Free flight dynamically similar model testing—the value of this programme toward solving the problems of transition and vertical flight was problematic but a study had to be made of the feasibility of obtaining usable data in this manner with reasonable expenditure. This portion of the programme could easily be eliminated by the prototype airplane itself flying with a normal landing gear.
5. Miscellaneous research programmes—these programmes were essentially mechanical in nature but had to be done prior to airplane first flight.
    a. Drop tests of the landing devices to determine permissible load factors.
    b. Proof tests of the prototype airplane structure.
    c. Hydraulic system mock-up and laboratory tests of boost components.
    d. Fuel system test to simulate load factors which would probably be experienced and unique attitudes in which this fuel system would be required to operate satisfactorily.
6. Power plant operational test—complete power plant operational test on a ground

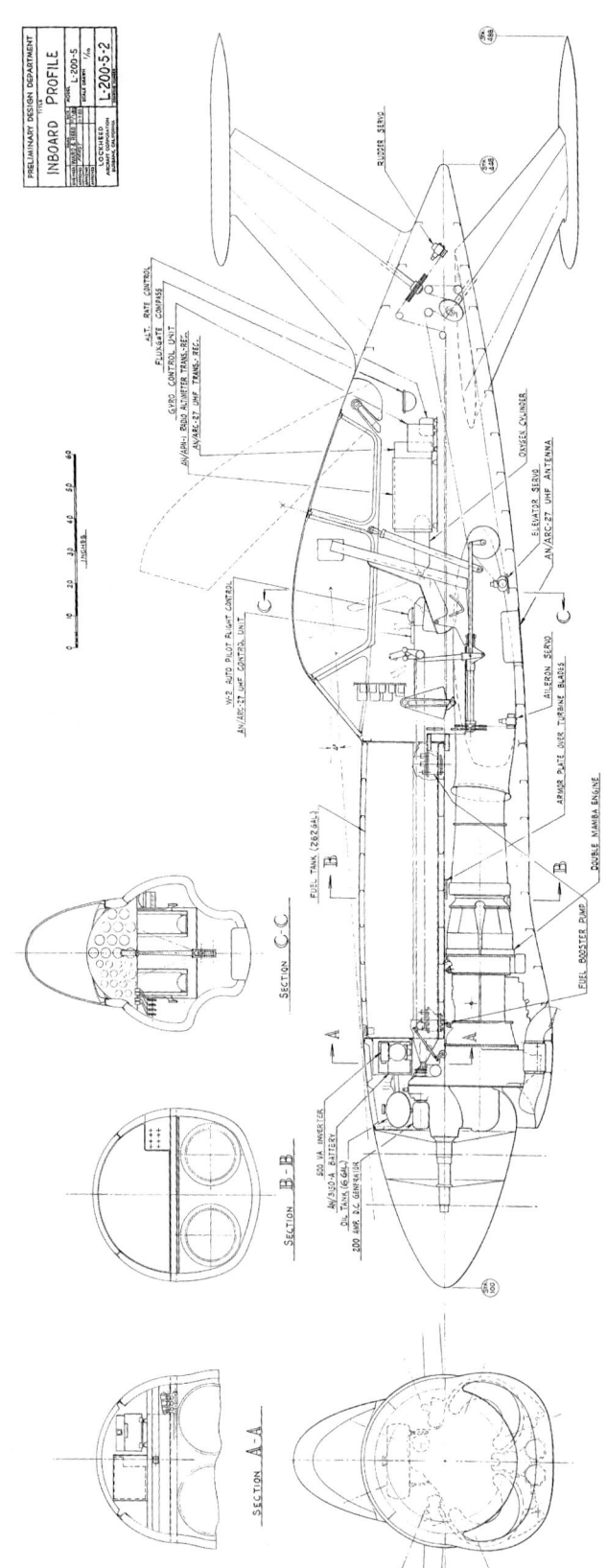

Inboard profile of the 0.766 scale prototype of the Lockheed Convoy Fighter powered by the Double Mamba.

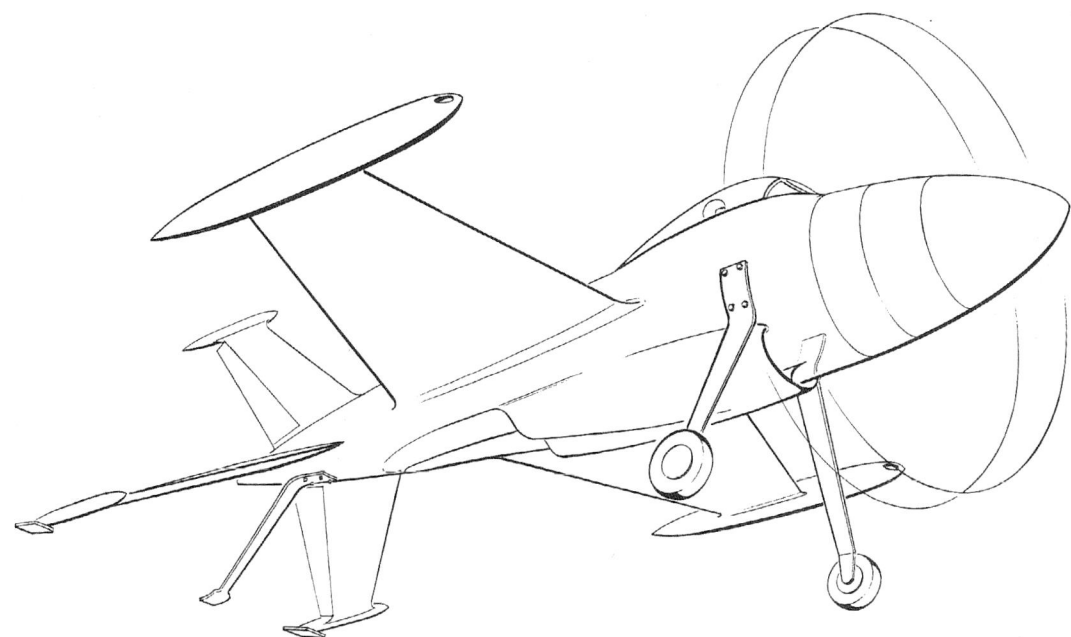

NORMAL LANDING GEAR FOR PROTOTYPE

Perspective diagram of a normal landing gear which could be installed on the L-200-02; it was similar but not identical to what was used on the actual XFV-1.

rig at the attitudes likely to be experienced. This programme might have easily included Item 5d above as part of this power plant ground test.

7. Tethered flight at vertical attitudes—Lockheed provided an artist's conception of a possible rig for simulating vertical flight under controlled conditions with the full-scale prototype. This procedure was used with success on the free flight models which had already been tested and appeared to offer considerable advantage and maximum safety. The purpose for tethered vertical flight testing was pilot familiarity with airplane attitude and checkout of the autopilot full-scale to determine whether or not carefully controlled airplane attitudes could be held. It was possible, however, that a careful approach to vertical flight at relatively high altitudes with the prototype and the normal gear could eliminate the entire portion of a flight test programme.

8. Flight tests with normal landing gear—having proven the ability of the airplane components to operate under the probable changes in attitude which would be required, flight research could begin with a normal landing gear. These tests would include:
    a. Transition flight development.
    b. Stability and control in normal flight (rough approximations only).
    c. Stability and control in vertical flight.
    d. Development of autopilot controls to maintain vertical flight.

9. Flight Research with Vertical Landing Provisions—Having satisfactorily demonstrated that the airplane could be flown through transitions and maintained under control in a vertical attitude, it was satisfactory to install landing provisions for vertical ascent and descent, thus eliminating the external fixed gear. This flight test programme would demonstrate the following:
    a. Vertical rising flight with ground castoring gear and possible externally mounted 'safe' struts to prevent tip-over.

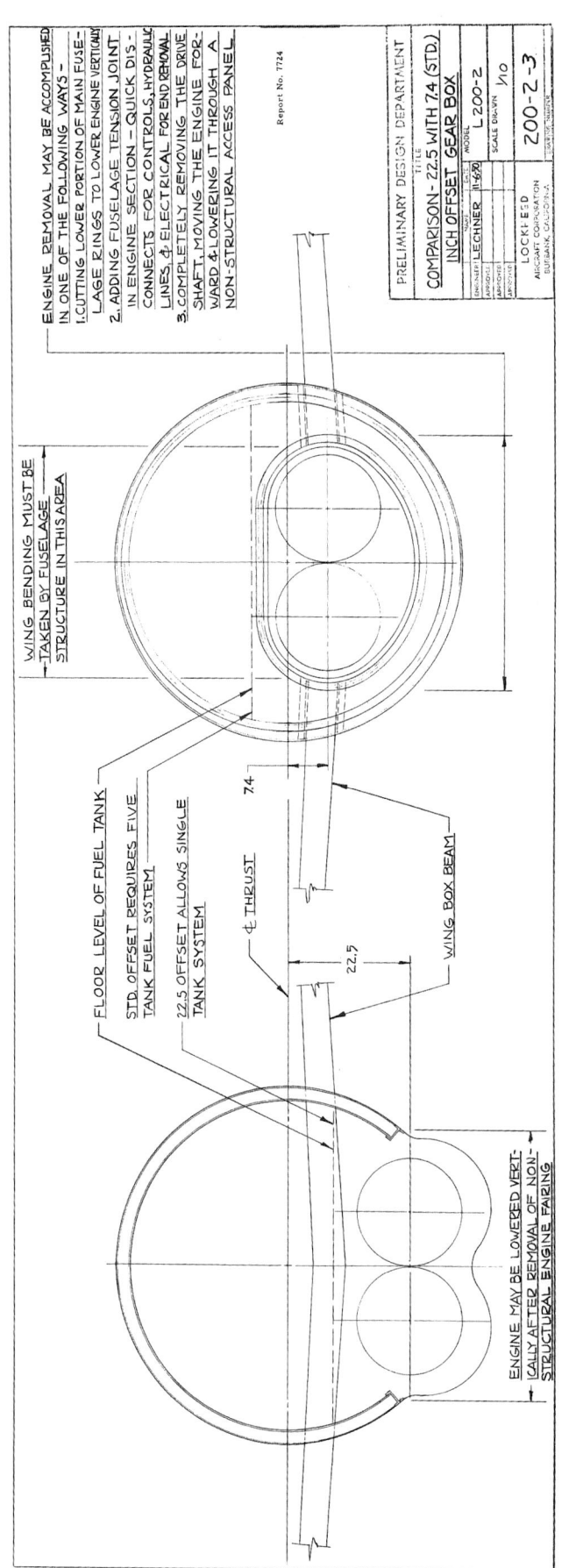

Drawing showing the effect of the low or high offset gearbox position on the basic fuselage structure of the Convoy Fighter prototype, with the latter being the preferred option.

b. Vertical rising flights without safe struts but including castoring wheel gear.
　　c. Complete flight research to develop accuracy of positioning on landing area and proper adjustment of autopilot flight controls.
10. Vertical rising flight research with final gear configuration and landing platform.
　　a. Detailed research on airplane performance and normal flight stability and control.
　　b. Detailed research on airplane systems installed in the prototype if applicable to final tactical airplane.
　　c. Equipment flight tests prior to tactical installation of electronic components, etc., if the full-scale airplane prototype was used.

Major comparison of T40 and Mamba time schedule

Lockheed calculated comparative timespan performance for the entire development programme using the two types of prototype airplanes. The company drew up two programmes: Programme A being that suggested in Navy letter Aer-AC-21 in which the Mamba prototype was developed, and Programme B being a Lockheed proposal based on the assumption that a gearbox could be made available for prototype flight testing with the maximum desired offset in 15 months.

As shown by comparing these two programmes, a production airplane could be expected nine months sooner if Programme B was followed. Even under the most severe conditions of gearbox availability, the programme with a T40-powered prototype would make possible a production airplane at the same time as with the Navy programme. Using the scheduled development of Gear Box No. 4 at 21 months, the T40 programme would make possible a production article three months earlier than Programme A.

Conclusions

Lockheed's analysis showed that a satisfactory prototype airplane for demonstrating vertical rising flight, transition between vertical and horizontal flight, and vertical descent to landing could be made, following either the Navy proposed programme or a programme suggested by Lockheed which used a full-scale Allison T40-powered prototype airplane. It was concluded that a reasonable series of research programmes could be outlined which would safely assure the success of the final programme, and that these could be consummated within time spans previously outlined.

It was the final conclusion of the Lockheed studies that a vertical rising convoy-based airplane was extremely advantageous to develop and would open many fields now closed to the use of an airplane in tactical problems.

COST PROPOSAL

Lockheed's Convoy Fighter cost proposal letter was dated November 28, 1950. In the document, it recommended that BuAer abandon the design and development of the 0.766 scale prototype airplane in favour of the company's alternate proposal, which was the design and development of full-scale prototype airplanes powered by the Allison T40-A-6 and their later conversion to tactical airplanes equipped with the more powerful T40-A-8. However, Lockheed was willing to undertake the Convoy Fighter project on the basis of either programme as determined by the Navy.

Lockheed presented two programmes to the Navy with differing schedules. Programme A involved the design and construction of two scaled-down prototype airplanes and two full-scale experimental convoy fighters in accordance with the requirements of OS-122:

1. Part One—scaled-down prototype airplanes—first scaled-down prototype airplane could be available for flight within 18 months after receipt of authorization to begin Phase One of Part One.
2. Part Two—experimental Convoy Fighter—first full-scale experimental Convoy Fighter could be available for flight within 41 months after receipt of authorization to begin Phase One of Part One, while the second airplane could be available for flight within 45 months.

Programme B involved the design and construction of two full-scale prototype airplanes and later conversion to tactical airplanes in accordance with Lockheed's alternate proposal for the prototype programme:

Diagram summarizing Allison's estimates of the types of gearboxes which could be made available along with the weights and times required; Gear Box No. 3 was preferred by Lockheed for both the prototype and tactical airplane.

1. Part One—full scale prototype airplanes—first full-scale prototype airplane could be available for flight test within 18 months after receipt of authorization to begin Phase One of Part One.
2. Part Two—converted prototype airplanes to tactical units—first full-scale tactical airplane could be available for flight test within 32 months after receipt of authorization to begin Phase One of Part One, while the second airplane could be available for flight within 36 months.

In presenting its price proposals Lockheed submitted estimates on both a cost plus fixed fee (CPFF) and a fixed price basis. Each proposal was broken down by phases and by items of work. In addition, its proposals were broken down into Prototype Programme Proposals A and B as described above. A summary of the costs by Programme is presented below, inclusive of a 2% fixed fee and 5% profit, as applicable:

BuAer Programme

Price Proposal A-1 — CPFF Basis	$7,596,066
Price Proposal A-2 — Fixed Price Basis	$8,992,400

Table comparing the weights of the full-scale and 0.766 scale prototype Convoy Fighters.

Lockheed Programme

Price Proposal B-1 — CPFF Basis	$5,679,575
Price Proposal B-2 — Fixed Price Basis	$6,723,613

LIGHT WEIGHT DESIGN POLICY

On February 23, 1951, Lockheed vice president and chief engineer Hal Hibbard sent a memo to his company's engineering staff regarding Lockheed's policy for light weight design; a copy was also sent to BuAer, perhaps to underscore the company's commitment to meeting the weight requirements of the Convoy Fighter and other programmes.

According to the memo, intense competition between Lockheed models and those of other manufacturers as well as the increasing demand for the highest possible performance for military needs necessitated a new general policy to be followed in order to improve the company's designs from the weight standpoint. In the past, Lockheed had followed a philosophy in structural, functional and equipment of design which resulted in decisions that minimized the risk in static tests, in production, etc., frequently at the expense of added weight.

Consequently, Hibbard wished to have the Engineering Branch of Lockheed adopt a more progressive judgment in the selection of materials, structure, equipment, etc. and be more courageous in choosing design alternatives in the interest of saving weight.

Naturally, this could be carried to unwise extremes and Lockheed's products would fail competitively in other ways. Therefore, to help engineers evaluate the degree to which this philosophy should be applied and to illustrate the kind of action required to implement this policy, the following examples were cited:

1. Lockheed abandoned the idea that a 'follow-on' version of an airplane had to be structurally identical in every respect to its

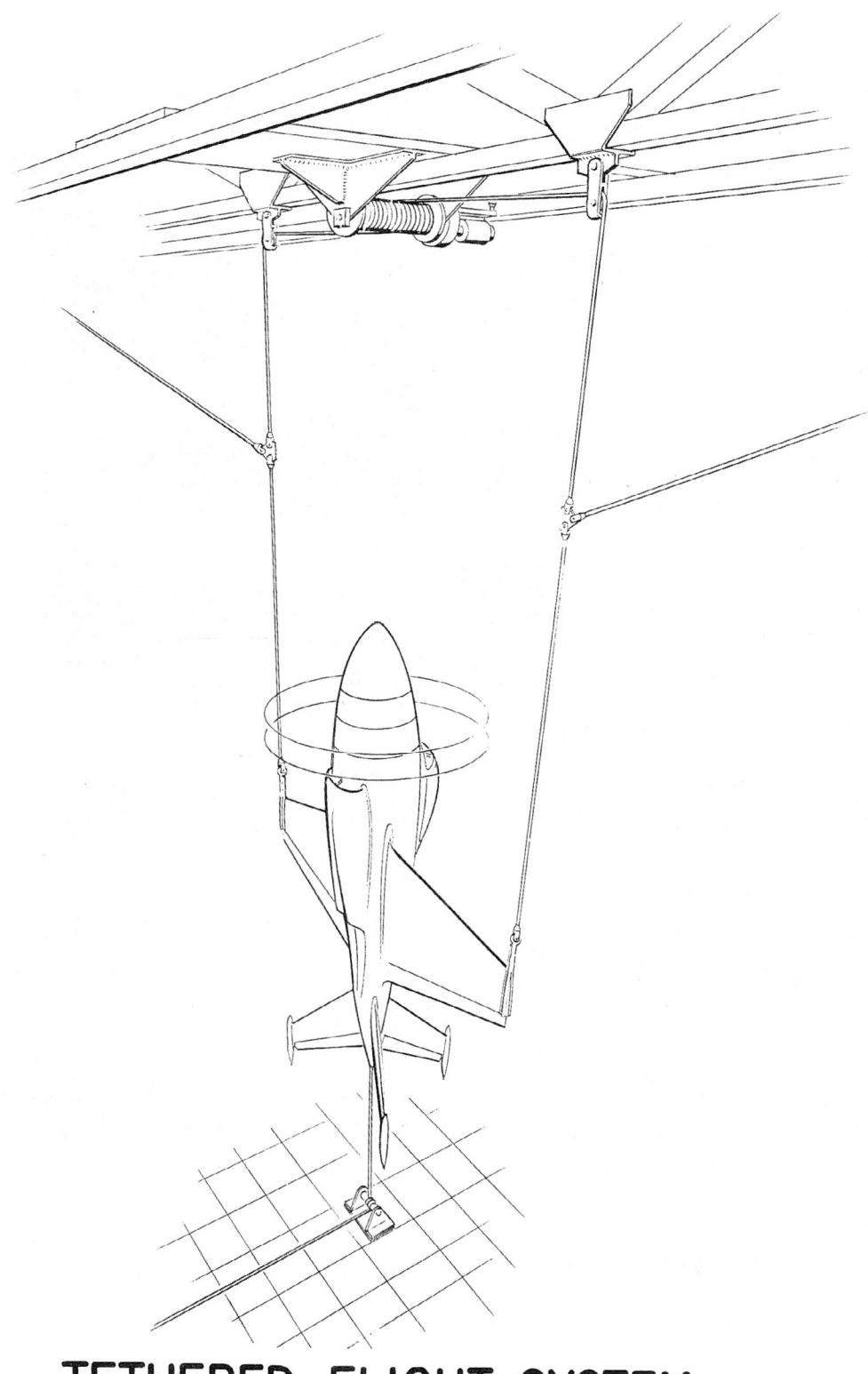

TETHERED FLIGHT SYSTEM FOR PROTOTYPE

Artist's impression of a possible rig for simulating vertical flight under controlled conditions with the full-scale prototype.

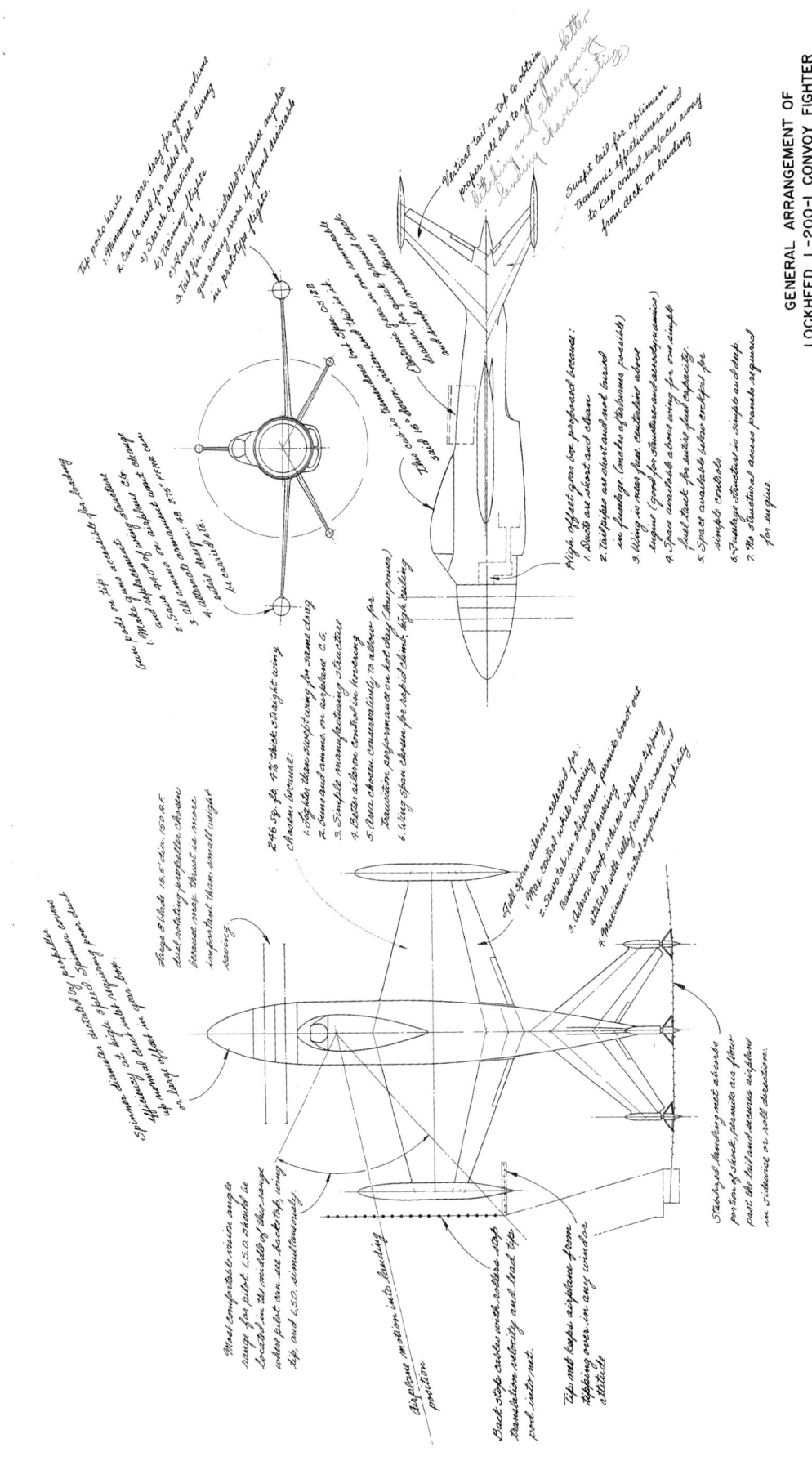

Annotated general arrangement of the L-200-1 Convoy Fighter which accompanied Lockheed's November 1950 correspondence with BuAer.

predecessor. It would design each model according to the best information at hand, making it lighter if the facts justified it.
2. It would design with the understanding that in static test some structural failures and subsequent changes were tolerable.
3. It would design structurally so that there were narrower margins of safety and fewer tests that carried well above 100%.
4. It would select design alternatives and equipment with closer adherence to the 'value of a pound' criterion.
5. It would make increased but judicious usage of magnesium, high strength, steel, titanium, sandwich construction, etc., fully aware that these decisions could involve a little more effort on its part.
6. It would give more judicious consideration to nebulous and often far-fetched contingencies regarding possible service troubles, possible production troubles and possible customer whims.
7. It would not go out of its way to 'dream up' compound improbabilities that necessitated structural or functional weight to provide for.
8. It would give more consideration in the selection of its aerodynamic parameters, design loads, functional test conditions and flight test extremes.
9. It would make more of any effort to induce its equipment vendors to reduce the weight of their equipment which was to be used on Lockheed airplanes.

The responsibility for the lightest practicable design was incumbent upon every individual and weight had to be given the fullest consideration.

CONTRACT AWARD

Lockheed was awarded a contract on April 19, 1951, for two full-scale prototypes under the designation XFO-1; this would later be changed to XFV-1 when BuAer's code for Lockheed was changed from O to V. BuAer's likely rationale in selecting both Convair and Lockheed's proposals for further development is covered in Chapter 7, along with some details of the XFV-1's preliminary development as well as a summary of its brief flight testing programme and subsequent cancellation.

A speculative colour profile of the L-200-1, the tactical version of the Convoy Fighter. The overall Glossy Sea Blue scheme was the standard for most US Navy aircraft of the early 1950s; markings are inspired by the actual XFV-1 prototype.

A speculative colour profile of the Lockheed Model 200-6, an alternate configuration studied in the preparation of the Convoy Fighter proposal. It featured a swept wing and inboard gun pod.

Speculative colour profile of the Model 200-7, a canard design with a shortened fuselage. Both it and the Model 200-6 were judged to be inferior to the relatively conventional straight wing configuration ultimately chosen.

Speculative colour profile of the Lockheed L-200-5, a reduced-scale prototype of the Convoy Fighter concept powered by an Armstrong Siddeley Double Mamba turboprop engine. Initially part of the Navy's requirements, the L-200-5 was ultimately dropped in favour of a stripped version of the full-scale Convoy Fighter, the L-200-2.

5

Martin Model 262 and 262P

MARTIN DEVOTED three and a half months of serious engineering effort toward a study of the problems associated with the development of a vertically rising fighter designed to be based on a merchant ship and to supply air defence for the convoy. These studies confirmed the feasibility of developing a Convoy Fighter that would meet the intent of the requirements set forth for the airplane, leading Martin to the conclusions drawn below.

Martin believed that a large amount of quantitative data, which were not available at the time of the competition, needed to be obtained before deciding on an optimum configuration which would ensure success. The required information was basically aerodynamic and was two-fold; first, the effect of a propeller absorbing unusually high horsepower on the horizontal flight characteristics of the airplane, and second, the aerodynamics of the airplane in essentially vertical (hovering) flight. These data could be obtained through a well-planned wind tunnel programme.

Martin believed that a satisfactory method for launching and recovering the aircraft from a merchant ship could be developed. The optimum method depended to a great extent on the information obtained during the flight testing of the prototype (scale model) airplane. These tests would supply data on hovering flight characteristics and control response, and permit the development of techniques upon which a sound recovery method could be evolved.

Martin believed that the utilization of this kind of aircraft was not limited to the role of Convoy Fighter. Its tactical potential in many other phases of naval warfare was tremendous, and Martin was intensely interested in becoming associated with its development.

General overview

During the course of preparing the proposal, Martin's primary objective was to develop as complete an understanding of all aspects of the problem as possible. In so doing, many different configurations were investigated. Of all the various possibilities studied, the Model 262 appeared to be the best solution to the problem based on the data available at the time the proposal was submitted.

Certain considerations, however, indicated the desirability of further investigation of modified arrangements. Martin proposed investigating the most promising of these modifications in the wind tunnel to ensure the selection of the optimum aerodynamic arrangement. These modifications are shown later in this chapter, along with a discussion of their outstanding features.

The basic Martin Model 262 was a swept-wing airplane powered by a 7,500 horsepower XT40-A-8

Colour cover to the proposal for the Martin Model 262 Convoy Fighter dated November 17, 1950.

turboprop engine driving 16ft contra-rotating propellers. The development of the gas turbine engine to this equivalent shaft horsepower output made it possible for the airplane to hover by supporting its 16,890lb of weight entirely on the thrust of the engine. A suitable device was designed to be mounted on a merchant ship to launch and recover the airplane in its hovering attitude.

The mission of the Model 262 was to protect convoy vessels from enemy air attack. To do this, the airplane was capable of making a transition from the hovering (vertical flight) attitude to a horizontal attitude where it was designed to operate close to sonic speed at attitudes up to 45,000ft. It carried a single pilot and was equipped with an ejection seat in a pressurized cockpit. Its armament consisted of four 20mm cannon. Fire control was provided either by a visual computing sight or by use of the radar mounted in the aircraft which was capable of both search and tracking operations. Automatic control devices were provided to assist the pilot. The airplane was designed to be readily adaptable to high rate production with special emphasis on ease of maintenance.

Launch & recovery

In studying the operation of the Convoy Fighter from a merchant ship, it became apparent that the problems involved in launching the aircraft were far less complex than those encountered in recovery. Martin's efforts were therefore directed toward solving the recovery problem. The method which Martin evolved permitted launching the aircraft from the same equipment that was used for the recovery.

Recovery

The feasibility of any recovery method depended primarily on the degree of control that the pilot had over the airplane in hovering flight. If it was assumed that he had perfect control with instantaneous response, it was possible to land the airplane on its tail on a platform area with little or no auxiliary equipment on the surface ship. If it was assumed that pilot control and response rate were poor, recovery methods involving elaborate deck equipment were required. Martin's proposed recovery method was about midway between these two extremes. The company believed that proof of the ultimate method depended on flight tests and techniques that would be developed on the prototype.

The landing area was defined by specification as being 200ft aft and 25ft above the pitch and roll centre of the ship. The specification also required that a landing be made while the ship was rolling at

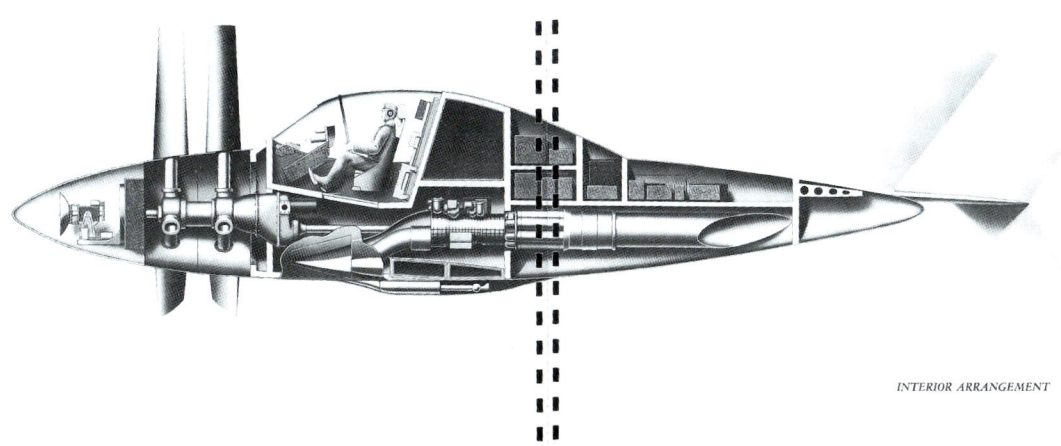

INTERIOR ARRANGEMENT

A simplified inboard profile of the Martin Model 262 showing the internal layout of this remarkable VTOL turboprop fighter.

PHYSICAL CHARACTERISTICS

DESIGN WEIGHTS

TAKE-OFF GROSS WEIGHT	16,890 LBS.
FUEL (500 GALLONS)	3,000 LBS.
COMBAT WEIGHT	15,690 LBS.
LANDING WEIGHT	15,090 LBS.
WEIGHT EMPTY	12,665 LBS.

DIMENSIONS

SPAN	31.5 FT.
LENGTH	44.7 FT.
HEIGHT	16.0 FT.
WING AREA	247 SQ. FT.
WING ASPECT RATIO	4
SWEEP BACK OF 40% WING CHORD	45°

+/-15° and pitching +/-4°. This pitch and roll induced a total vertical motion of the landing area of 28ft and a sideways motion of 13ft. Martin's recommended recovery method could be accomplished in the specified landing area or in an area immediately aft of amid-ships adjacent to the No. 4 hatch. This latter location was recommended since it was only 80ft aft of the centre of rotation in pitch where the vertical motion induced by pitch was only about 11ft.

Surface ship equipment

A platform was to be mounted vertically at the landing area of the ship. This platform was hinged about a fore and aft axis and was power operated through a rack and pinion. It was stabilized in such a manner that no sideways motion was imparted to the point 'A' by roll of the ship.

The point 'A' was in the centre of a contact area approximately 10ft on a side. The contact area consisted of a series of vee-shaped welded steel members which extended from the top to the bottom of the area. A series of shock absorbing arresting cables were stretched horizontally across and behind these vertical members. The remainder of the platform was covered with light steel grating with a reinforced region below the contact area. A protected station for a landing signal officer was situated on top of the platform.

A retractable spike was located just aft of the propeller plane and two retractable wing gears were located on each wing panel. The spike consisted of a short stroke shock absorbing unit with a pointed end equipped with a spring loaded latch which locked the spike in the contact area of the platform when it was pushed through the opening between

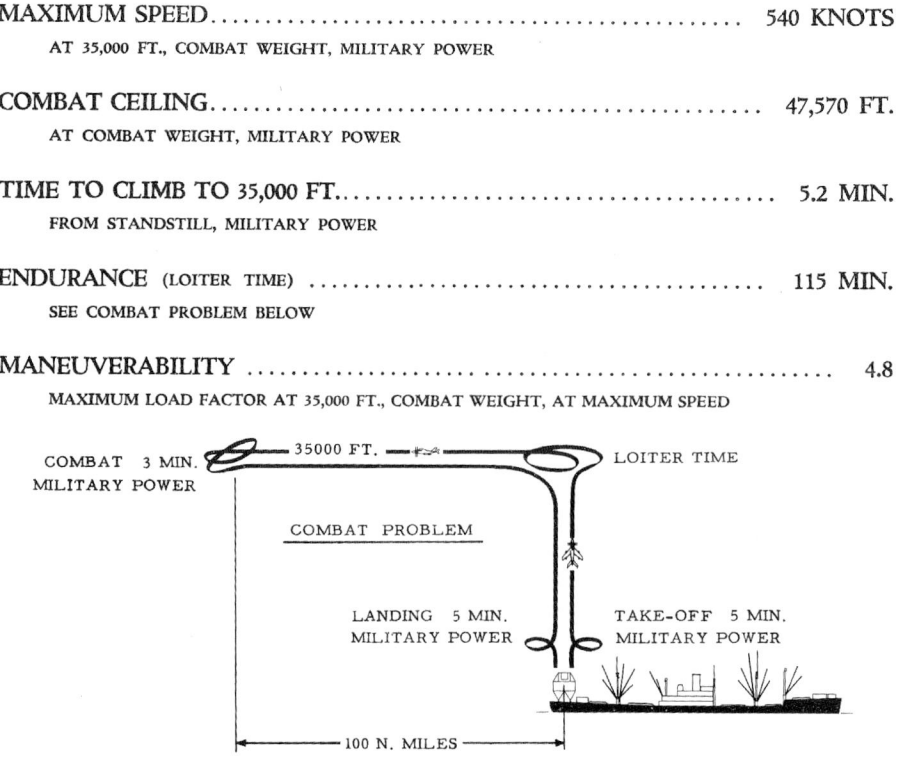

MAXIMUM SPEED	540 KNOTS
AT 35,000 FT., COMBAT WEIGHT, MILITARY POWER	
COMBAT CEILING	47,570 FT.
AT COMBAT WEIGHT, MILITARY POWER	
TIME TO CLIMB TO 35,000 FT.	5.2 MIN.
FROM STANDSTILL, MILITARY POWER	
ENDURANCE (LOITER TIME)	115 MIN.
SEE COMBAT PROBLEM BELOW	
MANEUVERABILITY	4.8
MAXIMUM LOAD FACTOR AT 35,000 FT., COMBAT WEIGHT, AT MAXIMUM SPEED	

The key performance figures of the Martin Model 262 along with a diagram of its principle mission.

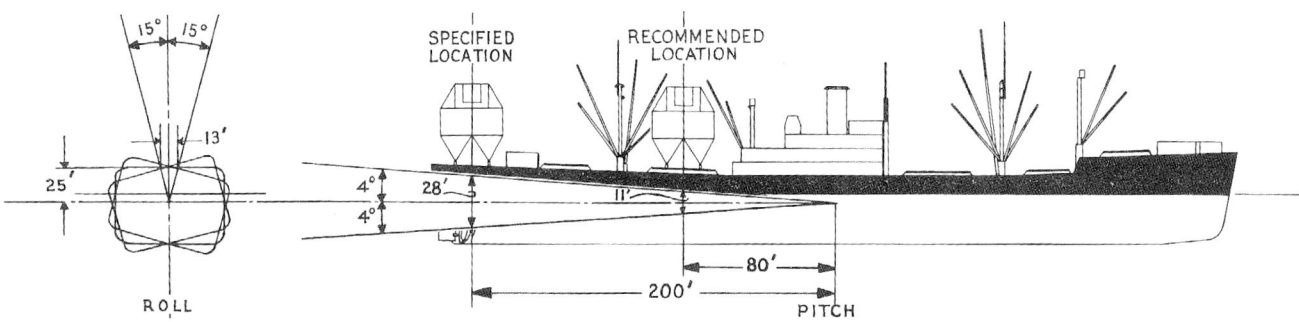

Martin proposed a hinged vertical launch/recovery platform for its Convoy Fighter; in a deviation from the original specification, the company recommended locating the platform in an area immediately aft of amidships to reduce the pitching moment of the vessel to make launches and recoveries of the aircraft safer.

the vertical vee-shaped members on the platform.

The pilot's seat was rotated through 45° when in the hovering attitude to place the pilot in a comfortable position for the vertical flight attitude and enable him to look down and over the sides of the airplane. A window in the floor of the airplane permitted him to see the end of the spike directly for final alignment prior to contact.

Recovery method

The landing was accomplished after making a transition from the horizontal to the vertical flight attitude, by an approach from aft of the ship with the wings approximately aligned with the fore and aft axis of the surface ship. Under these conditions, the pilot's view of the ship was ideal since the ship was out to the side and slightly below the airplane.

The pilot manoeuvred the airplane to a position just outboard of the landing platform. Since the platform was stabilized so as to transcribe no lateral motion to the point 'A' due to roll of the ship, and since the ship's speed was constant, the only significant motion of the platform relative to the airplane was essentially along the vertical axis of the airplane and was induced by pitching of the ship.

With this situation, the pilot was able to maintain a position prior to contact such that the spike was just outboard of the platform. Under normal sea conditions when the pilot was ready to make contact, he would pitch the airplane gently toward the platform, so as to fly the airplane toward the platform. The spike could strike the platform anywhere in the 10ft by 10ft contact area. The vee-shaped vertical members guided the spike into the slot between the members. When the spike had fully penetrated the slot, a spring loaded latch locked it in the slot and sent a signal which either cut the power automatically or told the pilot to do so. During heavy seas, the pilot would hover at constant altitude just outboard of the platform until he could anticipate the relative vertical motion of the platform. With the assistance of the signal officer, he would engage the contact area as the ship came up to meet him near the top of its vertical motion.

If the spike missed the contact area and struck the grating around it, it would tilt the airplane's nose away from the platform, since the spike was above the airplane centre of gravity. This would cause the airplane to move away from the platform and permitted a second approach.

With the power cut and the spike engaged, the airplane was allowed to move down the platform about 3ft as its vertical motion was reduced to zero by one of the horizontal arresting cables that was engaged by the spike. During this motion the wing gears contacted the platform, and the recovery was complete. The platform was then rotated down to a horizontal position and the airplane was repositioned for takeoff and secured.

Launch

To position the airplane for takeoff, the arresting cables were relaxed and the airplane was manoeuvred so as to slide the spike to the top of the slot where it engaged a quick-release mechanism.

The pilot entered the airplane, checked it out, and started the engines while the platform was in the horizontal position. Just prior to takeoff, the platform was rotated to its vertical stabilized position. As power was applied, the airplane raised off the

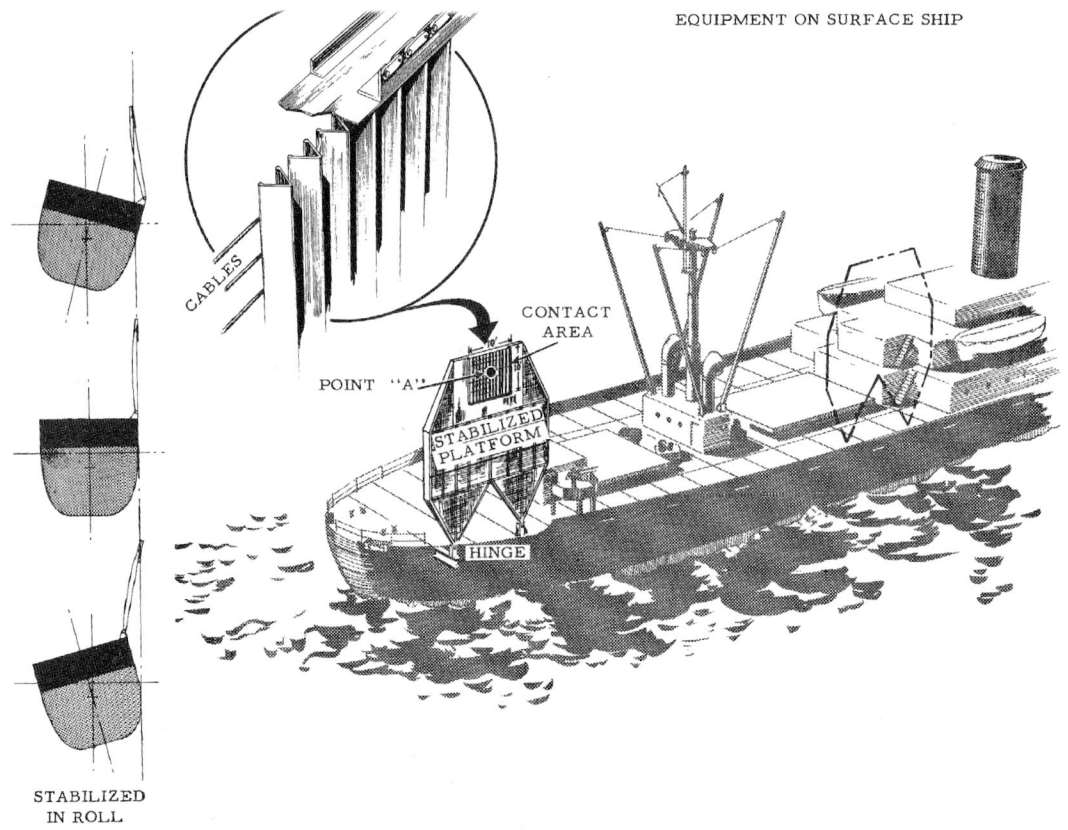

Illustration showing the platform mounted vertically at the landing area of the ship; it was hinged about a fore and aft axis and was power operated through a rack and pinion.

wing gears, the spike was released from the quick release mechanism by the pilot or signal officer, and the airplane left the platform.

When takeoff or recovery was made with the relative wind coming in from the side of the ship, the platform would be stabilized in a position slightly off the vertical in order to match the hovering attitude of the airplane.

To permit the airplane to be rolled around on the deck or dock areas, the spike was designed to accommodate a small wheel and jack assembly. When attached to the spike, this assembly permitted the airplane to be jacked up to support the weight on the wheel.

For training and shore based operations, the platform could be mounted on a trailer which served as its takeoff and recovery vehicle as well as for ground handling. Martin believed that this recovery method could be developed to fulfill the intended purpose, though additional thought and analysis was required. One contemplated refinement was to study the means of providing shock absorption in the platform that was more than just the vertical direction as provided by the horizontal arresting cables. If this were done, the shock absorbing units in the spike and wing gears on the airplane could have been shortened to reduce the alighting gear weight carried in the airplane.

AERODYNAMICS

During the preparation of the Convoy Fighter proposal, major emphasis was placed on the recognition of the aerodynamic problems involved and on studies of the many solutions to these problems. The mission of this airplane was such that it was extremely important to have an exceptionally sound basic aerodynamic design. The attainment of the optimum tactical configuration required a considerable amount of basic analytical aerodynamic studies, wind tunnel tests, and eventually flight tests of a prototype to solve those problems not previously encountered on an aircraft.

While it was feasible to design a vertically rising aircraft which was also capable of obtaining speeds approaching that of sound, a complete understanding of the various problems associated with this airplane was necessary at the onset if the design was to proceed on a sound basis.

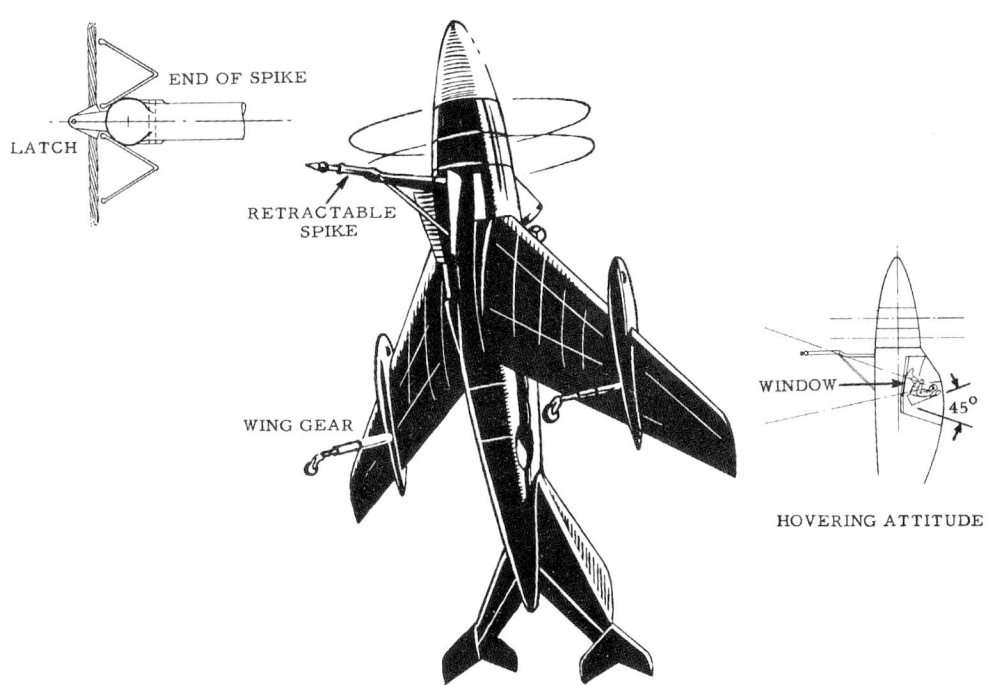

The Martin Model 262 was equipped with a retractable spike located aft of the propeller plane and two retractable wing gears located on each wing panel; this unusual undercarriage enabled the design to attach itself to the vertical landing platform.

Since the most critical feature of the airplane concerned its flying qualities, primary emphasis was placed on the stability and control characteristics of the airplane in hovering, transition, and in level flight attitudes. A summary of the more important aerodynamic considerations followed.

One means of obtaining the desired stability characteristics during some or all modes of flight was through the use of automatic electronic and mechanical devices. It did not appear advisable during the preliminary design stages to base a configuration entirely upon the use of automatic stabilization. Consequently, Martin's approach was to develop a configuration which was inherently stable aerodynamically and not to rely on the use of synthetic stabilizing devices. These devices could have been used, but they would not have been employed to refine the stability characteristics of the airplane and would not have been used in lieu of good aerodynamic design.

Hovering

During hovering, it was desirable to have the airplane tend to maintain a vertical attitude even in gusty air. To accomplish this, it was necessary that the airplane not tend to weather-cock (turn) into the relative wind. If the airplane had a large tail, a horizontal gust would rotate the airplane into the relative wind. The airplane would then begin to translate, tending to fly into the gust.

With a smaller tail, it was possible to make the airplane neutrally stable during hovering. In this case the airplane would not tend to rotate, but would only translate when subjected to a horizontal gust. After the gust subsided, the translation would also subside.

Thus, during hovering, no strong tendency to weather-cock into the wind was tolerated. However, in the normal level flight attitude, it was necessary that the airplane weather-cock into the relative wind if it were to be stable. Martin therefore concluded that a large tail effect was desired during level flight while a small tail effect was desired during hovering.

For longitudinal stability, two effects could be used to make the hovering and level flight characteristics compatible. These were:

1. The downwash behind the wing was much greater during hovering than in level flight. This resulted in a desirable decrease in the stabilizing contribution of the horizontal tail.

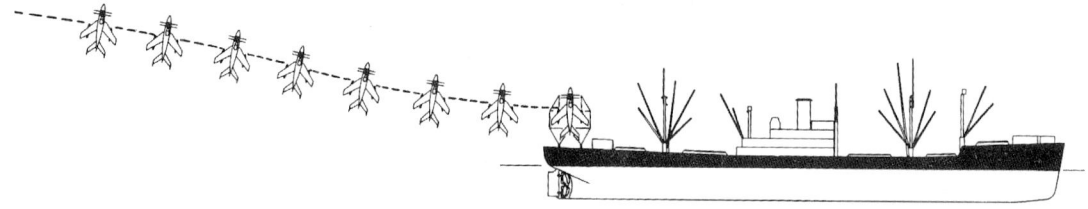

Landing the Model 262 was accomplished by transitioning from the horizontal to the vertical flight attitude, approaching from aft of the ship with the wings approximately aligned with the fore and aft axis of the vessel.

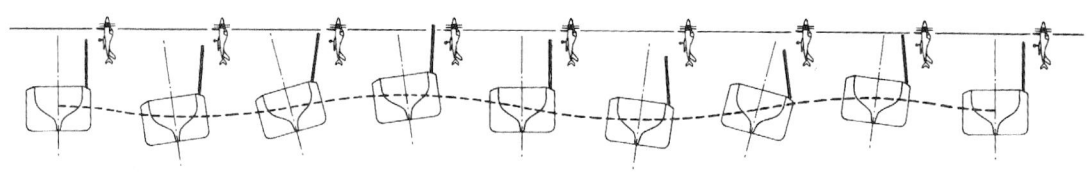

Stabilization of the platform minimized lateral motion caused by roll of the ship; in landing the aircraft, the pilot had to contend mainly with vertical motion of the platform caused by pitching of the ship.

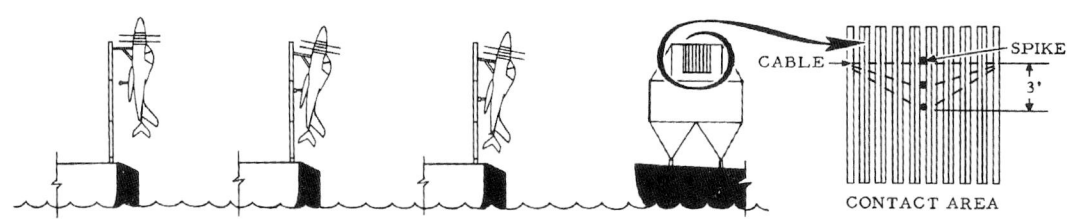

With the power cut and the spike engaged, the Model 262 moved down the platform about 3ft as its vertical motion was reduced to zero by one of the horizontal arresting cables that was engaged by the spike; at the same time, the wing gears made contact with the platform, completing the recovery.

After a successful recovery, the platform was rotated down to a horizontal position and the Martin Convoy Fighter was repositioned for takeoff and secured.

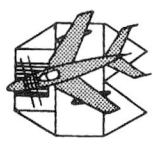 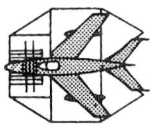 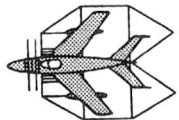 

In positioning the Model 262 for takeoff, the arresting cables were relaxed and the airplane was manoeuvred so as to slide the spike to the top of the slot where it engaged a quick-release mechanism.

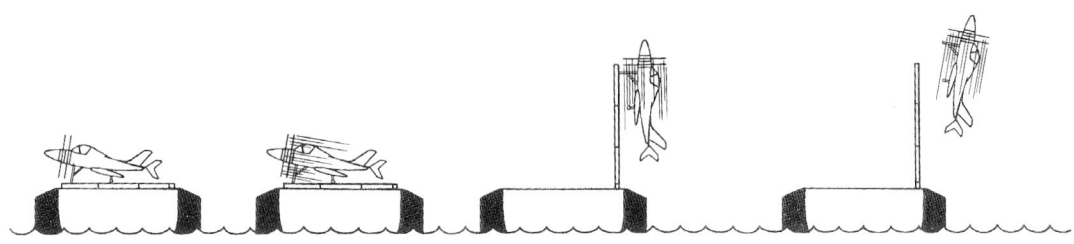

Graphic showing the takeoff sequence of the Model 262 leaving the hinged platform.

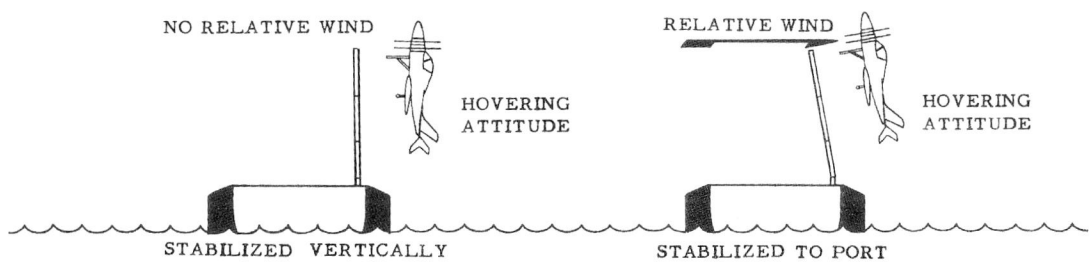

When takeoff or recovery was made with the relative wind coming in from the side of the ship, the platform would be stabilized in a position slightly off the vertical in order to match the hovering attitude of the airplane.

2. With a sweptback wing and the propeller in front of the wing, the slipstream acted only over the central portion during hovering. Since the wing was swept back, the centre of lift moved forward relative to its position in level flight. The result was a desired decrease in the tendency to weather-cock into the wind during hovering.

Actually, the above two aerodynamic effects may have been too great and resulted in an undesirably large tendency to weather-cock away from the wind during hovering. Wind tunnel tests were required to determine the proper balance.

In the directional case the above two aerodynamic effects did not exist and special consideration had to be given to the vertical tail design. It appeared very possible to make the level flight and hovering directional characteristics compatible by locating a portion of the vertical tail outside the slipstream. This was accomplished by the use of three vertical tails, one centrally located with the other two on the edge of the slipstream. Thus, in hovering, only the central vertical tail was effective while in level flight the entire tail was effective. If wind tunnel tests showed that large gusts caused the slipstream to drift over the leeward fin to an appreciable extent, it was possible to eliminate the contribution of this fin by the use of outwardly deflected tip rudders. This prevented the leeward fin from contributing to the tendency to weather-cock into the wing. These tip rudders, if required, would be deflected to a given position only during hovering.

During hovering, the control surfaces had to be

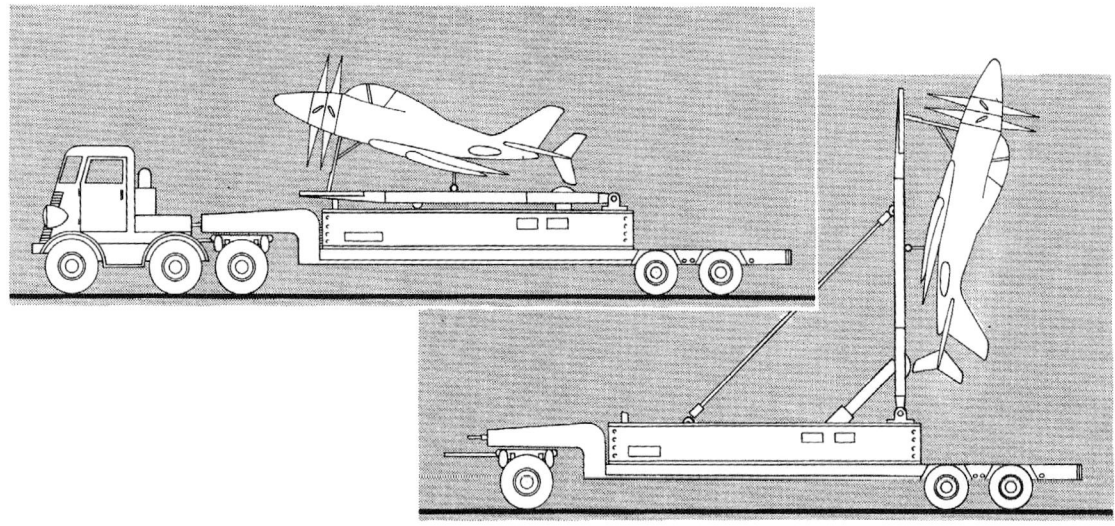

The platform could also be mounted on a trailer which served as a takeoff and recovery vehicle as well as for ground handling, enabling training and shore based operations.

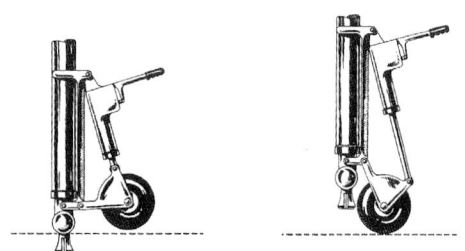

To permit the Model 262 to be rolled around on the deck or dock areas, the landing spike was designed to accommodate a small wheel and jack assembly.

located in the slipstream. The lateral control was seriously handicapped by this requirement since it was not possible to obtain a large moment arm. Thus, if inboard wing ailerons were used exclusively, it would have been necessary for them to develop large changes in lift. If the aileron were designed to produce the large changes in lift, however, it was expected that the resulting change in downwash at the tail would have tended to give a rolling moment opposing that of the ailerons.

While some available data indicated that the reversing effect of the tail did not exist, at least when the inboard ailerons produced a small rolling moment, it did not appear reasonable to ignore this possibility since other test data showed that the tail roll reversal effect did exist. Therefore, the latest control had been located on the tail. It was expected that the elevator, operated differentially, would have produced adequate roll control during hovering.

While the amount of control possible with the tail elevons may have been limited, the magnitude of the required roll control was not expected to be large. If future studies and tests showed that the proposed lateral control was marginal, some additional control could have been achieved by the use of inboard ailerons in conjunction with the tail elevons. Another possibility was the use of differentially operated all-movable tails.

Transition

The transition manoeuvre involved the change from vertical to horizontal flight attitude in takeoff and from horizontal to vertical in landing. A 'zoom' type transition involving a change in altitude in which the angle of attack was kept below the stall for the entire wing, or a level flight transition with no significant change in altitude, were the alternatives. Martin believed that the design philosophy should be directed toward the attainment of an airplane potentially capable of performing level flight transitions. Such an airplane could be capable of level flight at any speed from high speed down to zero speed and was superior to one that had to be manoeuvred in a predetermined fashion through a given range of speeds. If an airplane was capable of making level flight transitions it was also capable of making zoom transitions. Martin noted that the airplane had to be capable of level flight in the hovering attitude at least up to speeds of 35 knots even though zoom transition was used. This was necessary

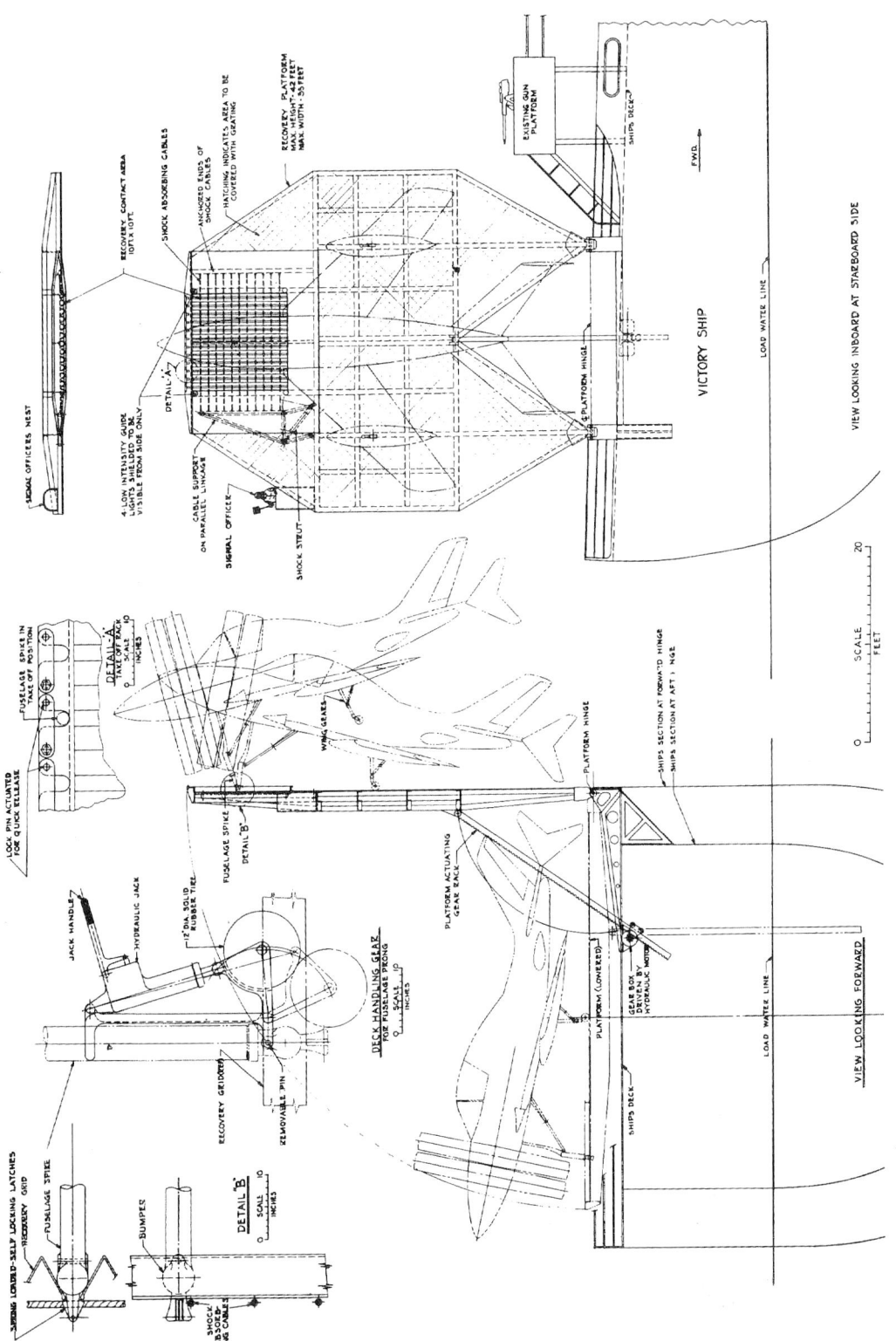

Detailed schematic of the hinged landing/recovery platform mounted on a merchant vessel with the Martin Model 262 attached.

in landing when the surface ship was heading into a strong head wind.

The design problem of making an airplane capable of level flight transition was more complicated since the angle of attack became large. The portion of the wing outside the slipstream would have been stalled during a level flight transition. No serious drag or stability complications were anticipated due to this fact. The primary effect would have been a possible buffeting of the airplane. Martin believed that this would not have been a serious disadvantage due to the low speeds involved. If necessary, several methods existed which could have been used to alleviate this condition. One method was the use of slats to keep the wing tips unstalled at least down to speeds where the aerodynamic forces became negligible. Another method consisted of varying the incidence of the wing tip. A third possibility was the use of a low aspect ratio delta wing which would have had only a negligible portion of the wing stalled.

An analysis of the airplane characteristics during a level flight transition was made for the proposed configuration including the effect of stalled wing tips. The angle of attack of the centre portion was below the angle of stall at all speeds due to the action of the slipstream, a desirable but not necessarily required condition. Although a reversal in the elevator position existed over a portion of the speed range, ample control was available. Wind tunnel tests might have shown that this elevator reversal could be eliminated by the use of wing and tail incidence in conjunction with propeller thrust moment.

In summary, it appeared that it was possible to perform the transition in essentially level flight. It followed that it was possible to perform zoom transitions, since the additional power required was not as high as that required for takeoff.

Level flight
The primary source of the aerodynamic problems in level flight for this airplane stemmed from the use of the propeller as the propulsive element. The power loading was much higher than normal and the usual power effects were greatly aggravated. This indicated that the propeller design had to be carried out to minimize the very large influence of the propeller on the trim and stability characteristics of the airplane, in addition to obtaining the highest possible propeller efficiencies.

Two propeller design philosophies were considered.

In the first, an essentially transonic propeller design was used. The propeller tip Mach number was maintained at approximately unity. This required a low propeller rpm, which in turn resulted in high blade angles. Two serious design problems resulted:

1. A gear-shifting mechanism was required to obtain the low rpm for high speeds and the required high rpm for takeoff and hovering.

2. The large size force developed by such a propeller caused a very serious stability problem. The effect on stability was similar to having a surface of approximately one-third the wing area located at the propeller position.

The above two problems were greatly alleviated by utilizing a supersonic propeller design philosophy. Most aerodynamic phenomena became critical at Mach 1, whereas improved flow condition existed at supersonic speeds. Therefore, it appeared desirable to design a propeller with a large portion of its blade operating above the speed of sound. To obtain this condition, it was necessary to increase the propeller rpm, making it feasible to eliminate the gear-shifting mechanism. The blade angle, of course, would have been reversed and it then became possible to solve the stability problem in a reasonable manner. Available test data indicated that the efficiency of a supersonic propeller was expected to equal that of the transonic propeller in the high speed range of this airplane.

Since the propeller fin effect was roughly proportional to the number of blades, a six-bladed propeller was preferred to an eight-bladed propeller. The changes in performance were not significant and Martin therefore strongly recommended a single speed gearbox driving a six-bladed supersonic propeller.

A new method was derived to obtain these relations since existing methods did not adequately cover compressibility effects. The derivation of the method was given in G.L.M. Engineering Report 4105, Considerations of the Dominating Propeller Interference Effects for a Convoy Fighter Design by H. Multhopp, November 16, 1950.

Even with a six-bladed supersonic propeller having minimum fin effect, relatively large tail sizes were required. This condition could be greatly relieved by locating the propeller in the centre portion of the fuselage near the centre of gravity of the airplane. A reduced frontal area, and improved visibility would have also been obtained with this

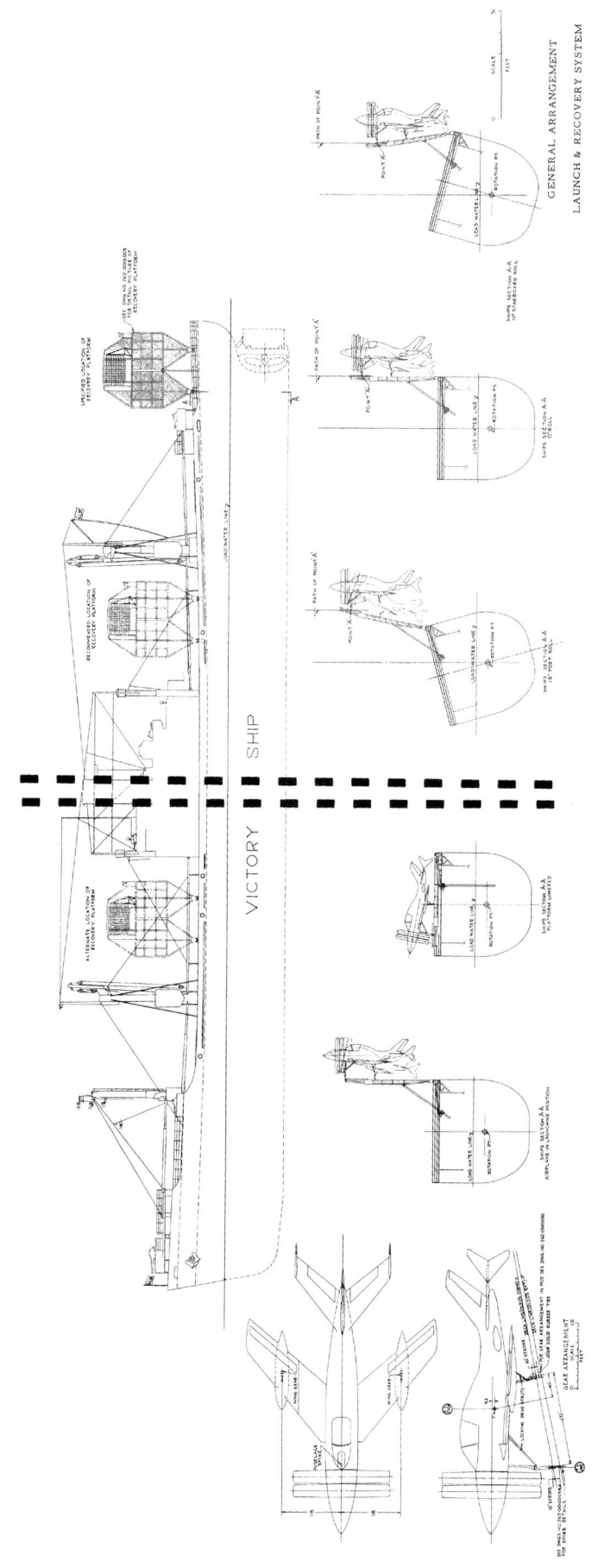

General arrangement of the Model 262's launch and recovery system showing various possible locations of the recovery platform on a Liberty Ship as well as how the platform could be adjusted to the rolling motion of the ship in rough seas.

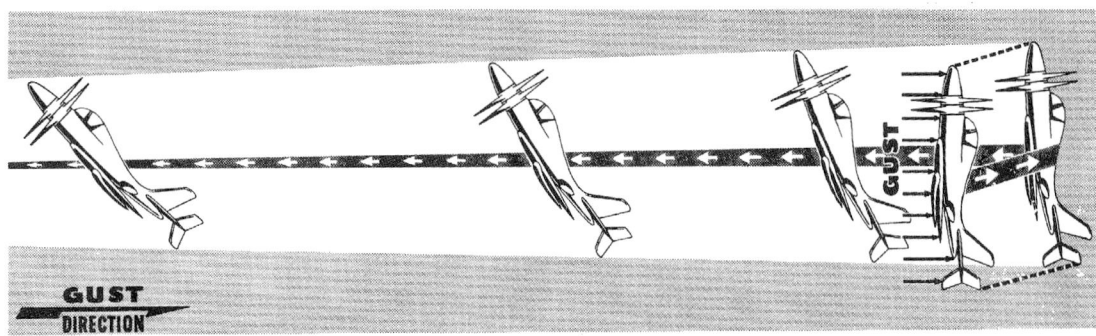

In gusty conditions, the tail arrangement of the Model 262 was designed to minimize the tendency of the vehicle to unintentionally turn into the relative wind while hovering.

arrangement. This required a propeller gearing arrangement which accommodated the structure necessary to connect the fore and aft portions of the fuselage.

In addition to the problem imposed by the propeller, the usual compressibility problems existed on this airplane. High speed rolling effectiveness, usually restricted by aeroelastic influences, was maintained in Martin's design by the use of spoiler ailerons which had but a fraction of the aerodynamic twisting moments of normal ailerons. This proposed lateral control for horizontal flight with its artificial feel had been used successfully at high speeds on the Martin XB-51. Use of a large amount of sweep and a thin airfoil section minimized the adverse effects of compressibility at transonic speeds. A low wing position was used to reduce the possibility of tail buffet at high speed. The low wing allowed a tail location above the wing and still within the slipstream during hovering.

To summarize, the mission of this airplane introduced design considerations not previously encountered. The time available for the preparation of this proposal had been expended in attempting to ascertain the nature of the basic problems and investigate possible forms of solution. The results of the study indicated that although analytical and wind-tunnel work was necessary, Martin was convinced that it was entirely feasible to develop a configuration which would perform the mission desired.

CONTROLS

Surface controls and boost

The surface control systems were in general of conventional design. Directional control was provided by foot-operated rudders with no power boost. Longitudinal control was provided by power-boosted elevators. The power boost was designed to cause the pilot to furnish a portion of the force which operated the control surface, thereby supplying aerodynamic feel and enabling him to control the pitch of the airplane in limited manoeuvres with the boost system inoperative. Lateral control was provided by differential movements of the elevators and by operation of wing spoilers. The lateral control surfaces were hydraulically operated and mechanically synchronized with the pilot's control, enabling him to operate the surfaces manually in the event of hydraulic failure.

Autopilot

Components of a conventional autopilot such as the Sperry E-4 were used with special sensing equipment to provide the freedom necessary for this aircraft. The servo system of the autopilot was used without modification. Trim servos were provided to keep the airplane control surfaces operating at zero hinge moment during autopilot control to prevent violent aircraft manoeuvres and sudden application of forces to the pilot's controls in the event of autopilot failure. A throttle servo was used to control thrust and to complete the set of autopilot functions necessary to provide for all modes of aircraft flight.

A formation stick type of autopilot hand control was located on the pilot's console. Use of this type of control permitted a pilot to manoeuvre the airplane through the autopilot if he so desired. In the event of power boost failure, the pilot could have combined his own efforts with those of the autopilot to control the airplane.

Automatic control

The autopilot could be used to hold the airplane in any attitude of horizontal or vertical flight and

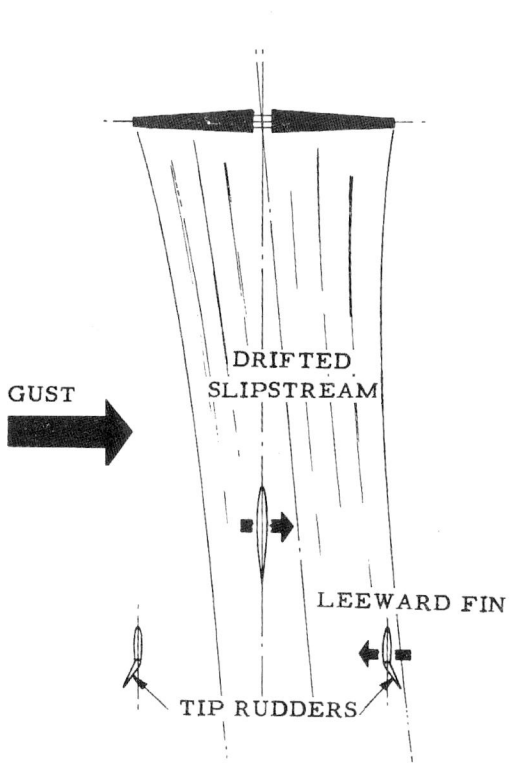

The Model 262's small tip rudders, if required, could be deflected to a given position during hovering to reduce the chance of the aircraft to 'weather-cock' in gusty conditions.

to make the automatic transition from vertical to horizontal flight. A controlled torque was applied to one axis of the autopilot's gyros which caused the gyros to process and operate the airplane's control servos for the transition. During transitional flight the power setting of the engine was changed by the throttle servo to give the proper power setting for each pitch angle.

Automatic transition from horizontal to vertical flight was provided as the reverse of the transition from vertical to horizontal flight. Another method of providing a transition from horizontal flight for use during a blind landing is discussed in later paragraphs.

Artificial roll stability was probably required for the airplane when it was in the vertical attitude. The stability signals obtained from a rate gyro operated the lateral control servo. The contemporary trend in fire control development was toward fully automatic flight control of the airplane during the attack phase of the mission. The basic components required to do this, the tracking radar, computer and autopilot, were installed in the airplane. Martin gave definite consideration to completely automatic attack during the development of the control system for the airplane.

Electronics installation

The electronics equipment installed in the aircraft included:

- AN/ARC-27—UHF transmitter-receiver
- AN/ARR-2A—navigation unit
- AN/APN-22—radio altimeter
- AN/APX-6—IFF unit

In addition, the aircraft carried an AN/APQ-42 radar and a small beacon furnished by Martin for use in blind landing.

In conformance with OS-122, both the AN/APG-25 and the AN/APS-25 radars were considered. The AN/APG-25 was a 50 kilowatt, monopulse, automatic tracking radar. It had no provisions for search and, owing to the relatively small amount of transmitted power, this radar could not provide adequate search range. For these reasons it was not recommended for use in this application.

The AN/APS-25 was a 250 kilowatt radar which provided for search, acquisition and automatic tracking. For automatic tracking, this radar employed the conical scanning technique. It was well known that radars employing this technique provided a lower degree of tracking accuracy against targets travelling at high angular rates than monopulse radars. Since the AN/APQ-42 provided the same features as the AN/APS-25 but employed the monopulse technique for automatic tracking, this radar was recommended for use in the Convoy Fighter. Nevertheless, the airframe was capable of accommodating either the AN/APS-25 or the AN/APQ-42 radars.

It was considered essential that the RF head, which contained the transmitter, be mounted directly behind the scanner in order to obtain the maximum available search range. If the RF head were mounted back in the fuselage, around 15ft of waveguide leading through the propeller shaft to the scanner would have been required. This resulted in a search range penalty of about 10%. The RF head was consequently mounted in the spinner. It was repackaged to fit into the 10" depth behind the scanner; this was the only unit of the AN/APQ-42 requiring repackaging.

Blind landing (electronics)

A fully automatic or pilot controlled blind landing system was developed to make the Convoy Fighter an operable weapon during times of poor visibility.

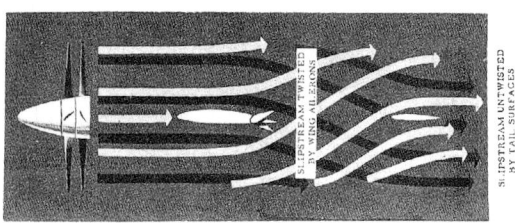

Diagram showing how the slipstream from the propellers would be twisted by the inboard wing ailerons, then subsequently straightened out by the tail elevons, possibly improving lateral control.

The only basic equipment required in addition to that already employed, the system consisted of a small radar on the surface ship and an airborne beacon, weighing less than 30lb, in the aircraft. The system provided the pilot or automatic control devices with all the necessary information to bring the aircraft through transition from horizontal to vertical flight which terminated directly above the ship, oriented the aircraft's wings parallel to the longitudinal axis of the ship, and executed a descent to the recovery devices on the deck of the ship.

To execute a blind or automatic landing, the pilot turned on his airborne beacon when he was in horizontal flight and his range was about eight miles from his ship. When the range decreased to five miles the shipborne radar detected the beacon and began to automatically track it in angle and in range. The ship's heading angle with respect to magnetic north (obtained from the ship's compass) was added to the azimuth angle of the radar sight line measured with respect to the longitudinal axis of the ship.

The resultant angular information was shifted through 90° to provide an azimuth angle, with respect to magnetic north, which represented a perpendicular to the radar sight line. This information was sent to the aircraft, via conventional radio link, where it was compared with the angle between the wing line and magnetic north. The resultant difference, which represented the misalignment of the aircraft wings with the perpendicular to the radar sight line, was used to rotate a line on the face of a 3" cathode ray tube or to feed the automatic control devices which controlled the wing alignment of the aircraft. In this manner, it was possible to maintain the orientation of the aircraft wings perpendicular to the radar sight line and the transition from horizontal to vertical flight could be controlled by means of pitch signals.

The pitch signals were obtained in the following manner. A synthetic elevation angle was generated by means of a motor-driven synchro. This angle was made to increase with time in accordance with the variation in elevation angle which occurred for a representative transition terminating with the aircraft hovering in a vertical attitude directly above the ship. The elevation angle of the sight line was compared with the synthetic angle and the resultant difference signal was transmitted to the aircraft. This signal, which represented a change in speed necessary to bring the aircraft on to the representative transition, was used to displace the line vertically on the face of the indicator or to feed the automatic devices which controlled the pitch and therefore the speed of the aircraft.

This technique made possible a completely controlled transition terminating with the aircraft directly above the ship. The throttle servo was controlled by signals sensitive to the pitch of the airplane such that the power supplied was that required for horizontal flight at each pitch angle of the airplane. The use of this technique did not restrict the pilot from making the transition in any fashion he desired and the presentation on the cathode ray tube continuously indicated that a change in pitch or wing alignment was necessary to bring the aircraft on to the representative transition selected.

As the sight line approached the vertical, the wing orientation information transmitted from the ship to the aircraft was changed so that when compared with the wing-line angle in the aircraft, the resultant error signal provided the rotation of the line on the indicator necessary to bring the aircraft wings parallel to the longitudinal axis of this ship. At this point an additional signal representing the position of the sight line measured about the transverse axis of the ship was also sent to the aircraft and was used to deflect the line on the indicator horizontally or to feed automatic devices which controlled the aircraft in yaw.

In this manner it was possible for the aircraft to be retained along a vertical line to the deck of the ship and for the wing line to be maintained parallel to the ship's longitudinal axis. Then, by the use of the radar range (height) information, the throttle could be controlled manually or automatically to permit descent to the landing platform of the ship. Since the pilot had to see the landing platform only during the very last part of the recovery, this method

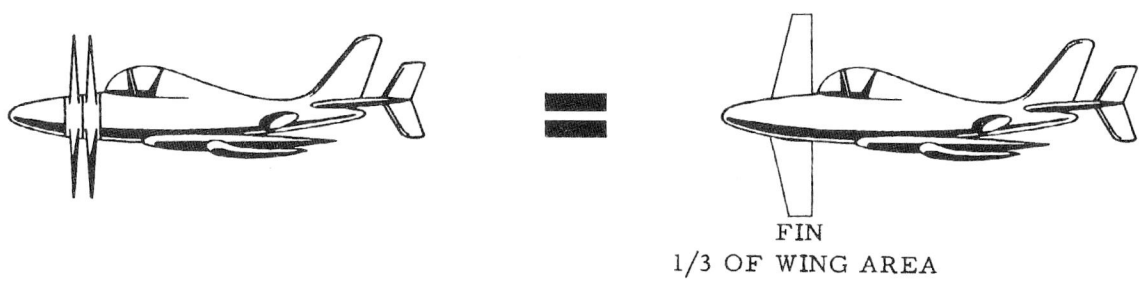

The transonic propeller specified for the convoy fighter produced a large side force, causing a serious stability problem. This issue was similar to having a surface of approximately one-third the wing area located at the propeller position. Martin proposed the development of a supersonic propeller to address this problem.

permitted recovery during conditions of visibility down to about 75ft.

Armament

Two 20mm cannon were mounted adjacent to each of the wing gears in underslung streamlined bodies just outside of the propeller disk. Ammunition was carried in compact boxes in the forward part of the body and loaded through doors on the top.

The fire control system consisted of the recommended AN/APQ-42 radar and the Aero X1A computer. A Mark 6, Mod. 1 visual sight unit was also provided. Since the airplane was equipped with automatic flight control devices, consideration was given to the problem of automatically controlling the aircraft during the attack phase. Developments at the MIT Instrument Laboratory had successfully demonstrated the ability of a radar, computer, autopilot combination to control an airplane on an attack course. Westinghouse was also developing a similar system. Preliminary study had shown that the basic Westinghouse control scheme could be applied to this aircraft. The same components of the autopilot system used for the takeoff and landing phase could be utilized for automatic control during the attack phase if the entire system was properly integrated. Future developments in the art of automatic flight control might have produced a system which could manoeuvre an attacking aircraft to a higher degree of accuracy than could a well-trained pilot. Martin recommended that the aircraft control system be designed in such a manner that automatic flight control during the attack phase could be incorporated.

The company also recommended that definite consideration be given during the development of the aircraft and its systems to the use of other weapons such as rockets, either controlled or otherwise, and to the installation of air-to-air missiles.

Power plant

The engine used was an Allison XT40-A-8 mounted in the fuselage. The engine reduction gear housing was mounted in the fuselage by means of vibration isolating (dynafocal) mounts which isolated propeller vibration from the power sections and aircraft structure. The dual power unit utilized a three-point mounting arrangement. The two upper mounting pads on the diffuser assembly were used together with the bottom compressor air inlet tie bracket.

The induction system employed separate air intakes located in the fillets between the wing and the fuselage and partly submerged in the fuselage. A revolving door which was designed to impose a minimum amount of drag on the airplane was incorporated in each intake to provide for single power unit operation. A boundary layer bleed was incorporated in the intake to remove boundary air. An almost identical air intake and door were used on the Martin XB-51 and the ram recovery was excellent. Provisions were made in the inlet ducts for passage of the drive shaft housing through the ducts.

The fuel system consisted of a service tank and five auxiliary tanks with a total capacity of 500 gallons of fuel. A constant fuel level was maintained in the service tank by means of constant level valve which controlled the fuel flow from the auxiliary tanks. The service tank was designed so that fuel would be continuously available to the engine at any possible sustained flight attitude by the use of two booster pumps located in opposite ends of the tank and two check valves to prevent the fuel from being pumped back into the tank. The pilot could direct fuel to either or both power units by means of two on-off

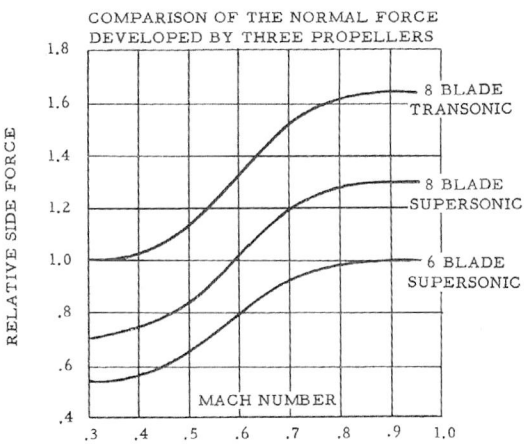

Graph showing the superiority of a six-bladed supersonic propeller over other options, as it produced less undesirable relative side force over the entire speed range. Martin endorsed a single speed gearbox driving a six-bladed supersonic propeller for its Convoy Fighter proposal.

valves located at the service tank. In the event that the service was shot out, flight could be continued by routing it through a separate line around the tank. Two lines and strainers were used between the service tanks and the power sections to increase engine reliability.

The service tank was pressurized by the vapour pressure of the fuel to 2psi across the tank walls to reduce fuel losses due to boiling. This pressure was maintained by means of a pressure relief valve in the service tank vent system. An automatic pressure release for landing and a manual release for combat operations or emergency were incorporated in the system. The auxiliary tanks were vented into the service tank and were also pressurized.

The lubrication system consisted of a 5½ gallon oil tank including a 1½ gallon expansion space and two 11" diameter oil coolers. The cooler duct inlets were located in the wing fillets and duct exit doors were employed to control the air flow through the coolers.

The engine combustion chambers, turbine sections, and tail pipes were enclosed in a fire resistant chamber of corrosion resistant steel.

Maintenance

Ample access and ease of removal and replacement of components were considered mandatory for this aircraft to provide for a high degree of availability when maintained by small crews on the merchant ship.

The engine reduction gear and dual power section were mounted separately in the fuselage. Access to the aft side of the reduction gear was obtained through the recovery gear compartment which was open when the spike was down. In this area the reduction gear supports and accessories, drive shafts, and radar nose section support were accessible. Release of the radar nose section at this point permitted the removal of the radar nose and exposed the propeller thrust nut for propeller removal. The reduction gear unit was pulled forward out of the fuselage centre section.

Removal of the dual power unit was accomplished by separation of the fuselage at the rear wing spar bulkhead. Access doors in the aft fuselage section permitted disconnection of the tail pipe from the engine. Access was provided to quick-disconnects at the fuselage break for tail surface controls and electrical lines. The fuselage was separated by removal of five tension bolts. The aft power unit supports were then accessible. The forward power unit support was mounted in a track which permitted the power unit to be withdrawn to the point where the forward support could be grasped for complete removal. Access, without removing the engine, to all accessories, supports, and lines on the forward end of the dual power unit was provided through a large non-structural access door on either side of the fuselage.

Electronic equipment in the upper portion of the aft fuselage section was fully accessible through large, quickly openable doors on the side of the fuselage.

Guns mounted in the under-wing pods were serviced and removed through large non-structural doors on the sides of the pods. An ammunition box access door was located on the upper surface of the forward end of each pod. Wing gears mounted in the underwing pods and between the guns were accessible through the gear doors on the lower surface of the pod.

Lower surface panels in the wing provided access to the wing fuel cells. Spoiler controls in the wing leading edge were accessible through a door in the leading edge. Controls in the wing above the pod were open to inspection through the wing gear well or could be exposed by removal of the upper cover in this area.

To simplify maintenance and spare parts problems, identical right- and left-hand assemblies were used for the tip fins, the horizontal stabilizers and elevons,

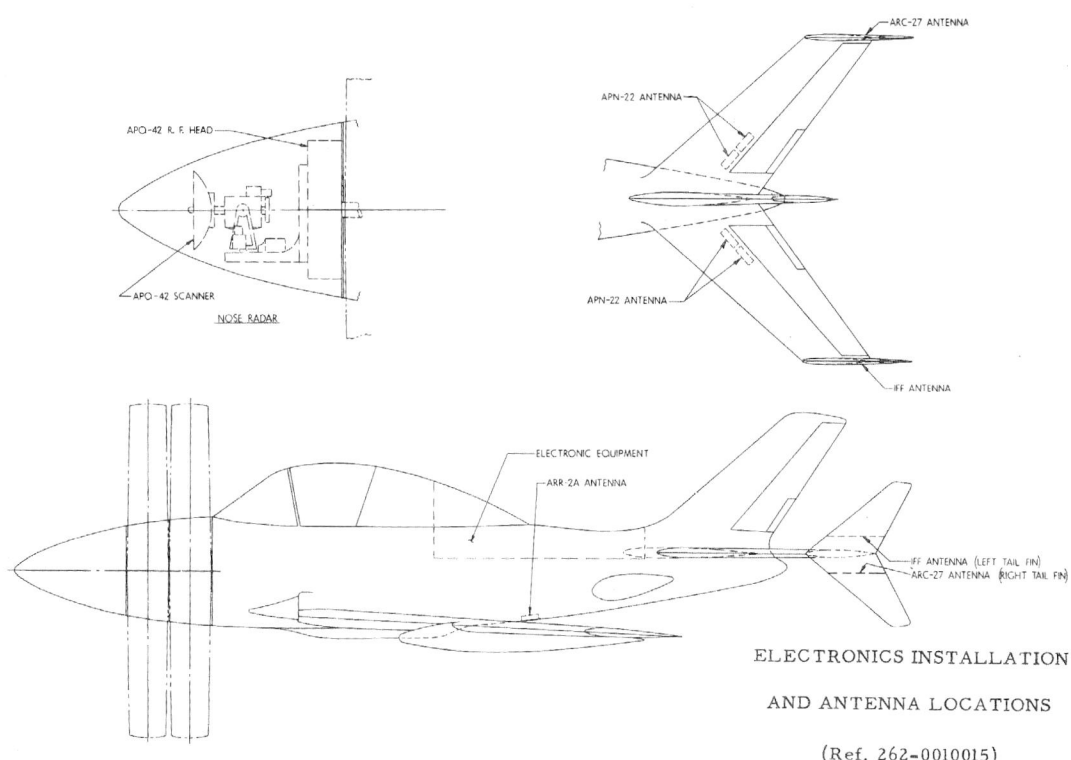

Diagram showing the location of the antennas and radar on the Model 262; the electronic equipment was located primarily in the upper mid-fuselage.

the wing spoilers, and wing gears.

Structure and producibility

Design of the airplane and its tooling were planned to minimize the elapsed time between the experimental programme and the delivery of production articles. This was accomplished by incorporating certain production features in the design of the experimental airplane and in the tools used to fabricate these airplanes. The degree to which these production aspects were incorporated was tempered by economic considerations.

The structural design was based on the use of shapes and fabricating methods which were readily adaptable to large scale production. Major production splices would have been incorporated in the experimental airplanes. The design would have employed forgings, castings, and extrusions in all applications where they were used in the production airframe except where excessive elapsed procurement time or tooling cost necessitated substitution. When this occurred, the part would have been made to conform as closely as possible to the configurations and physical properties of the production counterpart.

In considering design producibility for this airplane, several salient features were introduced which enabled adaptation of the airplane to high rates of production. The basic cross-sectional shape of the fuselage centre section was circular, and for its major portion this assembly was cylindrical. Although the wing fillet altered this shape locally, it permitted the use of radius cut frame tools which were symmetrical about their centrelines.

The cylindrical shape permitted multiple usage of frame sections between longerons as well as straight lengths for longerons and single curvature forming or rolling of skin and door panels. The fuselage aft section had basic cross sections which were circular, permitting the use of a reduced number of tools with multiple usage for circular frame sections. For the greatest possible extent this portion of the fuselage was made conical with straight line elements for single curvature forming of skin panels and straight longerons.

Prior to engine installation, the fuselage interior presented a clear opening which facilitated assembly of the units. The large fuselage access doors provided for maintenance of the engine and electronics equipment further facilitated this

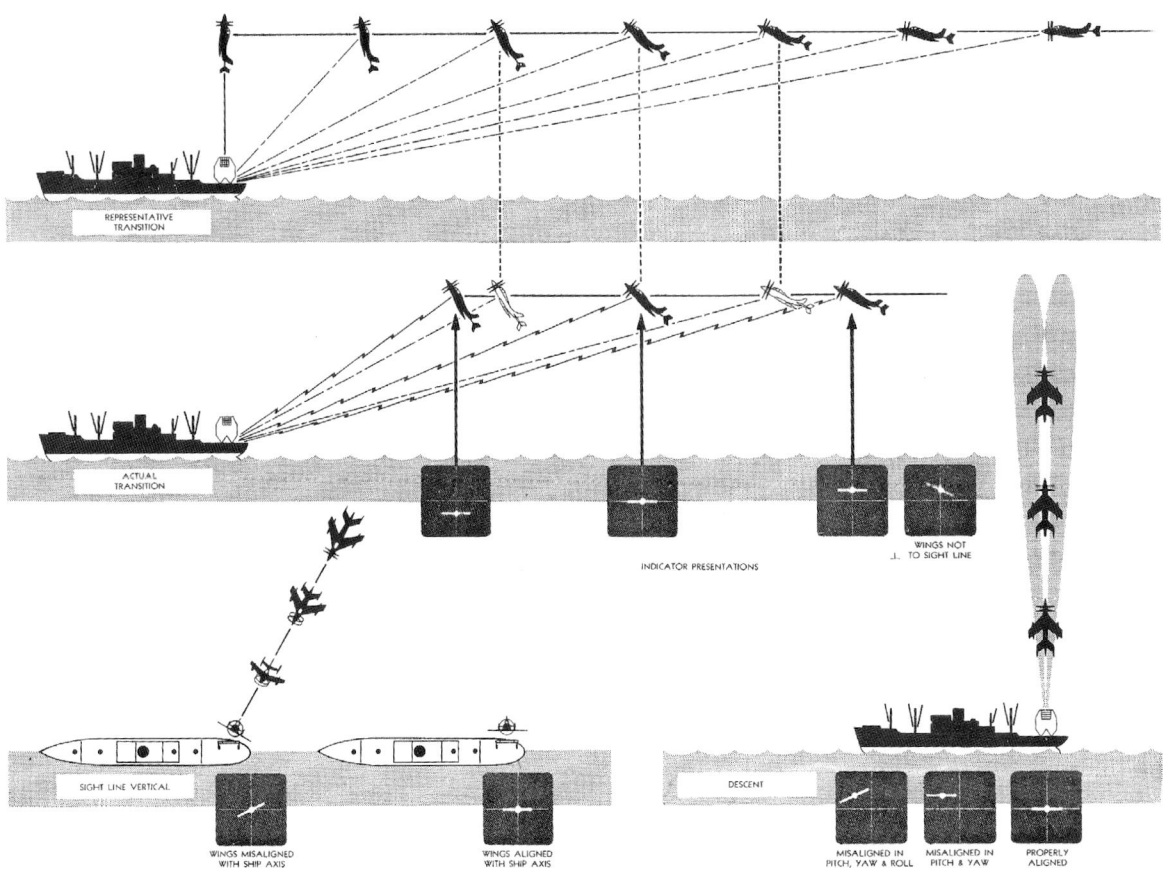

Illustration of the Model 262's blind landing system in operation; the system would have used an airborne beacon to help accomplish these hazardous recoveries.

assembly. The pressurized cockpit enclosure area was contained entirely within the fuselage centre section. This permitted complete sealing of the pressurized area and subsequent pressure testing of this fuselage section prior to its attachment to other major structural components.

The wing centre section between the outer panels was rectangular in planform and of constant section, thereby resulting in straight spar chords, rectangular spar webs, constant length spar stiffeners and single curvature surface covers. The fixed wing and tail surfaces were of the blanket type construction which permitted use of a minimum number of ribs and lent itself to a means of assembly that was not handicapped by the very limited access between the covers of the thin airfoil surfaces.

The wing spoiler aft of the rear spar was made with a flat surface and its ends were cut normal to its length. This eliminated all contour in this surface and permitted manufacture of identical parts for the right and left wing installations. Identical right and left hand assemblies were also being employed for stabilizers, elevons, tip fins, and wing gears. Thin trailing edge surfaces such as elevons, rudder, fin tips, and tabs were made of metal honeycomb construction, metal covered, for greatest economy eliminating numerous detail sheet metal parts.

Splices between the wing centre section and outer panel as well as the attachment of fixed tail surfaces to fuselage stubs were of the simplest practical interchangeable design employing a minimum of matching surfaces and attachments. If design considerations permitted, tension type splices would have been employed. The fuselage splice between centre and aft section was of the simple tension bolt type employing five tension bolts with intermediate self-engaging studs for shear load transfer.

Major assembly and subassembly tools were of the production type designed with provisions for all locators, drill plates, trim bars, and other refinements required for ultimate production. These assembly tools also contained temporary experimental tooling

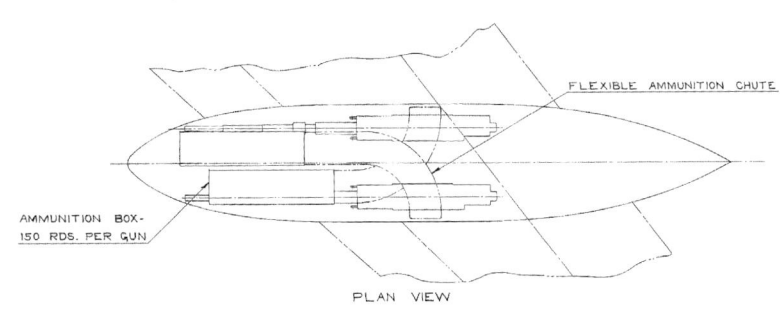
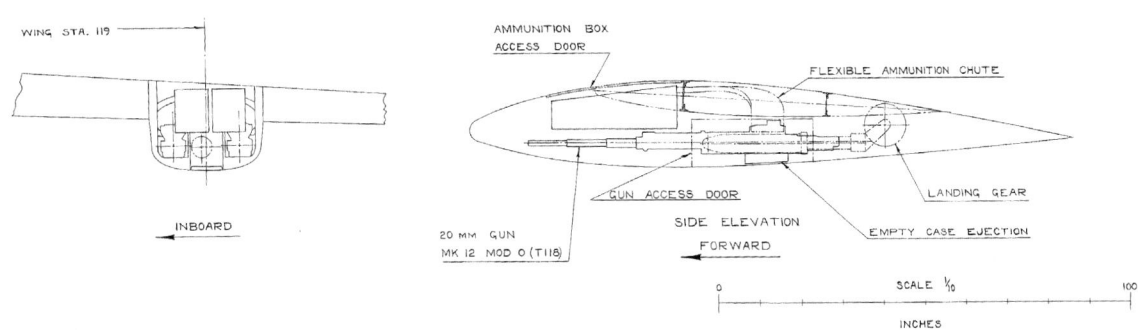

ARMAMENT INSTALLATION

The basic armament of the Model 262 consisted of four 20mm cannon, two in each underslung wing pod.

features such as contour boards, auxiliary locators etc., to take the place of minor subassembly tools. Tools used to fabricate detail parts and small subassemblies were of the experimental type generally suitable for the manufacture of up to ten airplanes. Wooden or masonite form blocks would have been used in lieu of steel dies to form sheet metal parts.

MODIFIED ARRANGEMENTS

Three modified arrangements were given serious consideration by Martin during its design studies. The company believed that certain aspects of each of these required further study, both analytically and in the wind tunnel, before deciding on the final configuration of the Convoy Fighter.

The manner in which this airplane was expected to be flown introduced aerodynamic problems which could not be analyzed by conventional and proven methods. Applicable methods of analysis had been evolved during Martin's design studies and were presented in accompanying engineering reports. These methods were derived primarily from theoretical considerations. They had not been proven experimentally since very little experimental data existed.

Martin found that the method of analysis and the evaluation of various aerodynamic effects had a profound influence on the selection of the optimum configuration. Therefore, the company planned to verify their methods of analysis by wind tunnel tests before proceeding with the construction of the airplane. The modifications which Martin envisaged as possible solutions to the Convoy Fighter problems involved a different propeller location for each configuration: one mounted the propeller on the nose, the second was centrally located, while the third had the propeller aft in a pusher installation.

General arrangement and inboard profile drawings are shown for two modifications using delta wings. In addition, the swept wing airplane modified to locate a pusher propeller between the wing and the tail is shown.

An alternate propeller gearing arrangement was also included which provided various speeds of propeller operation without a gear shift. This was presented for possible consideration in lieu of a

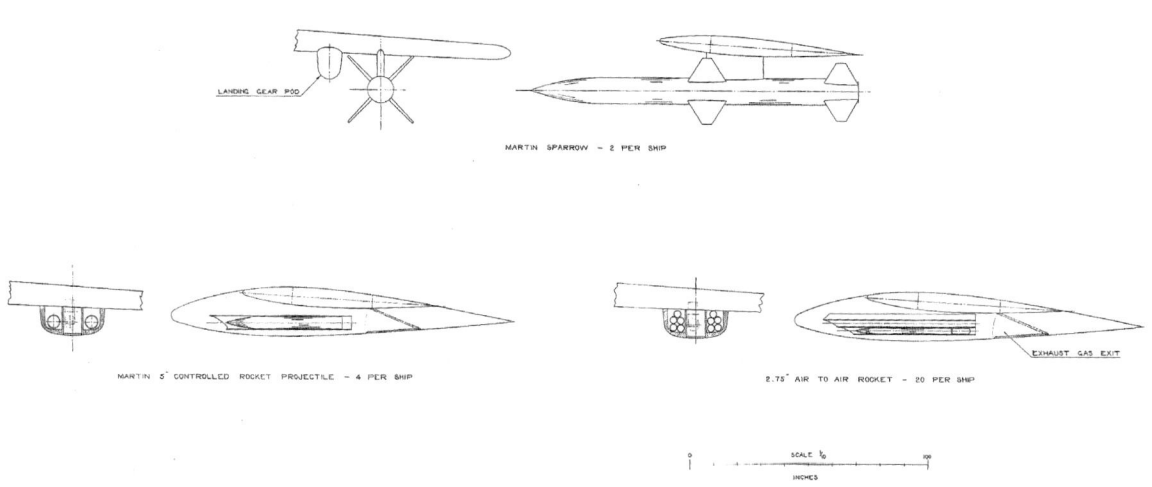

Martin also proposed these alternate missile and rocket installations for the Model 262.

gear shifting arrangement if two propeller speeds became mandatory to meet the propeller's performance requirements.

Modification A—delta wing
The delta wing planform offered several distinct advantages in the Convoy Fighter application. A considerable saving in wing and tail weight was achieved which resulted in increased loiter time and reduced airframe weight. It permitted the elimination of the horizontal tail whose contribution to stability in level flight was comparatively small. Tail buffet problems in high speed flight were eliminated. The ability to perform constant altitude transitions was much improved since very little of the wing was outside of the slipstream. The disadvantages of this planform were the reduction in rate of climb and ceiling due to the lower span and possible reduction in longitudinal control response with no horizontal tail.

The pilot was located immediately behind the propellers under a large canopy. The radar scanner was in the propeller spinner. Engine air was inducted through flush air intakes in the side of the fuselage. Engine exhaust was aft through the sides of the fuselage. Fuel was carried in the forward part of the wing and in a service tank in the fuselage.

The four 20mm guns were mounted in bodies of revolution on each wing tip. Their ammunition was carried in spanwise ducts in the wing.

Lateral control was supplied by differential operation of the elevons on the wing trailing edge. Longitudinal control was through symmetrical displacement of these elevons. Directional control was provided by a rudder on the central vertical surface. The additional vertical surface required for directional stability in horizontal flight was supplied by the fins mounted on the tip bodies which were not in the slipstream.

Modification B—central propeller
This modification was envisioned with a delta wing and consequently realized the same advantages and disadvantages as the 'A' modification. However, the central propeller location reduced the moment arm from the propeller to the centre of gravity of the airplane. The destabilizing effects of the propeller were then minimized. The pilot, the radar and the guns were mounted in the fuselage forward of the propellers. The pilot's forward vision was then unimpaired by the propeller and in an emergency the pilot could be ejected in a capsule rather than in an ejection seat. Increased gunfire accuracy and aerodynamic cleanliness resulted from the installation of the guns and radar in the fuselage nose.

Engine air was taken aboard through intakes in the wing root. Engine exhaust was through the sides of the fuselage in the rear. Fuel was carried in the forward part of the wing and in the fuselage. The control surfaces operated the same as in the 'A' modification.

Since this modification required supporting a part of the fuselage forward of the propeller, the gearbox and propeller hub assembly had to be adapted to permit this support. Preliminary discussions with the gearbox and propeller manufacturers indicated

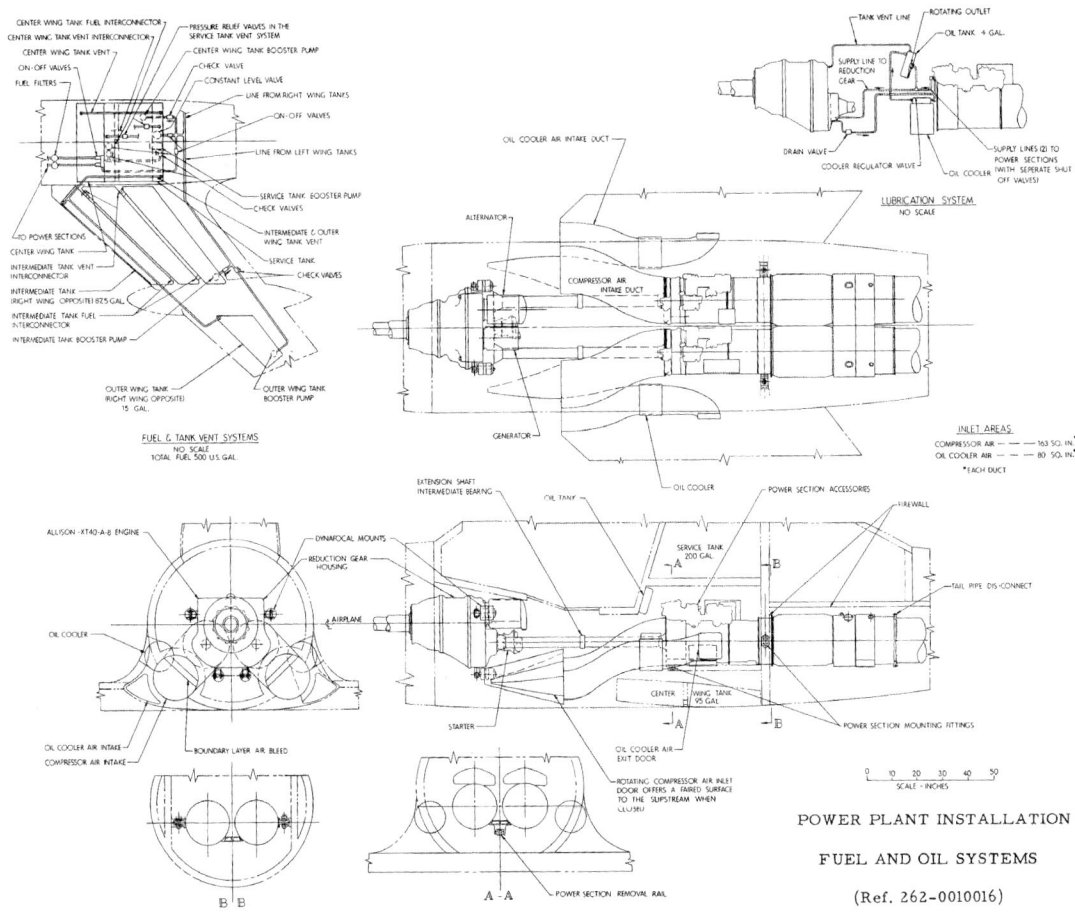

Blueprint showing the installation of the Allison XT-40-A-8 turboprop engine along with the associated fuel and oil systems in the Martin Model 262.

Modification C—pusher propeller

This modification placed the propeller aft of the centre of gravity in a pusher installation. The propeller was then no longer destabilizing in level flight but contributed to the stability with a consequent reduction in required tail size. With the propeller operating behind the wing, propeller stresses were greatly reduced since the downwash of the wing tended to cause the propeller to act at a small and constant angle of attack throughout the entire speed range.

The tails were located behind the propeller in the slipstream to provide control during hovering. Since the wing was ahead of the propeller, the possibility of a drag increment induced by the slipstream acting over the wing was eliminated. Engine exhaust was through eight louvers equally spaced around the fuselage to allow the exhaust to mix with the freestream before passing through the propeller disk.

Locating the pilot, radar and guns in the nose provided the same advantages in pilot's vision and armament installation as was found in the 'B' modification. With the pilot prone, the large canopy was eliminated, a small frontal area and favorable length-diameter ratio were realized.

The disadvantage of the arrangement was the extra effort which had to be borne by the engine and propeller manufacturers. The propeller drive shaft had to be move to the exhaust side of the engine. The propeller hub and gearbox again had to be capable of accommodating a support structure as in Modification 'B'—this time for the tail surfaces.

Alternate gearing arrangement

During Martin's studies of the pusher propeller configuration (Modification C), a propeller gearing arrangement compatible with the configuration was investigated. It consisted of a differential ring gear arrangement driving one fixed pitch propeller and one variable pitch propeller in contra-rotation. This arrangement provided a variation in propeller speeds without resorting to a gear shifting arrangement.

Use of the ring gear provided the space through

the centre of the gearbox for airframe structure and control lines. The differential feature split the torque equally between the two contra-rotating propellers and eliminated all rolling trim changes due to power. Their relative speeds were then a function of the blade angle on the variable pitch propeller which was governed by the engine rpm. Thus, for hovering, the variable pitch propeller was made to operate at low blade angles and consequently high rpm.

Since the torque was always equally divided between the two propellers, the variable pitch propeller then absorbed most of the engine power output and efficiently converted it to thrust. An increase in the diameter or number of blades may have been required to absorb the power. The large differences in the rpm of the two propellers produced a gyroscopic effect which should have contributed materially to hovering stability.

For the high speed flight regime, the pitch was increased so as to make both propellers turn at comparable (lower) rpms. Power was then divided about equally between the propellers. The fixed pitch propeller was profiled to produce maximum efficiencies at high speeds. A weight saving would have been realized by having one propeller of fixed pitch. Although this gear arrangement was conceived for use in the pusher arrangement, it could have also been adapted to the other arrangements.

WIND TUNNEL PROGRAMME

The proposed wind tunnel programme was designed to produce sufficient aerodynamic data to develop a successful prototype airplane which would have ultimately led to the optimum Convoy Fighter configuration.

Aerodynamically, the prototype powered by the Double Mamba engine was almost identical to the Convoy Fighter which was powered by the Allison XT40-A-8. However, this difference in power plants caused the placement of the pilot's canopy and the intake ducts to differ between the two. The initial wind tunnel tests would have been aimed at producing a successful prototype and therefore the wind tunnel model configurations would have adhered to the geometry dictated by the installation of the Mamba engine. In this way, an invaluable direct comparison between wind tunnel and flight test data would have been obtained for use in designing the Convoy Fighter.

Five separate types of wind tunnel tests were proposed. These, in their proper order, were:

1. *Isolated propeller tests*—Since the propeller became the prime lifting surface during part of the flight, its basic aerodynamic characteristics were needed just as the characteristics of the wing and tail were needed. Therefore, in addition to the thrust and torque properties, the propeller normal forces would have been obtained for angles of attack between zero and 90 degrees using a wide range of blade angles and advance diameter ratios. Although the propeller for these tests was to have been used later on the prototype wind tunnel models, the scope of the propeller tests would have been made sufficiently broad to encompass power loadings required for the Convoy Fighter. The data obtained would have provided the information to set correct blade angles and power input for the model tests and also a means of interpreting the effects of power on the airplane components.

2. *Low speed tests*—Power off runs were to have been made from zero lift to stall. Tests with power would have been run to cover the high speed, takeoff, transition and hovering phases of flight. A ground board would have been used during some of the takeoff studies. Special emphasis was placed on establishing the optimum wing and tail position relative to the thrust line. Although the prototype configurations would have been tested, power conditions compatible with the requirements of the Convoy Fighter would have been investigated. In addition to obtaining static stability and the effectiveness of all control surfaces, pressure measurements to establish air loads would have been obtained as well as velocity surveys in the vicinity of the engine intakes.

3. *Free flight tests*—The models would have been dynamically similar to the prototype equipped with remotely controlled surfaces and power plant. Only the hovering and transition near hovering phases of flight would have been investigated on these models. Each model would have

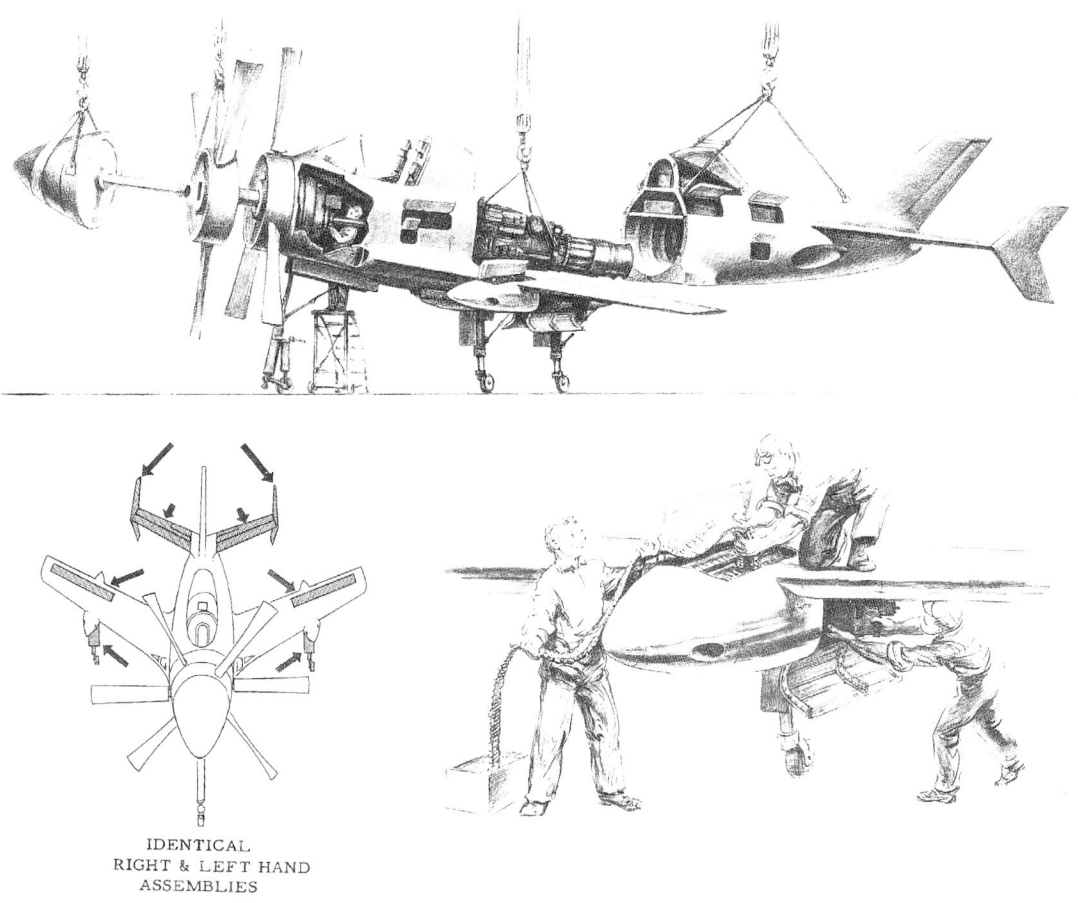

Martin designed the Model 262 to be easily serviced and maintained, with several identical right- and left-hand components to minimize the need for spares.

first been run in the hovering position and its response to control surface action recorded on film. Next it would have been subjected to horizontal winds of various velocities. The model would have been turned to get the effect of the horizontal wind coming from several quarters to represent fore and aft drift, lateral drift and intermediate drifts. Some tests would have been run close to the ground to investigate the response to control surface movement in manoeuvres representing the landing condition.

4. *Intake duct tests* — These models would have been unpowered, fairly large scale models; one to represent the prototype engine installation and one to represent the Convoy Fighter engine installation. Variations in lip shapes and internal duct lines would have been provided. The tests would have been run to obtain the refinement of lip shapes and internal geometry necessary to obtain a high critical Mach number entrance and an optimum duct efficiency for the range of inlet velocity ratios encountered in flight.

5. *High speed tests* — The high speed of the prototype airplane was not sufficiently great to require high speed wind tunnel tests, but that of the Convoy Fighter was. The wind tunnel tests would have therefore been performed on a model of the Convoy Fighter configuration. These tests could have been accomplished directly after sufficient low speed data had been obtained to design the prototype. In the event that the high speed tests showed the need for some redesign, corresponding changes would have been made in the flying prototype design to ensure that it was a true prototype airplane. It was therefore considered necessary to run the high speed tests before

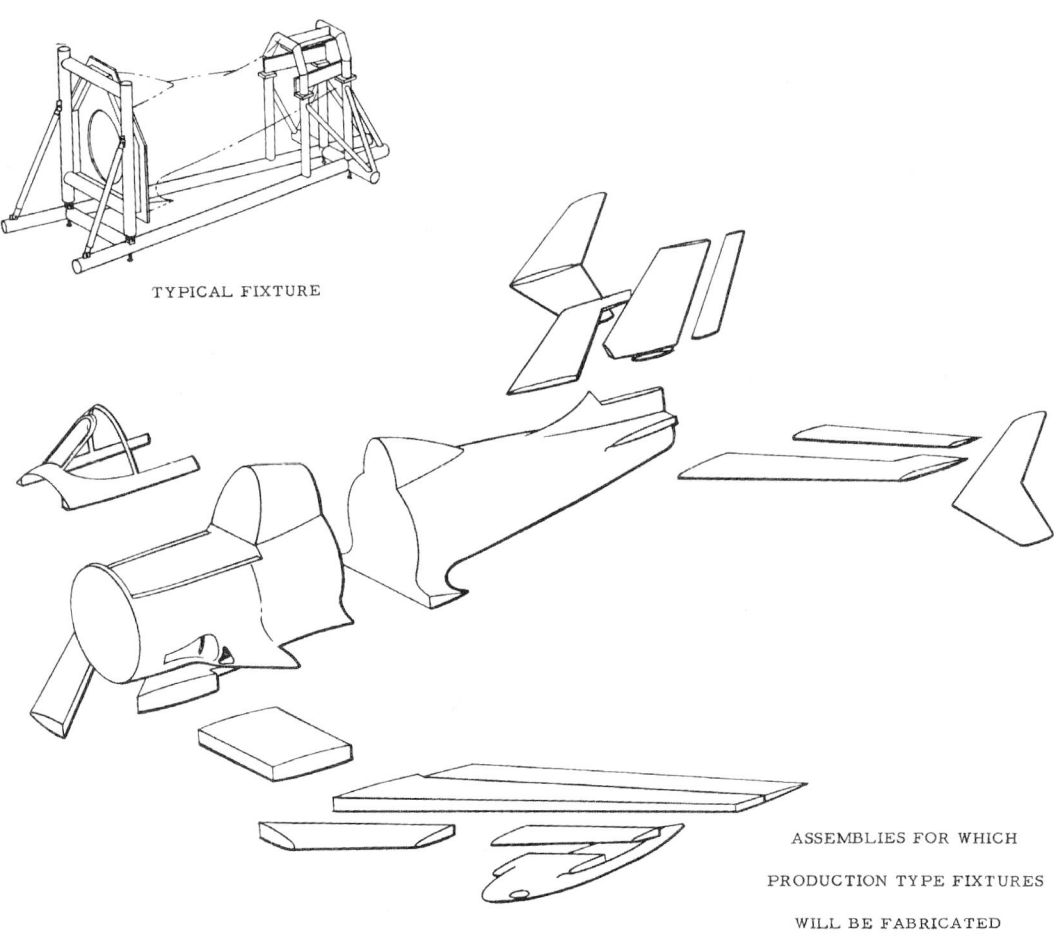

Exploded view of the Model 262, which was designed with large scale production in mind.

proceeding too far with the prototype construction.

Data on performance, stability and control, airloads and pressures adjacent to critical surface junctures would have been obtained by normal wind tunnel techniques. Studies indicated that it would have been unnecessary to equip the high speed model with a power plant. At high speeds the power coefficients were low and it was believed that by providing the model with a windmilling propeller having properly adjusted blade angles, the effects of the propeller on stability could have been adequately reproduced.

Modifications to the basic proposal showed promise of leading to a successful design. These could not have been evaluated until quantitative data was available from the wind tunnel. Therefore, the low speed static tests and the free flight tests to have been run on the proposed Model 262 configuration would have been paralleled by a similar series of tests to determine the merit of the modifications. The modification tested in the wind tunnel would have been selected after further analytical studies. The duplication of tests would have ceased as soon as one configuration emerged with superior characteristics. This method of testing was expected to lead to the optimum prototype in the shortest length of time.

ALTERNATE LAUNCH AND RECOVERY

The alternate recovery method described below probably required the least accurate control of the airplane of all those investigated, although the equipment required on the surface ship was rather elaborate. Martin believed that the flight testing of the prototype would have established the direction in which the recovery method development proceeded.

The equipment on the surface ship consisted of a stabilized platform which rolled on curved tracks whose centre of curvature coincided with the roll centre of the ship. The platform contained sockets

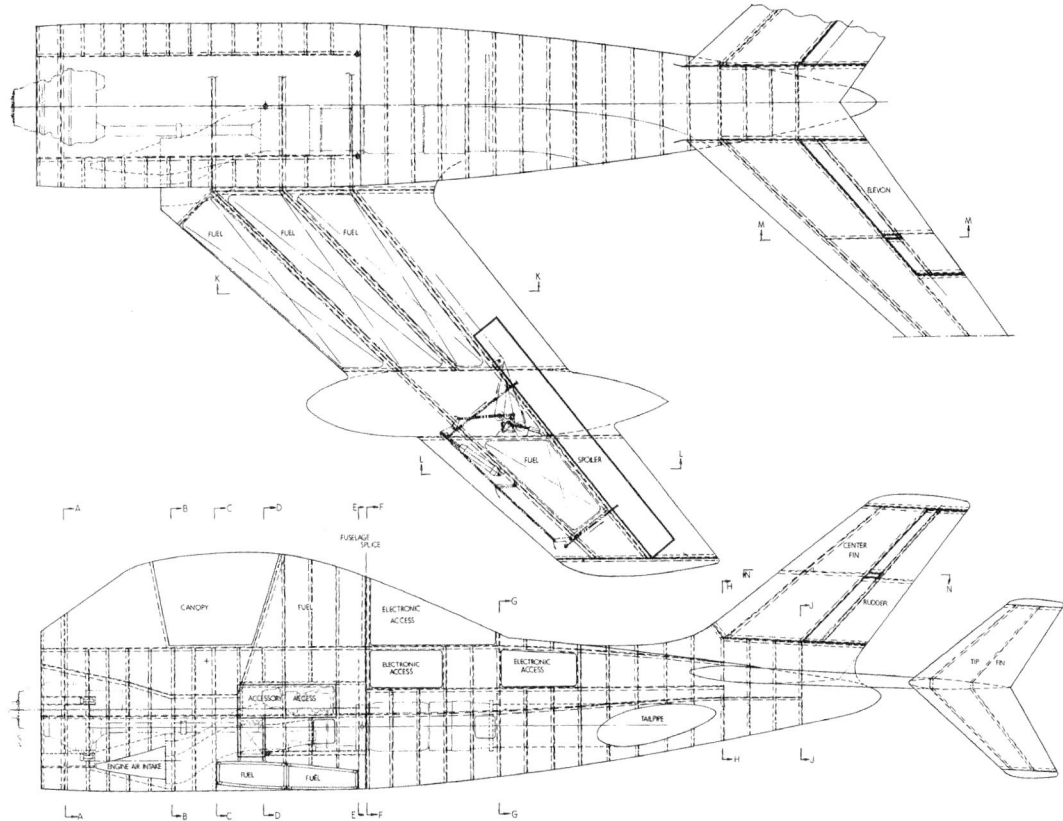

Structural diagram of Martin's VTOL turboprop fighter showing the major frames, longerons, spars, and so forth.

which accommodated recovery struts on the aft end of the airplane. Two lengths of flexible steel cable were wound on two winches on the platform. The airplane was equipped with recovery shock struts on its tail and two lengths of rope which paid out from each wing.

Recovery was accomplished as follows; after making a transition from horizontal to vertical flight, the airplane flew in a hovering attitude in a path which caused the ends of about 75ft of rope trailing from each wing to be dragged across the after deck of the ship. Deck hands fastened these ropes to the flexible cables located on the stabilized platform. The ropes were then reeled in by the airplane so as to pull the cables up to the airplane and latch them to it. During this operation, the relative vertical motion of the ship and the airplane was taken up by the winches on the platform which automatically reeled the cables in or out so as to keep the cables taut but with a very light load.

At the top of one upswing of the ship, the winches were locked. The airplane was then drawn down with the ship as the ship went on its downswing. When the pilot felt this, he applied full power. From then on the cables were kept continually in tension by the excess thrust available from full power operation of the engine and thus there was no longer any relative vertical motion between the ship and the airplane. The airplane was then reeled down onto the platform by the winches on the platform. Suitable devices on the platform then rotated the airplane to the horizontal attitude.

The airplane was launched by standing on its tail on the platform, applying full power, and operating quick-release mechanisms attached to the shock struts.

The cables might have only required use during extremely rough weather. During normal weather conditions, the airplane could have probably been landed on its tail on the platform similar to a helicopter. On shore bases no auxiliary equipment would have been required.

Two cables were used so as to automatically stabilize the airplane in yaw. The cables permitted the airplane to translate sideways but it could not yaw. This kept the thrust line vertical at all times in the plan view of the airplane. In the side view, the cables also tended to stabilize the airplane in pitch but

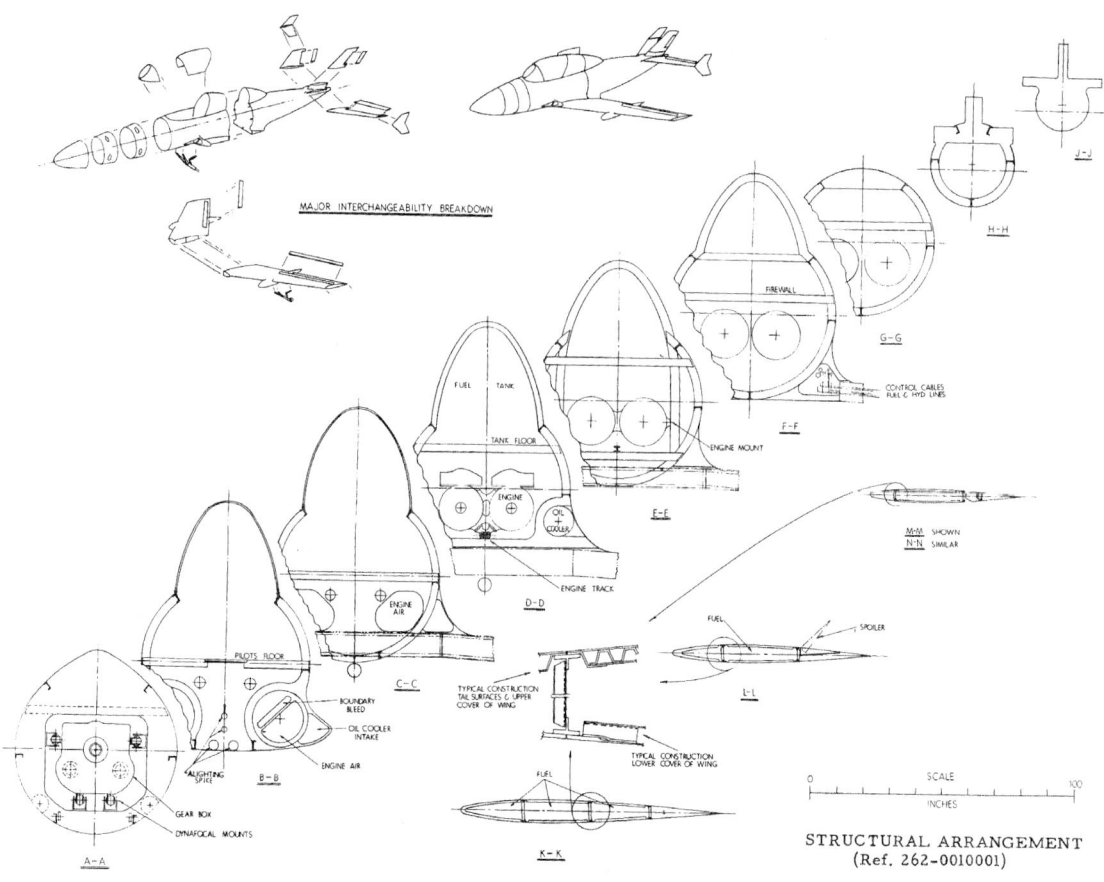

Structural arrangement of the Martin Model 262 showing major fuselage cross sections. The basic cross section of the centre fuselage was circular, which lowered the cost of producing the aircraft.

not to the extent that the airplane could not be controlled in pitch by the horizontal tail. The cables were mildly stabilizing in roll when reeled out to long lengths and strongly stabilizing in roll at the shorter lengths.

This recovery method influenced the horizontal tail configuration of the airplane. A three-way tail would have been used to provide for recovering the swept wing airplane on its tail. On the delta wing, modifications to the vertical tail extended above and below the fuselage to accommodate this recovery method.

MODEL 262P CONVOY FIGHTER PROTOTYPE

The prototype for the Convoy Fighter, designated Model 262P, was an inhabited, flying model (0.766 scale). Dimensional and dynamic similarity to the Convoy Fighter was maintained as closely as possible. The prototype was designed in accordance with BuAer Specification OS-122.

The prototype was designed to operate in both the hovering and normal flight regimes. It was aerodynamically similar to the Convoy Fighter airplane except for the aft location of the pilot and the engine air intakes. These differences were caused by the use of a British Double Mamba engine in place of the XT40-A-8 engine used in the Convoy Fighter.

The Model 262P was equipped with a retractable gear for launching and recovery in the hovering attitude which was similar to that of the Convoy Fighter.

A truck-mounted recovery platform was designed to accommodate the prototype. A conventional, fixed tricycle landing gear was provided for flight test purposes to permit conventional takeoffs and landings from a runway prior to vertical launching and recovery. Martin considered the temporary installation of a simple split flap to reduce the landing and takeoff speeds during the initial phase of flight testing.

Overall, Martin appears to have put less effort into the design of the Model 262P than the full scale Model 262; it may have determined early on that it was more cost effective for the Navy to build

Contemporary artist's impression of the Martin Model 262 Convoy Fighter proposal of 1950, one of the unsuccessful rivals to the Convair XFY-1 and Lockheed XFV-1.

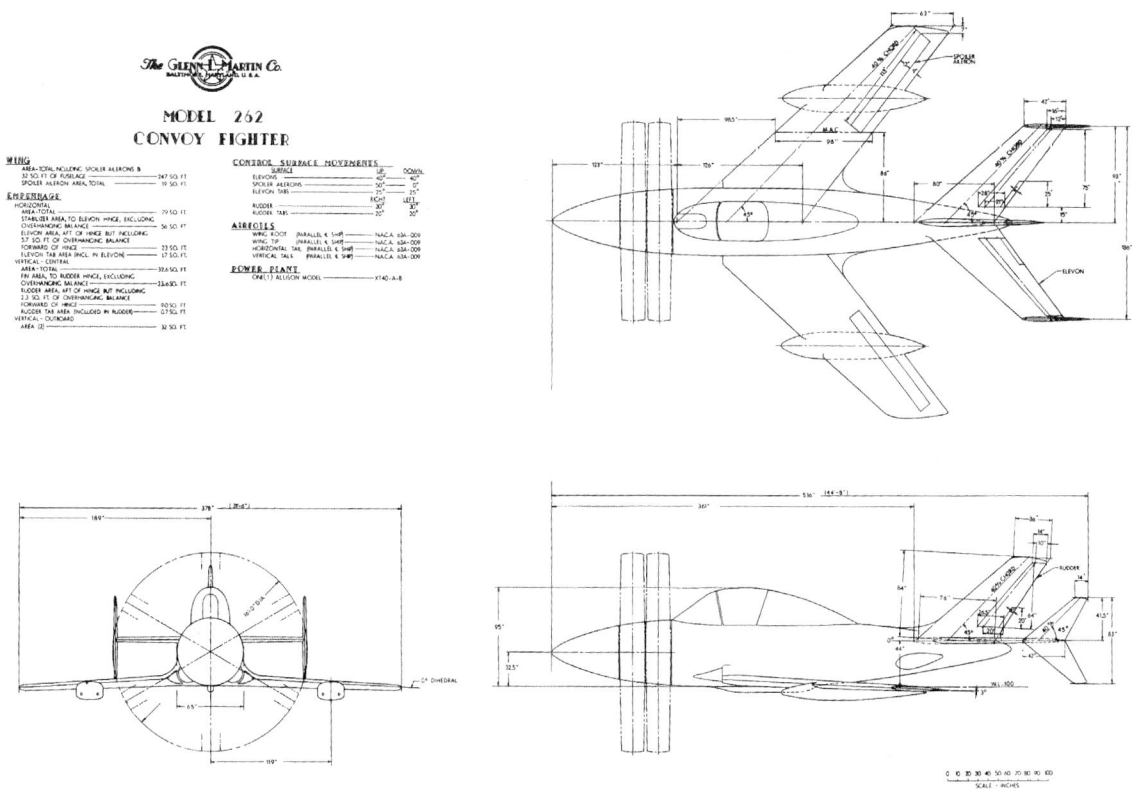

Three-view of the Martin Model 262 with detailed dimensional data added. Except for the contra-rotating propeller and triple vertical tails, the configuration was relatively conventional for a fighter of the early 1950s.

and test a stripped down version of the latter than bother with the three-quarter scale demonstrator.

COST PROPOSAL

Martin's Convoy Fighter cost proposal, dated December 1, 1950, was divided into two parts. Part I included an intensive wind tunnel test and aerodynamic study phase, the design and construction of two prototype scale airplanes, structural proof testing of one of the prototypes, flight testing of the prototypes, and design and furnishing of launching, retrieval and ground handling equipment. Part II included preliminary design data, construction of a mock-up, additional wind tunnel tests of the full

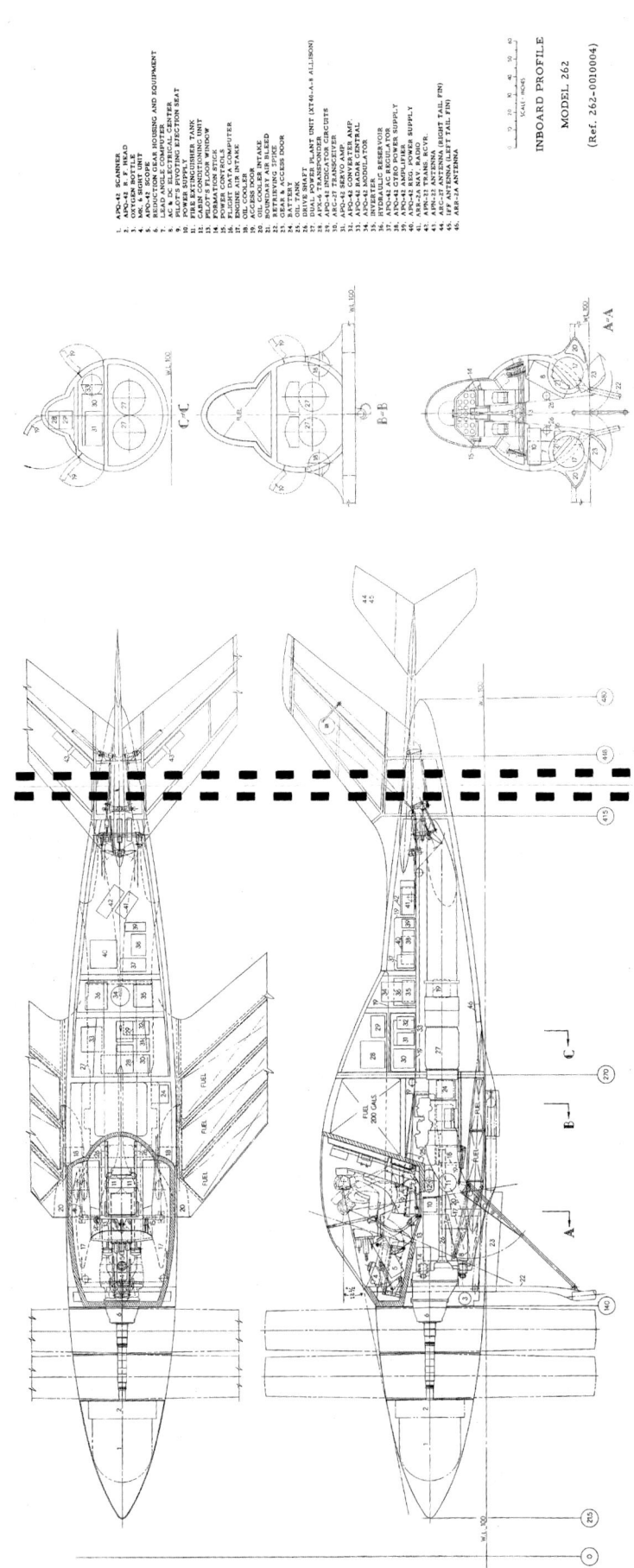

Inboard profile of the Model 262; note the rotating pilot's seat and ventral window under the cockpit to aid him in 'sticking' a landing on the recovery platform with the nose spike.

Artist's impression of Martin's Convoy Fighter in flight.

scale configuration, the design and construction of two complete flight articles and one static test article, a complete static test programme, a flight test programme and the furnishing of launching, retrieval and ground handling equipment.

Based upon the award of a cost plus fixed fee (CPFF) contract having mutually satisfactory provisions, Martin submitted the following quotation for the design and construction of the Model 262 Convoy Fighter and the prototype airplanes; all amounts are in 1950s dollars. For the scale prototype airplane, the cost was $4,661,159; for the experimental Convoy Fighter, the cost was $6,399,047. Total for the complete project was $11,060,206. This quote included a 7% fixed fee on estimated costs.

BuAer required contractors to also submit quotes on a fixed price basis. Martin felt that due to the highly experimental nature of the project and the time span involved, that protection against abnormal difficulties and rising costs of labour and materiel had to be included in a fixed price quotation. Accordingly, an additional amount was added on a percentage basis to the aforementioned estimated costs to arrive at a fixed price quotation. For the scale prototype airplane, the fixed price cost was $5,693,405; for the experimental Convoy Fighter, the fixed price cost was $7,808,933. Total for the complete project was $13,502,338 on a fixed price basis.

The above quotations included the following considerations:

1. A very intensive analytical and wind tunnel programme, aimed at solving the aerodynamic problems associated with the development of an optimum configuration was planned. Martin assumed that the high speed tunnel tests would be run in a government tunnel at government expense.
2. The estimate included funds for the development and complete flight testing of a reliable control system for automatically controlling the Convoy Fighter during level, transitional and hovering flight. This was considered important to the tactical utilization of the aircraft. Although maximum use of existing autopilot components was contemplated, the unusual flight attitudes in which this airplane operated required special sensing equipment.
3. The above prices included all services and

Martin looked at three additional configurations during the development of its Convoy Fighter proposal. Modification A of the Model 262 was a delta wing design which achieved significant reduction in wing and tail weight, increasing loiter time and reducing overall airframe weight.

Modification B was a delta wing design with a centrally located propeller. This reduced the moment arm from the propeller to the centre of gravity of the airplane, minimizing the destabilizing effects of the propeller. However, with this layout, the gearbox and propeller hub assembly had to be adapted to permit the support of the forward fuselage.

Modification C had a swept wing planform with the propeller located in the aft fuselage behind the wing and forward of the tail. In this location, the propeller was no longer destabilizing in level flight but contributed to overall stability and enabled a reduction in tail size. Note the prone position of the pilot.

data requested by BuAer with one exception to specification SR-38. With regard to the spin requirements, Martin considered that due to the unconventional characteristics of this airplane and its ability to maintain hovering flight, the spin demonstration requirements were considered too extreme. In lieu of complying with these requirements, Martin proposed investigation of the spin characteristics within the limits considered to be safe and practical as indicated by wind tunnel model spin tests and flight handling characteristics.

4. A launch and recovery platform would have been furnished for the prototype and the Convoy Fighter. The prototype platform would have been mounted on a suitable trailer to facilitate ground handling. The Convoy Fighter platform would have been suitable for mounting on a trailer or on a ship.

5. Included in the price of the static test article was the cost of an engine test stand for the Mamba engine suitable for checking engine operation, propeller and controls in horizontal and vertical positions. An operable engine test stand was also included in the cost of the first complete flight article.

6. The cost of the first complete flight article included basic tooling; the major assembly and major subassembly tooling would have been of the production type.

Martin had evolved a completely automatic blind landing system for recovery of the aircraft when the visibility was down as low as 75ft. The total estimated cost plus fixed fee to design and furnish a completely automatic blind landing system for the Convoy Fighter was $674,421. This cost was in addition to that quoted above since such a system was not included in the original requirements.

Martin estimated that the first flight of the prototype would occur 20 months after the authority to proceed was received; the first flight of the first experimental Convoy Fighter was estimated to occur 18 months after the authority to proceed was received. Although a prototype could have been built and flown in less time than quoted above, Martin felt that detail design and construction should not have begun until sufficient wind tunnel tests and aerodynamic studies had been accomplished to arrive at an optimum configuration. Approximately eight months were required to accomplish this.

After a review of the total programme as proposed, Martin concluded that careful consideration should be given to eliminating the construction of the two

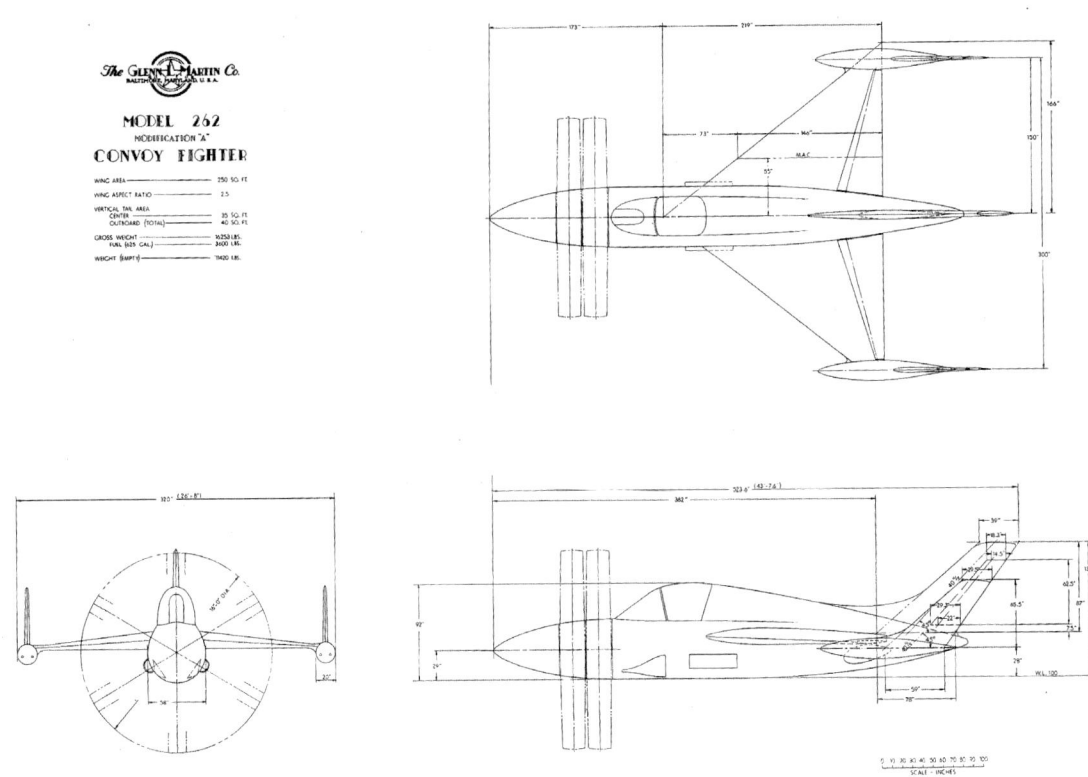

Three-view of the Martin Model 262 Modification A delta wing Convoy Fighter, the primary advantage of which was reduced airframe weight.

scale prototype models and proceeding directly to a full scale article powered by a T40-6 production engine rated at 5,500 ESHP. This power was sufficient to provide 5ft/sec² acceleration during the takeoff and transition at a takeoff weight of approximately 13,800lb.

The weight of the full scale prototype could have been held to this figure by omitting the armament, tactical electronic equipment and some fuel until such time as the 7,500hp T40 engine became available. This takeoff weight allowed sufficient fuel on board the airplane to permit flight testing. Since Martin's proposed design was based on the use of a single speed gearbox for the propeller, no gearbox availability problems were anticipated. Martin estimated that the time span for the development of the Convoy Fighter could have been reduced by approximately 14 months and the cost to a total of about $8,000,000 for the entire programme if the scale prototypes were eliminated.

The company appreciated that there may have been other considerations dictating the use of the prototype scale models of which it was not aware; however, if the alternate plan suggested had merit, a quotation on this basis would have been submitted upon request from the government. BuAer would ultimately agree to the elimination of the smaller prototype airplane, but did not award a contract to Martin for reasons discussed later in this book.

Martin definitely preferred a CPFF contract for this type of project, particularly when it was phased and extended over a multi-year period. It requested that any contract resulting from the proposal included provisions for partial payments to the maximum allowable by regulation and for passage of title to the government at the time of the first such payment.

The company claimed that it had the type of personnel and broad experience required to successfully develop the proposed Convoy Fighter and was intensely interested in undertaking the project.

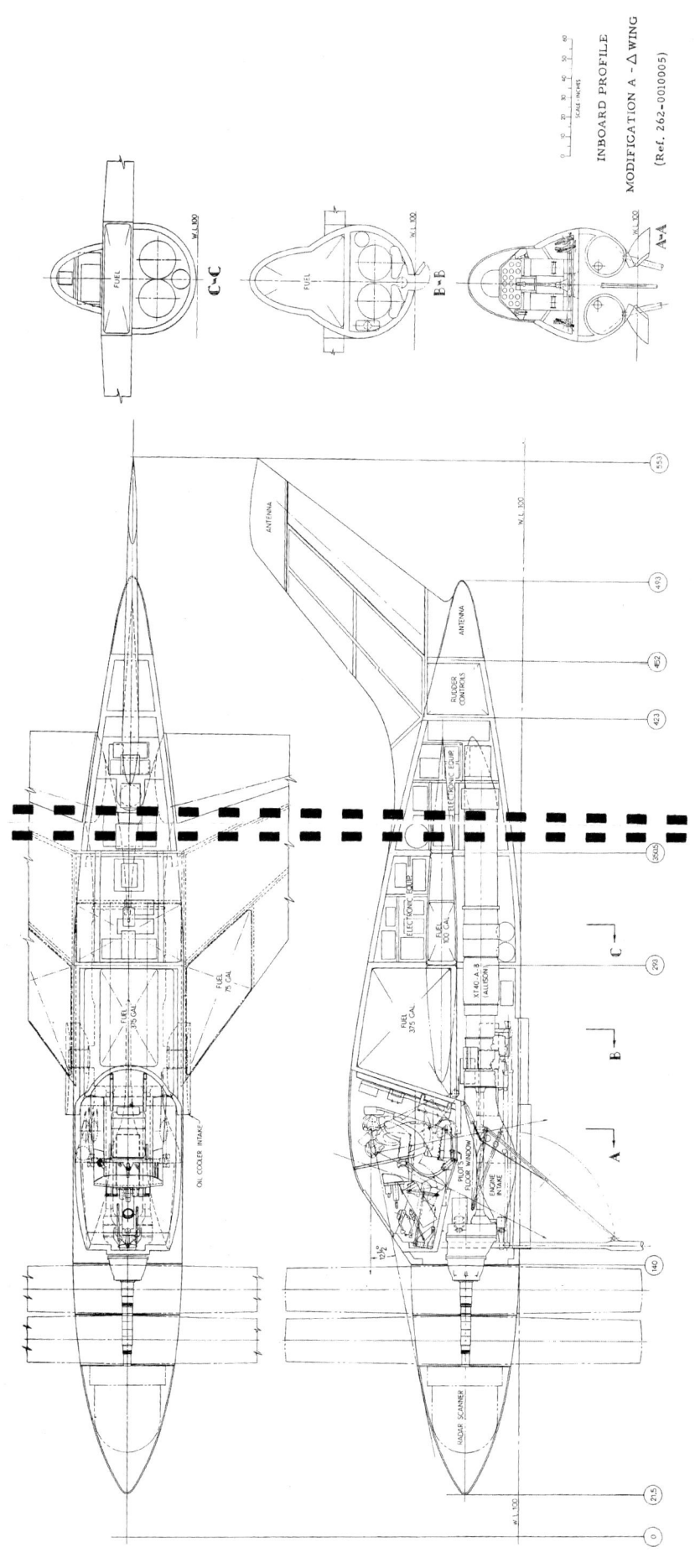

Inboard profile of the Modification A delta wing study, a lighter and simpler version of the Model 262.

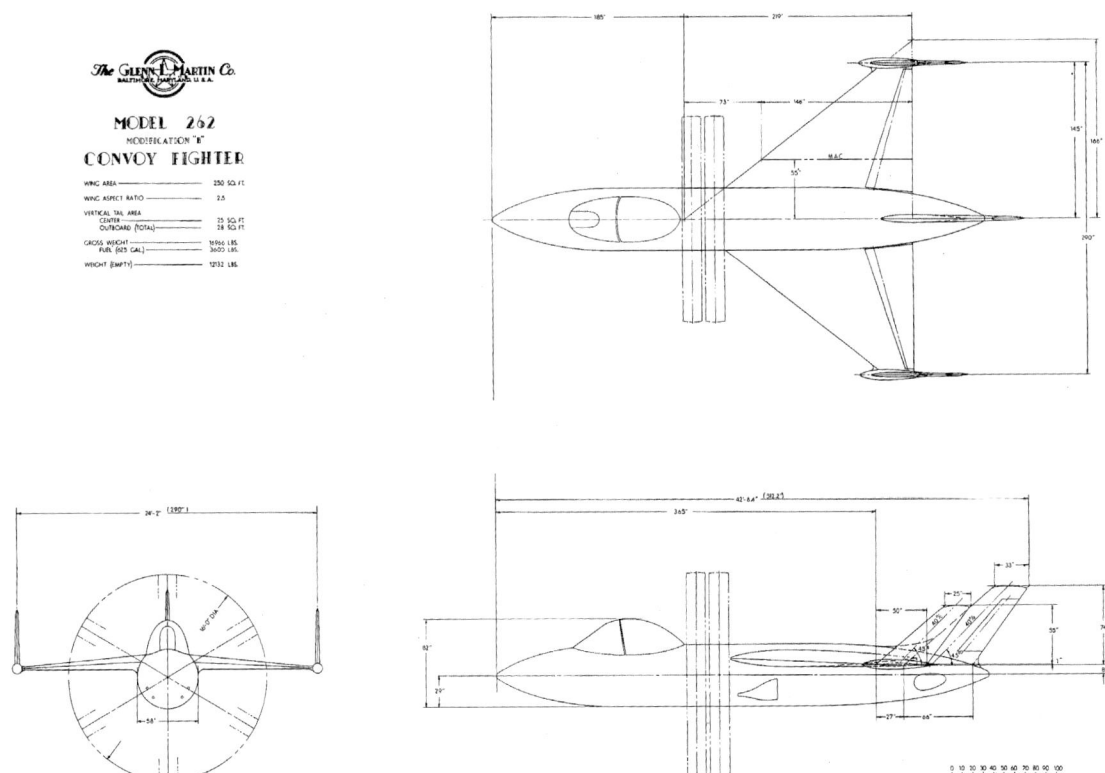

Three-view of the Martin Model 262 Modification B delta wing study with the propeller located mid-fuselage just behind the cockpit and forward of the wing.

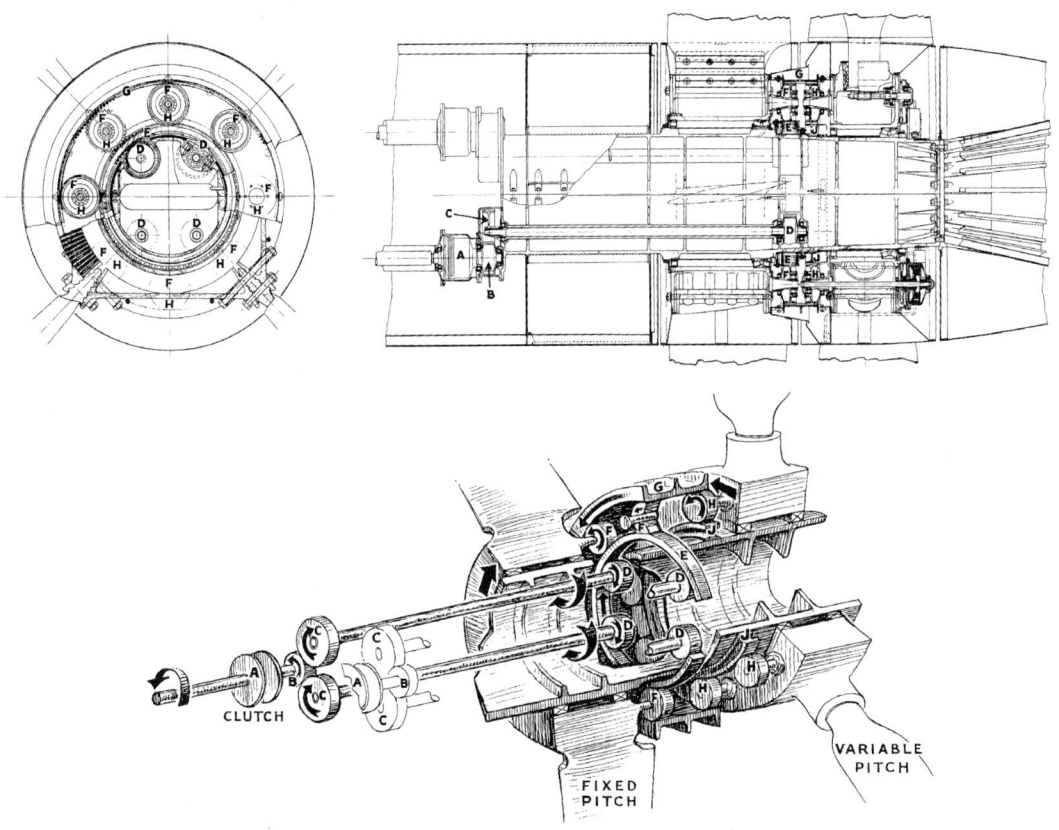

The alternate gear arrangement studied by Martin in connection with Modification C; it consisted of a differential ring gear driving one fixed pitch propeller and one variable pitch propeller in contra-rotation.

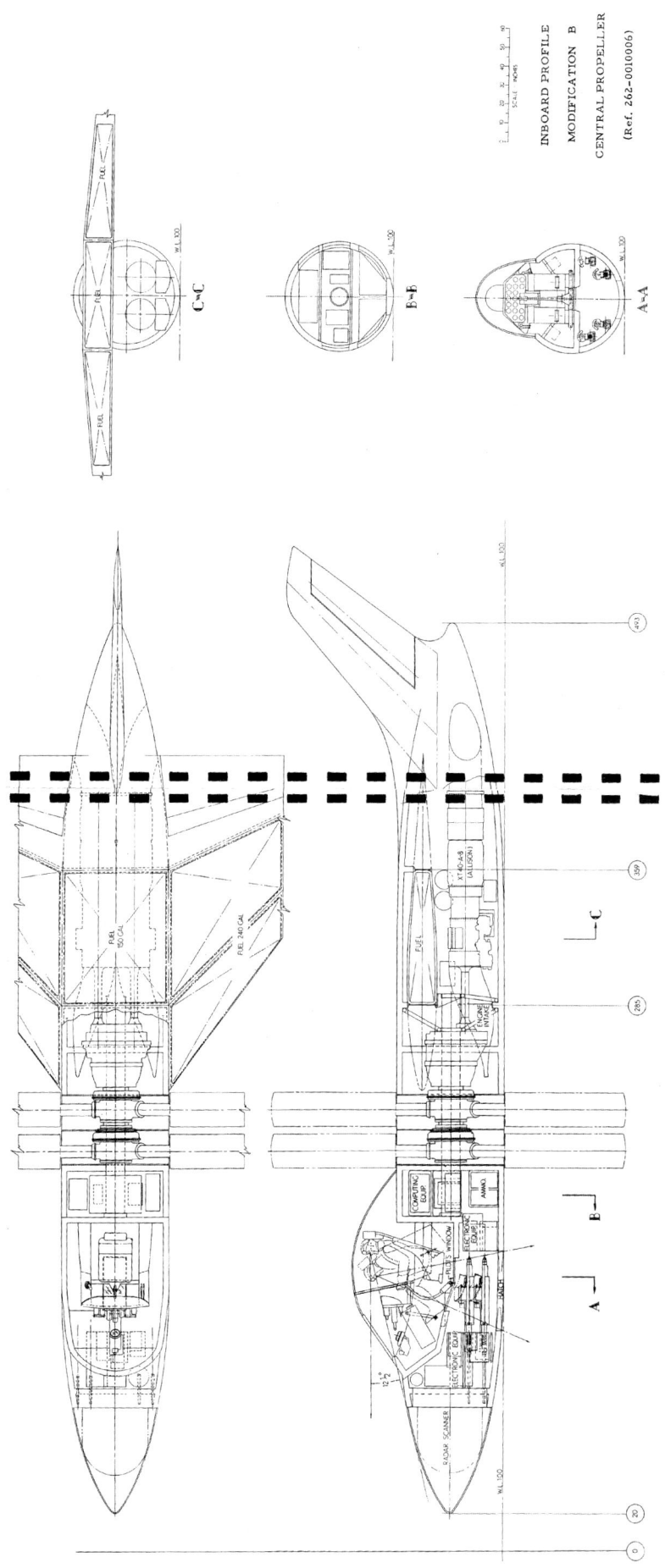

Inboard profile of the Martin Model 262 Modification B with a centrally mounted propeller. During an emergency, Martin suggested ejecting the pilot in a capsule rather than in an ejection seat due to the hazard posed by the unusual propeller location.

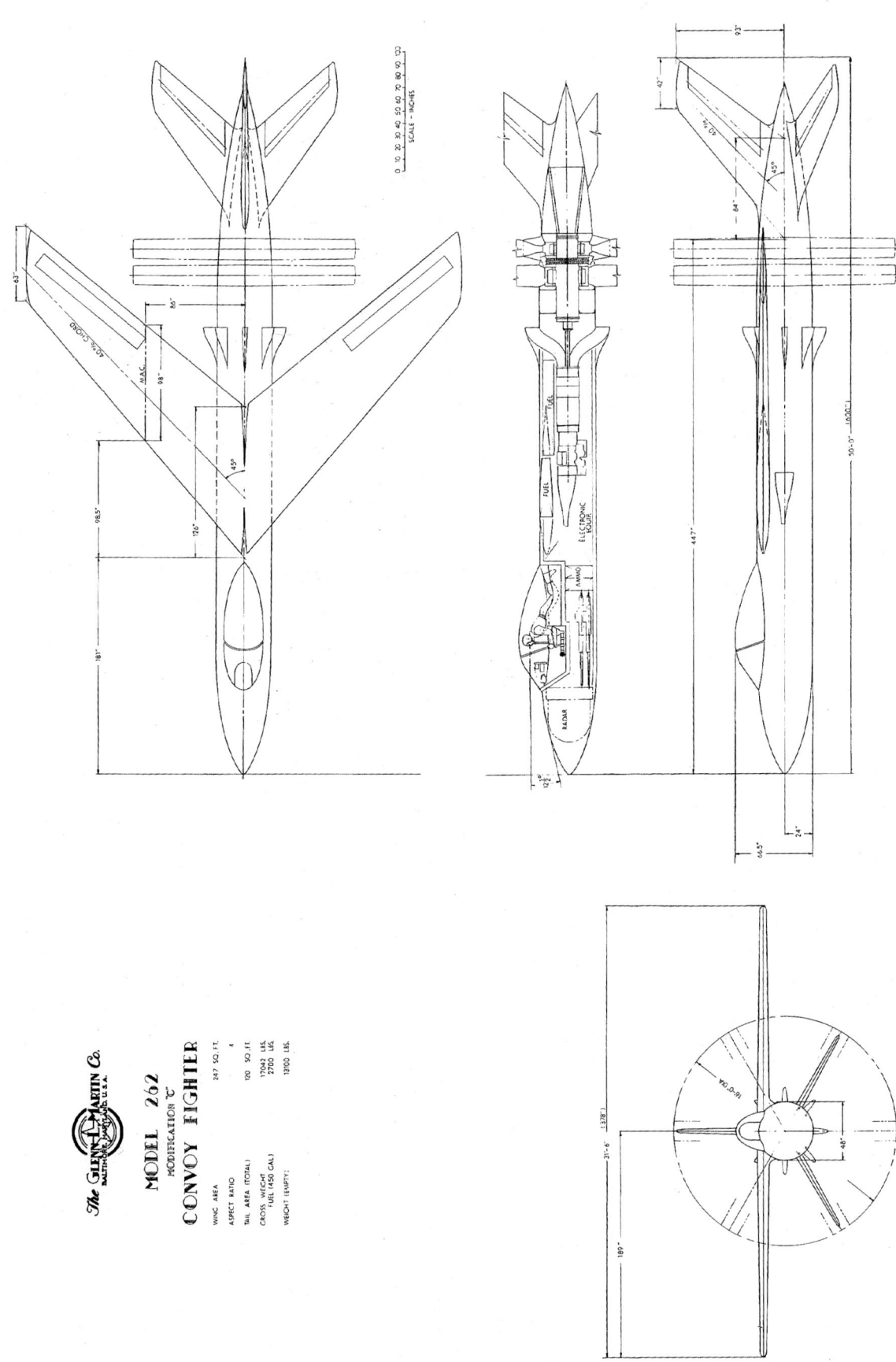

Three-view of Modification C with the contra-rotating propeller mounted in the aft fuselage; note the eight engine exhaust louvers just forward of the propeller.

QUALITATIVE COMPARISON

ITEM	MODEL 262 Swept Wing	MODIFICATION A Delta Wing	MODIFICATION B Central Prop.	MODIFICATION C Pusher
Performance				
Maximum Speed	base	comparable	superior	superior
Combat Ceiling	base	inferior	inferior	comparable
Time to Climb to 35,000 Ft.	base	inferior	inferior	comparable
Endurance	base	superior	superior	inferior
Stability				
Hovering	base	comparable	comparable	inferior
Level Flight	base	comparable	superior	superior
Arrangement				
Pilot Vision	base	comparable	superior	superior
Armament Installation	base	comparable	superior	superior
Gear Box Development	base	comparable	inferior	inferior
Weight Empty	base	superior (Lower)	superior (Lower)	inferior (Higher)

Table summarizing the pros and cons of the primary Martin Model 262 configuration relative to the alternate Modifications A, B and C.

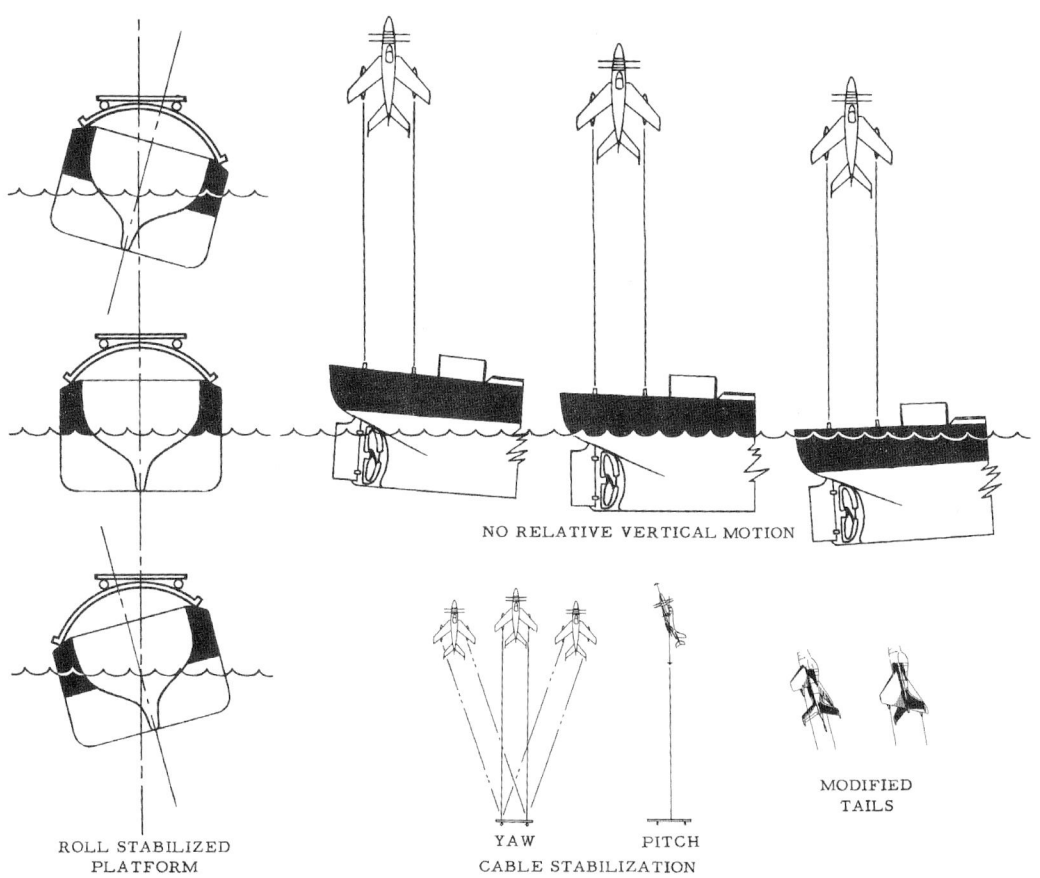

Illustration of an alternate recovery method for the Model 262, which involved the aircraft reeling itself in by lengths of cable to the ship, landing on a stabilized platform. This would have required modification of the tail, turning the aircraft into a true tailsitter.

Artist's impression of the Martin Model 262P Convoy Fighter prototype, a .766 scale demonstrator powered by a Armstrong Siddeley Double Mamba.

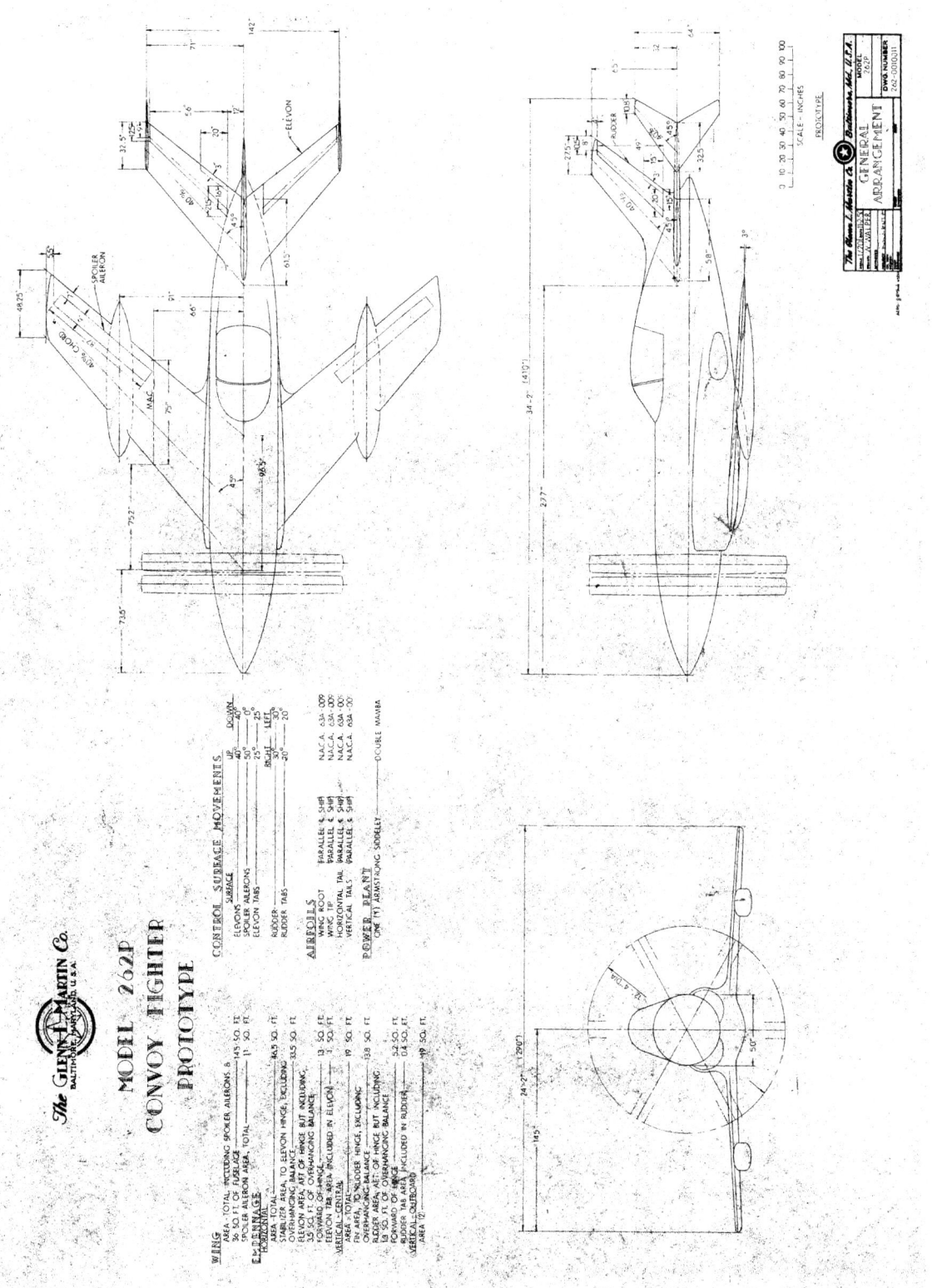

Three-view drawing of the compact Model 262P.

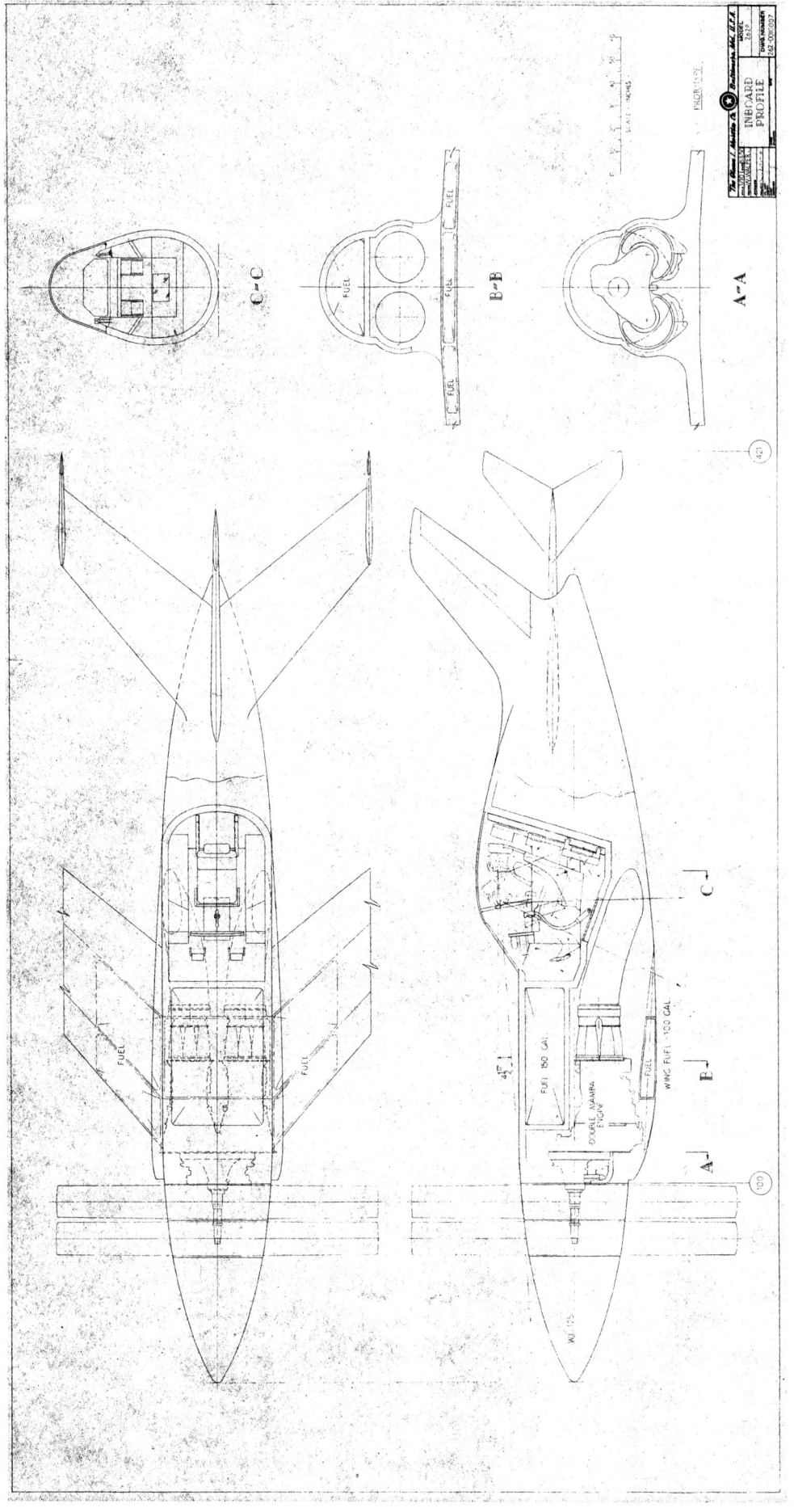

Inboard profile of the Martin Model 262P, which was aerodynamically similar to the full scale convoy fighter, except for the location of the pilot and air intakes.

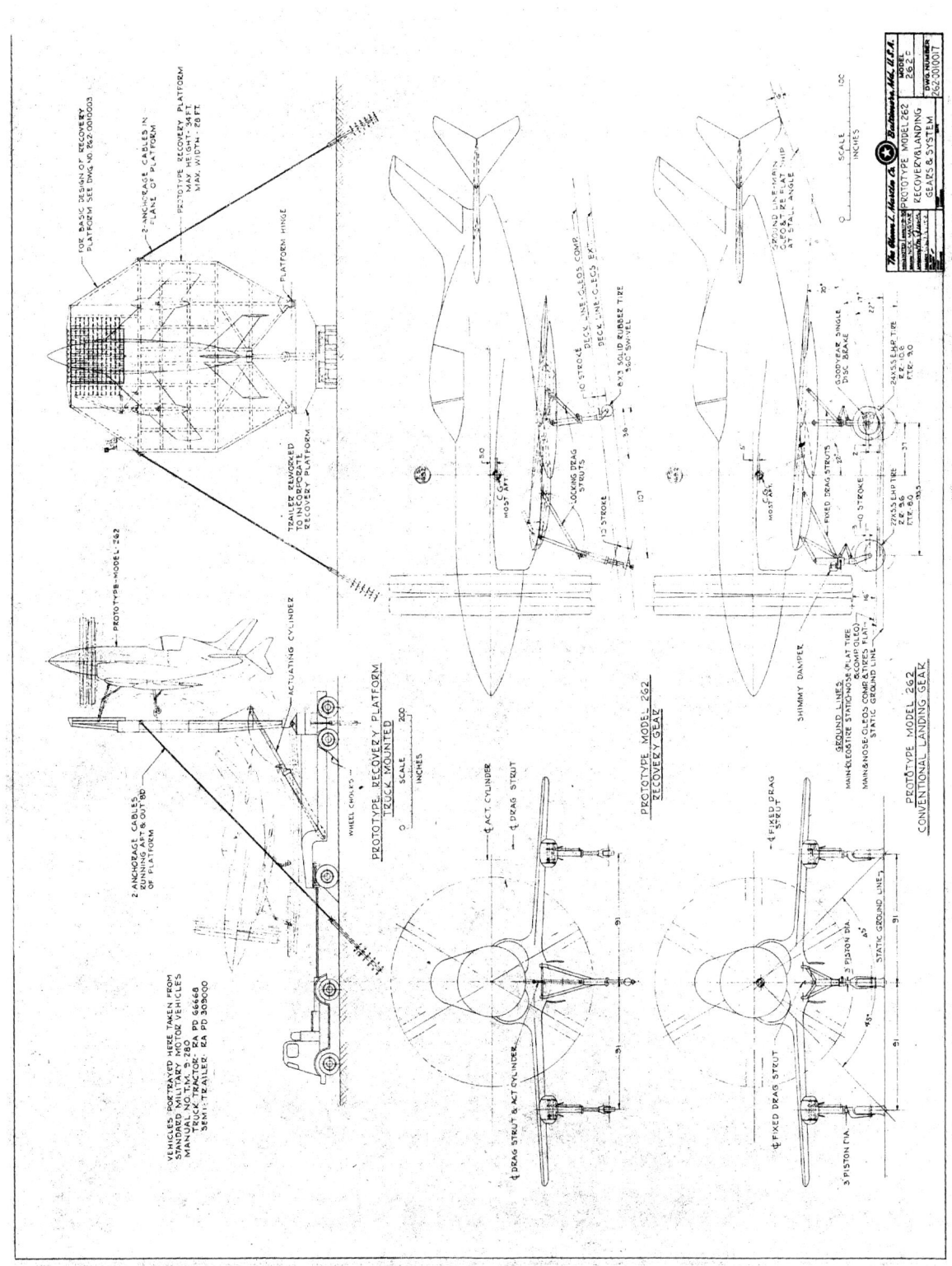

Blueprint showing the truck-mounted recovery platform designed to accommodate the Model 262P as well as the auxiliary fixed tricycle landing gear used for flight test purposes, permitting conventional takeoffs and landings from a runway.

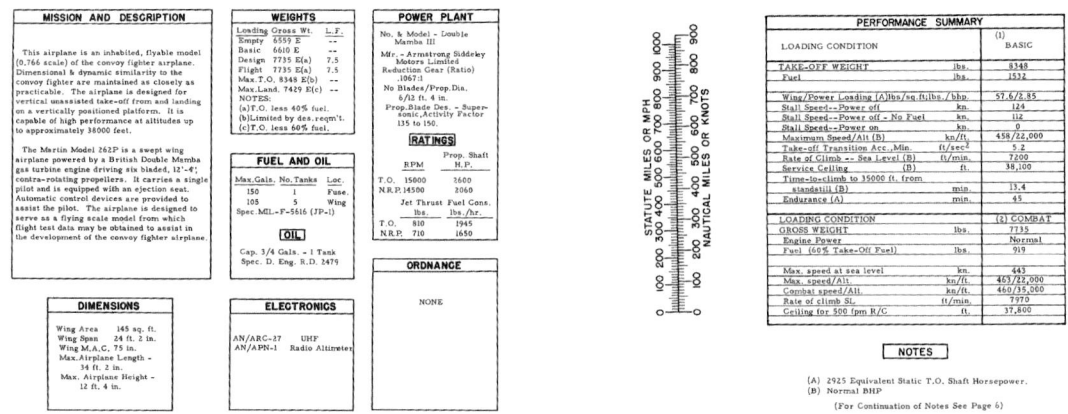

Standard Aircraft Characteristics charts for the Martin Model 262P presenting the physical and performance figures for the proposed aircraft.

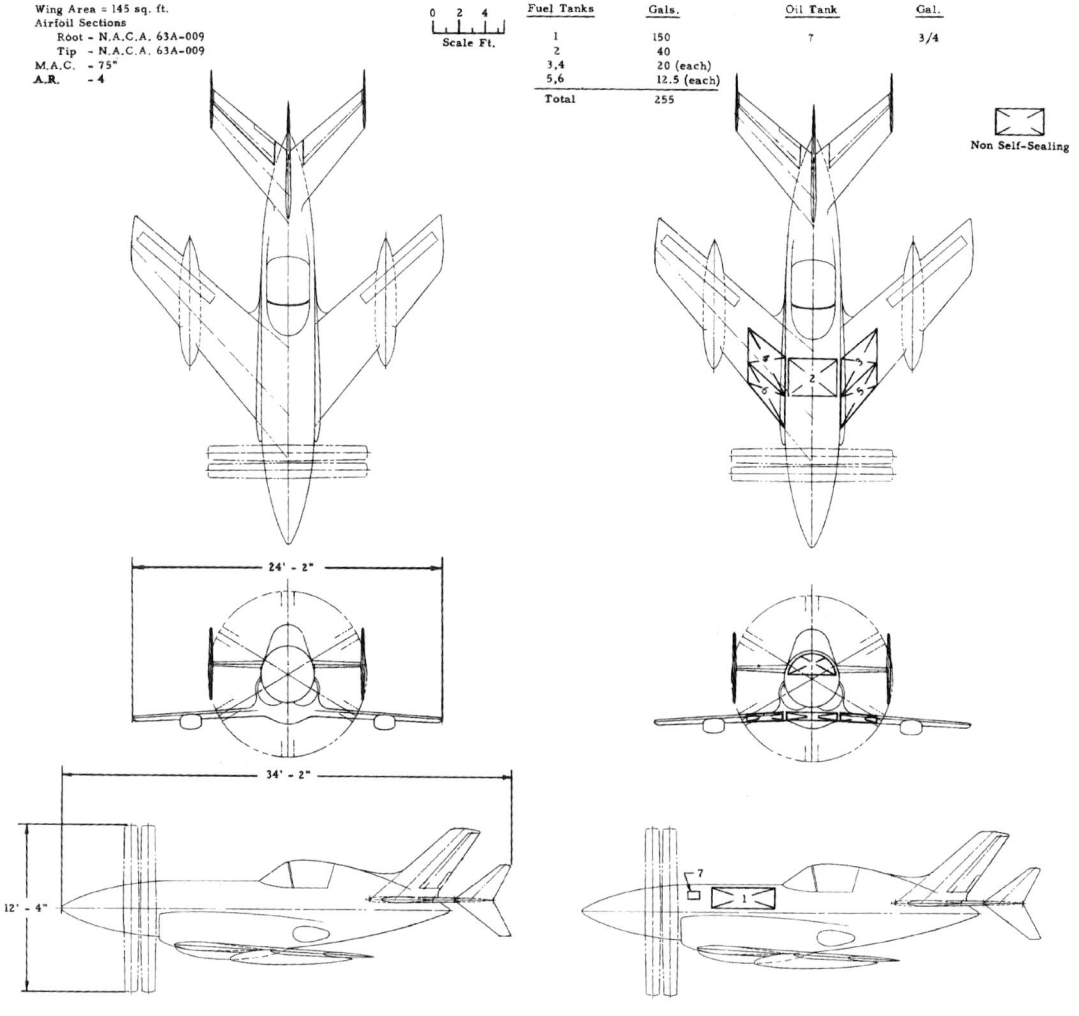

General arrangement drawings of the Model 262P taken from the Standard Aircraft Characteristics charts prepared for the type; the drawing on the right shows the location of the fuel and oil tanks.

A photo of the Glenn L. Martin Company of Baltimore, Maryland circa 1950.

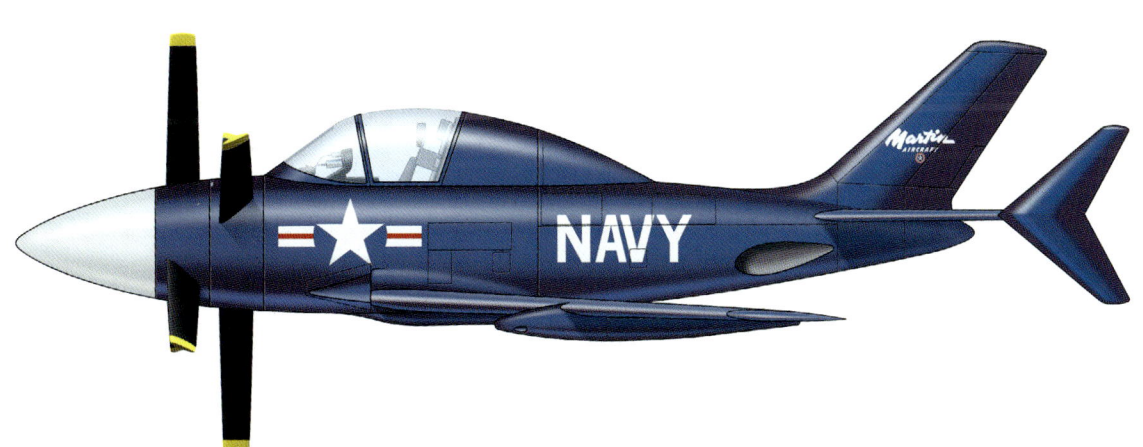

A speculative colour profile of the Martin Model 262 Convoy Fighter proposal of 1950 in the overall Glossy Sea Blue scheme which was standard for most US Navy aircraft in this period.

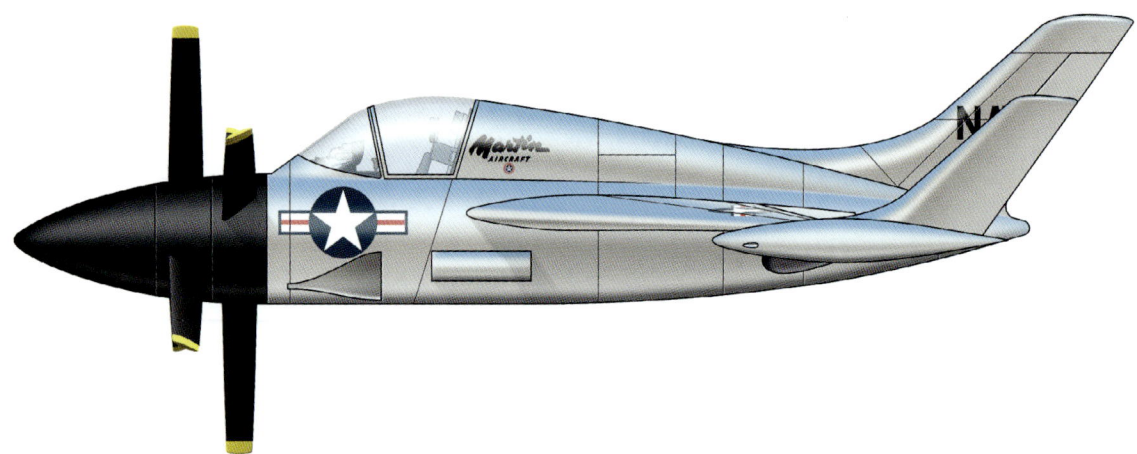

In addition to the primary Model 262 Convoy Fighter configuration, Martin presented three additional designs for the Navy's consideration. This speculative colour profile depicts the first of these, dubbed Modification A, which was a compact delta wing fighter that was lighter and had greater endurance than the baseline Model 262.

Modification B was a delta wing design with a centrally located propeller. This reduced the moment arm from the propeller to the centre of gravity of the airplane, minimizing its destabilizing effects. This configuration provided excellent forward vision for the pilot, increased gun fire accuracy, and overall aerodynamic cleanliness.

Modification C featured a swept wing planform with the propeller located in the aft fuselage behind the wing and forward of the tail. In this location, the propeller was no longer destabilizing in level flight but contributed to overall stability and enabled a reduction in tail size. A prone pilot position was envisaged for the design.

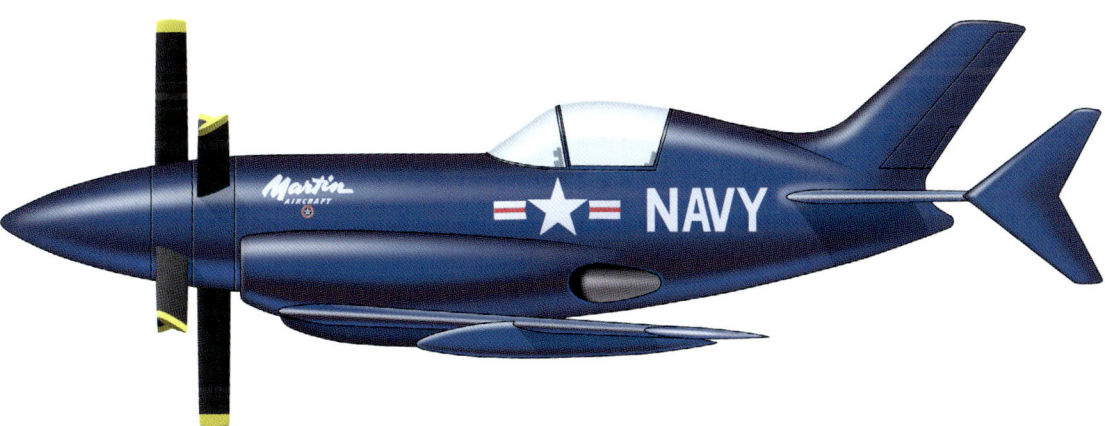

A speculative colour profile of the sleek and sharklike Model 262P in an overall Glossy Sea Blue scheme.

6

Northrop N-63 and N-63A

THE N-63 Convoy Fighter proposal produced by Northrop Aircraft, Inc. of Hawthorne, California was dated November 6, 1950. It was a single-place high performance fighter designed to protect convoy vessels from attack by enemy aircraft, and for vertical unassisted takeoff from, and landings on, small platform areas afloat or ashore. It was capable of high performance and manoeuvrability at altitudes from sea level to a combat ceiling of 47,000ft, and was controllable through a speed range from zero (hovering) to 528 kts.

The Northrop N-63 was a completely self-contained fighting weapon, requiring no auxiliary arresting or retrieving gear for successful takeoff and landing operations. It was designed for vertical takeoff from, and landing on, pitching and rolling decks under any sea and weather conditions in which normal operation of aircraft were expected. Shock absorbers at the aft extremity of the fuselage and at wing and vertical stabilizer tips, combined with widespread support points and a low centre of gravity, permitted vertical landing even under the most severe conditions. Only small positioning dollies and simple tie-down provisions were required to locate and secure the airplane. It was in the ready-for-takeoff attitude at all times. As a result, complete servicing and pre-flight operations were possible with the airplane in the vertical attitude.

Adequate control and stability during all phases of flight was assured by a full powered control system, with autopilot stabilization during transition and hovering. Inherent level flight dynamic stability assured safe flight in the event of failure of either the autopilot or powered control system. Excellent pilot vision for both combat and search resulted from the location of the cockpit well forward of the wing. Pilot comfort was assured during all phases of flight, with a rotatable pilot's seat permitted excellent vision down and aft during hovering and landing operations. Full control of the airplane was possible at any attitude, with or without autopilot operation.

An Allison XT40-A-8 turboprop engine drove an Aeroproducts 15.5ft six-blade dual-rotation propeller. Four 20mm aircraft guns with 600 rounds of ammunition were mounted in easily accessible wingtip pods. The Mk 6 gunsight and either APG-37, APG-25, or APS-25 radar provided accurate fire control.

Performance in all respects met and in many cases exceeded specification requirements. Whereas only safe horizontal flight was required with one power unit of the XT40-A-8 inoperative, the Northrop design permitted hovering and successful landing, under Standard NACA atmospheric conditions, even with such extremes of reduced power. Most performance requirements were achievable under 'Hot Day' (90°F, sea level) conditions. Manoeuvrability

Cover to the Northrop N-63 Convoy Fighter proposal brochure dated November 6, 1950.

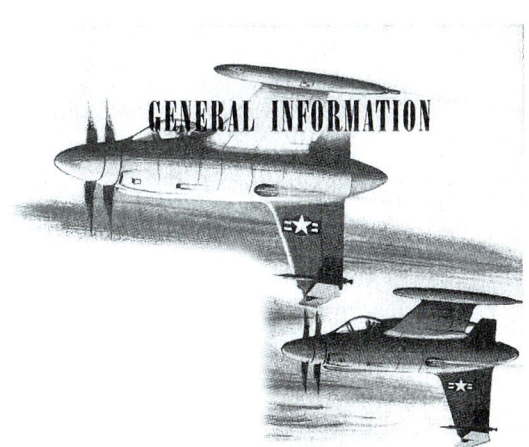

An artist's impression of a pair of Northrop N-63 Convoy Fighters in flight.

was sufficient to achieve a high probability of success in attacking a bomber travelling at 450 kts at altitudes between sea level and 47,000ft. Predicted standard performance characteristics included a 11,270ft/min rate of climb at sea level, combat ceiling of 47,000ft, 528 kts top speed, 2.78 hour endurance and 428nm combat radius.

Design history

The Northrop N-63 Convoy Fighter configuration evolved as the logical result of efforts to derive an optimum design to meet the various requirements set forth in the proposed specification. With the primary mission defined and certain components specified by Navy preference, analytical studies and years of Northrop experience designing full power control systems and stable unconventional aircraft

Title page from the Northrop N-63 brochure depicting the aircraft hovering over its takeoff/landing platform installed on a Liberty ship.

platforms were combined to produce the best possible vertically-rising convoy fighter.

Initial studies indicated that a tractor, dual-rotation propeller offered the most promise and that a two-speed reduction gear was necessary to achieve maximum propeller efficiency over a great range of speed. Preliminary analyses indicated that autopilot stabilization during hovering was required, although manual control could have been provided in the event of autopilot failure. Northrop experience indicated that a variation of control forces and reliability of operation made mandatory a fully powered control system.

Two general types of configuration were considered in the initial studies. One approach assumed an essentially conventional airplane to be recovered by a shipboard retrieving mechanism designed to meet the various requirements imposed by such a design. Studies indicated that such a retrieving gear would have been complicated and required an elaborate shipboard installation.

The other basic approach undertook the development of an unconventional airplane which was landed vertically, with simultaneous development of a matching retrieving gear which permitted the design of both components to be most effective. This retrieving gear was less complicated than the other and involved less elaborate shipboard installation.

Several wing planforms were considered, including swept, delta and thin straight wings. Theoretical studies substantiated by test of a Navy model at the David Taylor Model Basin indicated that wings with any appreciable degree of sweepback became highly unstable at low speed and high engine power. A thin straight wing was therefore selected in preference to the less stable swept or delta planforms.

Comprehensive studies of the effects of Mach number and power variations upon longitudinal stability were conducted. General results of these investigations led to the conclusion that (1) a propeller and wing arrangement could be designed with small changes in stability due to Mach number or power effects, but that (2) the addition of a conventionally located horizontal tail would cause large and erratic stability and trim shifts. The possibilities of buffeting or shake were greatly increased by the strong propeller slipstream. An unconventional location for the horizontal tail was indicated as the solution.

Intensive investigations of landing systems took into account various trapezes, barriers, nets, clamshells, booms and arresting hooks. The problem finally resolved itself into rather simple terms which led to the system finally selected as the most effective and simplest means of retrieving the airplane.

A successful landing system required compatibility with some reasonable positioning error of the airplane. In order to maintain stability in hovering during the final phases of the landing, it appeared extremely undesirable to attach any restraint to the airplane before its weight rested fully upon the retrieving gear. Thus, the problem of retrieving became merely one of transferring support of the airplane's weight from the propeller to the retrieving mechanism quickly and positively. It followed that the most desirable and effective landing system would simply have the airplane alight upright on a flat landing platform without any auxiliary arresting or retrieving mechanism. Loads imposed by

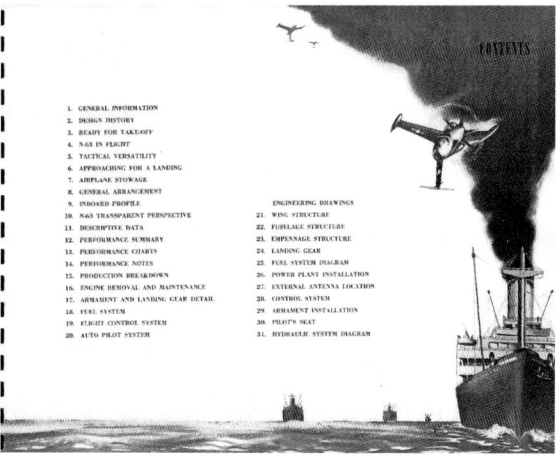

Table of contents page depicting the N-63 performing its primary mission—protecting a convoy of merchant vessels.

Illustration showing some of the earlier configurations evaluated by Northrop before settling on the final design of its N-63 proposal.

descent velocities, rolling and pitching of the ship, and misalignment of the airplane could be absorbed by properly designed shock struts, while upsetting after contact could be prevented by widely spacing the support points.

Aerodynamic and structural investigations indicated that a conventional empennage was undesirable. Relocation of the empennage permitted the airplane to be supported on its wing tips in the vertical attitude and eliminated the erratic effects of downwash on the horizontal tail. Directional stability was provided by the addition of a large ventral vertical fin, which also served as the necessary third landing support. The remaining problem of providing satisfactory longitudinal trim and control was solved by adding a movable horizontal surface at the bottom of the vertical fin, away from both wing wake and propeller slipstream.

As a result of these various studies, the desirable characteristics of an optimum configuration emerged. Stability and control requirements dictated a thin straight wing. Location of gun pods at the tips of the straight wing permitted good gun-platform characteristics without the dangers of aeroelastic twist and deflection associated with swept wings. The unconventional ventral fin and horizontal tail provided satisfactory stability and trim characteristics. Wide spread landing support points were provided at wing and fin tips, thus simplifying shipboard installations. Adequate control surface area would be located in the propeller slipstream for control during hovering flight.

Northrop N-63A

Northrop's design of the experimental prototype vertically-rising airplane involved two primary considerations. First, it had to be a flying aerodynamic model of the full-scale fighter. Second, it had to be equipped with an auxiliary landing gear to permit conventional takeoff and landing for exploration of unconventional transition and hovering flight characteristics at altitudes where takeoff and landing considerations would not be involved.

Extensive studies of the full-scale N-63 Convoy Fighter resulted in an optimum configuration for a vertically-rising airplane. Its design was in no way compromised by conventional landing requirements of the prototype airplane. The primary prototype requirement was satisfied by development of a design most nearly representing aerodynamic characteristics

READY FOR TAKE-OFF

The Northrop Convoy Fighter is a self-contained fighting weapon requiring a minimum of deck or ground handling facilities. It can operate from a simple flat deck over the stern of a ship, or from any equivalent flat area. Only small positioning dollies and simple tie-down provisions are required for locating and securing the airplane. It requires no special hoisting apparatus since it is in the take-off attitude at all times.

N-63 IN FLIGHT

Excellent visibility, aerodynamic cleanness and good gun-platform stability make the Northrop Convoy Fighter an effective fighting weapon. The pilot's cockpit has been placed much farther forward than in more conventional fighters. As a result visibility forward and down, to the sides, aft and overhead are exceptionally good.

Simplicity of design and configuration fineness permit a wide range of forward speeds from zero (hovering) to 528 knots. Four 20 mm. guns mounted in pods at the tips of the straight wings provide heavy fire-power coupled with the ultimate in accuracy of fire control. The line of fire of each gun is well outboard of the propeller circles.

TACTICAL VERSATILITY

The Northrop Convoy Fighter, though designed primarily for convoy protection, is well suited to the additional demands which might be made of it by Amphibious Ground Forces. Its unique tail landing gear gives real meaning to the heretofore academic phrase: "Close Ground Support."

APPROACHING FOR A LANDING

The Northrop Convoy Fighter is designed to land vertically on a flat deck over the stern of a ship, or on any equivalent flat area afloat or ashore. The airplane is placed in the hovering attitude over the platform, then lowered until it rests on the deck. Fine positive control in the hovering position makes the entire landing operation routine. Landing is accomplished with a minimum of deck gear or equipment. No arresting or retrieving gear is necessary.

The wide tread of the tail landing shock absorbers permits completely self-contained landing on pitching and rolling decks under the worst possible weather conditions in which the airplane might be operated. This completely self-contained operation of the airplane is an outstanding tactical advantage.

The airplane has been designed to be stowed or serviced in the vertical attitude to eliminate bulky handling equipment. In this attitude, the stability of the airplane at rest is greater than it would be if it were secured on equipment provided for horizontal stowage. This characteristic is directly attributable to the wide tread of the tail landing gear and the proximity of the center of gravity to the deck.

Tie-down cables, easily attached and removed, are provided to secure the airplane. Covers for propellers, canopy, etc., are provided for use in unfavorable weather. A dolly under each of the three landing struts is used for maneuvering the airplane on deck. The main central shock absorber is locked in the static position when the airplane is to be moved about.

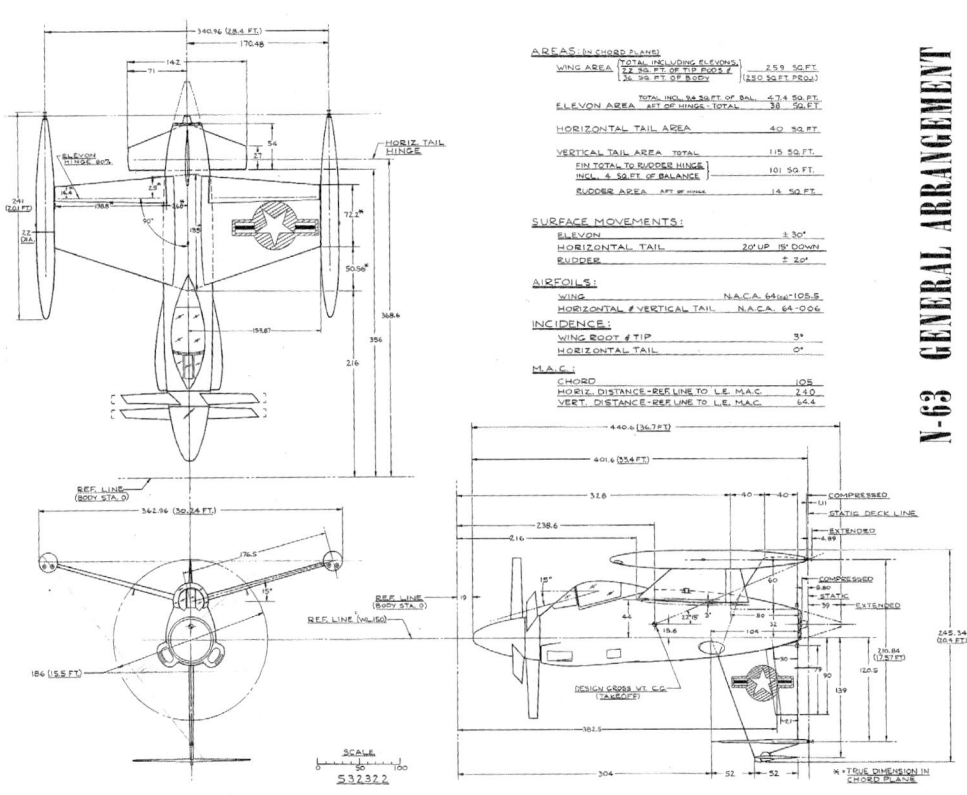

NORTHROP N-63 AND N-63A

N-63 TRANSPARENT PERSPECTIVE

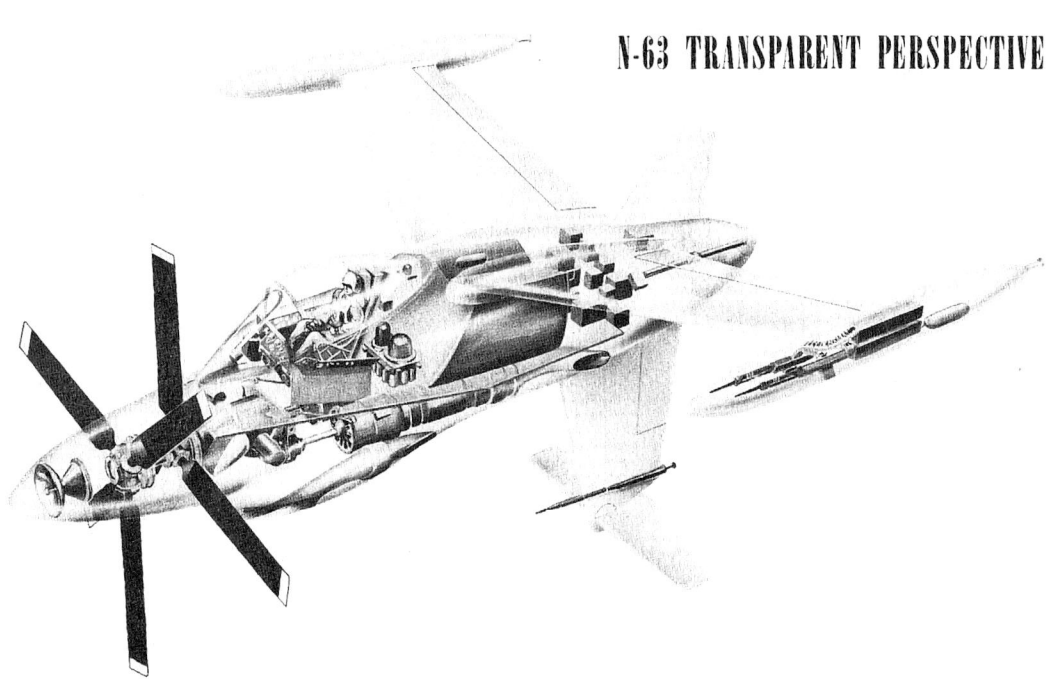

DESCRIPTIVE DATA

MISSION AND DESCRIPTION

The primary mission of the Northrop N-63 airplane is to protect convoy vessels from air attack by enemy aircraft.

The Northrop N-63 is a proposed U.S. Navy Class VF (Convoy Fighter) Airplane based on BuAer Outline Specification OS-122. It is designed for vertical unassisted take-off from, and landing on, small platform areas of convoy vessels. It also is designed for high performance at low and medium altitudes and for operation under all weather conditions.

The airplane is essentially a conventional single-engine tractor monoplane configuration except for features appropriate to vertical take-off and landing. Longitudinal control at low speed and lateral control for all conditions are obtained by elevons, which are drooped to improve ceiling and maneuverability. Longitudinal control during normal flight at higher speeds is obtained by an all-movable horizontal tail. Directional control is provided by a rudder on the vertical tail. Aerodynamic braking is provided by the propeller alone. The alighting gear consists of a main central shock absorber in the tail and stabilizer shock absorbers on the wing tips and lower fin tip. Construction is all-metal. Crew consists of a pilot.

WEIGHTS

Loading	Pounds	L.F.
Empty	12,283 (E)	
Basic	12,801 (E)	
Design (flight)	15,600	7.5
Combat	15,454 (E)	7.5
Max. Take-off	16,780 (E)*	
Max. Landing	14,795 (E)	

*Limited by Space

FUEL AND OIL

Location	No. Tanks	Cap.
Fuselage	1*	552.5 gal.
Total		552.5 gal.

Spec. MIL-F-5572
Grade 100/130
*Not self-sealing

OIL

Capacity (1 fus. tank) 7.5 gal.
Spec. MIL-O-6086
Grade M

DIMENSIONS

Span	30.2 ft.
Length	33.4 ft.
Height	20.4 ft.*
Wing Area	250. sq.ft.

*In horizontal position

ELECTRONICS

UHF Trans-receiver	AN/ARC-27
Homing Receiver	AN/ARR-2A
Radio Altimeter	AN/APN-22
Gun-laying Radar	AN/APG-37
Radar Identification	AN/APX-6

POWER PLANT

ENGINE

Number & Model	(1) XT40-A-8
Manufacturer	Allison
Type	axial-flow turbo-prop.
Augmentation	None
Length	190 in.
Width	43 in.
Height	35 in.
Reduction gear	
Take-off (& landing)	10.95:1
Normal	15.67:1
Specification	272-B (31 May 1950)

PROPELLER

Manufacturer	Aeroproducts
No. blades/dia.	6/15.5 ft.
Propeller No.	A086564F

RATING

(Sea level static guarantees)

	Engine Speed rpm	Shaft Power bhp	Jet Thrust lb	Fuel Cons. lb/hr	Time Limit min
Take-off	15,700	6825	1685	4325**	5
Military*	14,300	6955	1363	4095**	30
Normal	14,000	5790	1225	3642**	-

*Military rating used for take-off and landing.
**Using Grade JP-3 (Spec. MIL-F-5624) fuel.

ARMAMENT

No. & cal of guns	(4) 20-mm
Amm. per gun	150 rds.

IN TRANSITION BEFORE LANDING

DESCRIPTIVE DATA

PERFORMANCE SUMMARY

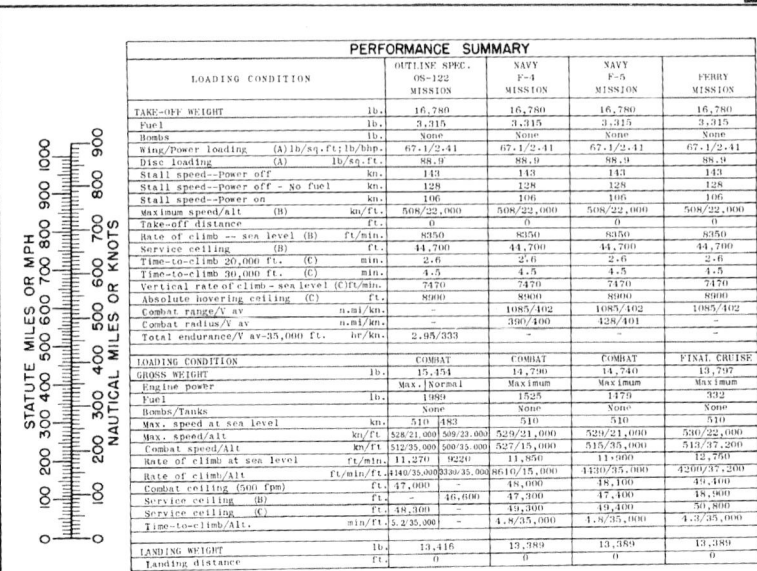

PERFORMANCE SUMMARY

PERFORMANCE CHARTS

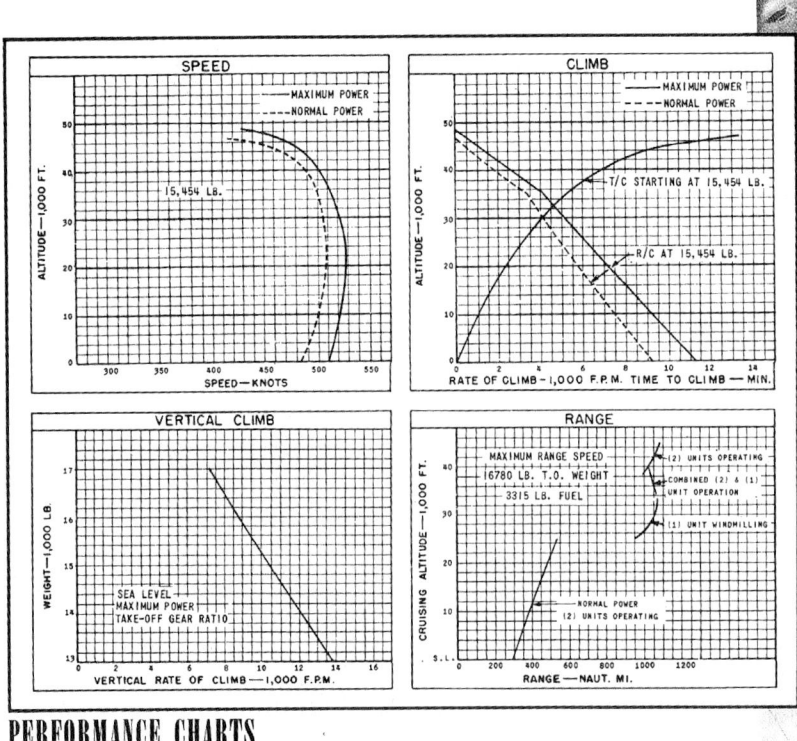

PERFORMANCE CHARTS

PERFORMANCE NOTES

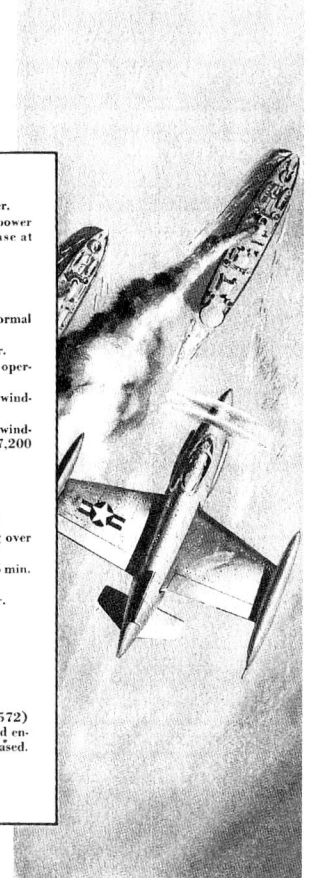

NOTES

(a) Performance basis:
   (1) NACA standard conditions, no wind, single aircraft.
   (2) Calculated airplane and propeller characteristics; estimated and guaranteed engine characteristics.
   (3) Blade thickness/chord ratio = 0.0497 at 0.7 radius

(b) Endurance, range, and radii are based on Allison Specification 272-B (as revised 31 May 1950) fuel consumption data pertaining to Grade JP-3 fuel (Specification MIL-F-5624) with a density of 6.5 lb/gal.

(c) Fuel consumption data are increased 5 per cent.

(d) Mission flight plans:
   (1) Navy Outline Specification OS-122 mission:
      1. Take-off fuel allowance ≡ 5 min. at maximum power at sea level.
      2. Climb to 35,000 ft. at military power.
      3. Loiter for 130 min. at 35,000 ft. at maximum-endurance speed (one power unit windmilling).
      4. Cruise out 100 n. mi. at 35,000 ft. at maximum power.
      5. Combat 3 min. at 35,000 ft. at military power.
      6. Cruise back at maximum-range speed, starting at 37,400 ft. and ending at 37,700 ft. (one power unit windmilling).
      7. Approach and landing fuel allowance ≡ 5 min. at static military power at sea level.
   (2) Navy Basic Fighter F-4 mission (radius):
      1. Take-off fuel allowance ≡ 5 min. at normal power at sea level.
      2. Climb to 44,300 ft. at maximum power.
      3. Cruise out at long-range speed, arriving over target at 45,500 ft.
      4. Descend to 15,000 ft. and combat for 10 min. at maximum power.*
      5. Climb to 36,600 ft. at maximum power.
      6. Cruise back at long-range speed (one power unit windmilling), arriving over base at 37,800 ft.
      7. Reserve ≡ 10% of initial fuel.
   (3) Navy Basic Fighter F-4 mission (range):
      1. Take-off fuel allowance ≡ 5 min. at normal power at sea level.
      2. Climb to 44,300 ft. at maximum power.
      3. Cruise at long range speed (two units operating).
      4. Descend to cruise altitude for one unit windmilling.
      5. Cruise at long range speed, (one unit windmilling), arriving over destination at 37,200 ft.
      6. Reserve ≡ 10% of initial fuel.
   (4) Navy Basic Fighter F-5 mission (radius):
      Same as Navy F-4 mission (radius) except:
      3. Cruise out at long-range speed, arriving over target at 45,500 ft.
      4. Descend to 35,000 ft. and combat for 15 min. at maximum power.*
      5. Climb to 36,500 ft. at maximum power.
   (5) Navy Basic Fighter F-5 mission (range):
      Same as Navy F-4 mission (range).
   (6) Ferry mission:
      Same as Navy F-4 mission (range).

(e) When Grade 100/130 fuel (Specification MIL-F-5572) is used, fuel consumption is reduced 2 per cent and endurance, range, and radii are correspondingly increased.

* No fuel nor distance allowed for descent.
   No distance allowed for combat.

PRODUCTION BREAK DOWN

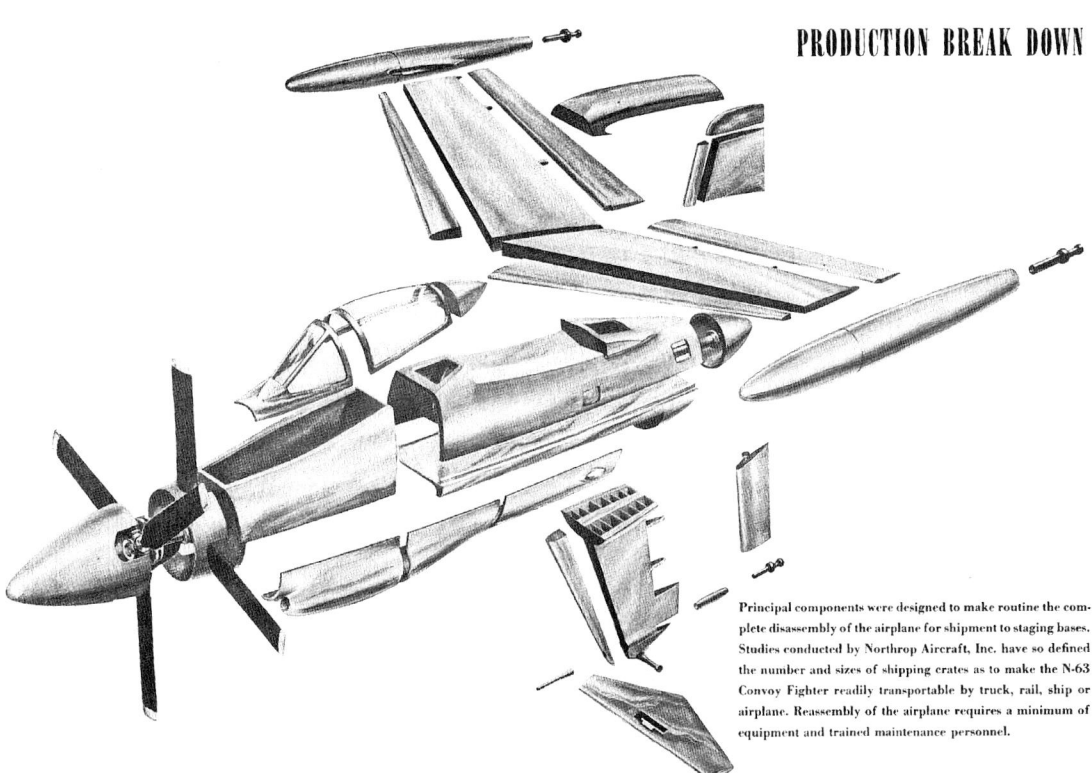

Principal components were designed to make routine the complete disassembly of the airplane for shipment to staging bases. Studies conducted by Northrop Aircraft, Inc. have so defined the number and sizes of shipping crates as to make the N-63 Convoy Fighter readily transportable by truck, rail, ship or airplane. Reassembly of the airplane requires a minimum of equipment and trained maintenance personnel.

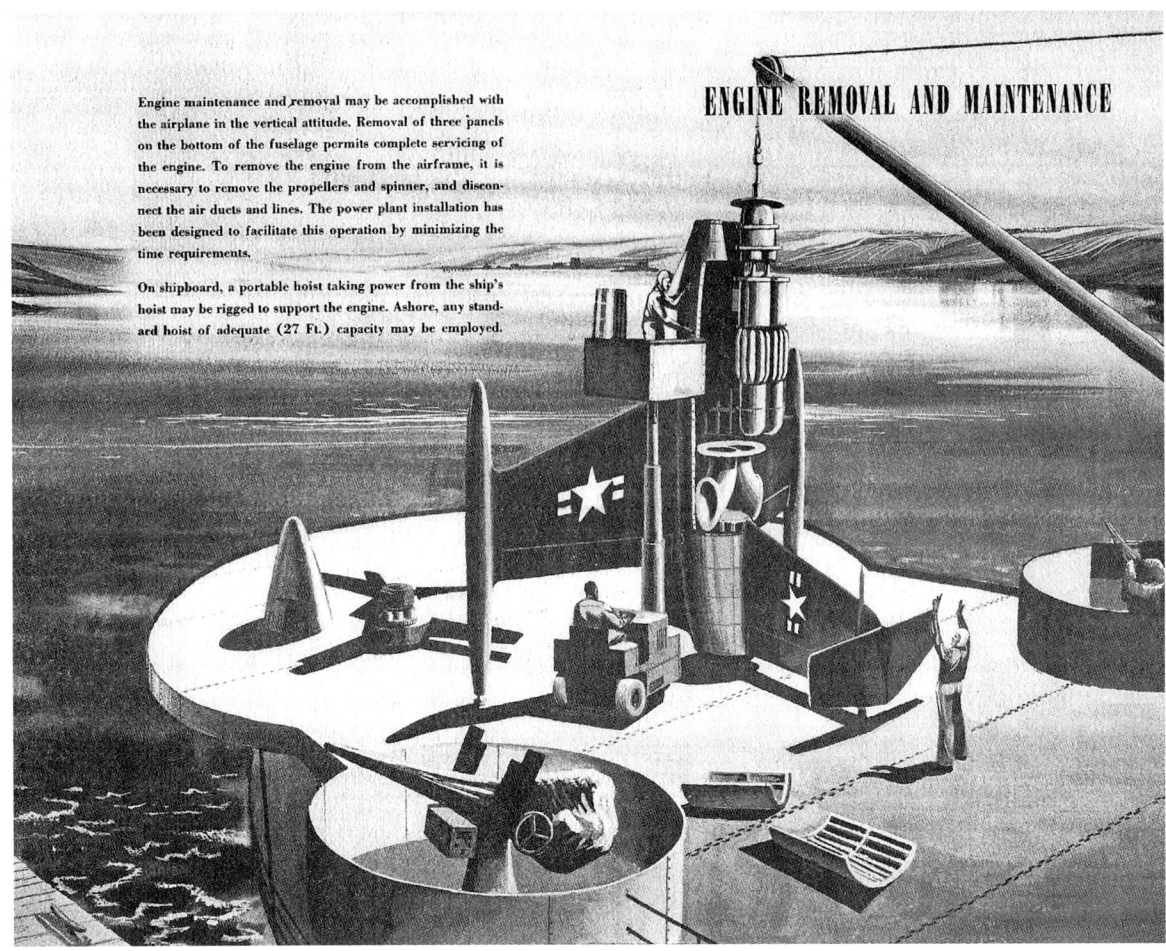

ENGINE REMOVAL AND MAINTENANCE

Engine maintenance and removal may be accomplished with the airplane in the vertical attitude. Removal of three panels on the bottom of the fuselage permits complete servicing of the engine. To remove the engine from the airframe, it is necessary to remove the propellers and spinner, and disconnect the air ducts and lines. The power plant installation has been designed to facilitate this operation by minimizing the time requirements.

On shipboard, a portable hoist taking power from the ship's hoist may be rigged to support the engine. Ashore, any standard hoist of adequate (27 Ft.) capacity may be employed.

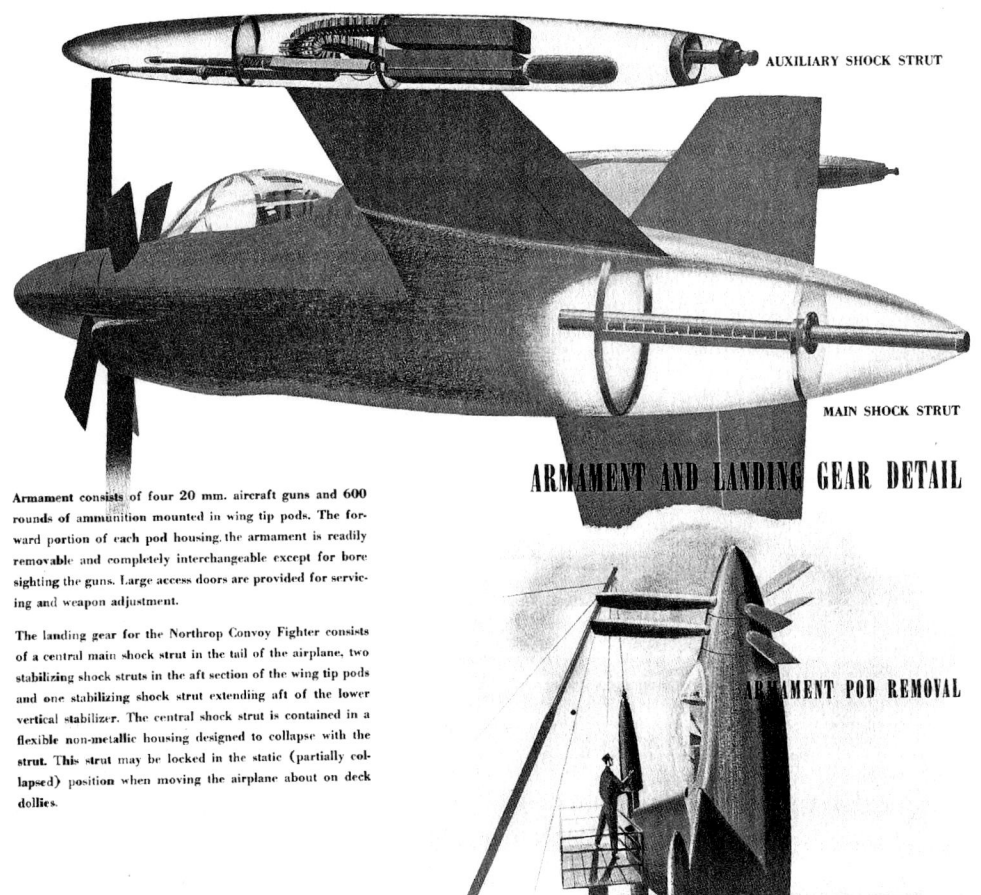

ARMAMENT AND LANDING GEAR DETAIL

ARMAMENT POD REMOVAL

Armament consists of four 20 mm. aircraft guns and 600 rounds of ammunition mounted in wing tip pods. The forward portion of each pod housing, the armament is readily removable and completely interchangeable except for bore sighting the guns. Large access doors are provided for servicing and weapon adjustment.

The landing gear for the Northrop Convoy Fighter consists of a central main shock strut in the tail of the airplane, two stabilizing shock struts in the aft section of the wing tip pods and one stabilizing shock strut extending aft of the lower vertical stabilizer. The central shock strut is contained in a flexible non-metallic housing designed to collapse with the strut. This strut may be locked in the static (partially collapsed) position when moving the airplane about on deck dollies.

FUEL SYSTEM

The N-63 fuel system consists of a single fuselage tank and sump with a total usable capacity of 552 gallons. Boost pumps in the tank and sump supply 90 per cent of the usable fuel to the engine at all level flight attitudes and 99 per cent of the usable fuel during hovering.

The tank system is vented to prevent excessive bursting or collapsing bladder cell pressures. At flight altitudes where the fuel normally boils, a suction-pressure relief valve in the vent line closes. Fuel boiling thus serves merely to pressurize the system and keeps fuel losses to a minimum.

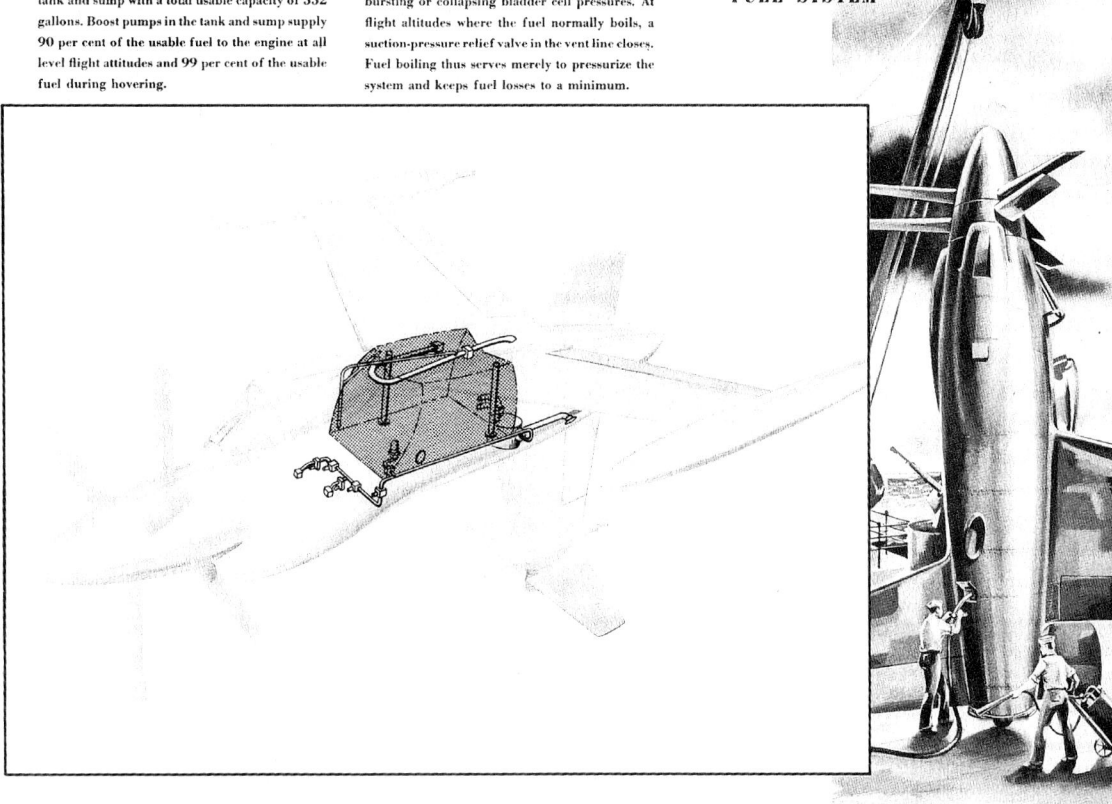

REFUELING

FLIGHT CONTROL SYSTEM

Control during hovering, transition, and take-off and landing is provided by elevons and rudder located in the propeller slipstream. Roll and pitch control are provided by the elevons, while the rudder contributes yaw control. In horizontal flight, elevator action of the elevons is eliminated and they become conventional ailerons. Directional control is provided by the rudder, while longitudinal control is provided by the horizontal tail located safely away from the wing wake and propeller slipstream. The Northrop fully powered system assures full control during all phases of flight.

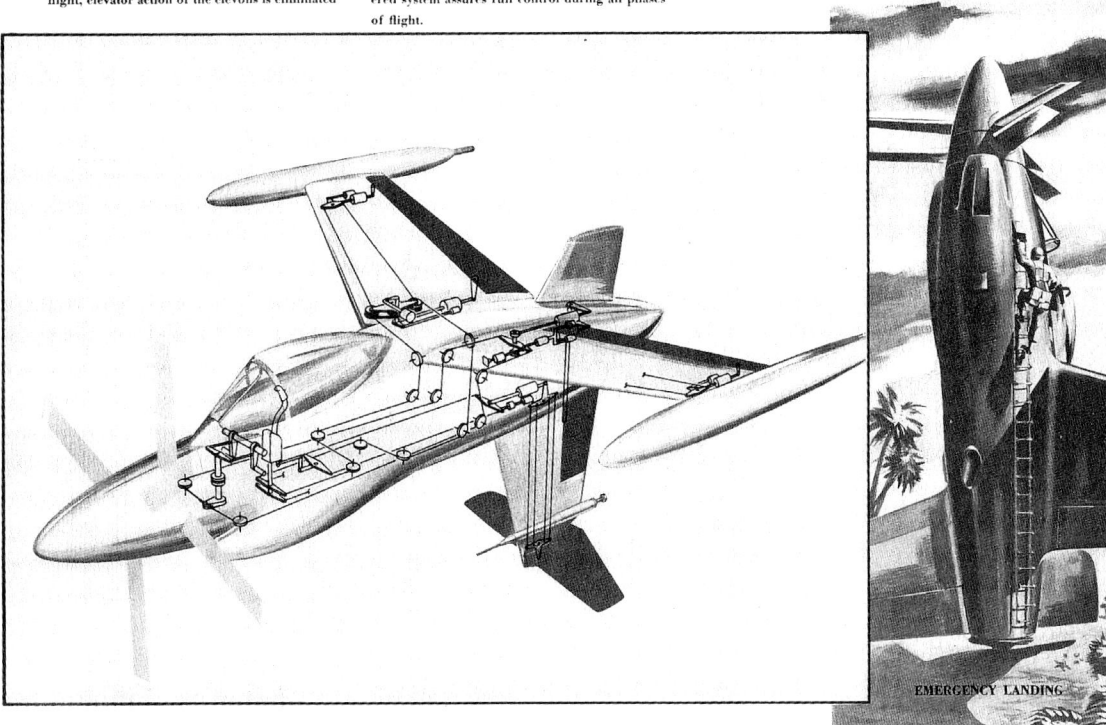

EMERGENCY LANDING

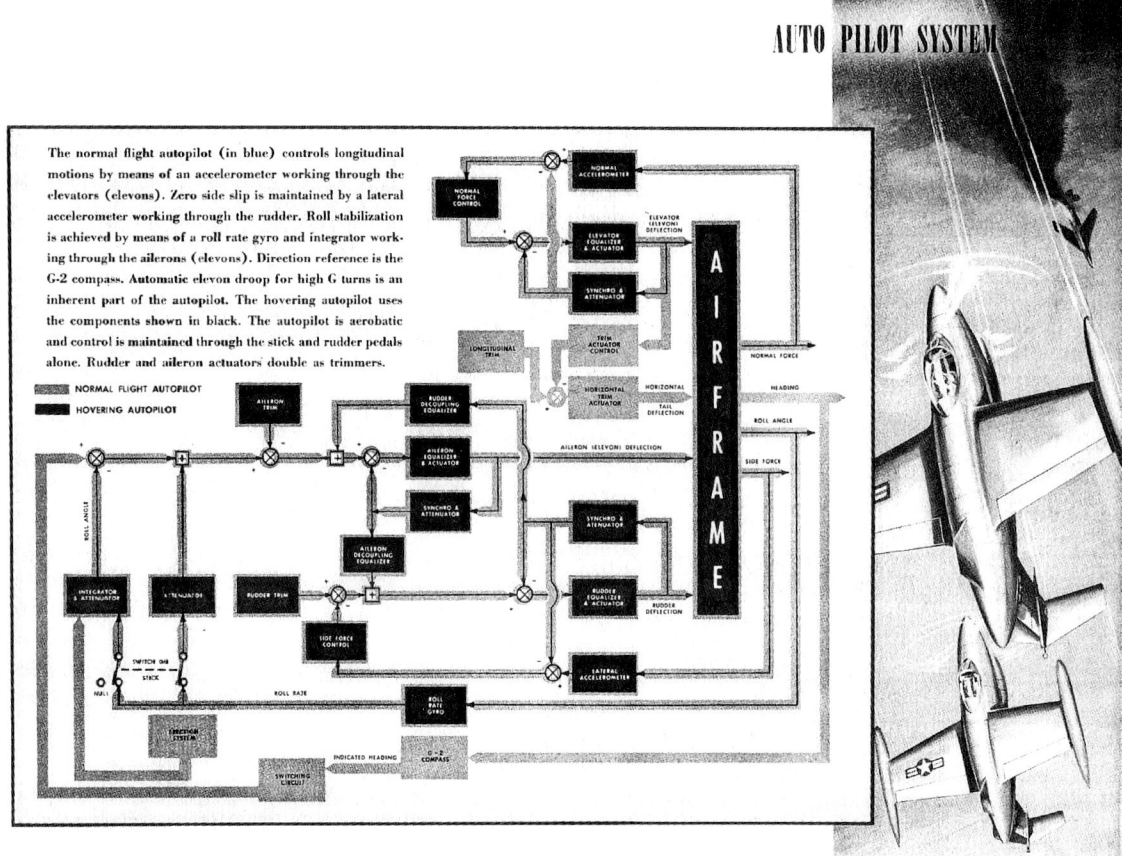

of the full-scale fighter. An auxiliary landing gear was therefore provided for pilot training.

The proposed prototype design permitted exploration of intermediate steps between conventional and vertical flight. Initial takeoffs could be made conventionally to explore level flight and hovering characteristics at safe altitudes. Simple modifications of the basic configuration permitted perfection of vertical takeoff and landing techniques. Short progressive steps from conventional to vertical flight were prime features of the Northrop proposal.

The Northrop N-63A prototype airplane was fitted with a conventional tail and landing gear for initial flights. Normal takeoff and landing procedures would have been employed for familiarization with level flight characteristics, after which transition and hovering manoeuvres would have been performed at safe altitudes without danger of takeoff or landing complications.

Replacement of the conventional tail by one more representative of that of the full-scale fighter and removal of the landing gear permitted investigation of vertical landing and takeoff characteristics. Initial hovering tests would have been made with the airplane tethered for control of test conditions, after which free-flight takeoff would have been accomplished. Aerodynamic characteristics of the full-scale fighter would have been investigated with this ultimate prototype configuration.

Thus the proposed Northrop prototype airplane permitted progressive transition from conventional to vertical flight. A properly designed autopilot could have been modified to provide correct flight characteristics as appropriate for each step in the transition from a simple pilot trainer to the ultimate aerodynamic model of the Convoy Fighter.

Various steps in the proposed programme could have been eliminated at the Navy's discretion, with corresponding savings in cost. For example, since the configuration with a conventional tail and landing gear was merely a pilot trainer, it may have been desirable to eliminate the first stages of the programme and commence with the tethered ultimate configuration. Sufficient pilot familiarization could have probably been achieved, with more accurate representation of full-scale aerodynamic characteristics.

Cost proposal

Northrop's cost proposal for the N-63 was dated November 20, 1950, with additional cost information contained in a letter to BuAer dated November 24.

PILOT'S SEAT

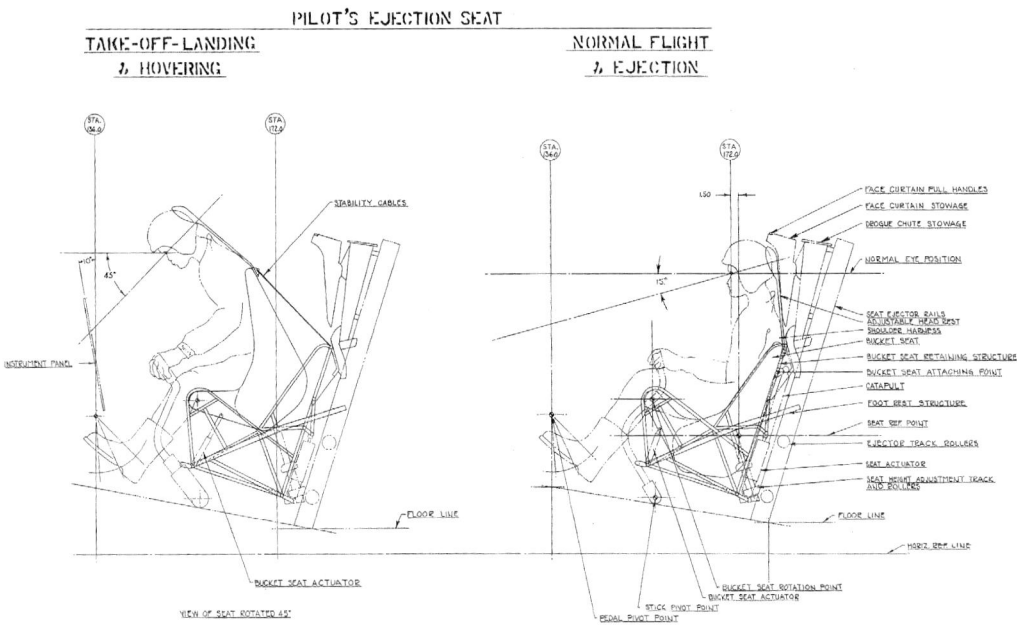

The pilot's seat of the N-63 could rotate 45°, permitting vision down and aft during hovering and landing operations.

The estimated total cost less fee for the Model N-63A scale prototype airplane was $7,103,153.92. (This and all subsequent figures are in 1950s dollars). This covered two complete flight articles; design data and structural tests; and demonstration. Northrop also proposed and recommended the following alternate programmes:

1. A flight simulator was proposed in lieu of auxiliary horizontal landing gear and the auxiliary tail configuration required for conventional flight testing. A net saving of $727,831.40 would have been realized by use of the flight simulator.
2. Northrop also recommended an abbreviated proof test programme, resulting in a net saving of $486,918.74 if this proposed test programme was adopted.

The cost of the N-63 experimental Convoy Fighter was $9,463,929.20, which included two complete flight articles and one static test article; design data and structural tests; and demonstration. Northrop also proposed and recommended a proof test programme which resulted in a net price increase of $1,271,034.69. This proposed programme was based on a careful analysis of the test requirements considered necessary to ensure complete operational safety and proper system functioning prior to first flight. It had been Northrop's experience that full scale test stands were necessary for the development of power control systems, fuel systems, and power plant installation.

The combined total cost less fee was $16,567,083.12. The cost of the Northrop suggested alternate programme was $784,115.95; this additional estimated cost brought the grand total to $17,351,199.07. The proposal was submitted on both a cost and a fixed-price basis; because of the unconventional, and therefore highly experimental nature of the programme, Northrop was willing to undertake it only on the basis of a cost plus fixed fee (CPFF) or equivalent type contract. Accordingly, the proposal was based upon estimated cost without fee with the intent that the fee would be negotiated separately. Progress payments would have been required in accordance with contemporary CPFF practice.

Alternate prototype

While Northrop was well aware of the logic behind the desire of BuAer to initiate the programme with a scale prototype airplane, it recommended the

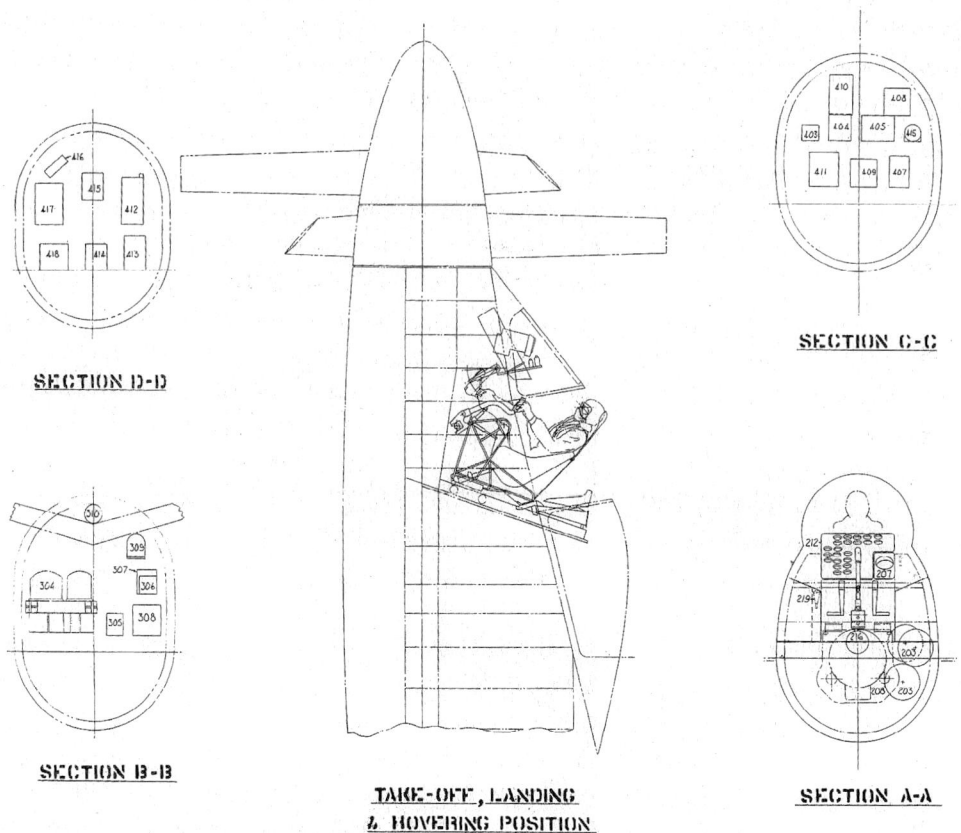

Inboard profile of the Northrop N-63 Convoy Fighter showing the location of the major internal components.

following alternate programme. Informal studies indicated that an experimental prototype configuration using the full-scale N-63 airframe would provide test data and experience far more useful to the Convoy Fighter programme than would the proposed N-63A, at an appreciable reduction in cost, time, manpower and material required for the overall programme.

The existing programme involved the design of three rather different airplanes: an optimum convoy fighter configuration had to be determined; a tail-landing prototype had to be developed to represent aerodynamic characteristics of the fighter; and the prototype had to be modified to permit conventional takeoff and landing, unless the proposed flight simulator was adopted.

All three designs involved separate and somewhat unrelated activity, including separate wind tunnel programmes, autopilot developments, proof test programmes, manufacturing planning, tooling, and construction.

From a technical standpoint, Northrop asserted that the great advantage of a full-scale prototype lay in its direct applicability to the ultimate fighter design. Initial aerodynamic studies, wind tunnel tests, load analyses and autopilot design could have been conducted with a single airframe in mind. Similarly, design studies and drawings, manufacturing planning, tooling, and production learning would have been applicable to one airplane. Other savings to BuAer not reflected in Northrop's studies were the development of a single engine-propeller combination. Overall savings, for Northrop work alone, on the order of 25 to 30% ($4.5-5.5 million) appeared probable.

The advantages of using the full-scale prototype were summarized as follows:

a. Northrop design of a single airframe instead of two or three.
b. Direct applicability of wind tunnel, laboratory and flight test data to the ultimate fighter.
c. Northrop procurement, tooling, and production for a single design.
d. Navy procurement of the XT40-A-6 engine, which was available at the time of the proposal.

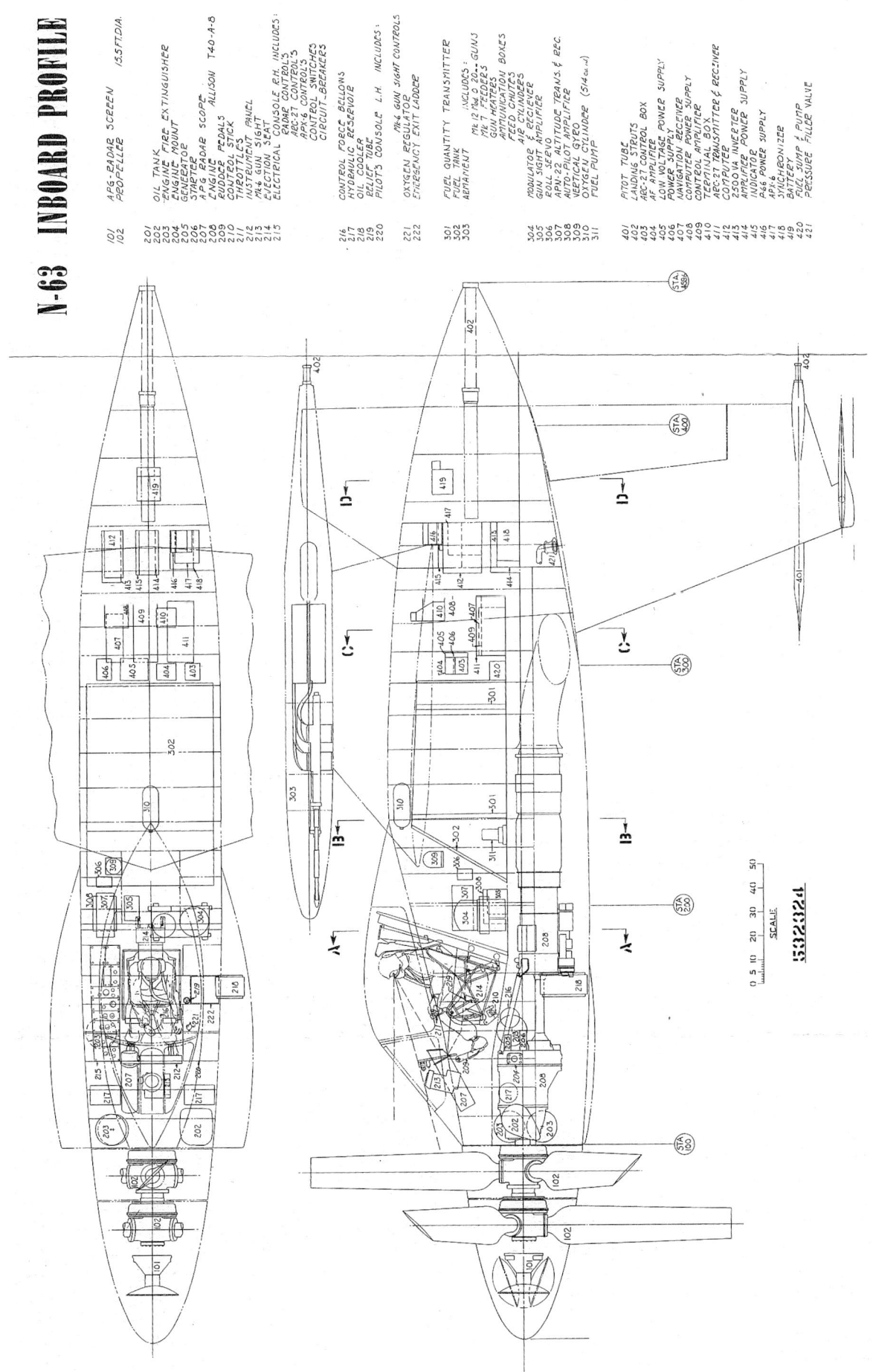

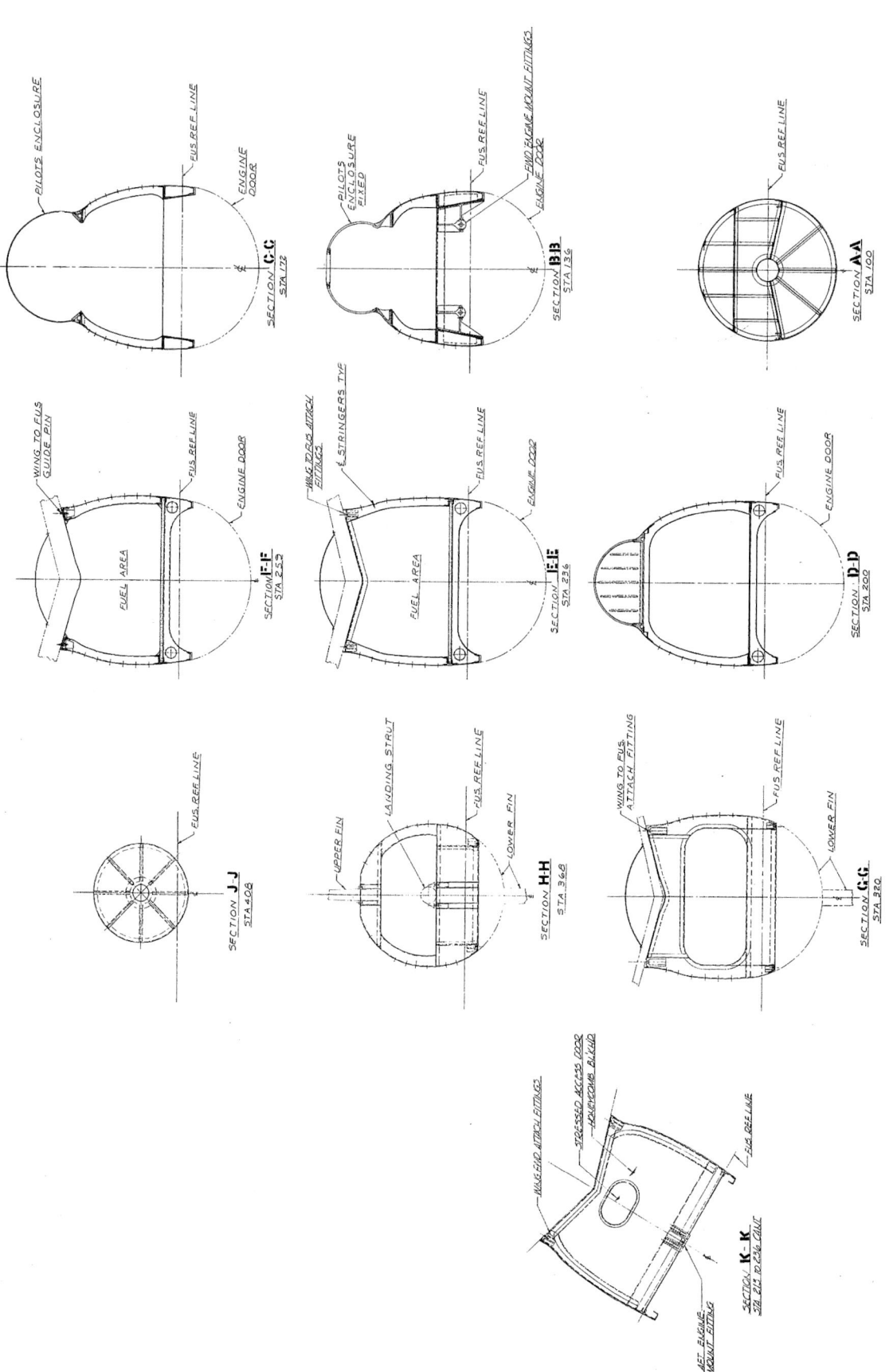

Structural diagram of the N-63's aluminium alloy fuselage, which was of semi-monocoque construction.

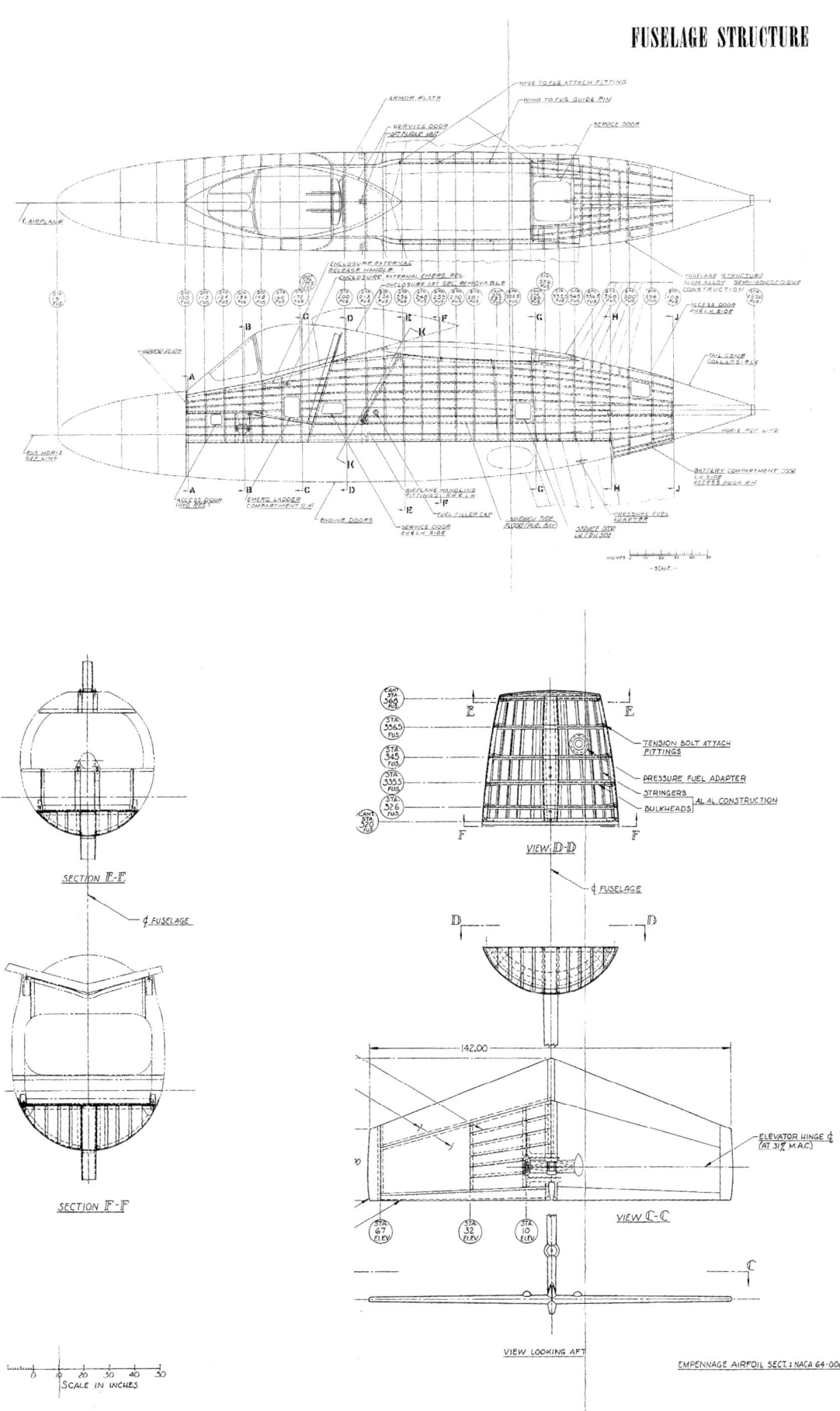

Schematic of the Convoy Fighter's empennage structure, which featured a large ventral T-tail as well as a small dorsal tail. The horizontal tail was positioned to avoid the wing wake and propeller slipstream.

e. Navy procurement of the same two-speed gearbox for the prototype and experimental articles.
f. Navy procurement of a single propeller, eliminating the need for a separate prototype propeller.
g. Reduction in overall elapsed time for the complete programme.

The disadvantage of the above was that the first prototype flight article would have been more costly than the first scale prototype flight article initially planned. This apparent disadvantage, however, would have been more than compensated by the elimination of the propeller and engine procurement costs for the scale prototype airplane. Informal studies were only made on the proposed alternate prototype airplane, with Northrop offering to submit a detailed cost analysis if the proposal was viewed favorably by BuAer.

Production facilities

The Northrop Aircraft facilities were ideally suited for the production of an airplane of the size and type of the Convoy Fighter, as exemplified by the Northrop P-61 night fighter produced during the Second World War and the F-89 all-weather fighter which was then in production. The company provided BuAer with photos of these aircraft on the Northrop assembly line to illustrate their production capability.

WIND TUNNEL TESTS AND DESIGN REFINEMENT

Northrop subsequently submitted an unsolicited progress report dated December 22, 1950 to BuAer, in which the company summarized its continued work on the Convoy Fighter design since the submission of its proposal the month before.

Introduction

During the period following submission of the Northrop Convoy Fighter proposal, the company continued preliminary design efforts directed towards refinement of the proposed configuration, and conducted wind tunnel tests with a low-speed stability and control model of the fighter airplane. Results of these efforts were informally submitted to BuAer, with the thought that they may contribute to the general fund of knowledge regarding this interesting problem. Northrop undertook these activities at its own expense, probably hoping to increase its chances of winning the contract.

Design refinements

Additional preliminary design studies of the proposed Convoy Fighter configuration had been concerned with minor design refinements. Detailed hovering stability calculations indicated desirability of some minor rearrangement of internal equipment and secondary structural components, in order to bring the centre of gravity into closer alignment with the propeller thrust line. Accordingly, the changes indicated by the accompanying drawings and sketches were made. More detailed duct analysis and refinement of external fairings were included in these minor changes.

The changes indicated were expected to in no way affect the earlier estimates of cost or time for production of the airplane, but improve its performance and stability characteristics. External refinements were expected to improve top speed, although no estimates were made of the possible magnitude of improvement, pending more reliable propeller data.

Wind tunnel tests

The configuration tested in the wind tunnel produced data which were in excellent agreement with predicted characteristics of the proposed Northrop Convoy Fighter configuration. The model was an approximate 0.16 scale model of the fighter airplane. Low speed tests were run in the Northrop wind tunnel to check basic aerodynamics regarding general stability and control characteristics, validity of applied theories, etc. Among the items studied were wing lift curve slope, wing and fuselage pitching moment, stability contributions of the propeller and horizontal tail, and drag increments of various components.

Test runs were made with various components of the model assembled in different combinations until the entire configuration was represented. Initial runs used the wing alone. Tip bodies representing the armament pods were then added. The basic fuselage was included in the next runs, without a propeller.

A windmilling counter-rotating propeller of the approximate desired characteristics was added next. Subsequent tests, as additional components were added, were run with the propeller both installed and removed. It was thus possible to establish the propeller contribution to the characteristics noted.

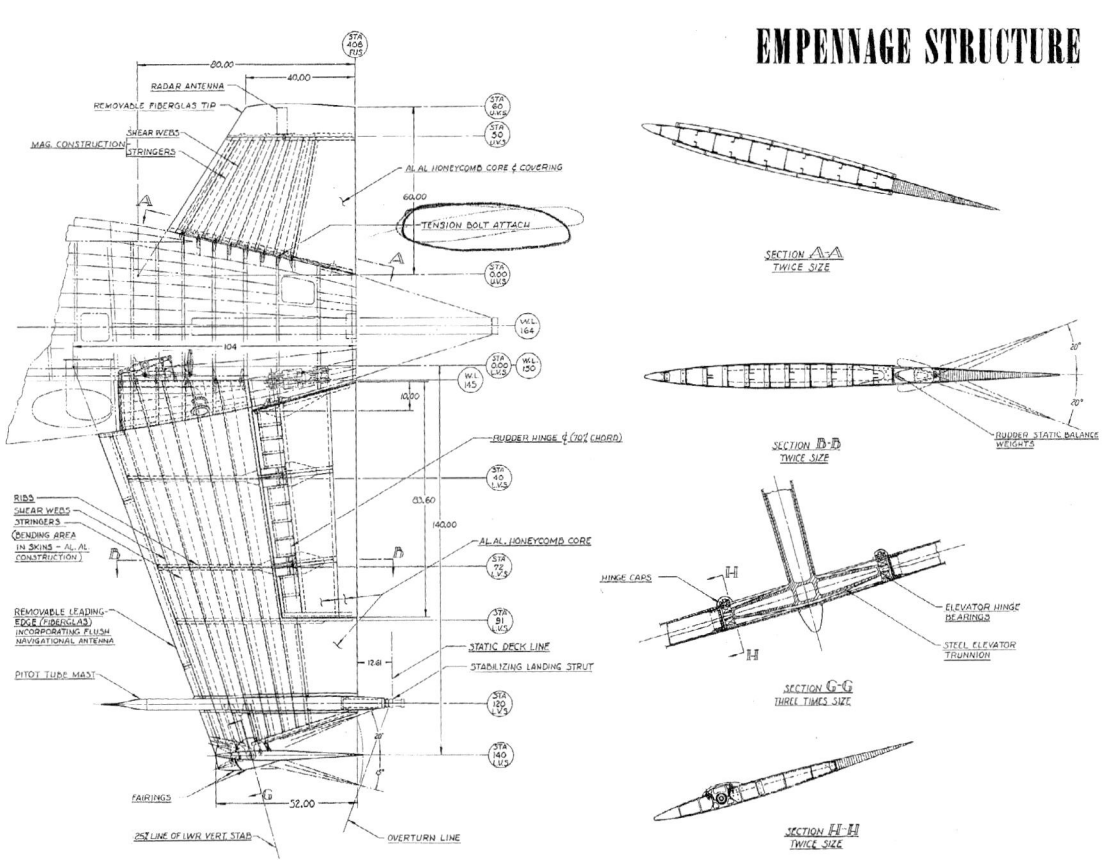

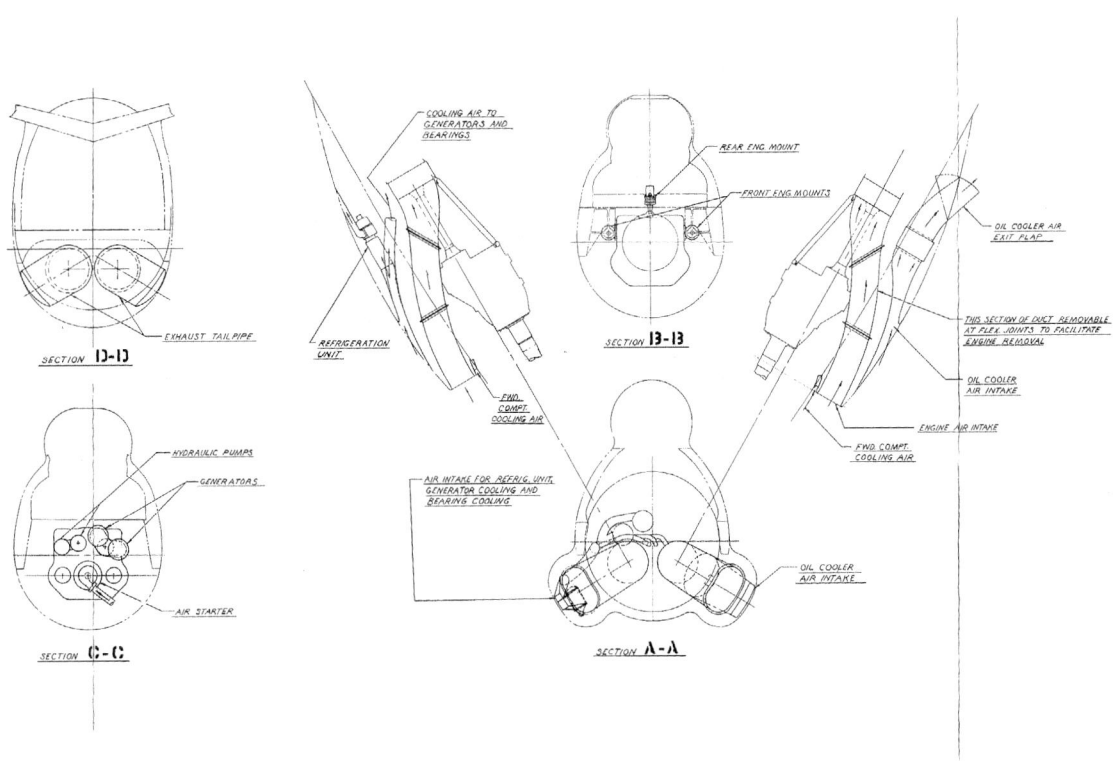

Blueprint of the N-63's power plant installation, which consisted of an Allison XT-40-A-8 turboprop engine driving an Aeroproducts 15.5ft six-blade dual-rotation propeller.

The ventral vertical fin was added for the next runs. This was then removed and runs were made with the dorsal vertical fin in place, after which both fins were installed for tests of their combined effects. Final runs were made with the horizontal trimming surface installed, completing the configuration.

As noted in the preliminary analysis outlined in the report, the tests showed excellent agreement with predicted aerodynamic characteristics. Purposes of the tests were completely satisfied by the indication of reliability of the general theories used in development of the configuration. More detailed discussion of the results was presented in the report. It was Northrop's opinion, however, that the work so far accomplished verified the entire feasibility of design of a high speed vertically-rising fighter of the proposed N-63 configuration.

Summary and conclusions
Northrop produced brief preliminary stability data from low speed wind tunnel tests of a simplified model of the N-63 airplane with windmilling propellers. The model differed in some details from the configuration of the airplane presented in the original proposal. The propeller blades used on the model were borrowed from another model, so they did not have the proper planform. No canopy or ducts were used on the fuselage. The vertical location of the wing corresponded to an advanced airplane arrangement which was completed after the Convoy Fighter proposal was completed. The flight surfaces had very simplified airfoil sections which were incorporated in order to speed construction of the model.

The results of the tests showed that the general behaviour of the airplane and the stability contributions of portions of the airframe were very close to the predictions presented in an earlier Northrop report after corrections were made for the effect of configuration differences.

From these tests Northrop concluded that:

1. The horizontal tail size was probably marginal, and an increase in size seemed warranted, subject to a study of stability and trim drag characteristics.
2. Although the values of the pitching moment coefficient obtained in the tests were not representative of the airplane, care was taken to obtain as small a value as possible.
3. Directional stability appeared to be somewhat too large, and a smaller vertical tail may have been adequate.
4. The effective dihedral of the complete airplane was larger than predicted or desired. The cause of this effect was not definitely isolated and further investigation was required.

Description of tests
As part of the preliminary investigation of the N-63 airplane configuration, a series of wind tunnel tests of a simplified model was initiated. The model was designed to be rapidly built and still yield data which were representative of the configuration. The data desired concerned only the stability and control characteristics of the airplane, and no special effort was made to obtain representative information on drag or stalling properties. The model had windmilling propellers with the blade angles set to correspond to high speed flight. This condition represented one of the critical conditions for both longitudinal and directional stability. Powered model tests were required to find the stability with power on and to determine which conditions were most critical.

Tests were conducted in the Northrop low speed wind tunnel using the standard fork mount system. A special support system would have normally been desirable for this model, but this would have required lengthy calibration. The test period was December 5-21, 1950. The wind tunnel programme and analysis of data were being continued in order to obtain a complete understanding of the stability properties of this aircraft type.

Support system and technique
A simplified model of the N-63 design was tested in the Northrop 10ft wind tunnel. The fuselage consisted of a body of revolution and was constructed from laminated mahogany. A removable nose section allowed the testing of windmilling propellers or a smooth nose. Aluminium alloy propeller blades of the correct diameter were clamped in steel hubs to provide blade angle adjustment. The dual rotating hub bearings were lubricated with an air-oil mist to ensure consistent windmilling velocities.

An aluminium alloy plate was cut to the correct wing planform and bent to the correct dihedral angle. A constant chord bevel was made along the trailing edge on the upper surface. A leading edge

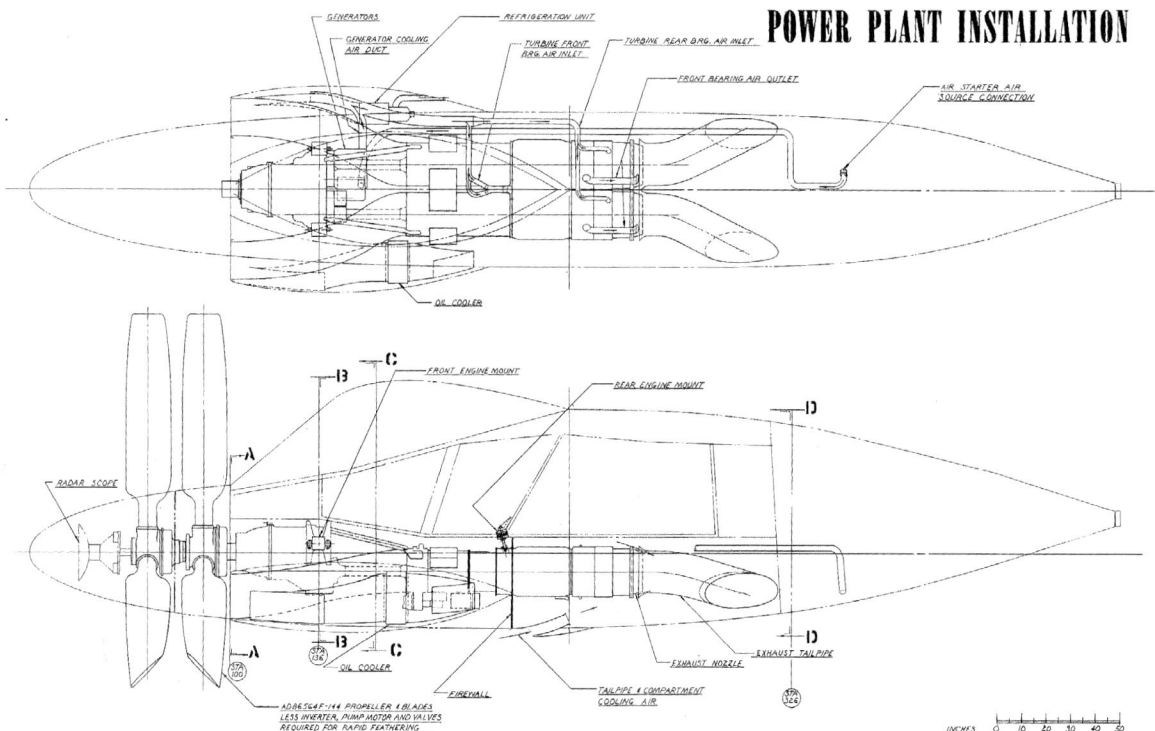

Schematic of the Northrop Convoy Fighter's fully powered control system. The elevons and rudder provided control during hovering, transition, and takeoff and landing. The elevons also provided roll and pitch and control, while the rudder contributed to yaw control.

radius of ½" was maintained across the span. At the wing mean aerodynamic chord location, the wing thickness ratio and camber corresponded with full scale. Consequently, at the root, thickness and camber were too small and at the tip were too large.

Tip bodies were turned from mahogany. These bodies had an elliptical nose, straight centre section, and a straight tapered afterbody beginning at the wing trailing edge.

The horizontal and vertical surfaces were fabricated, in a manner similar to the wing, from aluminium and plywood sheet. Variable incidence of the horizontal surface was provided by means of a tongue and groove arrangement of the horizontal in the vertical. A slight amount of negative camber was incorporated in the horizontal surface section to prevent premature separation due to the subcritical Reynolds number of that section. The incidence range was -20° to 15° by five-degree increments.

The model was mounted inverted on the fork support system to minimize empennage and strut interference. Tare corrections to the data due to the interference of the supporting system were estimated from previous data. Consequently, the zero lift and zero angle values of the six components were somewhat arbitrary.

Corrections to drag and angle of attack due to wind tunnel wall effect were applied, but no correction was made to the tail contribution.

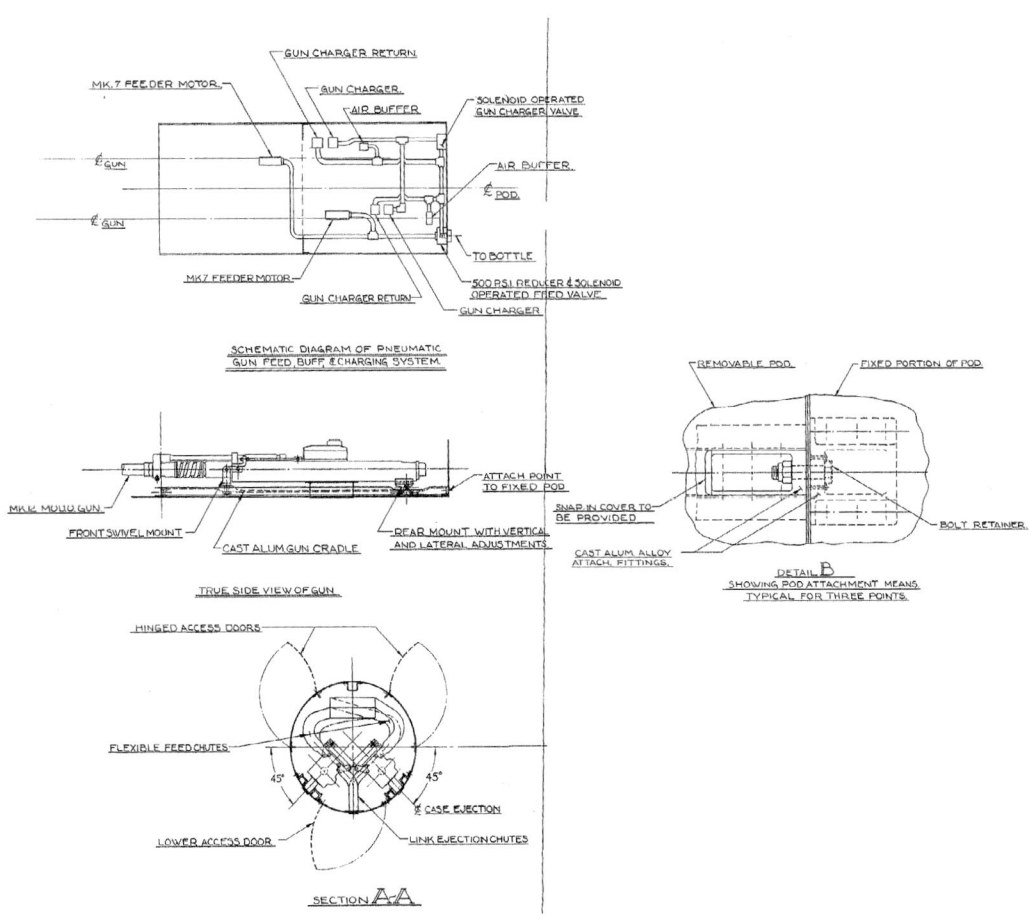

The N-63 was armed with four 20mm aircraft guns and 600 rounds of ammunition mounted in wing tip pods.

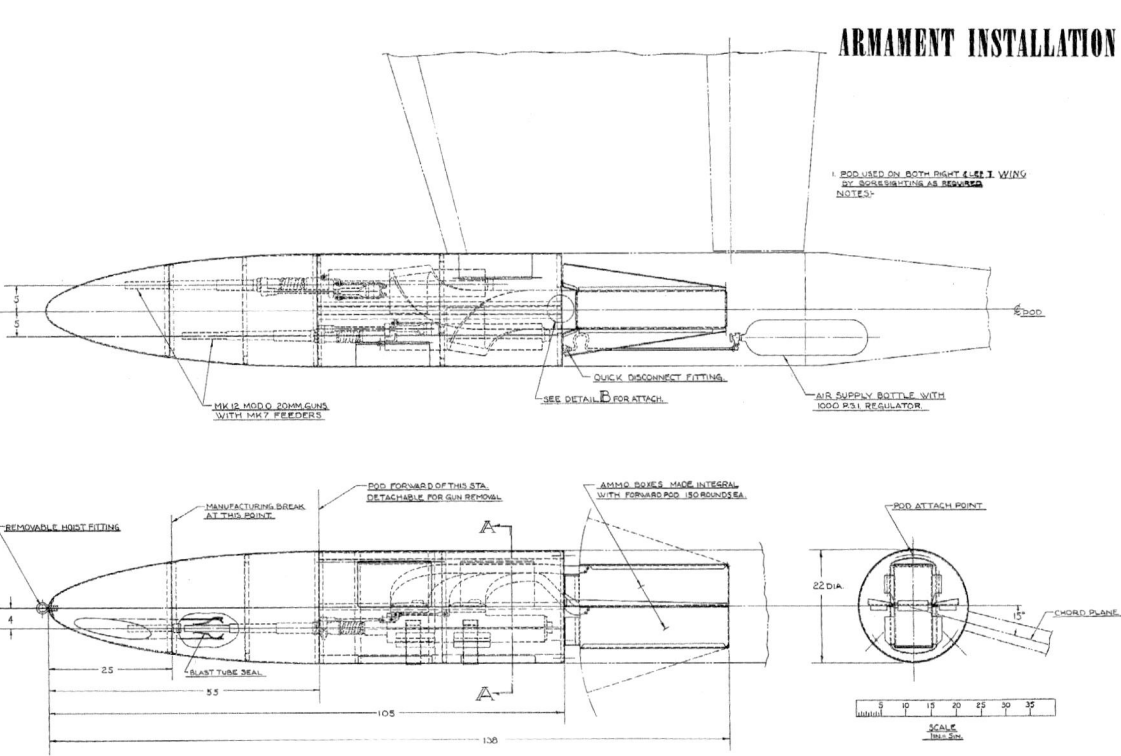

EXTERNAL ANTENNA LOCATION

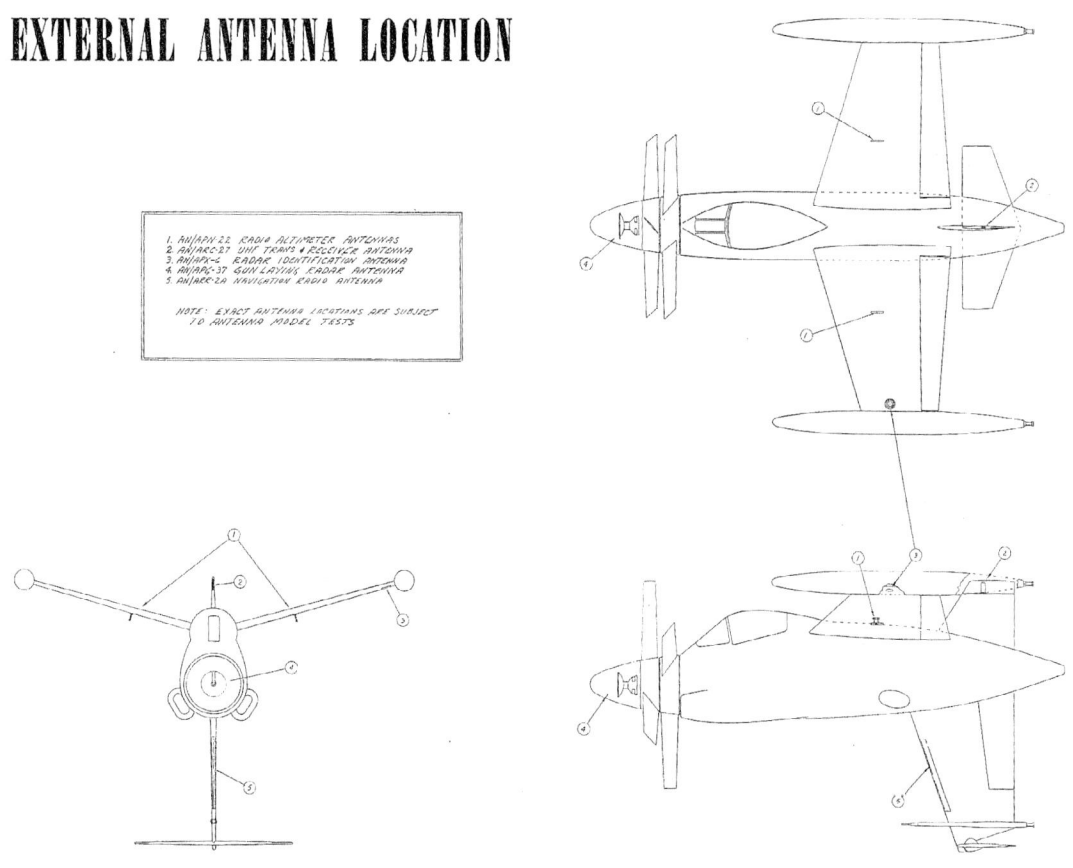

Diagram of the Convoy Fighter's external antenna locations, the largest of which was the AN/APG-37 gun laying radar mounted in the nose.

FUEL SYSTEM DIAGRAM

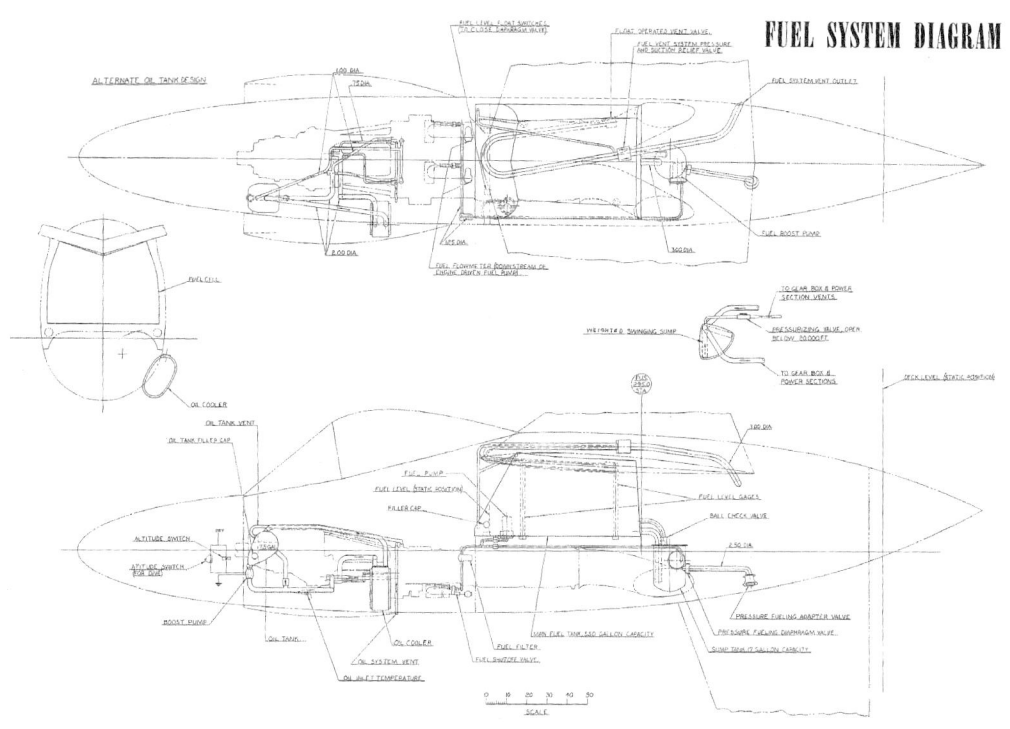

The N-63's fuel system featured a single fuselage tank and sump with a total usable capacity of 552 gallons.

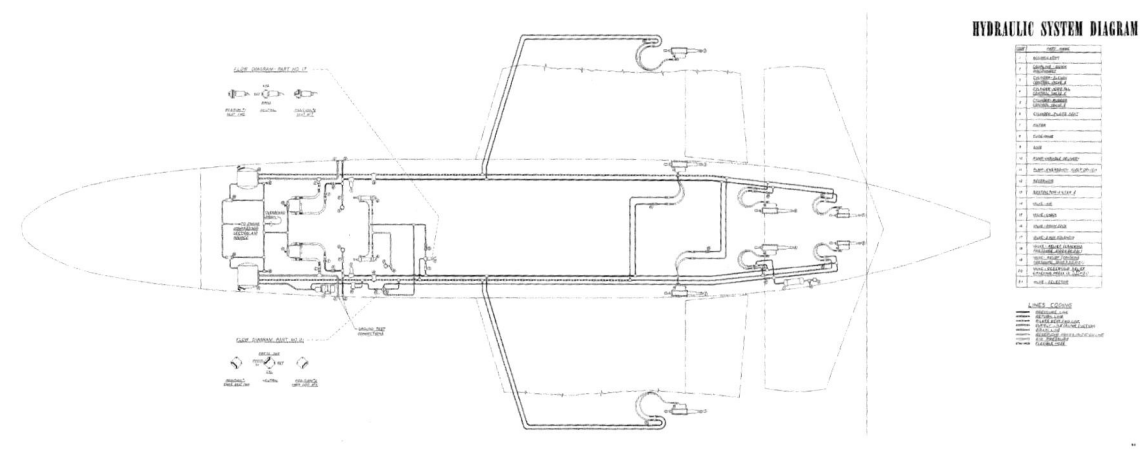

Blueprint of the aircraft's hydraulic system.

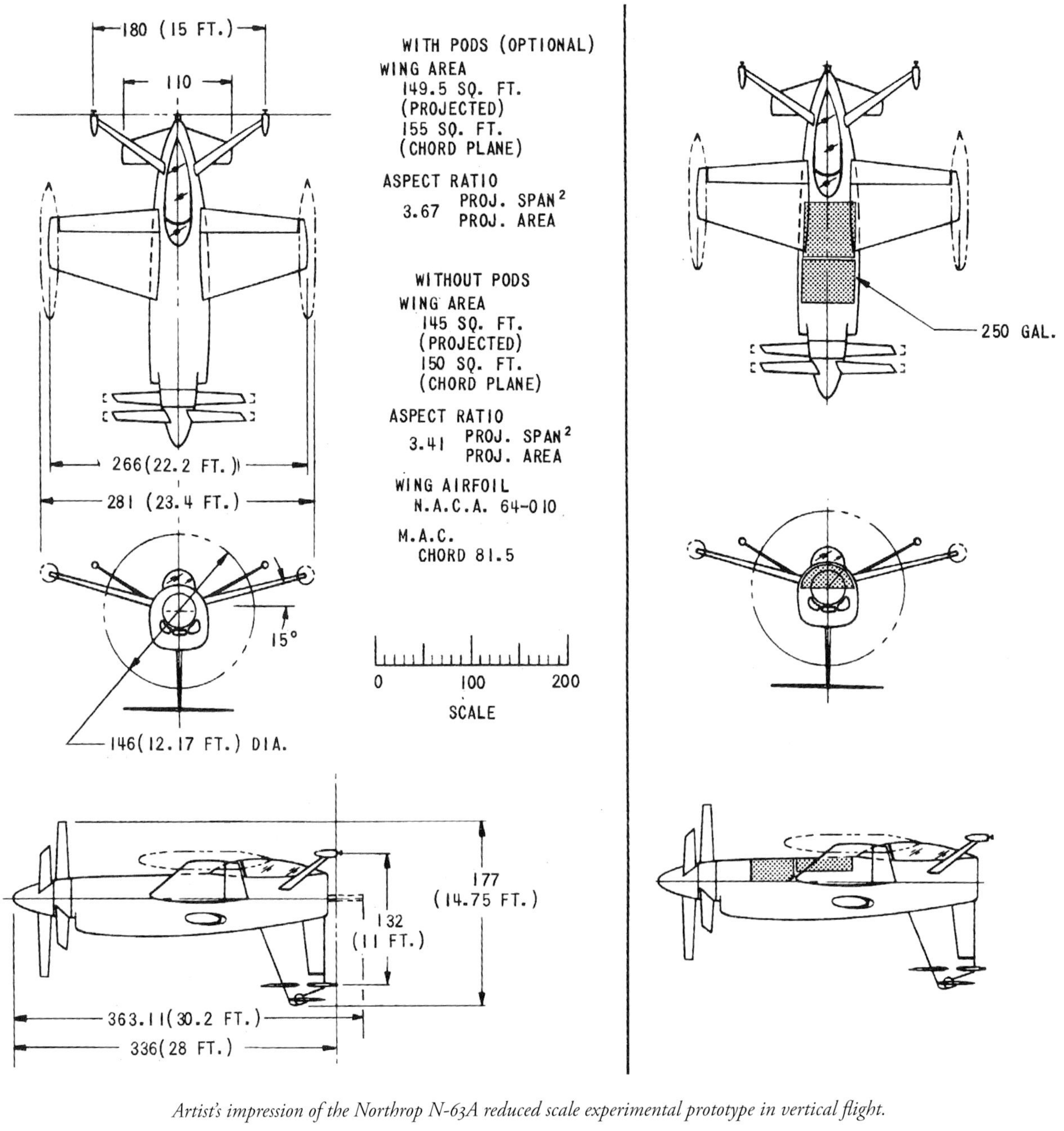

Artist's impression of the Northrop N-63A reduced scale experimental prototype in vertical flight.

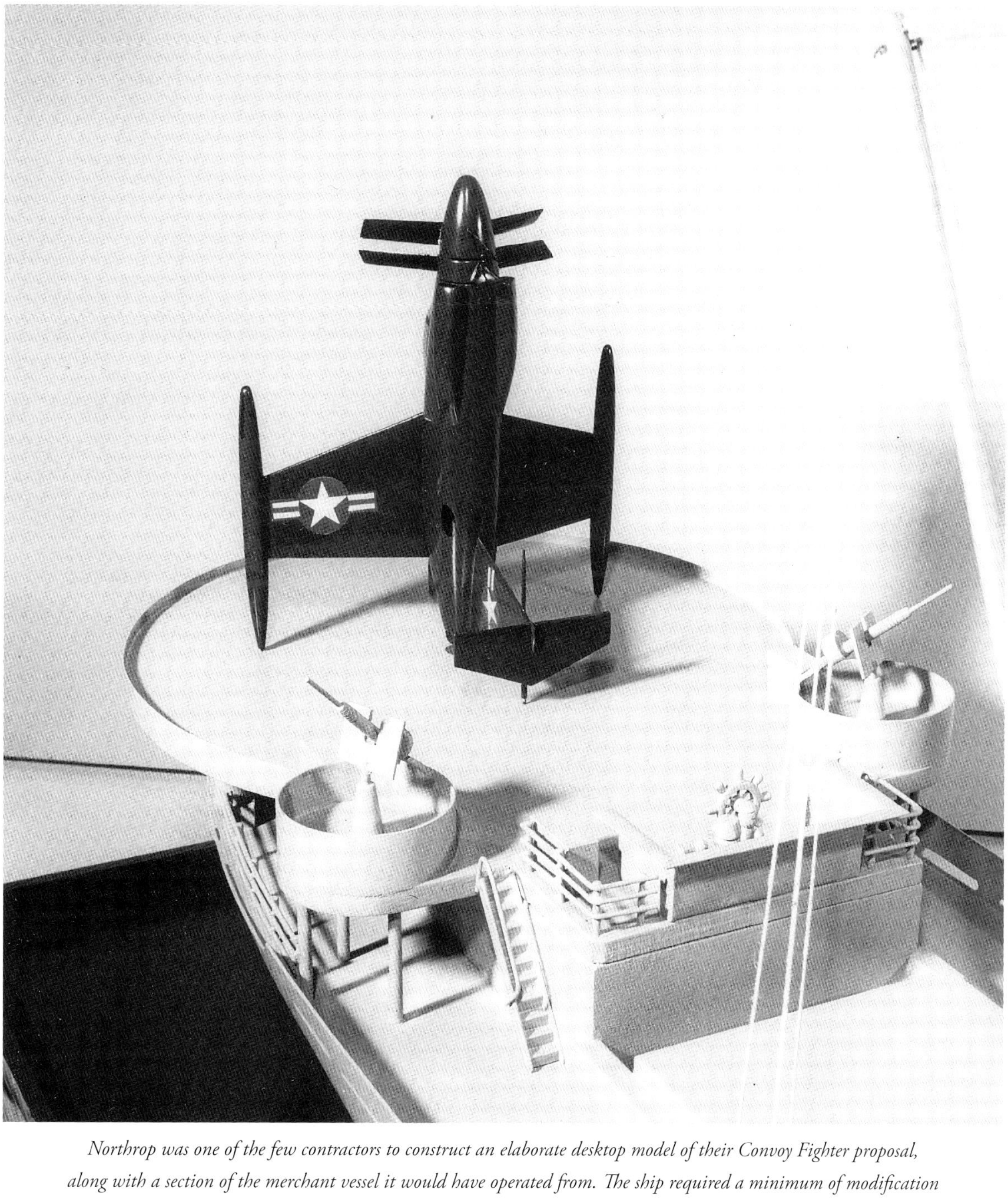

Northrop was one of the few contractors to construct an elaborate desktop model of their Convoy Fighter proposal, along with a section of the merchant vessel it would have operated from. The ship required a minimum of modification to operate the N-63, which did not need a hoisting apparatus since it was in the takeoff attitude at all times.

Artist's impression of a pair of Northrop N-63A tailsitters in flight taken from the type's Standard Aircraft Characteristics document.

Illustration of a pair of N-63A prototypes in a steep climb over the fleet.

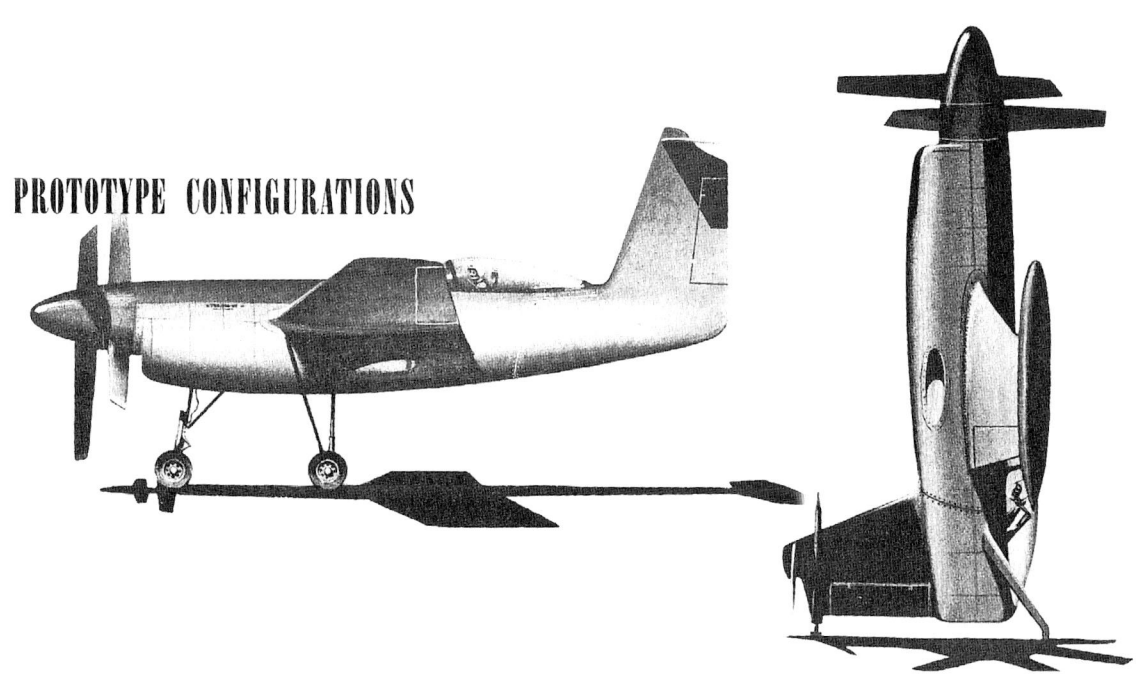

The Northrop N-63A scale prototype could be fitted either with a conventional tail for normal takeoff and landing or a special tailsitter unit for vertical testing.

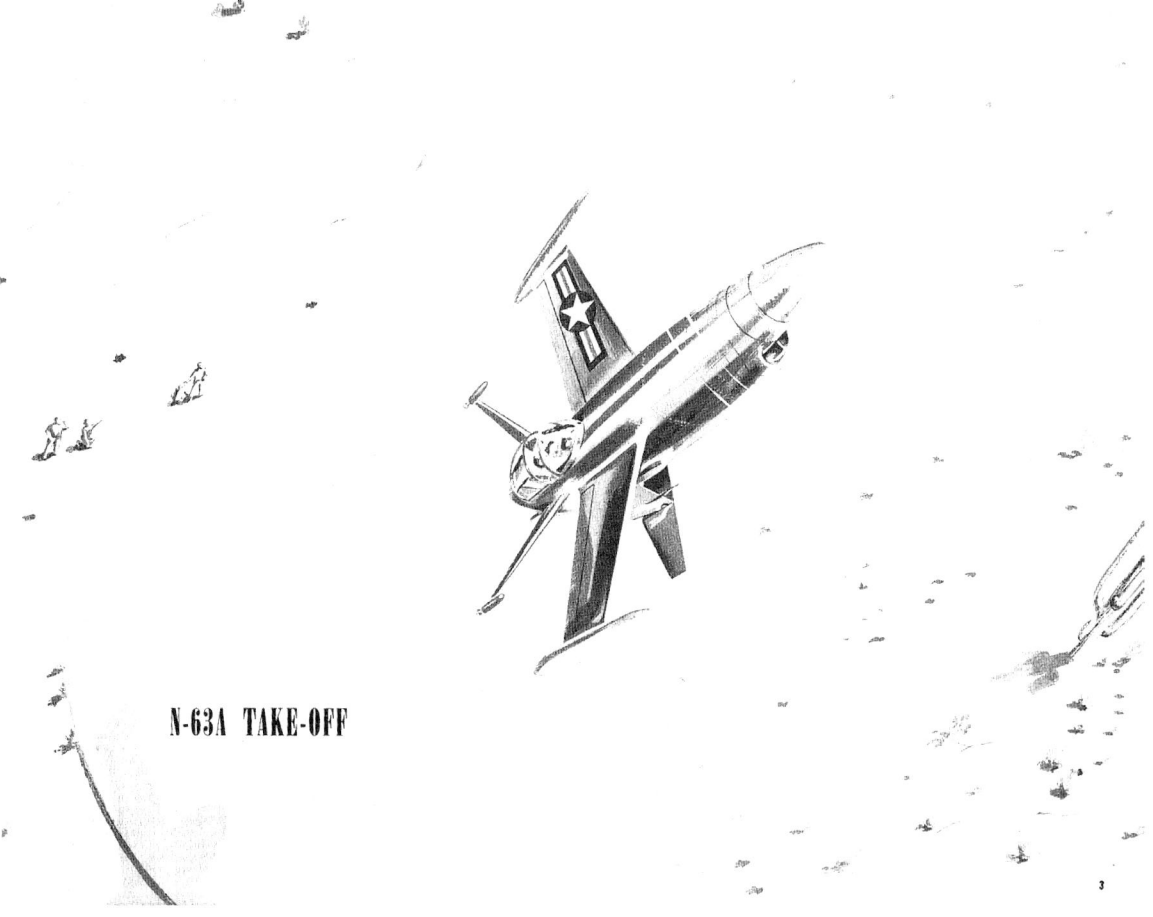

Artist's impression of the N-63A taking off vertically from a desert location.

An illustration of the N-63A just prior to touchdown on the deck of a merchant vessel; note the open canopy and tilted position of the pilot, both of which would have aided him in accomplishing these difficult landings.

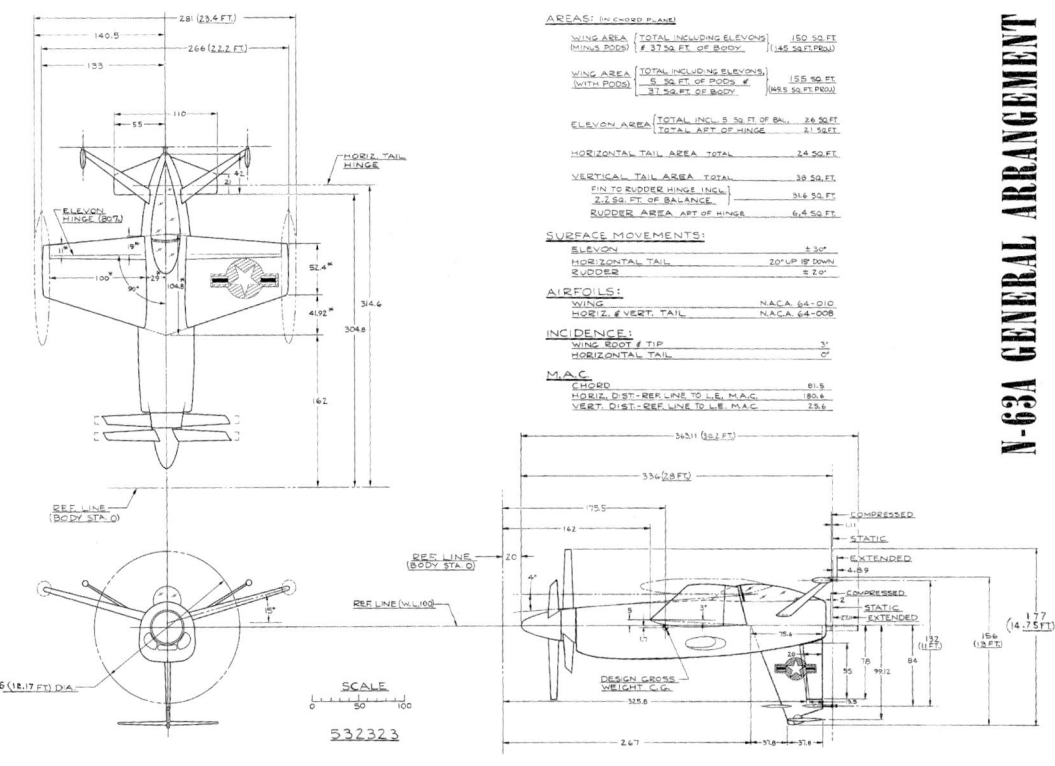

A three-view of the Northrop N-63A configured for vertical takeoffs and landings.

Artist's impression of the N-63A undergoing engine removal and maintenance in a desert location. Maintenance could be accomplished with the aircraft either in the vertical or horizontal attitude, depending on how it was configured.

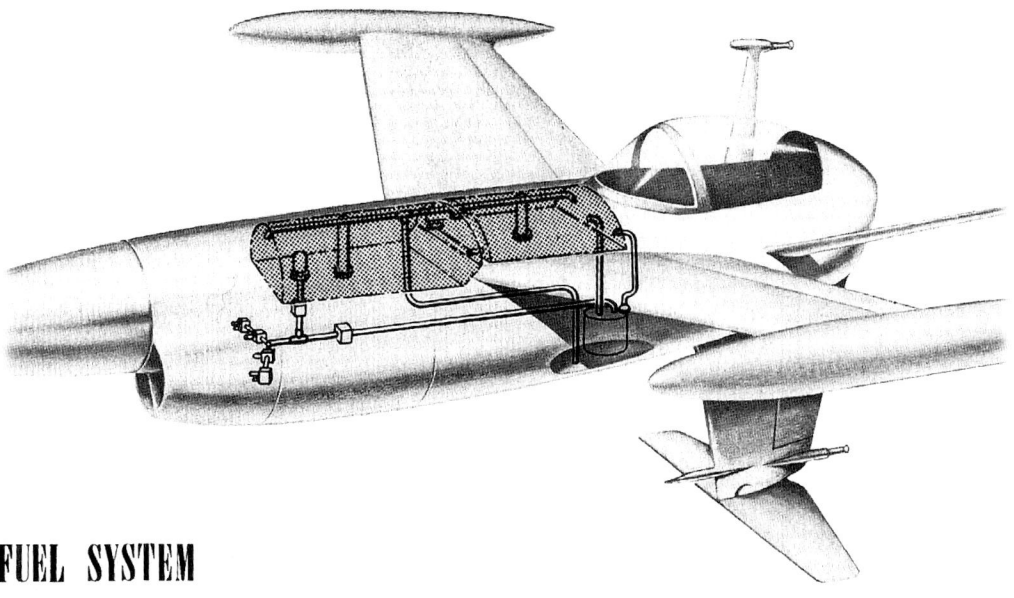

Diagram of the Northrop N-63A's fuel system, which had a capacity of 250 gallons.

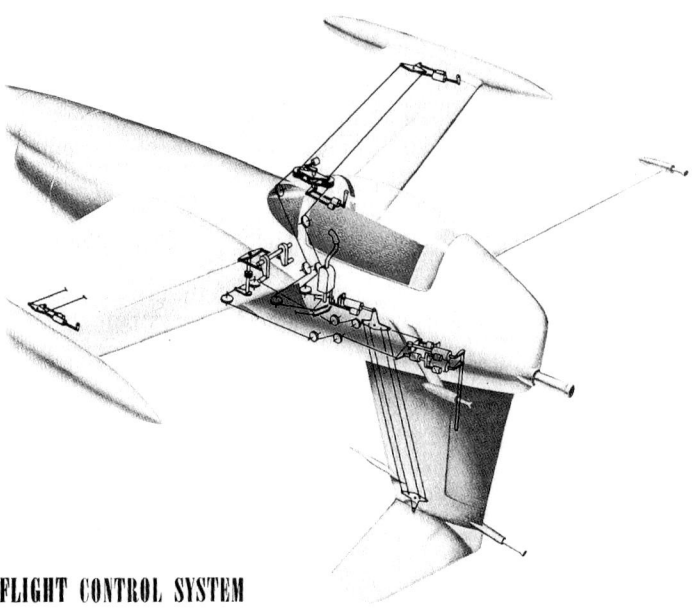

FLIGHT CONTROL SYSTEM

Illustration of the flight control system for the N-63A configured as a tailsitter.

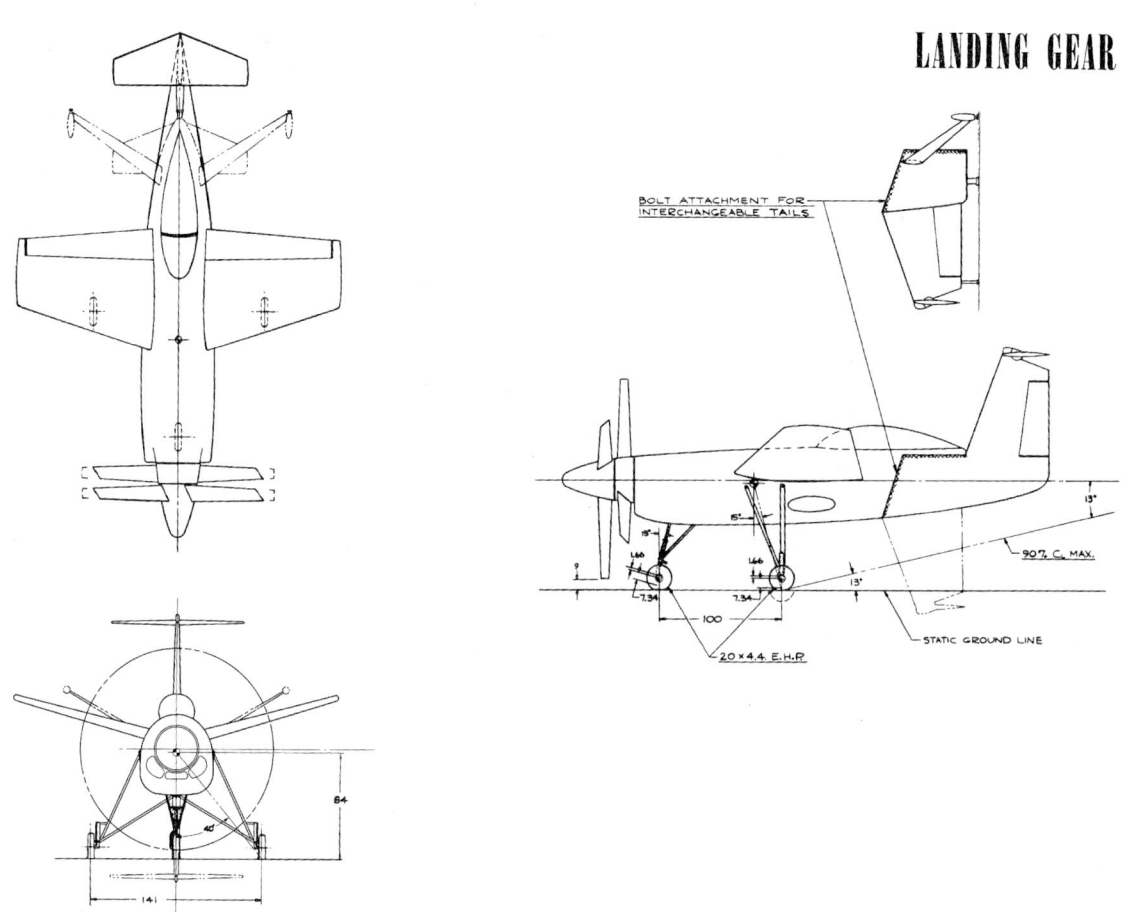

LANDING GEAR

A three-view of the Northrop N-63A modified for conventional takeoffs and landings with a bolt-on tricycle landing gear and T-tail.

NORTHROP N-63 AND N-63A

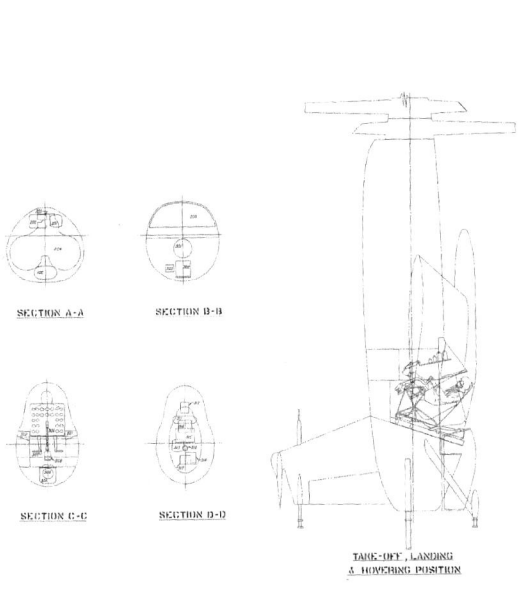

(Above and below). Inboard profile of the Northrop N-63A scale prototype, which was powered by a British Armstrong Siddeley Double Mamba turboprop engine.

Artist's impression of the Northrop N-63A reduced scale experimental prototype in vertical flight.

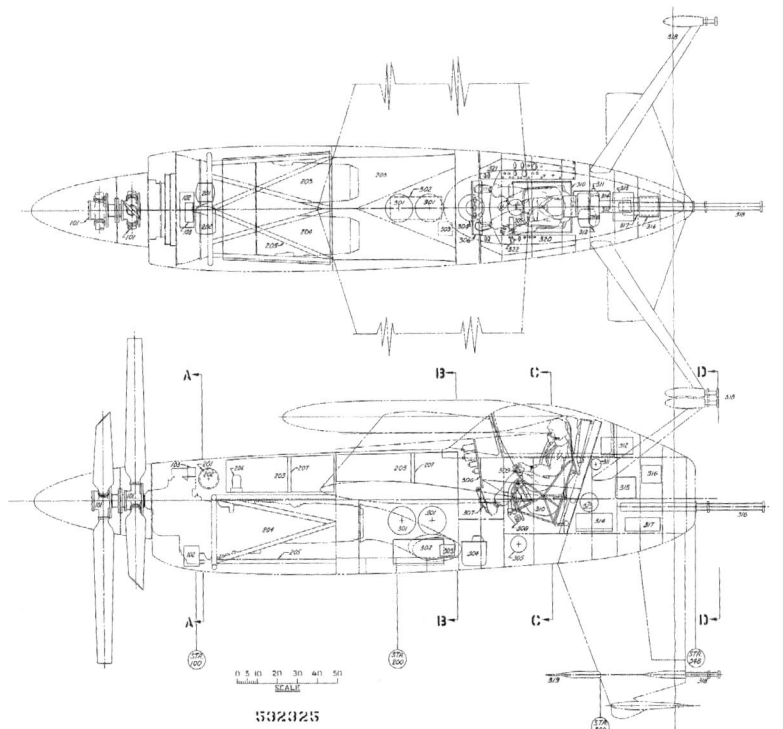

N-63A INBOARD PROFILE

251

MISSION AND DESCRIPTION

The primary mission of the Northrop N-63A airplane is to provide flight research data pertaining to the U.S. Navy Class VF (Convoy Fighter).

The Northrop N-63A is a proposed U.S. Navy Class VF (prototype for Convoy Fighter) airplane based on BuAer Outline Specification OS-121. This prototype is an inhabited, flyable model, similar to the Northrop N-63, a U.S. Navy Class VF (Convoy Fighter) airplane. It is capable of vertical unassisted take-offs from, and landings on, small platform areas. It is also capable of performance indicative of the characteristics of the Northrop N-63 airplane.

The airplane is essentially a conventional single-engine tractor monoplane configuration except for features appropriate to vertical take-offs and landing. Longitudinal control at low speed and lateral control for all conditions are obtained by elevons, which also are drooped to improve ceiling and maneuverability. Longitudinal control during normal flight at higher speeds is obtained by an all-movable horizontal tail. Directional control is provided by a rudder on the vertical tail. Aerodynamic braking is provided by the propeller alone. The alighting gear consists of appropriate shock absorbers on two supporting members and the fin tip. Construction is all-metal. Crew consists of a pilot.

WEIGHTS

LOADING	POUNDS	L F
Empty	6,477 (E)	
Basic	6,543 (E)	
Design (flight)	7,785	7.5
Combat	7,785 (E)	7.5
Max. take-off*	8,465 (E)	
Max. landing	7,445 (E)	

* Limited by space

POWER PLANT

ENGINE
No. & model (1) Double Mamba III
Mfr. Armstrong Siddeley Motors Ltd.
Type axial-flow turbo-prop.
Augmentation None
Length (w/exhaust cone) 102 in.
Width 53 in.
Height 44 in.
Reduction gear 10.38:1
Specification Installation folder Issue No.3

PROPELLER
Hub Manufacturer Rotol
Blade manufacturer Curtiss
No. blades/dia. 8/12.17 ft.
Blade design No. 630-1C2 & 631-1C2

RATINGS

	Engine speed rpm	Shaft Power bhp	Jet Thrust lb.	Fuel Cons. lb/hr
Take-off	15,000	2640	810	2160
Normal	14,500	2095	710	1833

FUEL AND OIL

LOCATION	No.Tanks	CAP.
Fuselage	1*	250 gal.
Total		250 gal.
Spec.		MIL-F-5616
Grade		JP-1

* Not self-sealing

OIL
Capacity (1 fus. tank) 3 gal.
Spec. MIL-O-6086
Grade M

DIMENSIONS

Span	22.2	ft.
Length	30.2	ft.
Height	14.75	ft.*
Wing area	145	sq.ft.

* In horizontal position

ELECTRONICS

UNF Trans-receiver	AN/ARC-27
Radio Altimeter	AN/APN-1

Standard Aircraft Characteristics charts for the N-63A. While the performance was not spectacular, it was sufficient to gather data on the feasibility of the VTOL tailsitter concept at minimal expense. Northrop would subsequently suggest abandonment of the N-63A in favour of a stripped down version of the full-scale N-63 to reduce the overall programme cost.

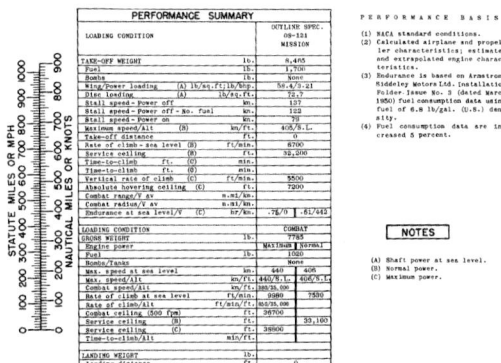

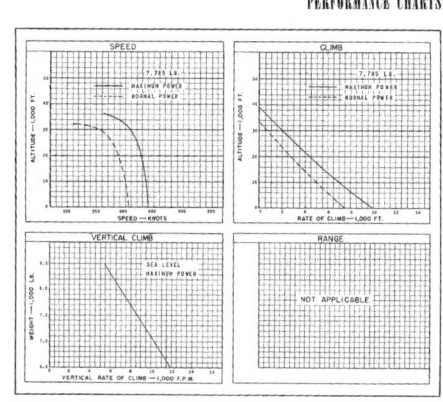

Cost Proposal
AND MANUFACTURING PLAN

Part of the cover to the cost proposal for the Northrop N-63 Convoy Fighter, which used the same artwork as the cover of the main proposal brochure.

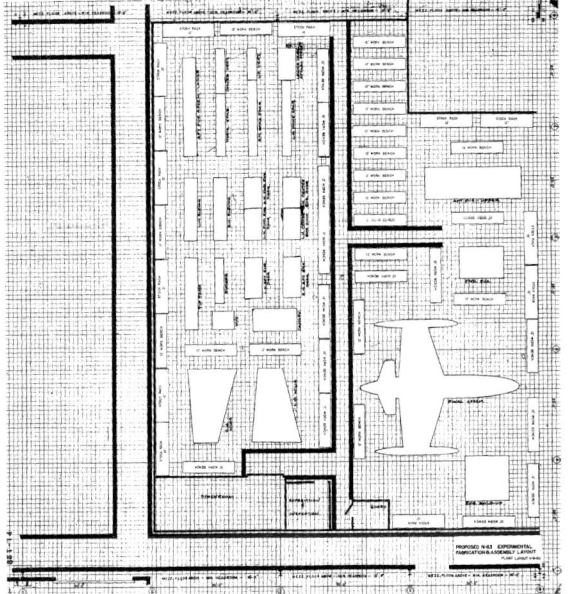

Northrop's proposed experimental fabrication and assembly layout for the N-63 Convoy Fighter.

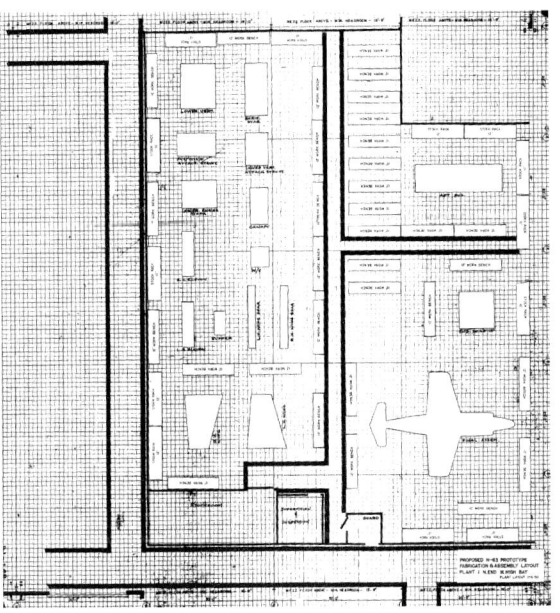

The proposed experimental fabrication and assembly layout for the N-63A prototype airplane.

Spot illustration of the N-63 Convoy Fighter in flight.

Cover to the Northrop N-63 Progress Report No. 1 dated December 22, 1950.

This view of the N-63 in flight emphasizes the type's unusual configuration.

Title page to the Northrop N-63 progress report which recycled artwork from the original November proposal brochure.

Painting of the N-63 in a steep climb.

A simplified model of the Northrop N-63 Convoy Fighter, which was tested in the company's 10ft wind tunnel December 5 through 21, 1950. The windmilling propellers were borrowed from another model and were not representative of those that would have been fitted to the actual aircraft.

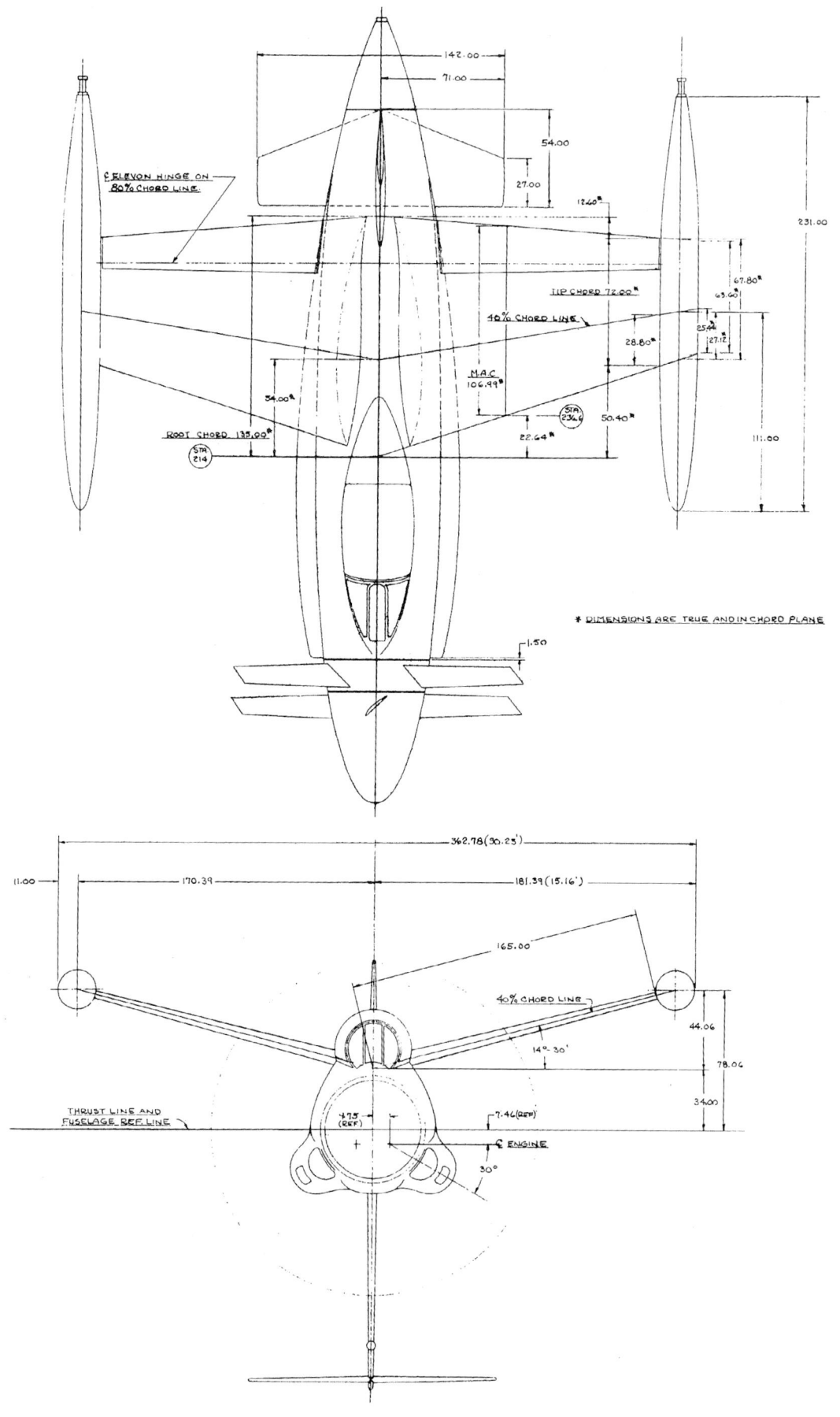

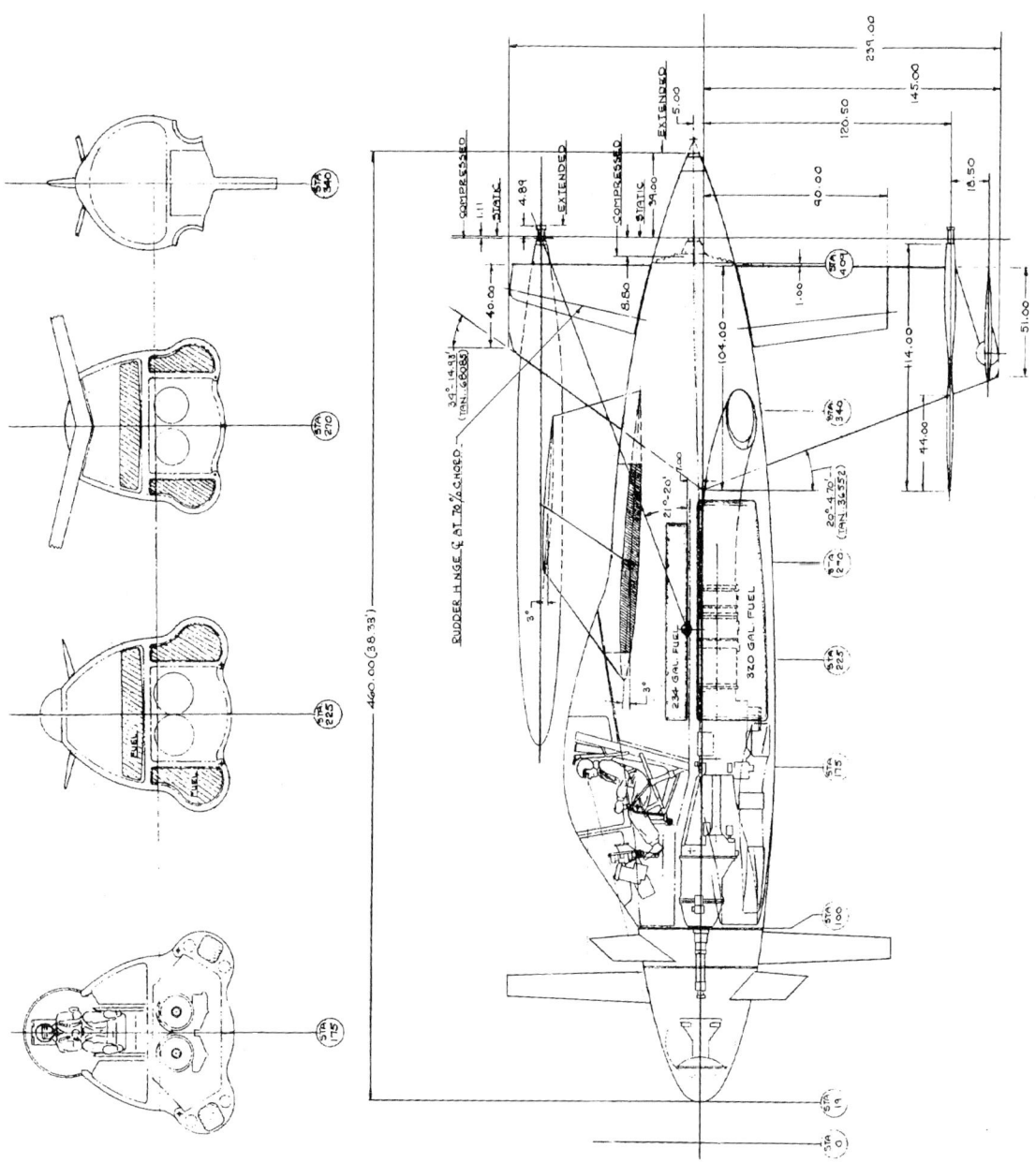

A comparison between this three-view of the N-63 from the December progress report and the previous one from the November proposal shown earlier in this chapter reveals several subtle changes to the design. These include a lengthening of the fuselage by 1.63ft; an increase in the frontal area and volume of the fuselage; recontouring of the air intake fairings, which now extend much further aft; and a larger dorsal vertical stabilizer with rudder added.

Aft views of the simplified N-63 wind tunnel model, which was constructed of mahogany, plywood sheet, and aluminium. It was designed to rapidly obtain data on the basic stability and control characteristics of the airplane at minimal cost.

NORTHROP N-63 AND N-63A

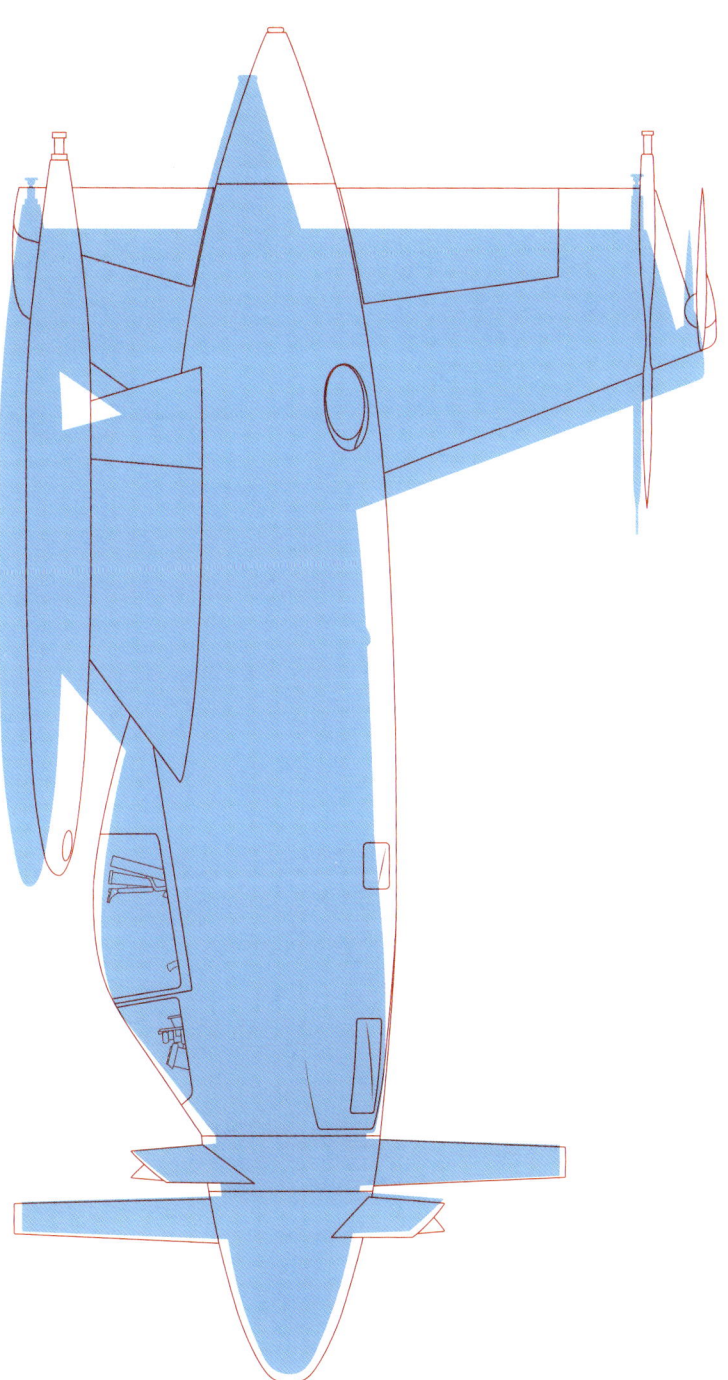

A side view comparison between the N-63 configuration of November 6, 1950 (shown in blue silhouette) and the later one of December 22, 1950 (shown as red line work), both to the same scale. The December design was larger and more aerodynamically refined thanks to wind tunnel tests undertaken by the company.

A speculative colour profile of the N-63 in the overall Glossy Sea Blue scheme which was standard for most Navy aircraft in the early postwar period; it portrays the original configuration dating from November 6, 1950.

A speculative colour profile of the Northrop N-63A scale prototype airplane configured as a tailsitter. It was a flying aerodynamic model of the full-scale Convoy Fighter powered by an Armstrong Siddeley Double Mamba turboprop engine.

The N-63A with a T-tail and fixed landing gear, which permitted conventional takeoffs and landings for exploration of unconventional transition and hovering flight characteristics at safe altitudes.

The full-scale Northrop N-63 Convoy Fighter configuration as it stood in late December 1950. Basic wind tunnel tests resulted in several subtle changes to the design, such as a longer fuselage, recontoured air intake fairings, and a larger dorsal vertical tail incorporating a rudder.

7

Winners and Losers

THE CONVOY Fighter competition was won by Convair, with Lockheed placing second and Martin third. The Lockheed and Martin proposals were scored very closely, but the former was rated slightly higher than the latter, and Martin lost out on receiving a contract. Based on a surviving table prepared by BuAer, takeoff gross weight was likely the key factor in determining the outcome:

CONVOY FIGHTER PERFORMANCE SUMMARY

Notes:

1. All performance figures were BuAer estimates, and were based on the same propeller (eight bladed—15.5ft diameter) and the same gear ratio for all designs.
2. Contractors' estimated weights.
3. BuAer estimated gross weight equalled contractor gross weight plus "BuAer Weight Increment".
4. Goodyear gross weight was different in its performance and weight reports, weight

	OS-122 Reqs.	Convair Model 5	Goodyear GA-28B	Lockheed L-200	Martin Model 262	Northrop N-63
T.O. Gr. Weight[2] (lbs)	16,000	16,000	17,200	15,600	16,9905	16,780
Fuel Wt. (lbs)		2,950	3,280	3,048	3,000	3,315
Time to 35,000 ft (mins)	4.5	5.9	5.9	5.3	5.9	5.8
Loiter time (hours)[6]	2.0	1.24	1.8	2.07	1.78	2.35
Combat weight (lbs)[2]		14,820	15,890	14,380	15,790	15,454
V_{max}–Sea Level (knots)		536	526	537	528	524
V_{max}–35,000 ft (knots)	540	511	518	516	525	506
Rate of Climb–Sea Level (ft/min)		11,320	10,250	11,910	10,360	10,950
Combat Ceiling (ft)	45,000	45,000	45,500	48,700	46,800	47,700
BuAer Weight Increment (lbs)		+724	+1,0344	+1,213	+773	+586
Adj.T.O. Gr. Weight (lbs)[3]		16,724	18,0284	16,813	17,453	17,576
Adj. Time to 35,000 ft (mins)		6.2	6.3	5.9	6.4	6.1
Adj. Loiter time (hours)[6]		1.12	1.5	1.88	1.70	2.23
Wing Configuration		Delta	Delta	Straight	Swept	Straight
Wing Area (sq ft)/Span (ft)		346/27.5	345/31	246/29	247/31.5	259/30.2

increment shown was overweight shown in the weight report.
5. Martin gross weight increased by 100 lbs to allow for a two-speed gearbox.
6. Loiter time was based on: a. five minutes of military power for takeoff, transition and acceleration; b. military power climb to 35,000ft; c. loiter at 35,000ft at speed for maximum endurance (single power unit operation permitted if less than normal power required); d. military power cruise out for 100 nautical miles at 35,000ft; e. three minutes' combat at 35,000ft; f. cruise back at speed for maximum range at 35,000ft (single power unit operation); g. five minute military power allowance for landing; h. no reserve.

The takeoff gross weight figures provided by Convair, Lockheed and Martin were 16,000, 15,600 and 16,780lb respectively. Aircraft contractors are frequently optimistic in calculating the weight of their proposals, as it greatly affects overall performance, and this case was no exception. The weight increment calculated by BuAer for the Convair, Lockheed and Martin designs was +724, +1,213 and +773lb, making the adjusted takeoff gross weights 16,724, 16,813 and 17,453lb, respectively. The lighter the design, the higher it placed in the competition. With the addition of the BuAer weight increment, the takeoff gross weight of the Goodyear and Northrop designs was 18,028 and 17,576lb respectively, effectively knocking them out of contention.

Focusing on the top three contenders, Convair was also clearly superior in terms of cost. The following figures, taken from the cost proposals of the manufacturers, were for two stripped-down full-scale prototypes plus a static test article and calculated on a CPFF basis:

Convair	$4,756,741
Lockheed	$5,679,575
Martin	$8,000,000

Surviving notes show that BuAer made little effort to evaluate the 0.766 reduced scale prototypes offered by each contractor and originally specified in the OS-122 requirements. Judging by the evidence, BuAer had come to the conclusion that it was faster and cheaper to procure stripped-down versions of the full-scale Convoy Fighter designs before the contractors had even submitted their proposals on November 1, 1950. Furthermore, BuAer likely communicated this to all the contractors at some point before they submitted their proposals, as the majority of them offered this as an option.

The following sections focus on the winners of the Convoy Fighter competition—the Convair XFY-1 and Lockheed XFV-1. The emphasis is on the preliminary design and configuration evolution, as much of this has not been widely published before. This is followed by a summary of the respective flight test programmes and subsequent cancellation of both aircraft. An in-depth examination of the actual XFY-1 and XFV-1 experimental aircraft and their testing is beyond the scope of this publication; hopefully it will be undertaken by another author in the near future.

CONVAIR XFY-1 POGO
Convair was awarded a contract in April 1951 for two full-scale prototypes and a static test article under the designation XFY-1; it retained the 'Pogo' nickname that the company had used in its earliest configuration studies.

Early blueprints

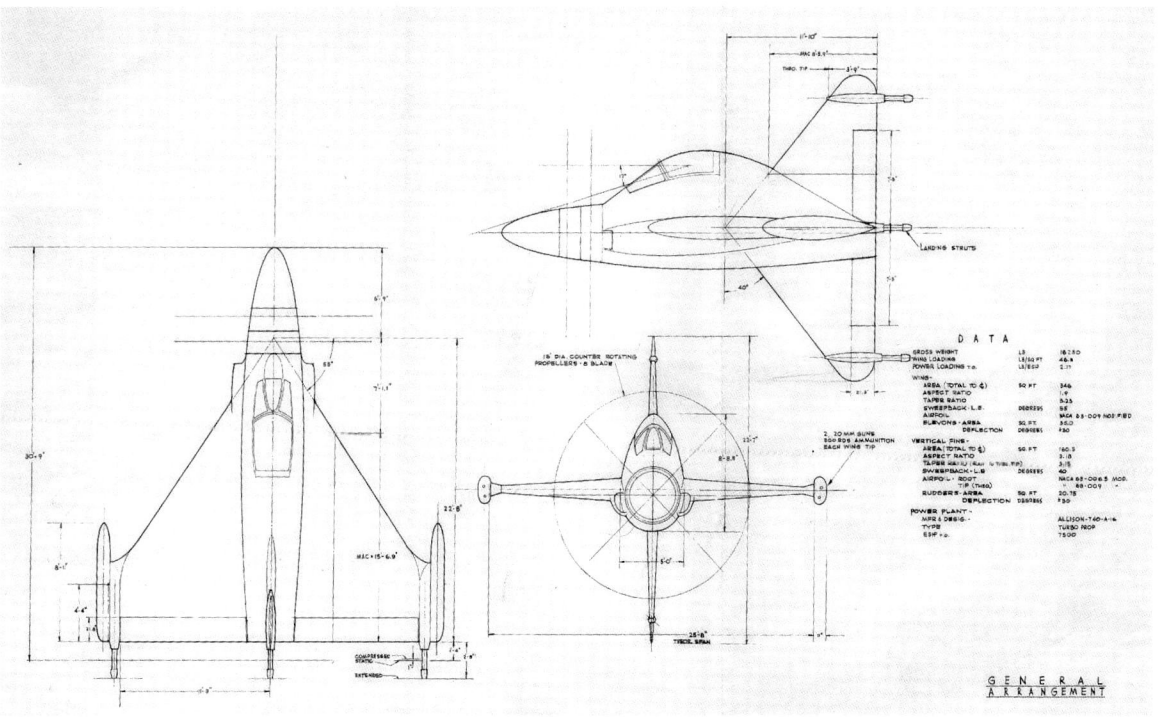

A three-view of the XFY-1 dating from mid-1951; at this point in its development, the configuration was still quite close to the initial proposal, though with several changes, including the addition of rounded fairings to the tips of the vertical tails and a smoother transition between the wing tip and gun pod, among other modifications. The subsequent mock-up appears to be based on this and the following drawings, also produced in mid-1951.

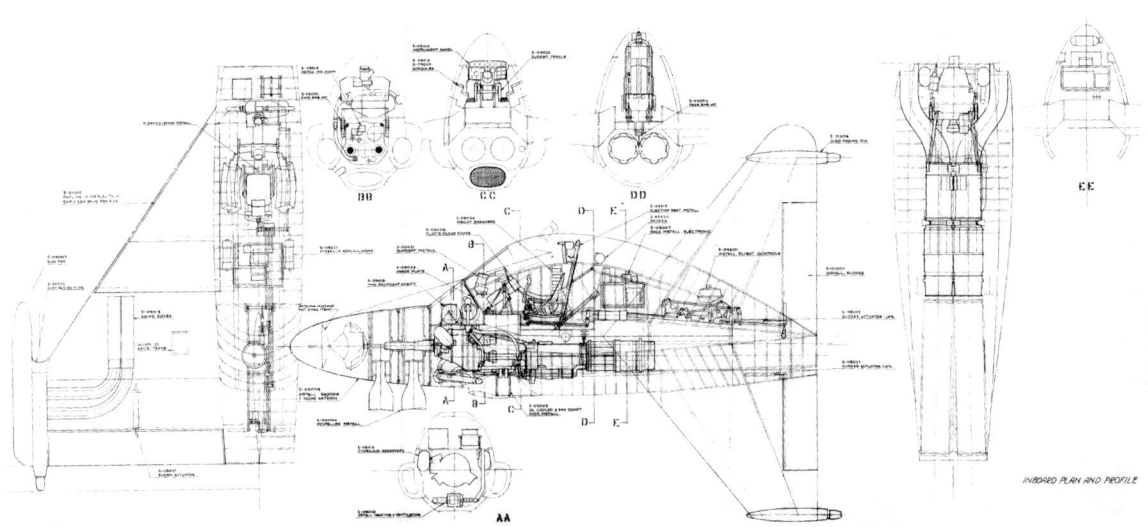

An inboard profile of the XFY-1; some minor changes in the interior equipment arrangement were made compared to the inboard profile of the original proposal.

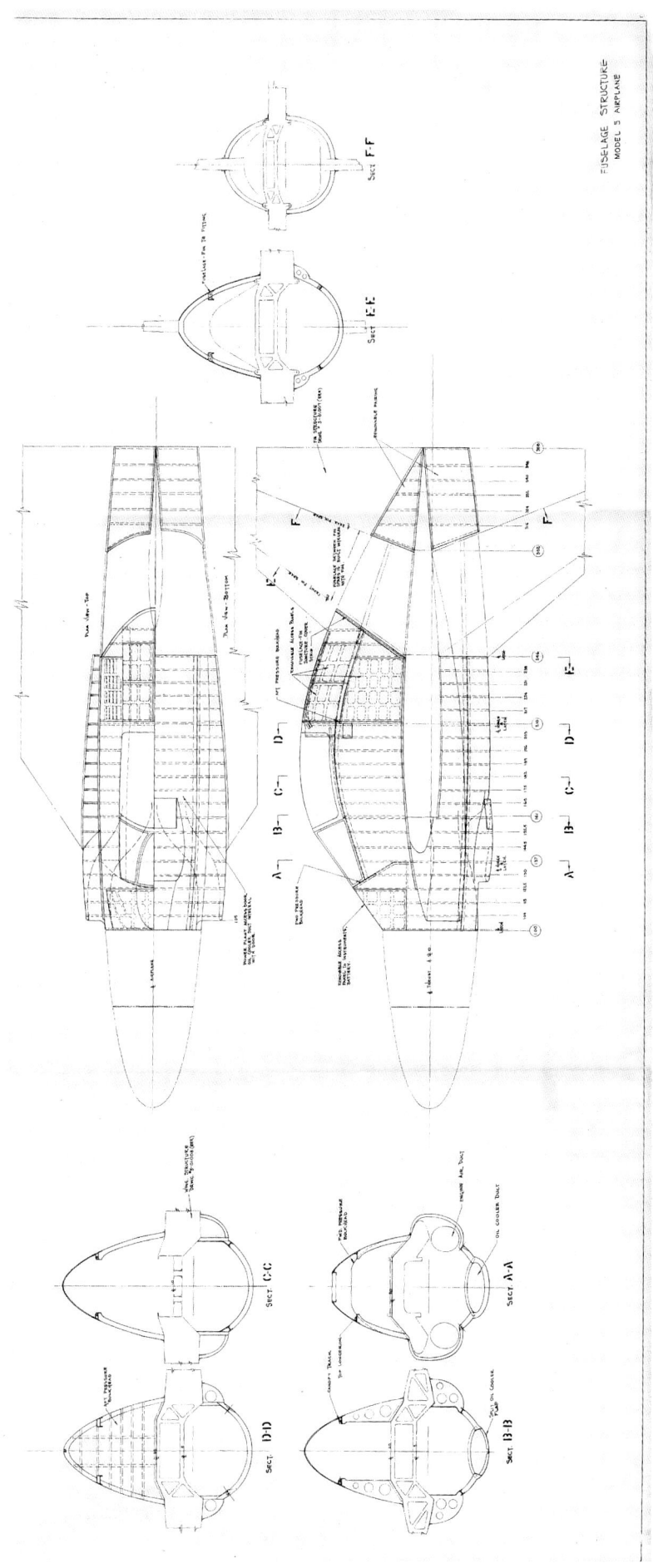

An early structural diagram of the XFY-1 drafted a few months after the contract award in May 1951.

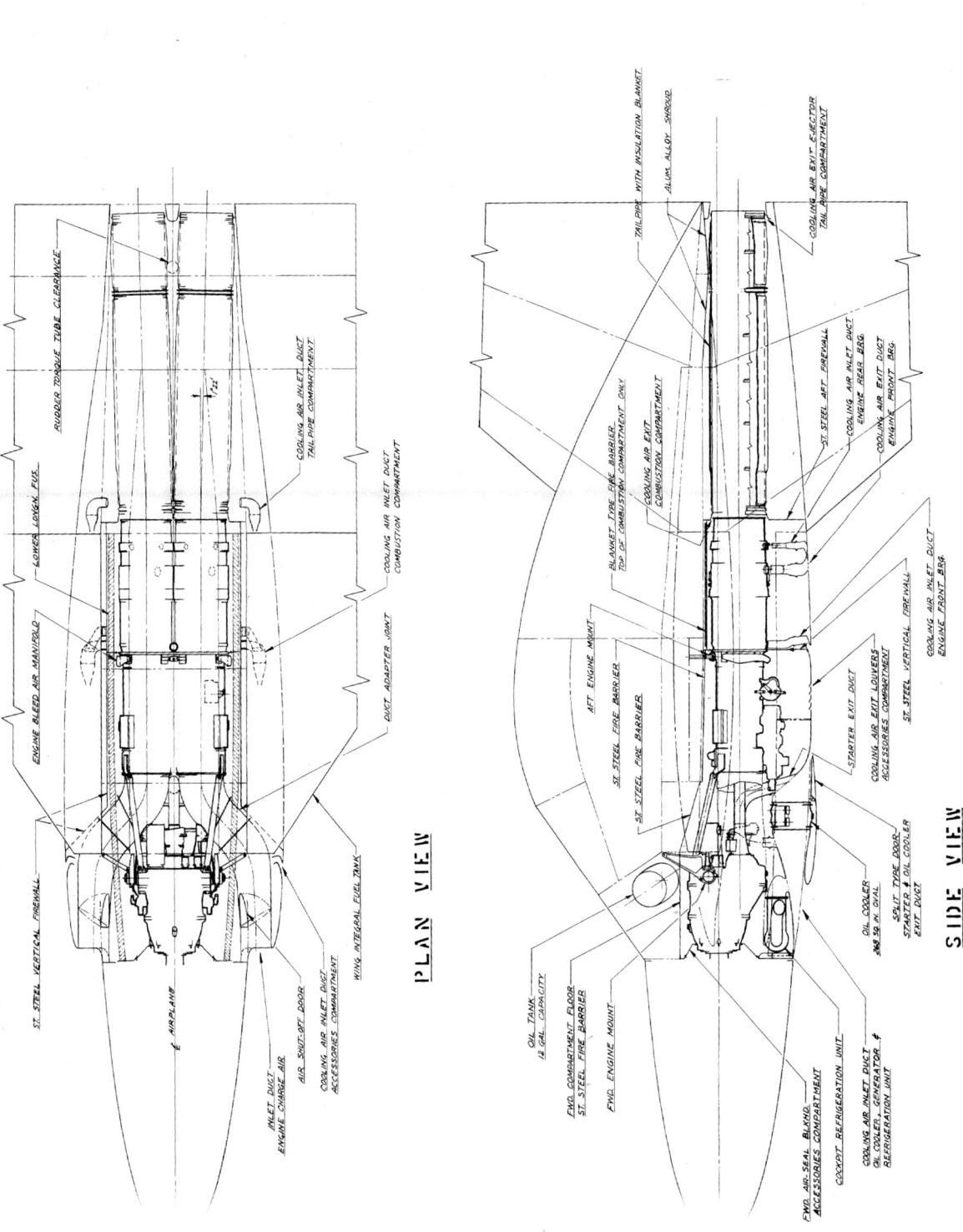

Blueprint of the Allison XT40-A-6 turboprop installation in the XFY-1 dating from mid-1951.

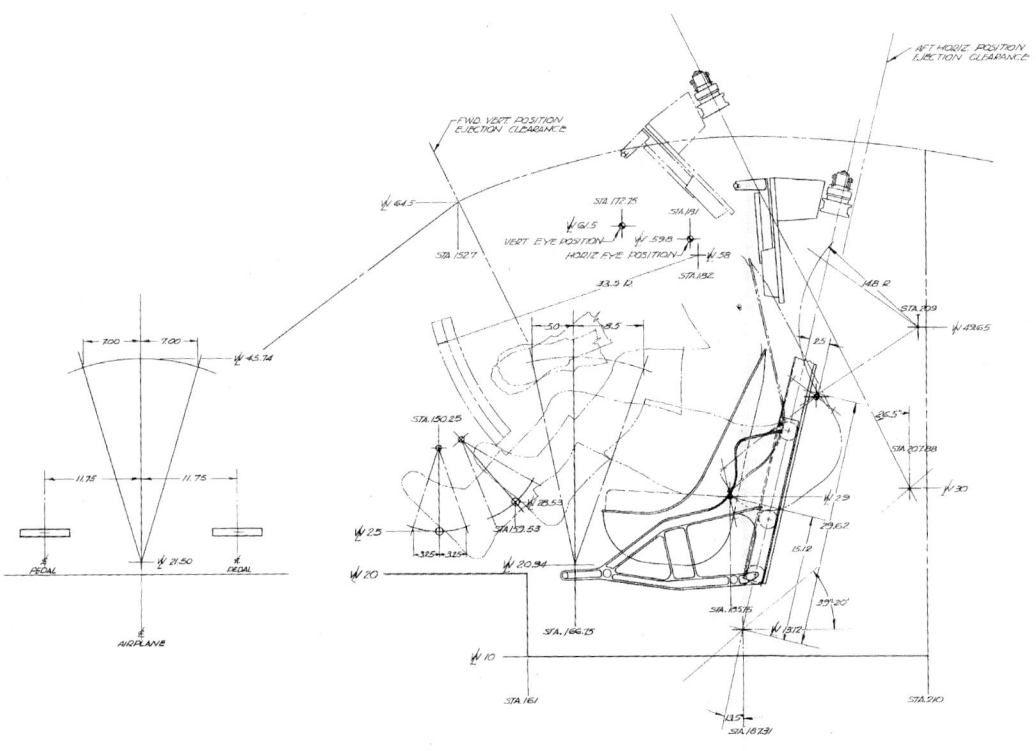

Basic diagram of the early Pogo's cockpit dimensions showing the seat rotation for takeoffs and landings.

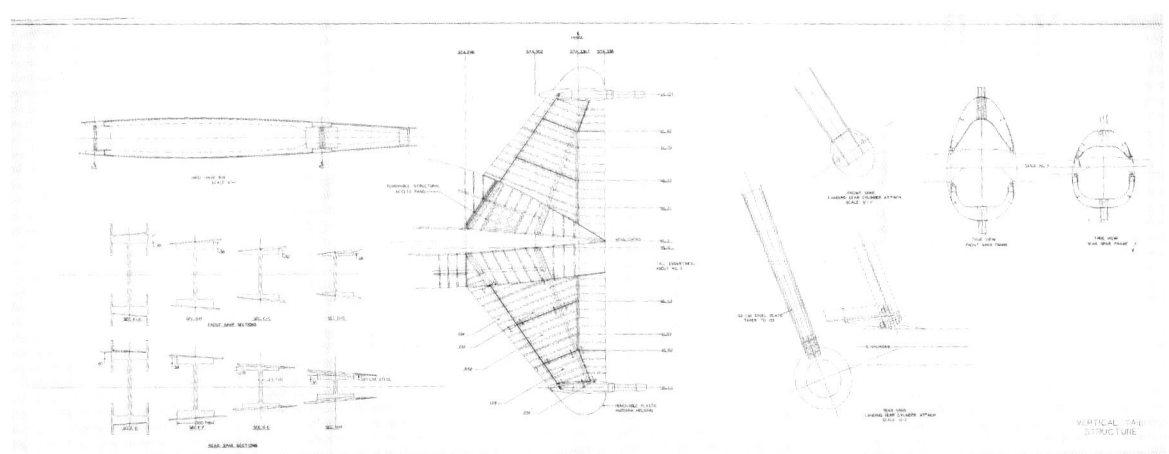

Early structural diagram of the XFY-1's vertical tails showing the addition of rounded tips to these surfaces; also note the thicker oleo struts and associated fairings.

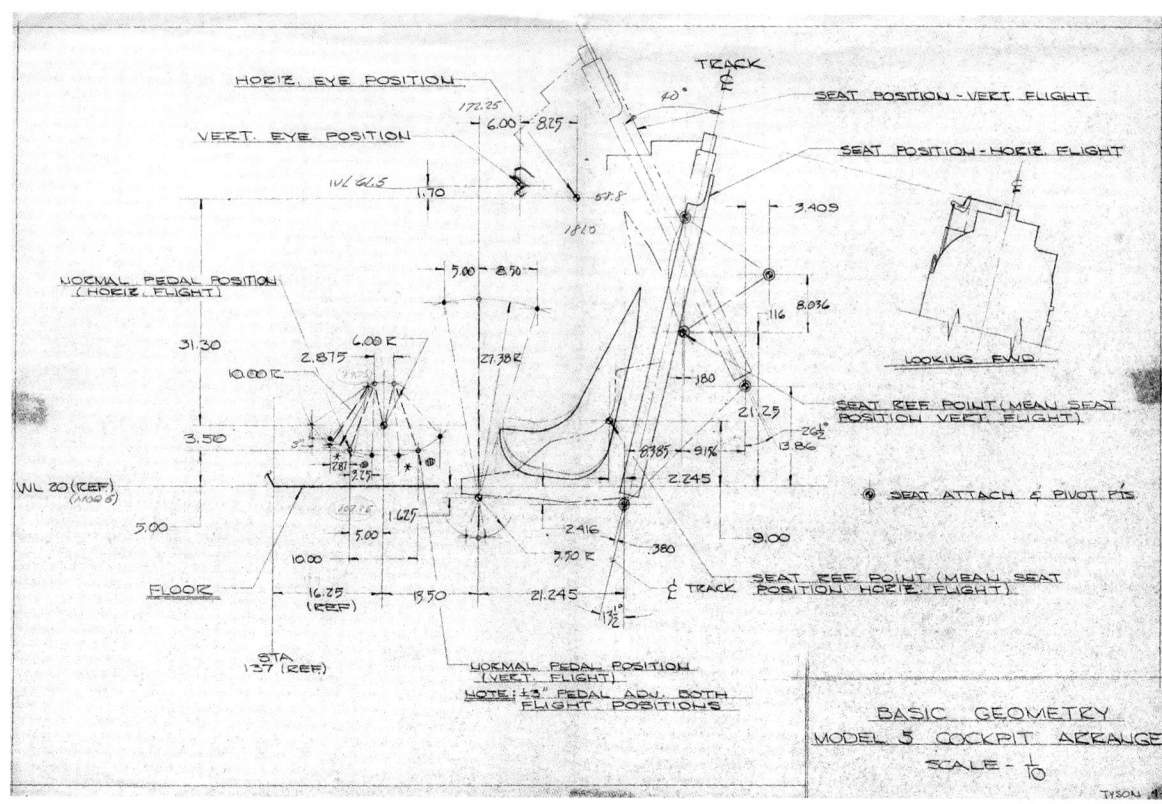

A drawing of the basic geometry of the XFY-1's cockpit arrangement dating from mid-1951.

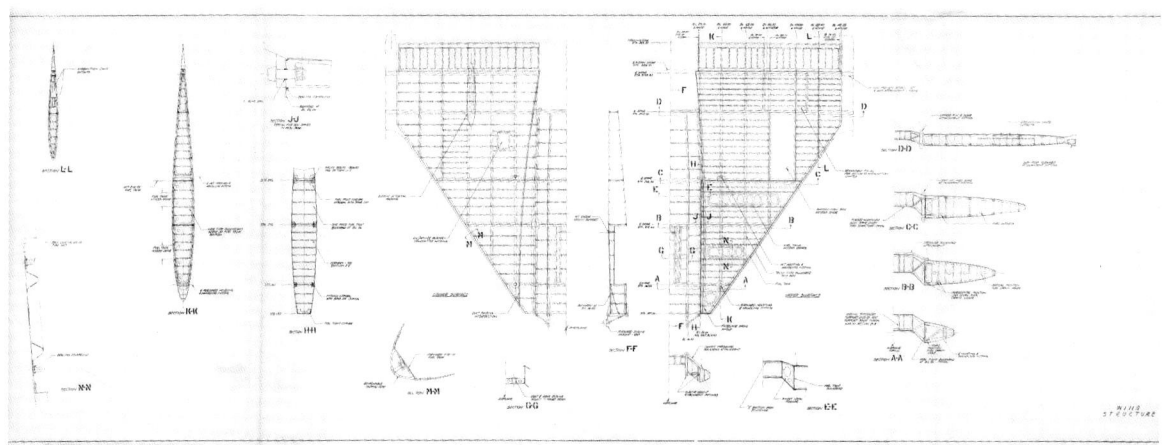

Blueprint of the early XFY-1 wing structure, which carried major loads from other major components of the aircraft, such as the power plant and pilot's compartment.

Included in this chapter are some early blueprints of the XFY-1 which are believed to date from mid-1951, having been drafted after the May contract award. The configuration depicted is still quite similar to the initial proposal, with the following differences: the addition of rounded tip fairings to the dorsal and ventral tail surfaces; a wider canopy; a more streamlined fairing between the wing tip and gun pod; and thicker oleo struts. The mock-up, presented to the Navy in June-July 1951, is believed to have been based on these drawings.

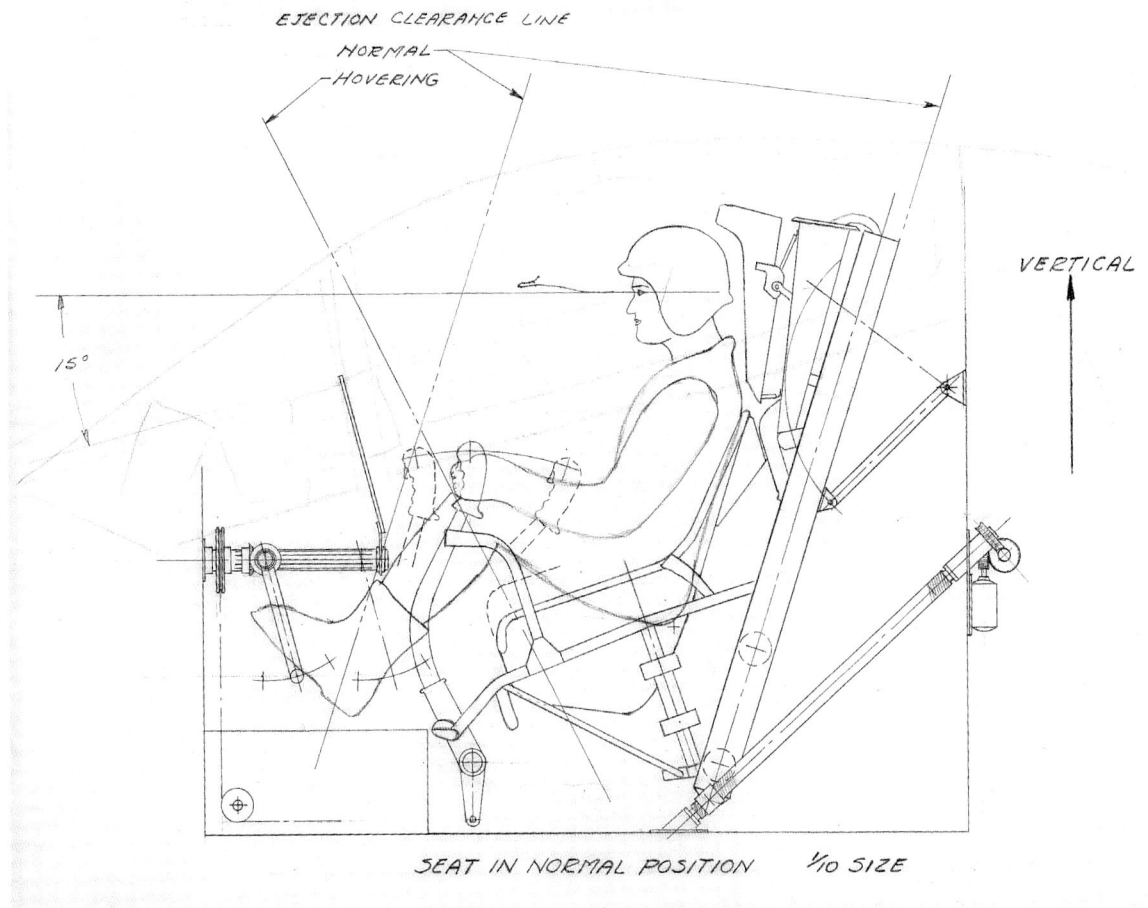

A drawing of the seat in the normal position for horizontal flight.

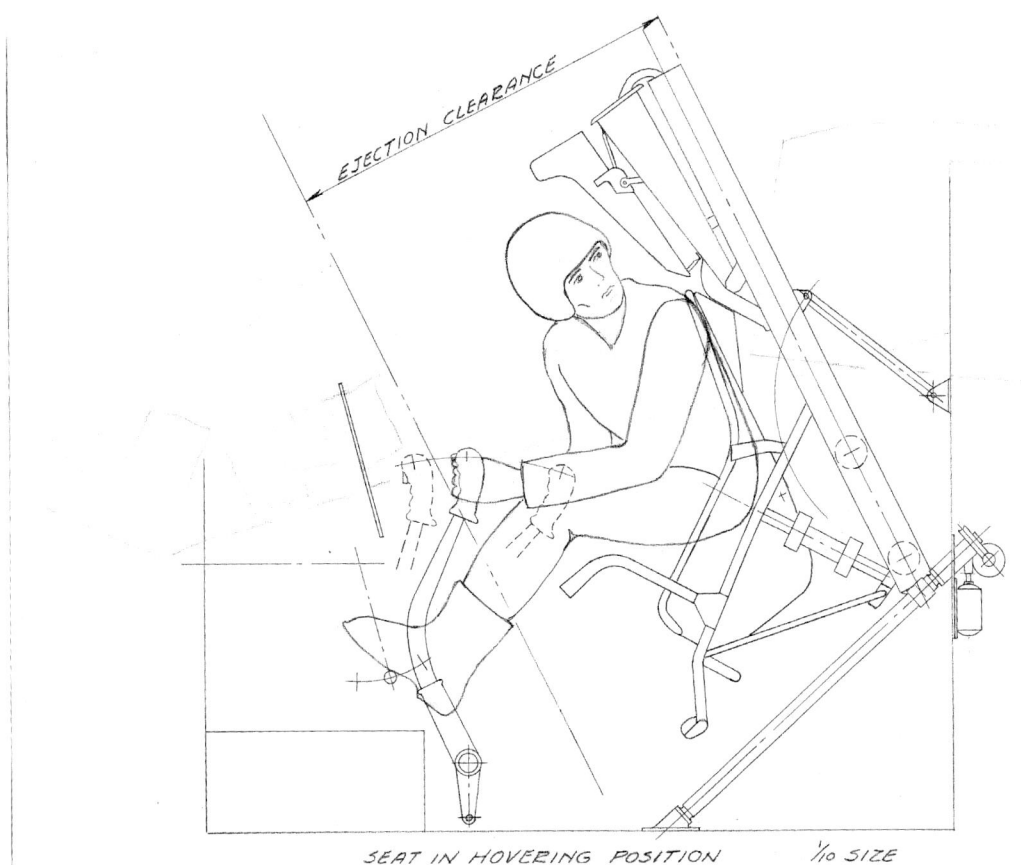

Depiction of the seat tilted forward during hovering to improve the vision for the pilot during landing attempts.

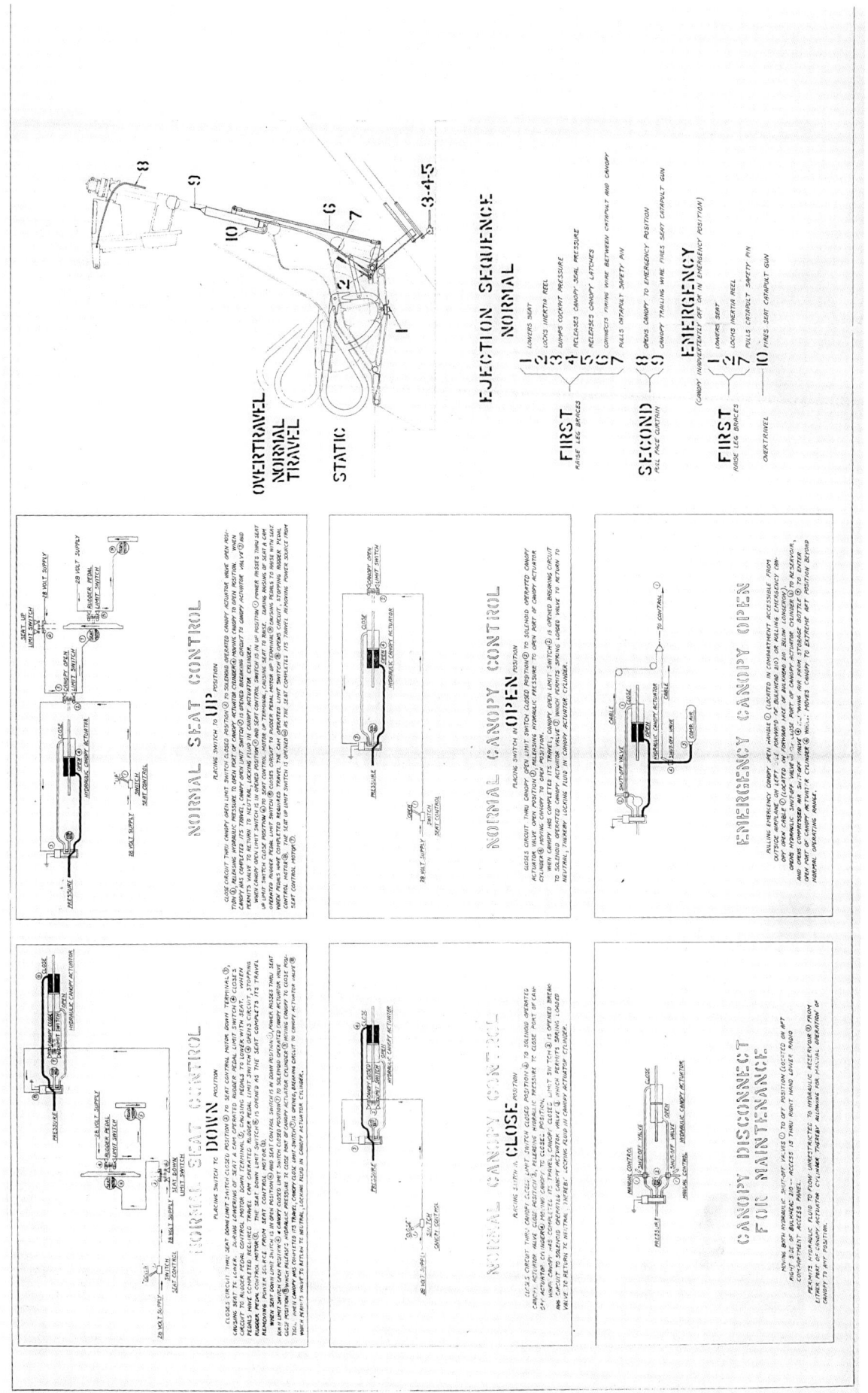

A detailed early blueprint covering seat and canopy operation of the XFY-1, including emergency ejection provisions.

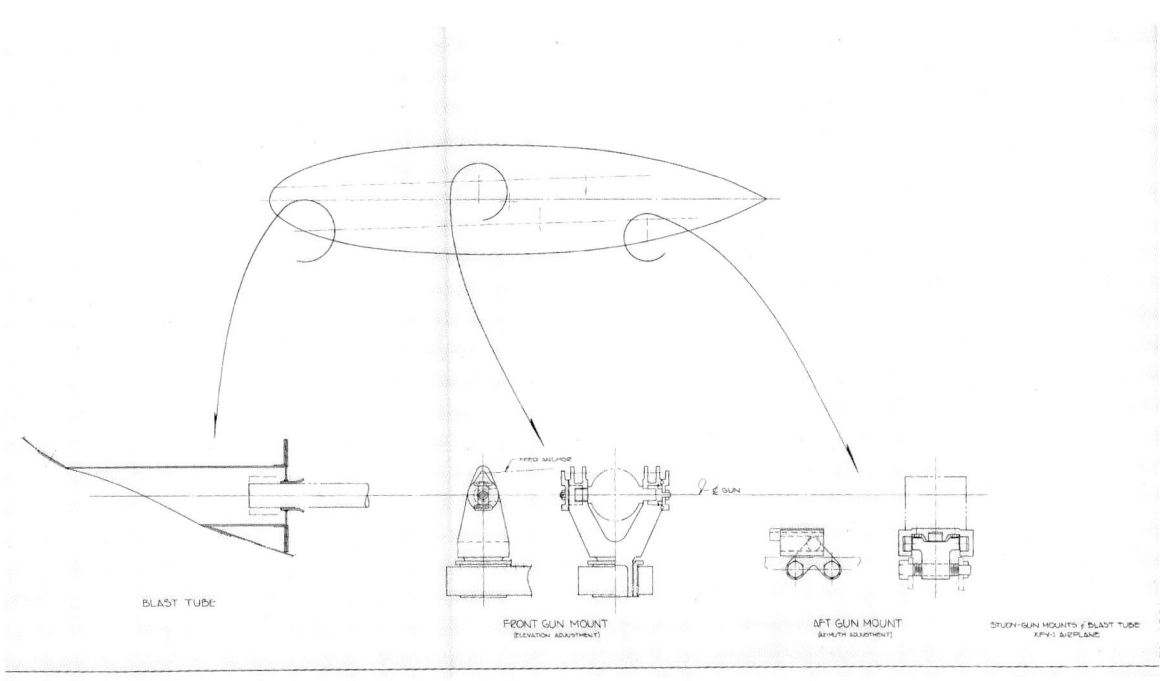

Schematic of the gun mount and blast tube components of the wing tip gun pod from mid-1951.

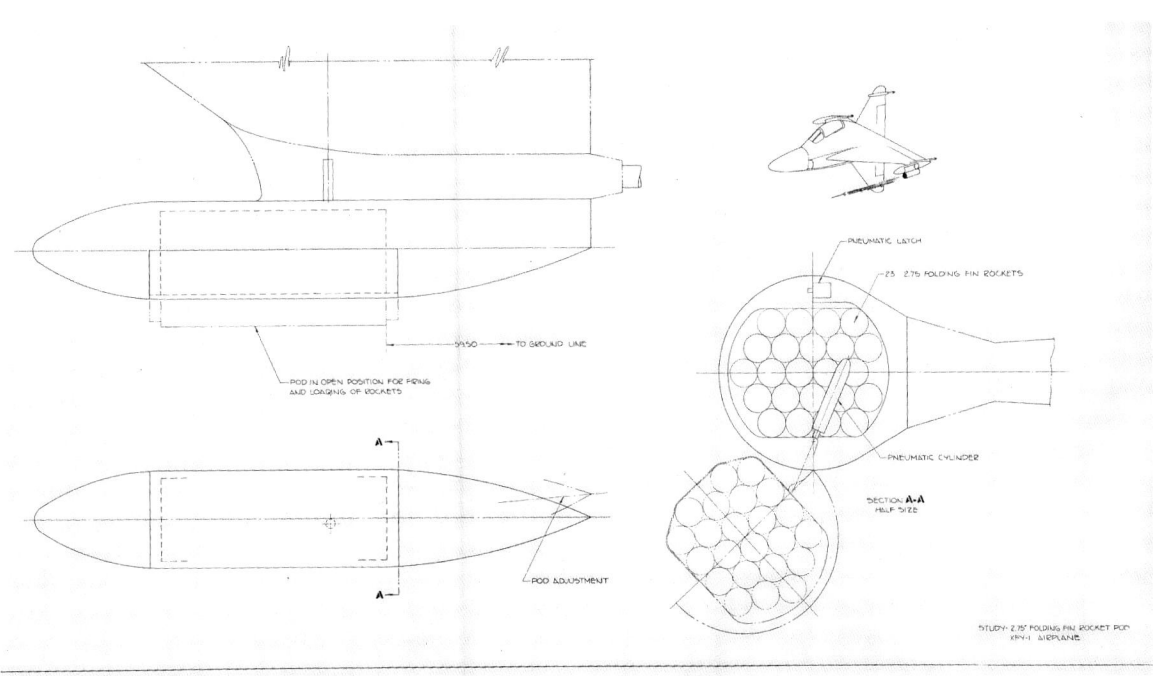

Blueprint of the early XFY-1 wing tip rocket pod, each of which could hold twenty-three 2.75" folding fin rockets. This drawing also shows the revised streamlined fairing between the wing tip and pod compared to the less aerodynamic one depicted in the November 1950 proposal.

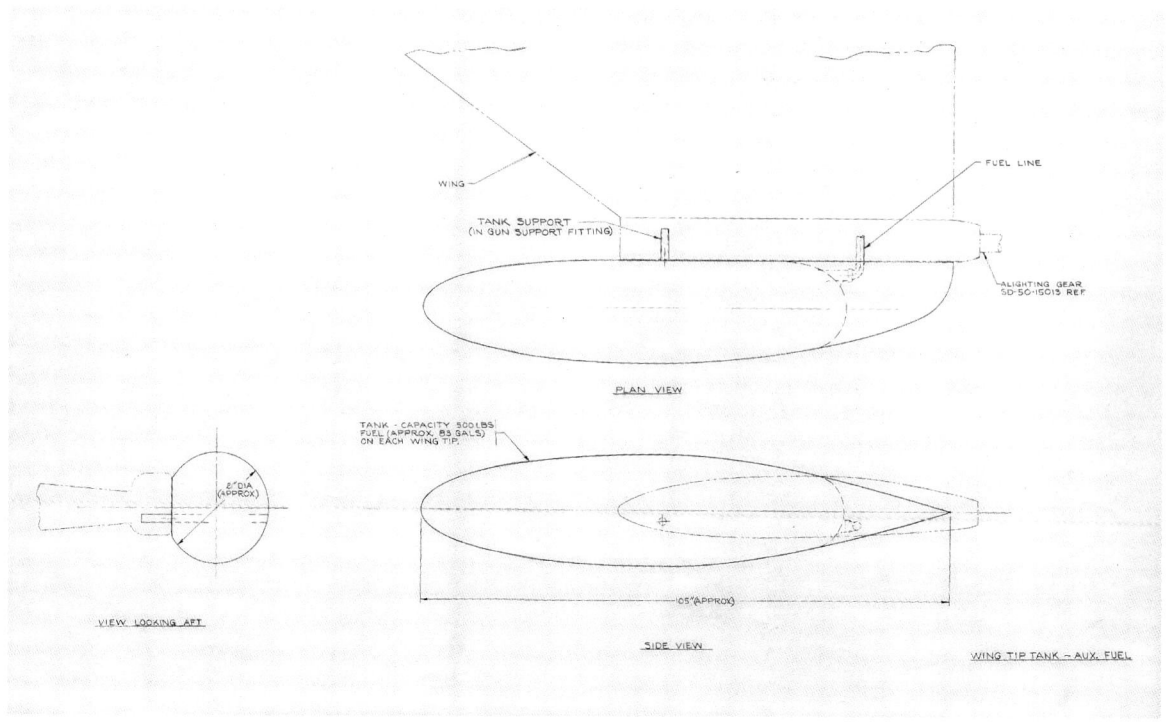

A drawing of a wing tip-mounted fuel tank, each of which could carry 500lb (83 gallons) of fuel.

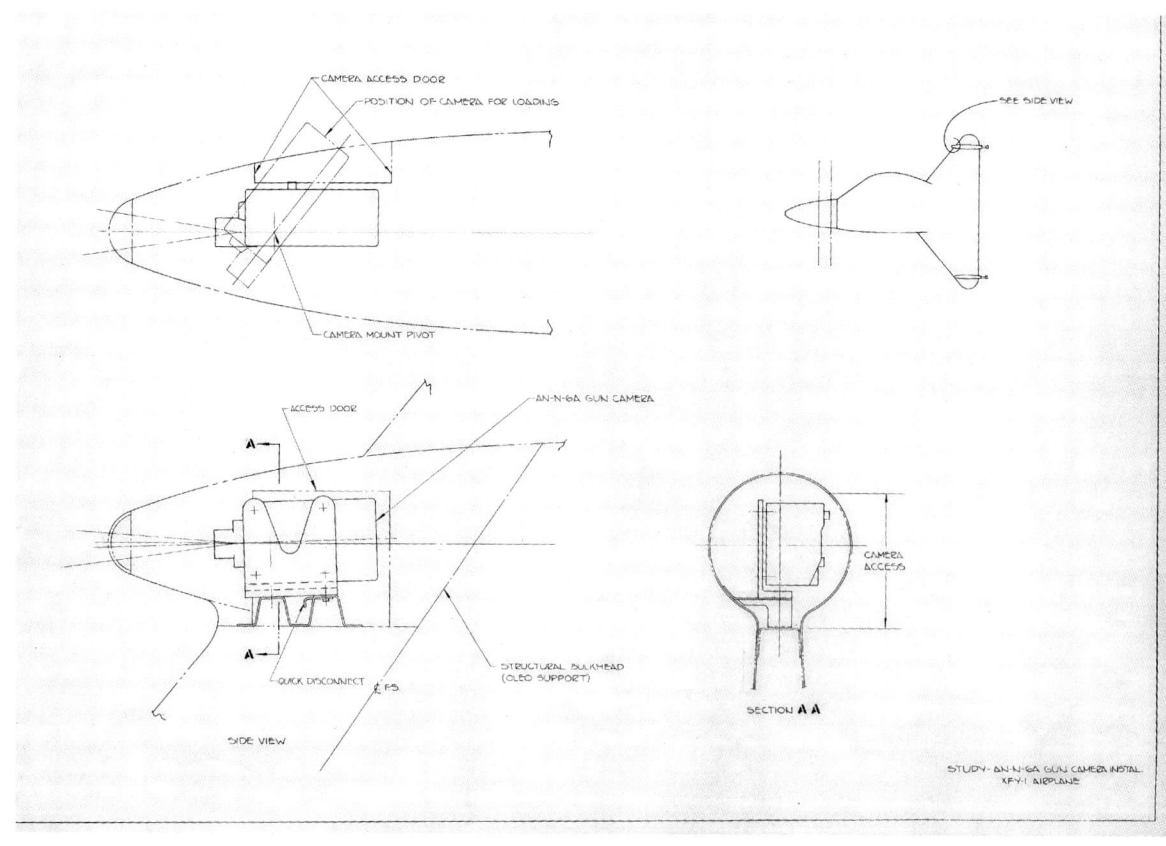

A study of an AN-N-6A gun camera installation at the tip of the XFY-1's upper alighting gear fairing.

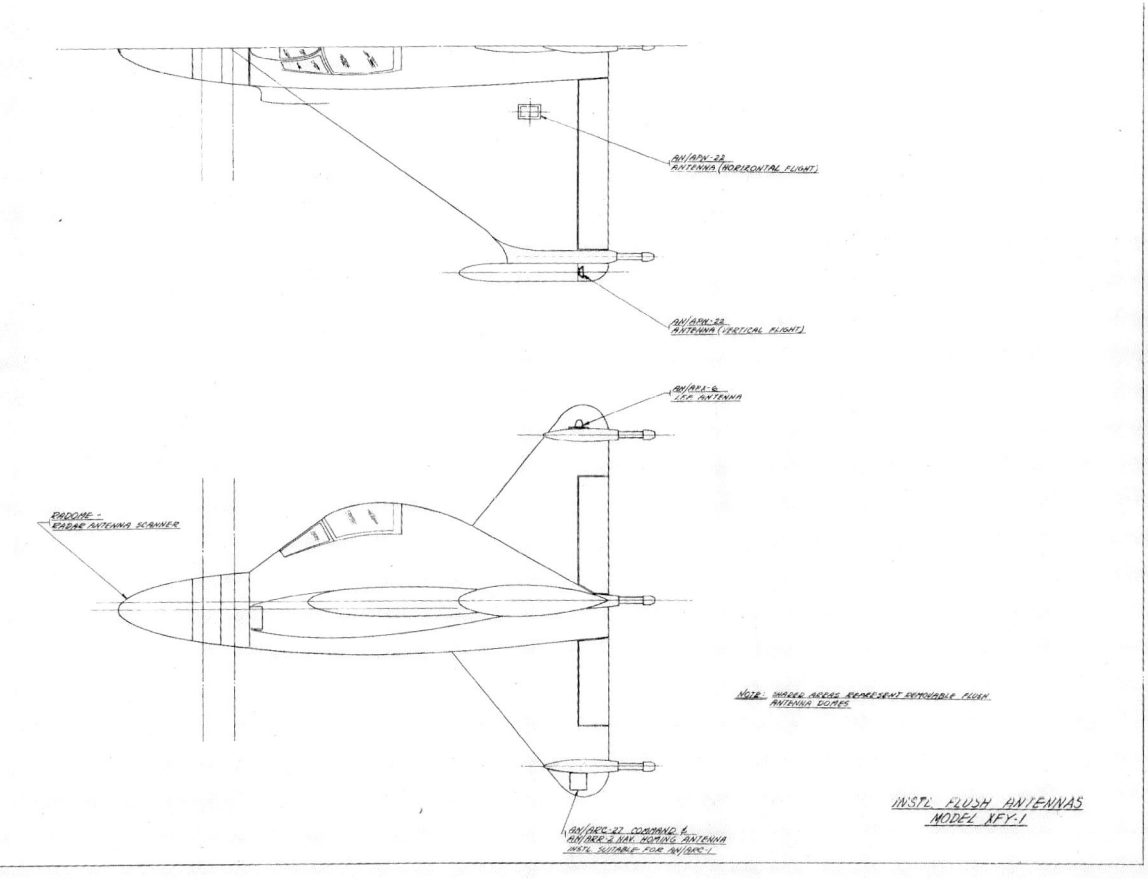

A blueprint of the installation of flush antennas on the Pogo from mid-1951.

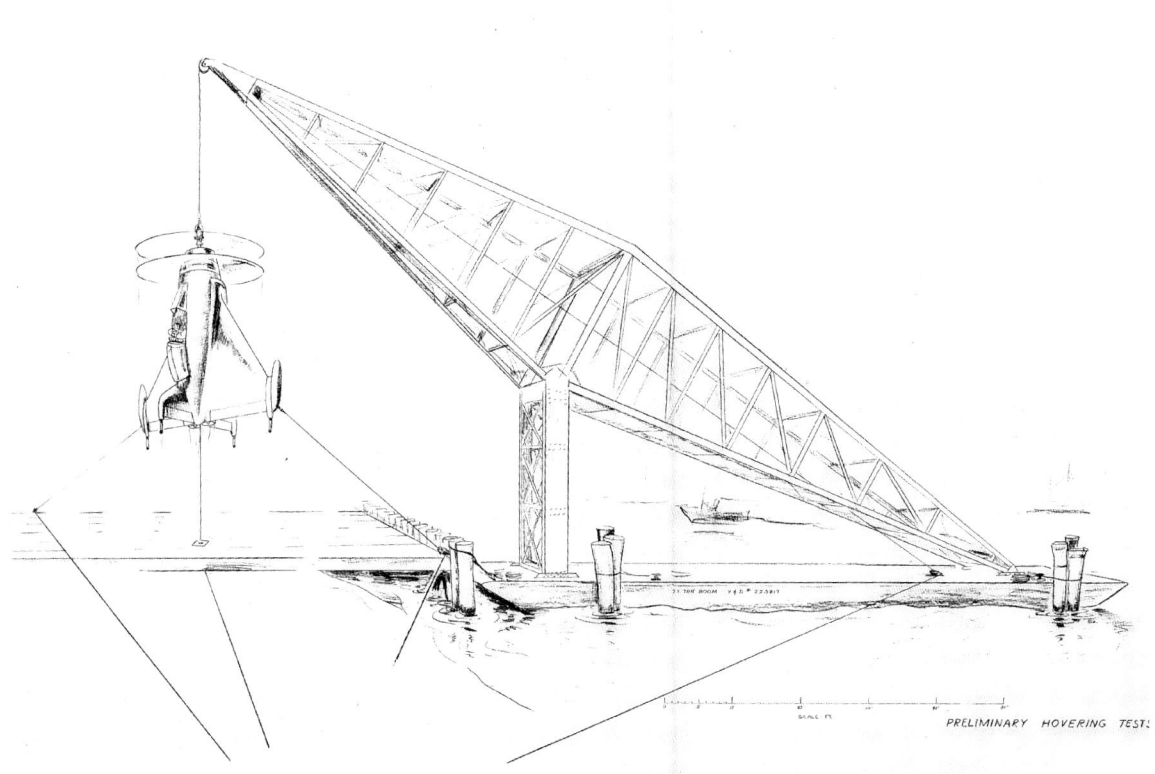

Convair considered performing preliminary hovering tests with the XFY-1 on a dock using a large 75-ton boom mounted on a barge. This concept would be rejected in favour of holding the tests inside of the naval airship hangar at Moffett Field.

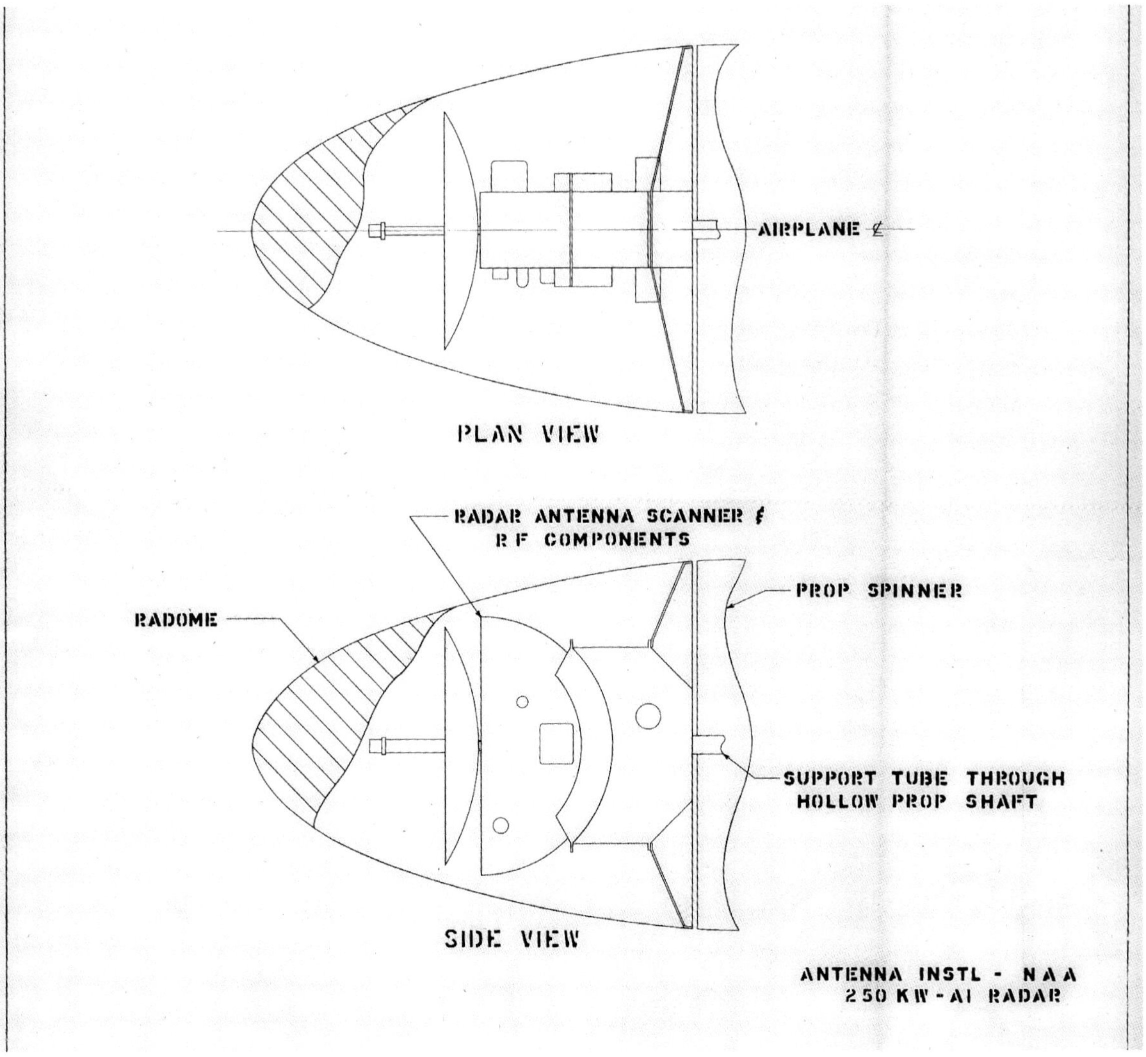

An early diagram of the radar antenna scanner and RF components in the propeller spinner of the XFY-1.

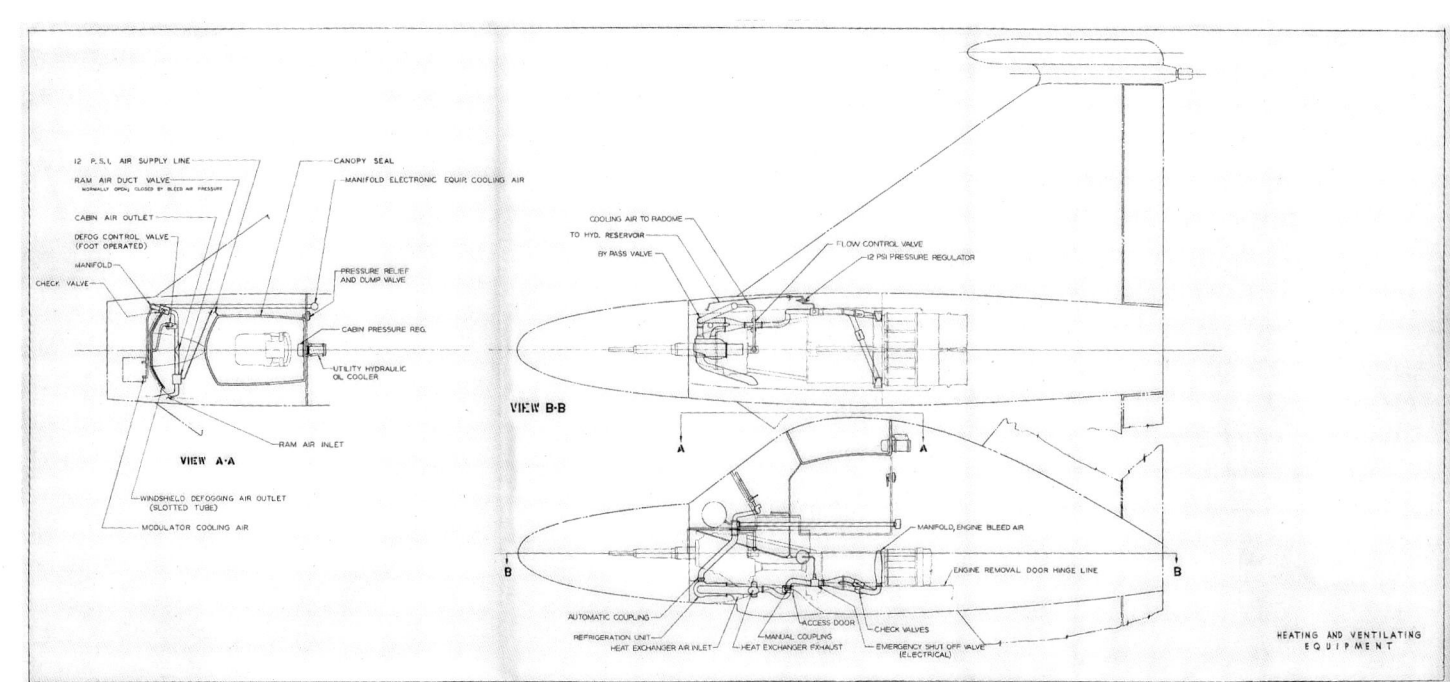

Early blueprint of the XFY-1 heating system.

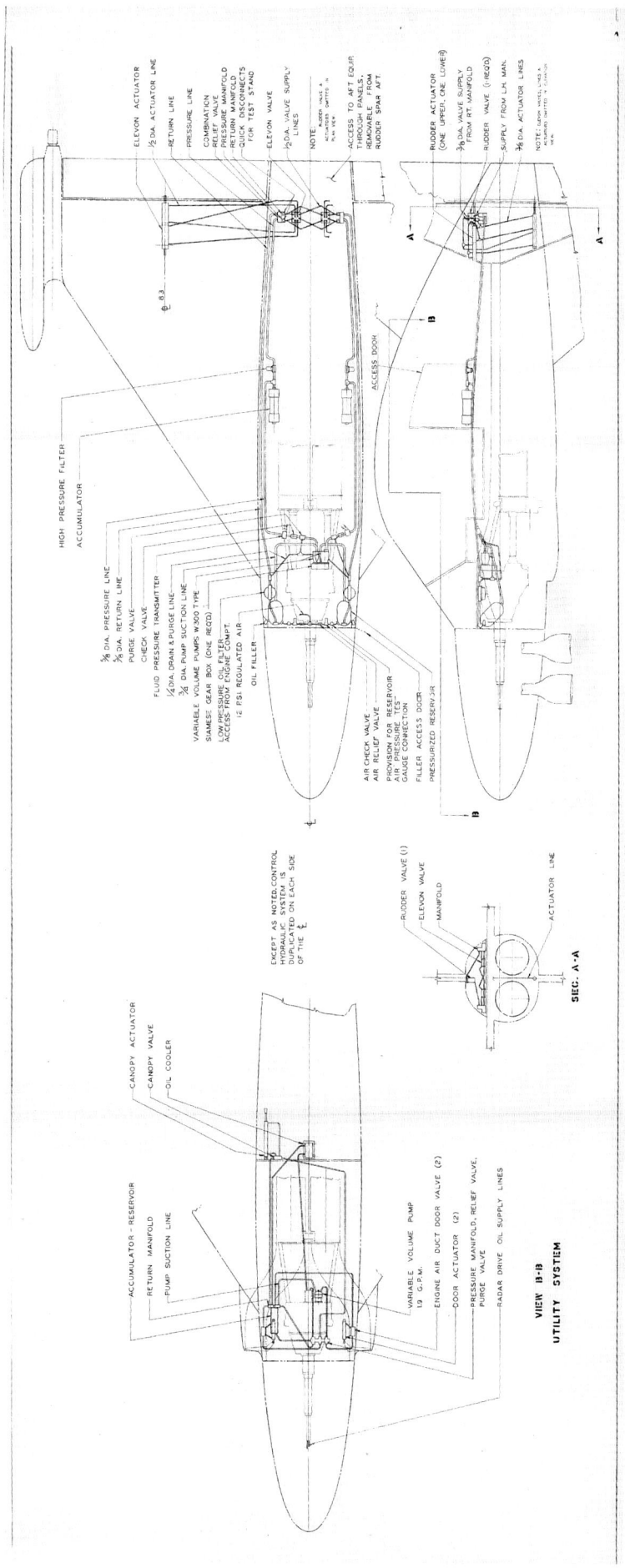

Schematic of the XFY-1 utility system from mid-1951.

XFY-1 MOCK-UP

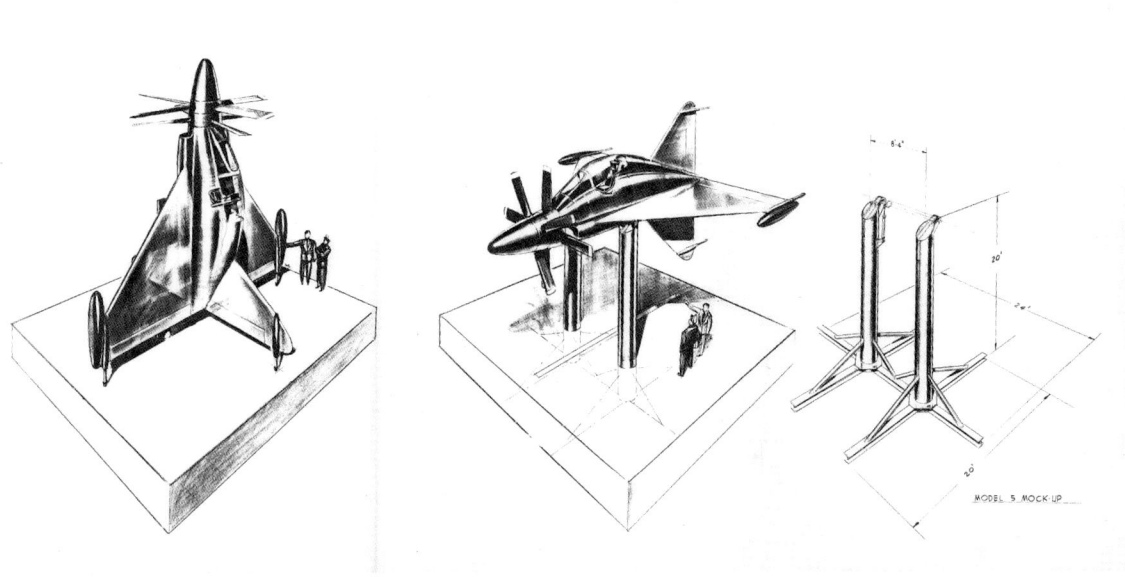

The Convair XFY-1 mock-up was displayed in both vertical and horizontal positions using the special twin post steel support stand shown here.

The XFY-1 mock-up in the horizontal flight position; it appears to be rather haphazardly painted an overall Glossy Sea Blue. The mock-up underwent official inspection on June 29-30 and July 2-3, 1951 at the Convair plant at Lindbergh Field, San Diego, California.

Side view of the Pogo mock-up, which was approved by the mock-up board on September 24, 1951 with the condition that several relatively minor changes be made.

This view of the Pogo mock-up emphasizes the unusual alighting gear mounted on the trailing edges of the wing and tail surfaces. The solid pegs at the ends of the oleos would be replaced with more standard wheels on the actual prototype.

Aft view of the XFY-1 mock-up in the horizontal flight position emphasizing the twin exhaust pipes of the Allison XT40-A-6 turboprop engine.

Top view of the mock-up in the vertical position.

A view straight down into the cockpit of the XFY-1 mock-up.

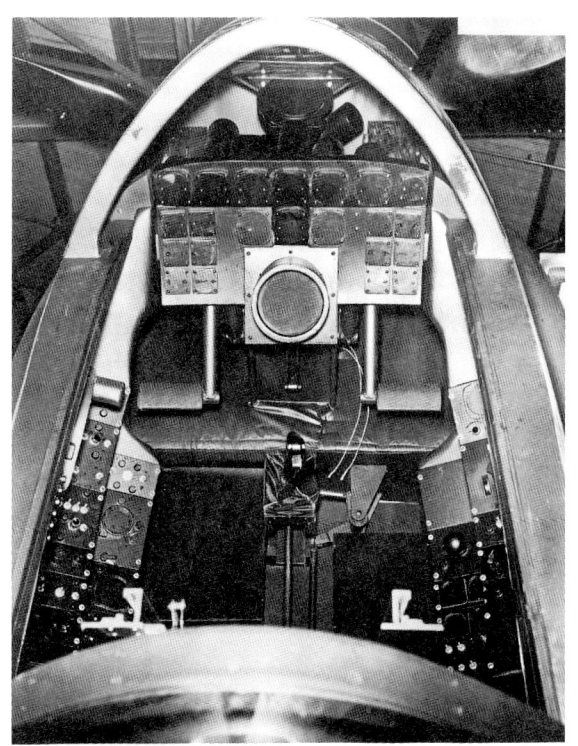

A good view of the instrument panel, rudder pedals, and the forward side consoles.

A similar view with the instrument panel removed showing the equipment behind the panel, including the gunsight, as well as more detail of the rudder pedals.

A view of the left hand console inside the cockpit of the Pogo mock-up.

A view of the right hand console.

A view of the engine compartment showing the installation of the Allison XT40-A-6 turboprop. The aircraft was designed to have the engine serviced in the vertical position, but the mock-up board subsequently asked the company to study the feasibility of removing the engine in the horizontal position using a large track or dolly.

A photo of the engine compartment with the turboprop engine removed.

A view of the upper right aft fuselage with the skin removed showing actuators and control valves.

The lower left aft fuselage panel and tail pipes of the XFY-1 mock-up were removed to show another aspect of the actuators and control valves.

A photo of the battery compartment behind the spinner and in front of the cockpit.

A panel removed from the left side of the fuselage spine revealing the controls compartment.

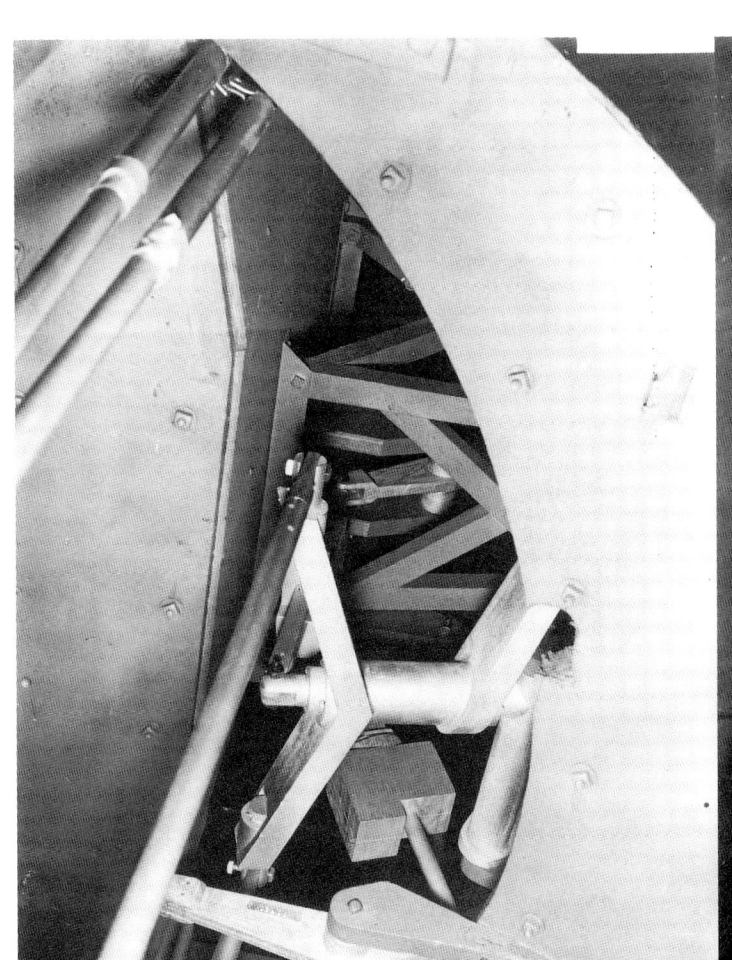

Another view of the controls compartment looking underneath a fuselage bulkhead.

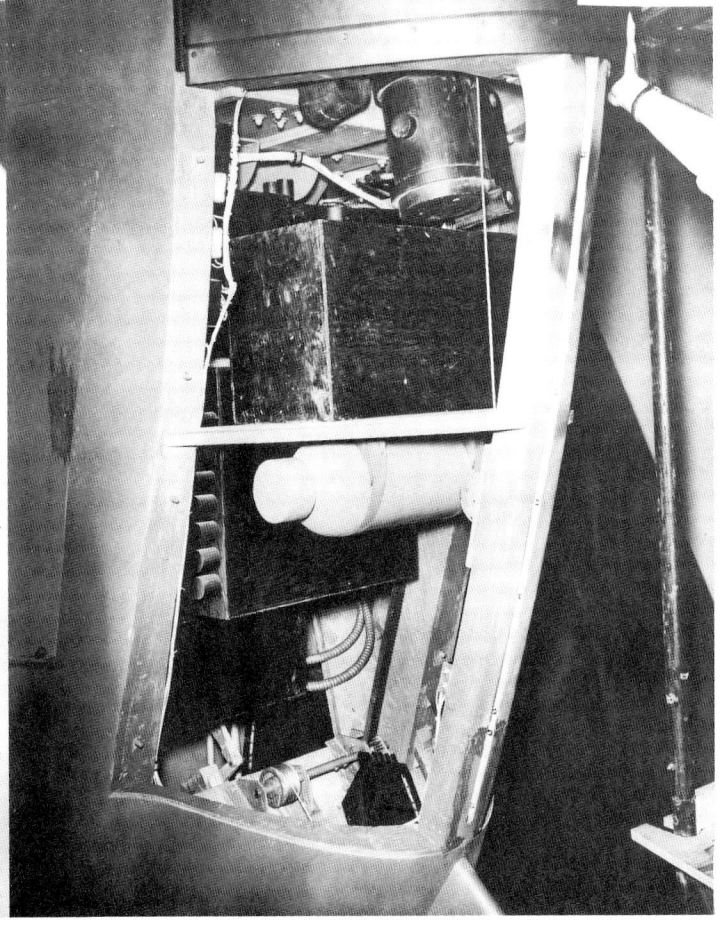

Port view of the electronics compartment located in the aft fuselage spine forward of the upper vertical stabilizer.

Starboard view of the same electronics compartment.

Another electronics compartment was located in the forward fuselage below the canopy on the port side of the aircraft.

A very similar electronics compartment was located in the forward fuselage below the canopy on the starboard side of the aircraft as well.

Ammunition stowage compartments in the port wing.

A close-up of the main ammunition stowage compartment in the port wing.

A photo of the port gun pod, which could be swapped out for rocket pod/fuel tank, depending on the mission.

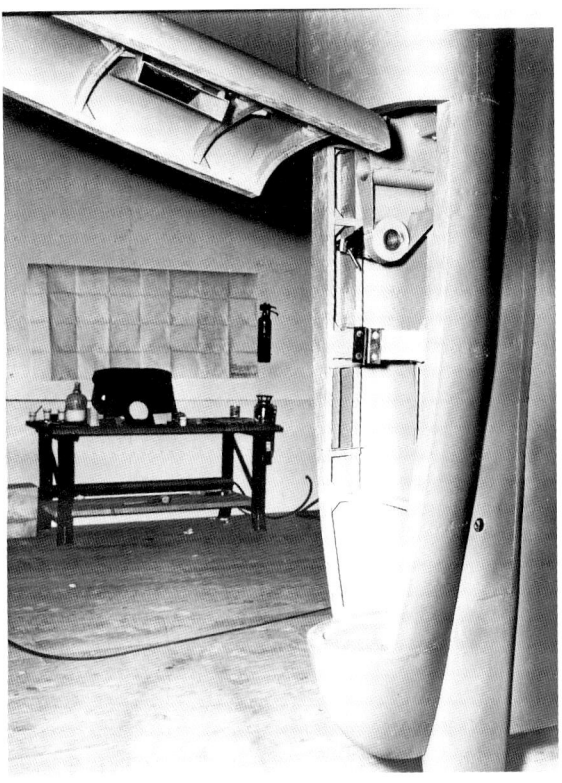

The port gun pod with the access door open.

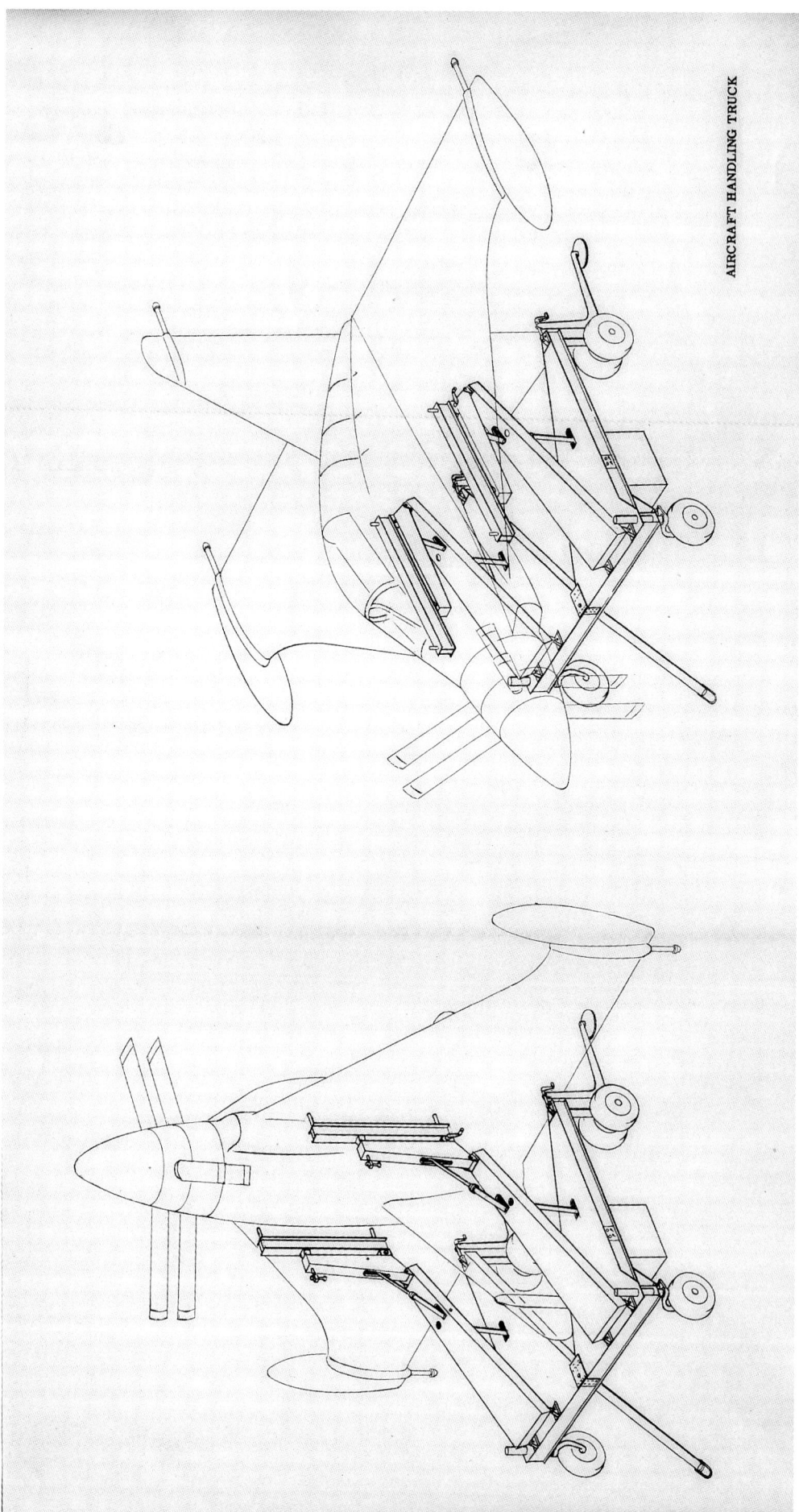

Convair designed an aircraft handling truck for ground handling of the XFY-1 and performance of certain maintenance functions. It incorporated a hydraulic system which actuated a mechanism for lifting the aircraft into the vertical position sufficiently to provide clearance for the fins upon rotating the aircraft to the horizontal position. Work stands for maintenance and inspection were under development; these would be readily attachable to the handling truck after the aircraft was placed in the horizontal position. The truck would be equipped with brakes actuated manually by the operator on the deck level adjacent wheels. Wheels would be equipped with pneumatic tyres and the truck would be capable of being towed over slightly rough terrain at a nominal speed.

Convair developed a large work platform with three levels to facilitate maintenance, servicing and rearming. The stands consisted of four units which encompassed the aircraft and securely attached to adjacent units, forming a complete working dock for all phases of maintenance. Platforms were equipped with safety rails, with the posts on the top platform being removable to provide clearance for the propeller. The top platform provided for the installation of a davit and the use of block and tackle for the installation and removal of equipment not adaptable to manual handling because of weight or size.

A sequence of photos showing how the engine bay could be easily accessed for maintenance by means of this large platform.

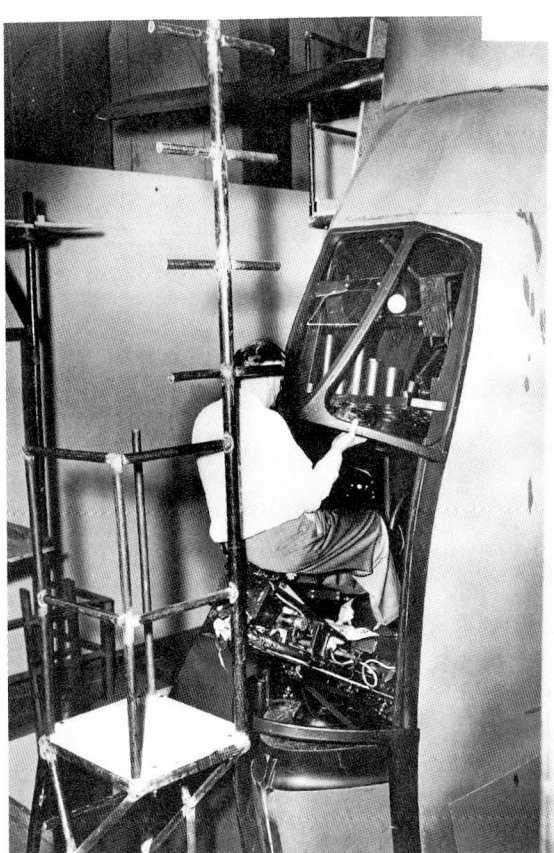
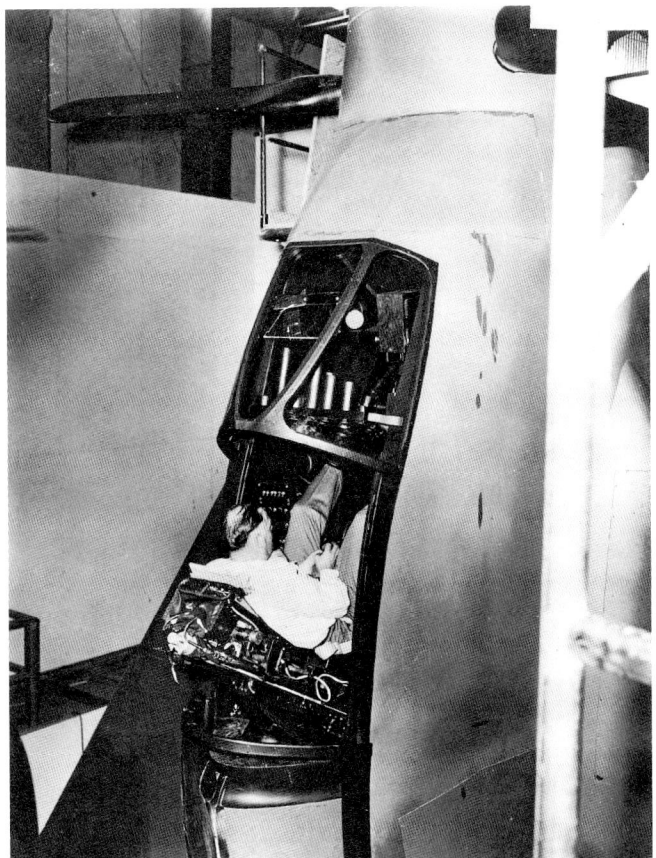

A sequence of photos showing how the pilot could enter the cockpit with the XFY-1 in the vertical position by means of a specialized tall, single person access stand.

The same access stand could also be used for refuelling the aircraft.

The single person access stand being used to place/remove an engine intake duct cover.

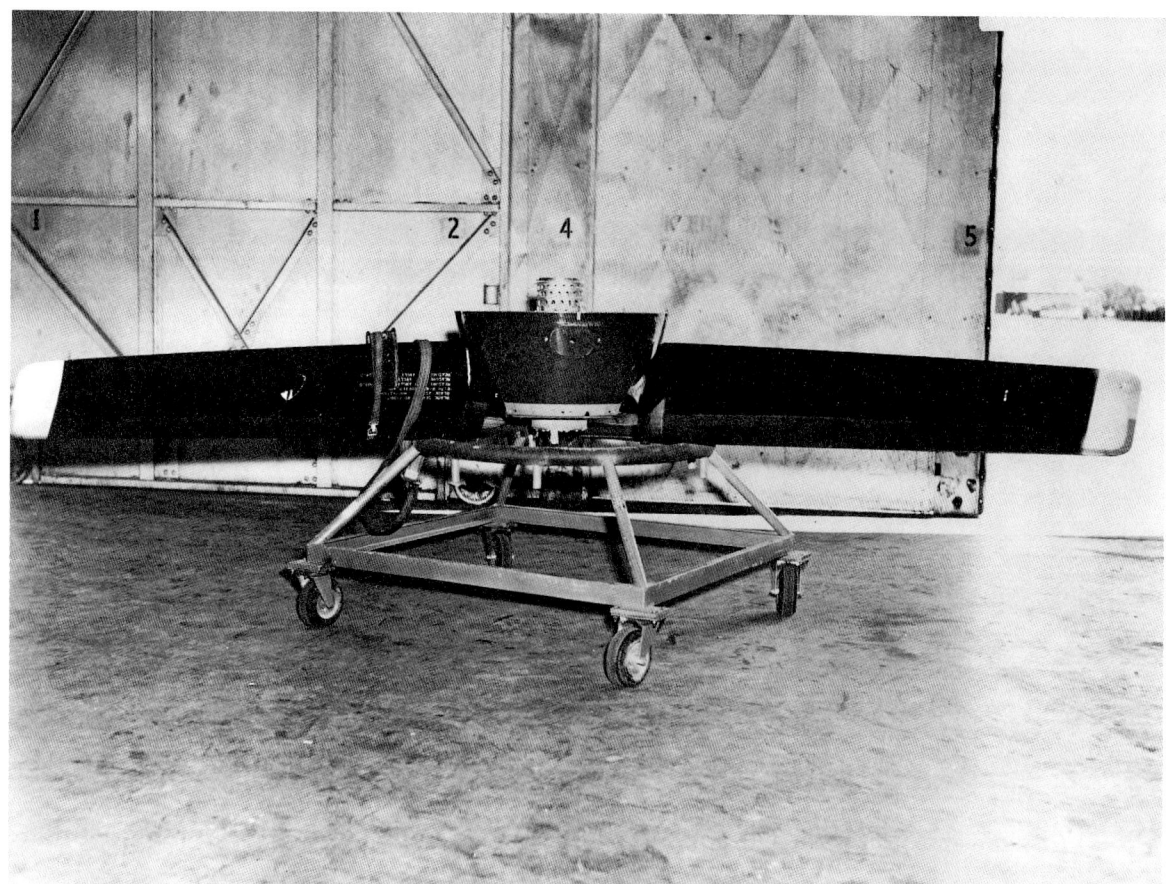
Convair designed a propeller horizontal storage and handling dolly for horizontal storage and limited handling of the propeller.

This vertical propeller storage stand provided for safe storage of the propellers during the change of the engine and propeller.

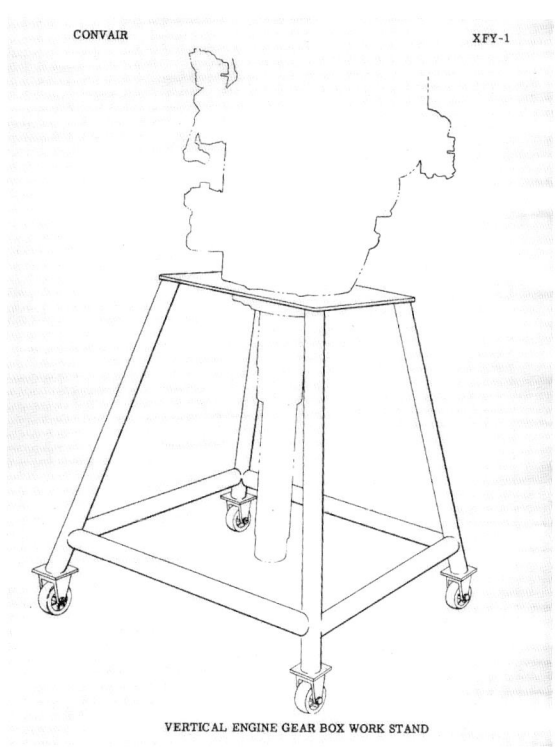

Convair also designed a vertical T40 engine gearbox work stand for servicing and storage of the gearbox in the vertical position.

Propeller removal was accomplished by means of a gantry and a conventional sling for three-bladed propellers. However, Convair recommended that BuAer consider the Sweeney Propeller shaft wrench, which also included a lifting assembly for removal and installation of the propeller. This assembly was particularly adaptable for removal of the propeller in the vertical or horizontal position.

A photo of the vertical engine hoist used for removal and servicing of the Allison XT40-A-6 turboprop engine.

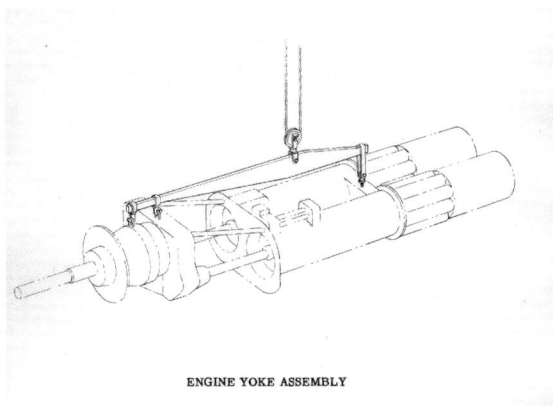

The engine yoke assembly was used for hoisting and transferring the engine from the rotating stand to the engine dolly.

The engine dolly was used for storage and handling of the Allison XT40-A-16 turboprop engine.

The XFY-1 airplane mock-up inspection was held at the Consolidated Vultee Aircraft Corporation Plant at Lindbergh Field, San Diego, California on June 29-30 and July 2-3. 1951. The cockpit lighting mock-up inspection was held on October 15-16, 1951. The official mock board had 11 members, 10 from the US Navy and one from the Marine Corps. These were accompanied by 12 official observers, mostly from the Navy, with some observers from the Marine Corps and Air Force.

Convair's mock-up report focused extensively on maintenance, as the tailsitter configuration presented unique challenges in this regard. The company developed special ground handling equipment and special tools, enabling personnel to perform specific maintenance functions in the approximate time required to perform similar functions on conventional type aircraft.

Items of major interest discussed during the mock-up inspection of the XFY-1 are summarized below.

1. In general, the seat rotating arrangement was considered satisfactory; however, it was apparent that additional rotation of that part of the seat supporting the neck and head was necessary from both a visibility and comfort point of view.
2. Installation of the North American Aviation (NAA) 250 KW AI radar including the NAA autopilot was approved. This radar was considered the most suitable because of its lower weight and size, and adaptability to the spinner installation. Considerable discussion centred around the advisability of including the autopilot as government furnished equipment (GFE), however, the majority board decision was in favour because of Convair's reluctance to undertake the sponsorship of such an important item of equipment and the problem of split responsibility on the integrated radar-autopilot twin, all of which was an NAA development. Convair was also requested to study the MK-11 sight as an alternate to the MK-6 because of easier installation and less interference with over the nose vision.
3. The board also approved the deletion of provisions for pressure fuelling and refuelling because of the considerable weight involved, and because the deck stands required for most maintenance and pilot ingress and egress were readily usable for fuelling too. This decision was further influenced by the expected operation off merchant ships and advance shore bases where a minimum of fuelling gear was carried.
4. Considerable attention was given to maintenance problems. Convair demonstrated the practicability of performing maintenance work with the aircraft in the vertical attitude, including radar accessibility, engine and engine accessory maintenance, armament servicing, and other points. Engine removal including the propeller and spinner was also shown. The workstands provided by Convair were generally satisfactory and no major objections were registered to the proposed maintenance set-up. However, the contractor was requested to study the alternate feasibility of engine

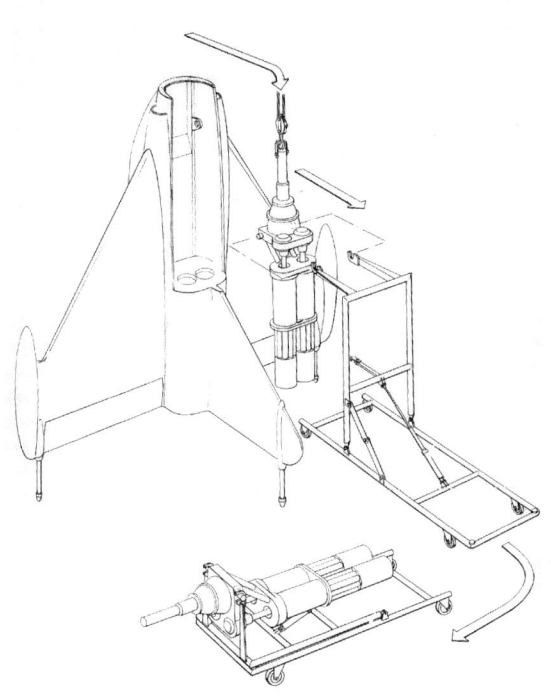

An illustration showing the removal and handling of the XT40-A-6 engine. The engine was secured to an adjustable component of the dolly while in the vertical position and suspended by overhead hoist by means of hoisting eye threaded on engine shaft. The vertical component of the stand with the engine attached was then rotated to the horizontal position, placing the engine in an accessible place for inspection and maintenance.

removal with the airplane horizontal on a large track or dolly.

5. The board approved the deletion of a separate utility hydraulic system. The utility system incorporated the radar drive, air duct door operation and canopy operation. Convair was authorized to tie these items into one of the dual power control hydraulic systems. However, adequate cut-offs and fuses were required in the lines to these utility components, so they could be removed from the system in the event of failure of the component or attendant piping. Since the XFY-1 was controllable on either hydraulic system alone, it was considered that the large weight penalty involved in providing a separate utility system was not warranted.

The various changes approved by the Mock-up Board resulted in a weight decrease of 147lb. The XFY-1 airplane mock-up was approved subject to subsequent action on the changes which were requested by the board.

XFY-1 WIND TUNNEL REPORTS

Convair conducted numerous wind tunnel tests to refine and improve the configuration of the XFY-1, with special emphasis on improving the controllability of the aircraft in the hovering and transition modes of flight. Several of the wind tunnel reports from 1952-54 are summarized here; the accompanying drawings show a variety of interesting modifications to the Pogo, some of which might have eventually been applied to the actual aircraft had the programme continued beyond 1955.

The first available report presented data from low-speed wind tunnel tests of a powered .15 scale model of the XFY-1; tests were conducted in the 8 x 12ft test section of the Convair wind tunnel in four periods during 1952, from late February to early July. These tests were accomplished in a total of 44 wind tunnel hours.

The purpose of these tests was to obtain basic stability and control data for the XFY-1 model during normal flight, transitional flight, and hovering flight conditions. Hovering flight was investigated both in free air and in close proximity to the ground. The basic configuration consisted of the complete model with wing, body, vertical tails, propellers, gun pods, oleos, and oil cooler. A different vertical tail with a 50° swept leading edge was tested.

Other modifications tested included an anti-roll parachute attached to the right-hand wing tip; leading edge spoilers applied above or below the wing; wing dams (fences); drag cups applied to the wing tips; leading edge slats; triangular wing tip extensions with a dihedral of 50°; and adjustment of the elevons to permit a 50° deflection. Transitional and hovering flight runs were made with the pilot's canopy open.

Additional tests were conducted in November 1952 with the same model employed in the earlier wind tunnel tests described above. The purpose of these tests was to investigate the effect of split and extended chord elevons, and various wing leading edge stall-control devices on longitudinal and lateral control in pitch during transitional and hovering flight conditions. In addition, a modified vertical tail and rudder was tested which was designed to reduce directional stability in normal flight and to

Side view of the engine dolly carrying the Allison XT40-A-6.

Another view of the engine on its dolly with a servicing cart in the foreground.

The turboprop engine in the vertical orientation on a special stand, which could be rotated 90° to the horizontal for transport and servicing.

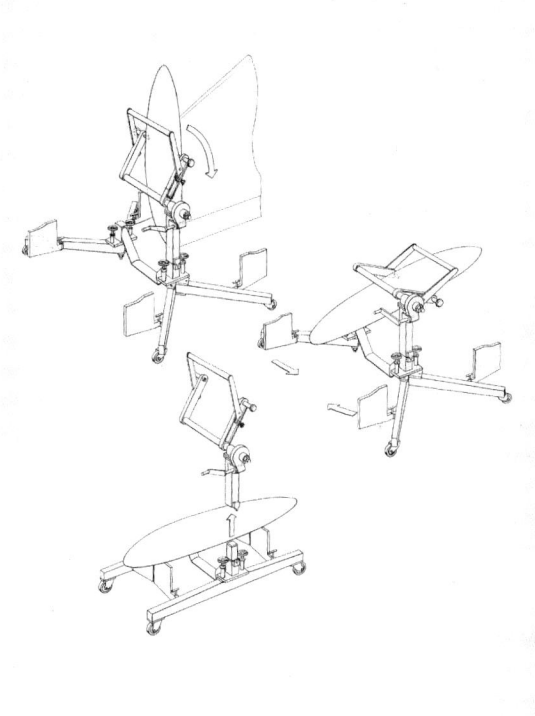

This dolly for the gun and rocket pod was designed for the rapid installation and removal of the gun pod, rocket pod and auxiliary wing tip tank.

A dedicated dolly for the storage and handling of the gearbox.

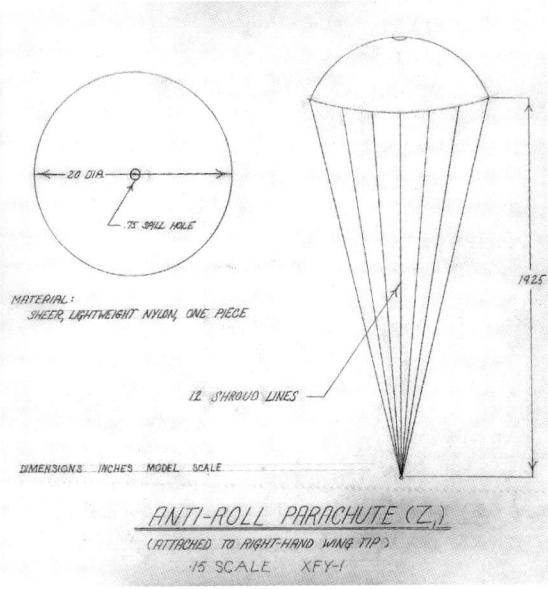

A drawing of a nylon anti-roll parachute attached to the right-hand wing tip of a .15 scale powered Convair XFY-1 wind tunnel model tested from February–July 1952; it may have been intended as a device to aid in landing the aircraft or as an emergency measure in situations where the aircraft was rolling uncontrollably during descent.

increase directional control in hovering flight.

More tests were performed by the company on the .15 scale powered XFY-1 model in its wind tunnel during February through March 1954; a total of 46 hours of wind tunnel time were expended. The purpose of the tests was to determine the following:

1. Maximum lift characteristics with and without power in normal flight.
2. Trim effects of a cambered wing leading edge, an elevon-type trimmer, and delta wing-tip trimmers in normal flight.
3. Slipstream direction with various crosswind velocities in hovering flight.
4. Duct pressure recovery characteristics in normal flight to hovering flight.

Several modifications to the wing were tested including cambered wing leading edges, an elevon-type trimmer, and delta wing-tip trimmers. The cambered leading edges were made of pine with wax fillets on the joints and rough spots. The elevon-type trimmers consisted of the outboard 25% of the actual elevon, deflected individually as a trimmer. The delta wing-tip trimmer was formed of sheet aluminium.

There were almost certainly other wind tunnel tests performed to refine and improve the Convair XFY-1 configuration, but thus far these additional reports have not been located.

XFY-1 STANDARD AIRCRAFT CHARACTERISTICS CHARTS

Several sets of Standard Aircraft Characteristics charts for the XFY-1 were produced during the existence of the programme. Charts from October 1951, May 1952, and July 1954 are reproduced here; these provide a valuable record of the evolution of the Pogo configuration, as well as minor changes in the performance and physical characteristics of the aircraft.

SUBSEQUENT HISTORY OF THE CONVAIR XFY-1 POGO

Three XFY-1s were built, with the first being used for engine testing; the second used in actual flight tests (serial no. 138649); and the third for static testing. It was powered by the Allison YT40-A-14 turboprop providing 7,100shp; production aircraft would have used the more powerful Allison YT54, but neither the aircraft nor the engine were built in quantity.

The first tethered flight of the Pogo was made on April 19, 1954 by Lieutenant Colonel James F. 'Skeets' Coleman inside a naval airship hangar at Moffett Field in Mountain View, California. The tether set-up was similar to the illustration shown in the original proposal. Over the course of several weeks, Coleman flew the tethered aircraft nearly 60 hours, with the first outdoor flight occurring on August 1, 1954. On the second test flight that day, Coleman flew the XFY-1 150ft into the air. This was followed by 70 takeoff-landing drills at the Naval Auxiliary Air Station in Brown Field, California. Coleman made the first transition from vertical to horizontal flight on November 5, 1954.

Further testing revealed the aircraft's inability to decelerate and stop efficiently after flying at high speeds due to its light weight and lack of air brakes. Landing could only be accomplished by the most experienced of pilots, as it required the pilot to look back over his shoulder to keep the vehicle stable. Even with the Allison YT54, the Pogo's top speed would have been under Mach 1, putting it at a disadvantage against jet fighters capable of twice that speed. These issues ultimately led to the type's cancellation on August 1, 1955.

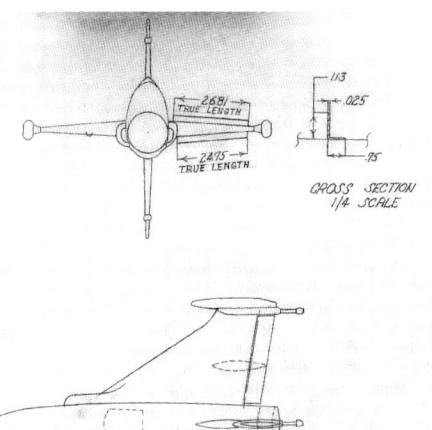
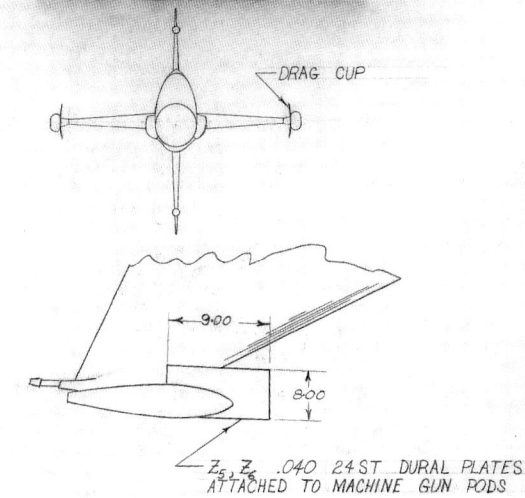

Narrow leading edge spoilers were applied both above and below the wing of the XFY-1 model, likely to improve roll characteristics.

Rectangular endplate 'drag cups' which curved inwards towards the aircraft centreline, were added to the forward gun pods, likely to control wingtip vortices and reduce drag.

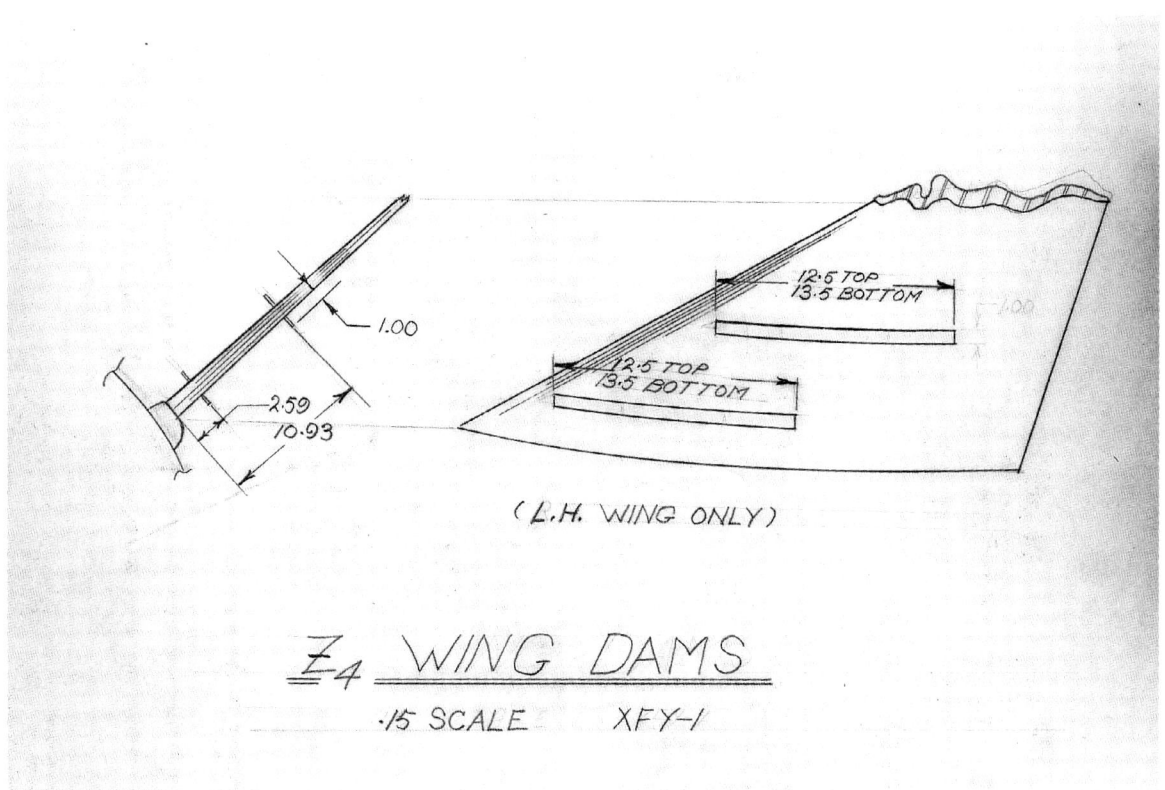

A pair of wing dams (fences) were added underneath each wing in an effort to block spanwise airflow along the wing and prevent the entire wing from stalling at once.

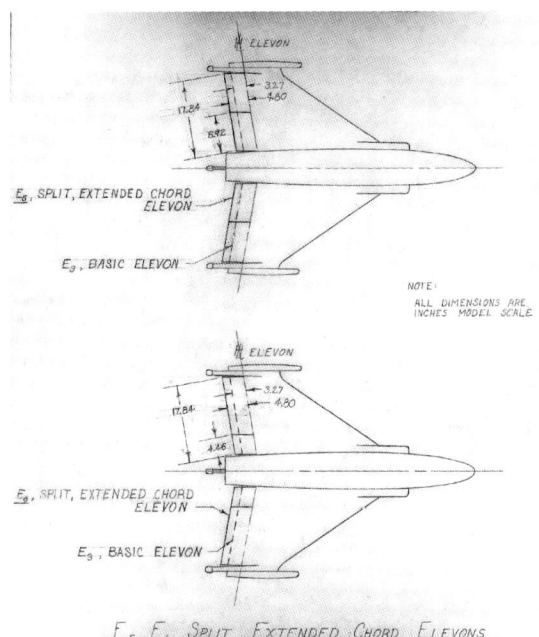

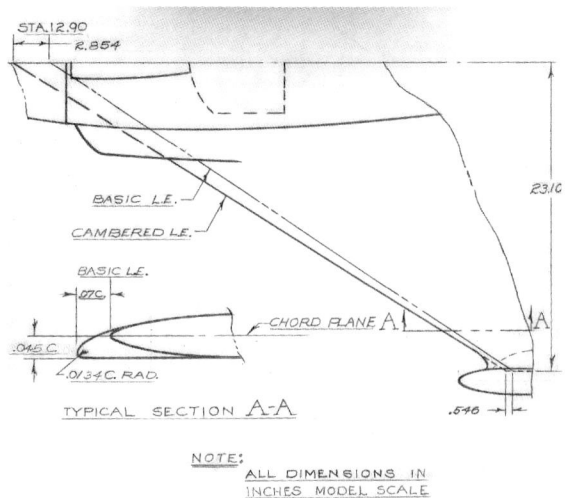

CAMBERED WING LEADING EDGE

In November 1952, Convair performed additional wind tunnel tests with its .15 scale powered XFY-1 model. Split and extended chord elevons were tested which a chord extension of 50% of the basic elevon chord. Each elevon was split into inboard and outboard sections at 25% and 50% of the elevon length. Tests were made with various combinations of split elevon deflections to determine longitudinal and lateral control effectiveness for transitional and hovering flight conditions.

Convair tested a cambered wing leading edge on its XFY-1 wind tunnel model to study trim effects in normal flight during February through March 1954.

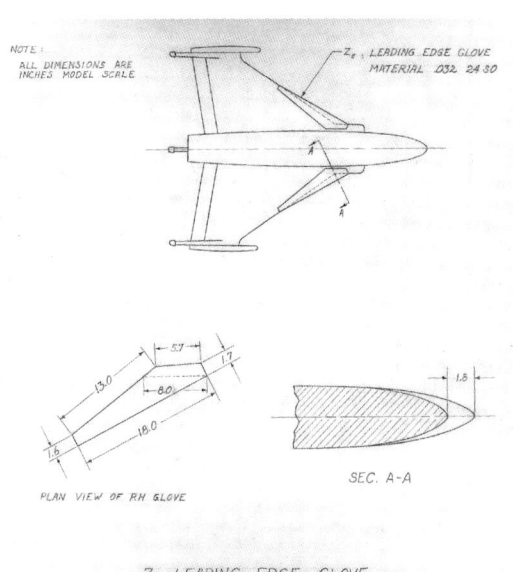

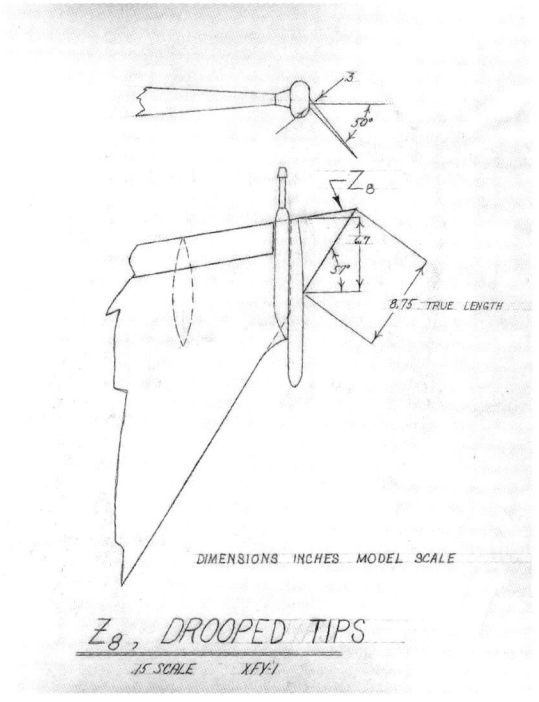

Another stall control device tested on the XFY-1 wind tunnel model was this leading edge glove, which increased the sweepback of the inboard wing section.

Drooped triangular wing tips with a dihedral of 50° were added outboard of the gun pods, likely to reduce drag and improve the Pogo's handling characteristics.

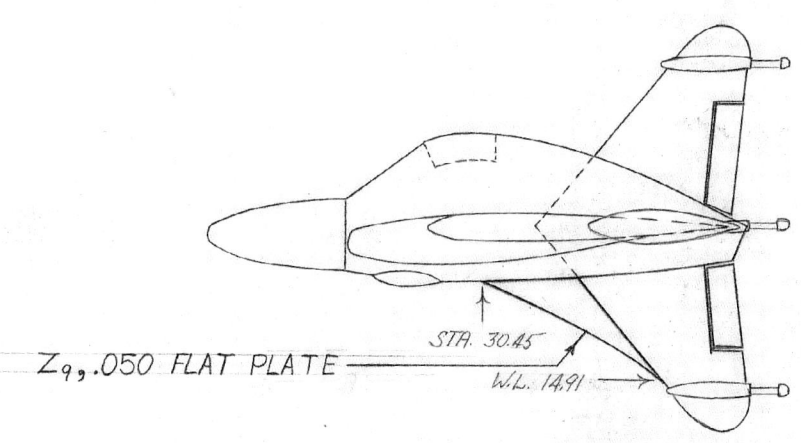

DIMENSIONS INCHES MODEL SCALE

Z_9, VENTRAL FIN

.15 SCALE   XFY-1

A prominent leading edge extension, swept at 50°, was added to the ventral vertical stabilizer of the XFY-1 wind tunnel model, likely to improve control, stability and trim in yaw at low speeds.

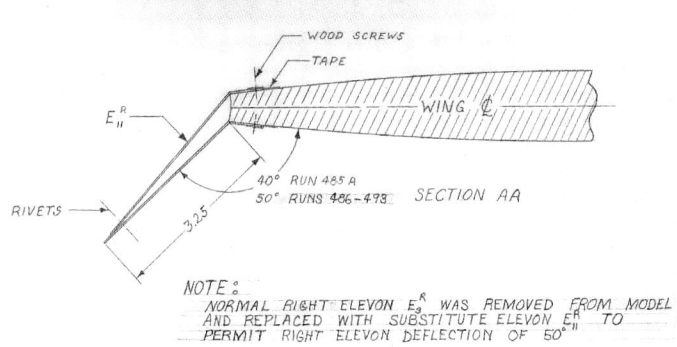

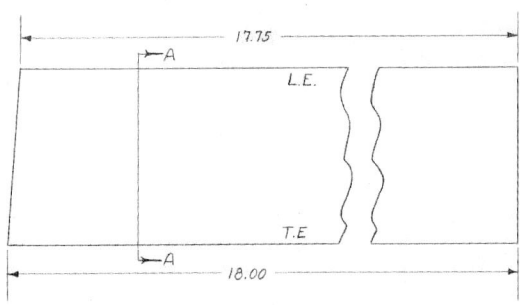

E_{11}^R, RIGHT ELEVON

E_{11}^R MADE TO PERMIT FIXED 50° ELEVON DEFLECTION
SAME PLANFORM AS E_9^R

Modifications were also made to the elevons of the XFY-1 wind tunnel model to permit deflections up to 50°, improving roll characteristics at low speeds.

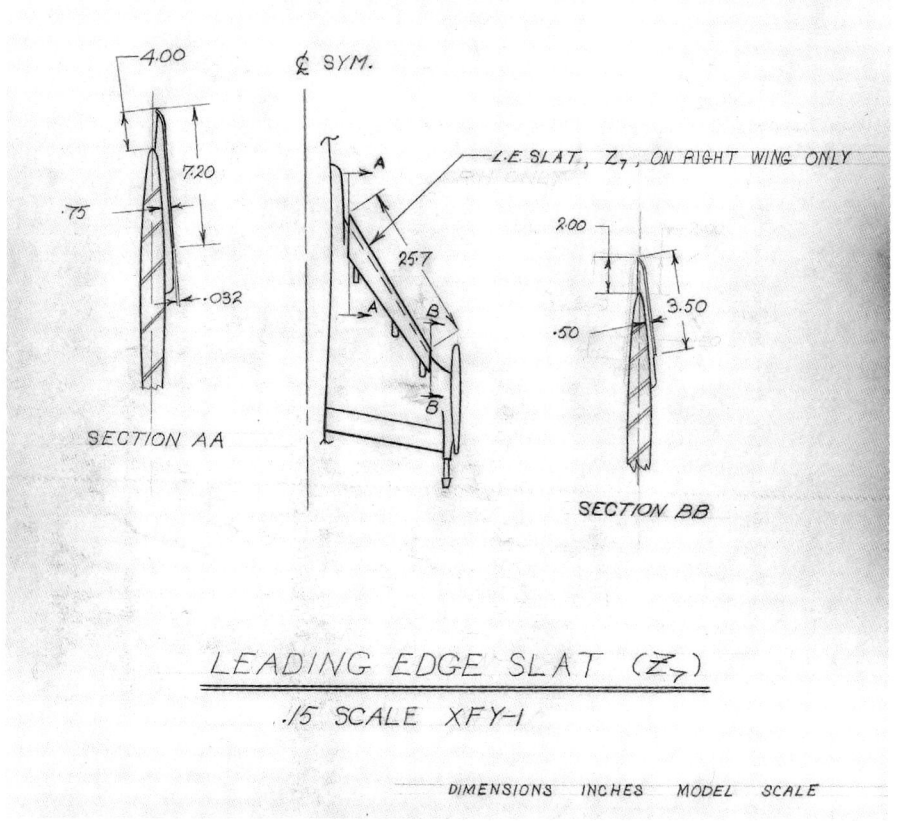

The .15 scale powered XFY-1 model was also tested with large leading edge slats to enable the wing to operate at a higher angle of attack and improve controllability at low speeds.

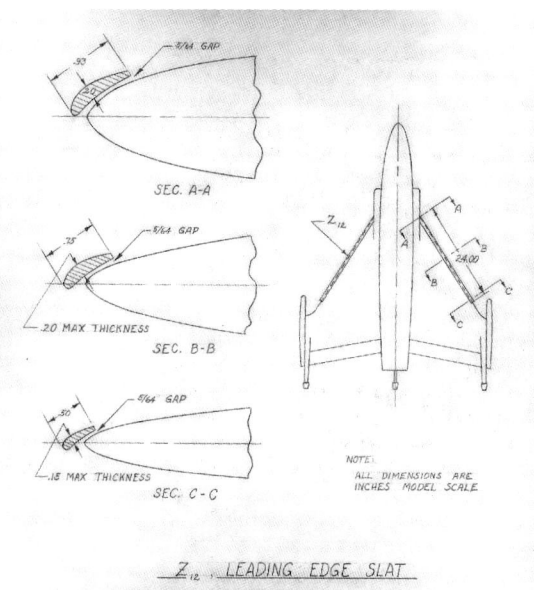

Narrow leading edge wing slats were also tested to improve the stalling characteristics of the aircraft.

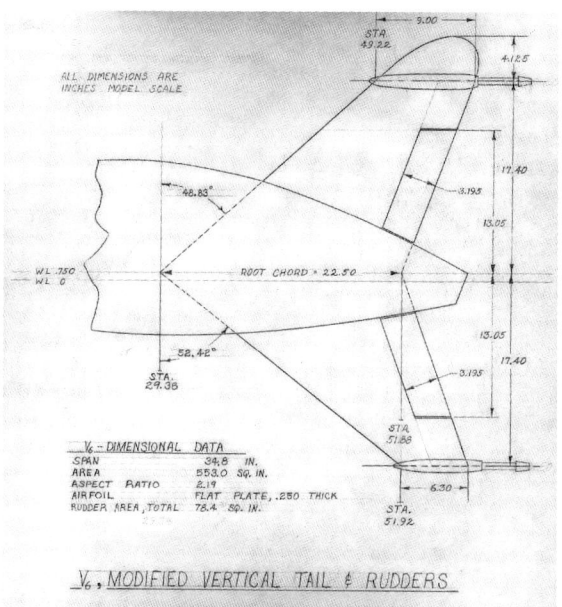

A modified vertical tail and rudder were tested on the XFY-1 model which increased leading edge sweepback and total rudder area than the basic configuration. The total fin area was smaller than the basic fin and was located farther forward to decrease stability during normal flight and increase stability during hovering flight.

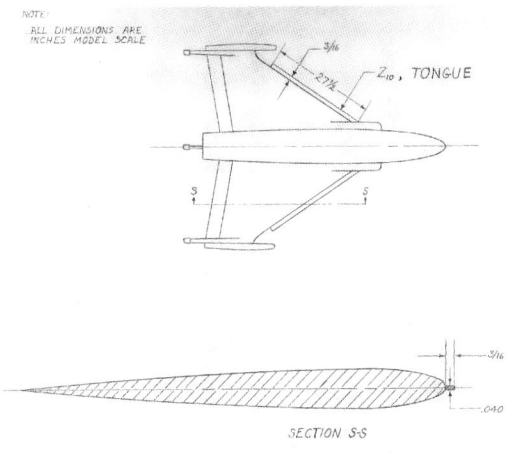

A narrow leading edge tongue was added to the wing leading edge of the Pogo model as a stall control device during transitional and hovering flight.

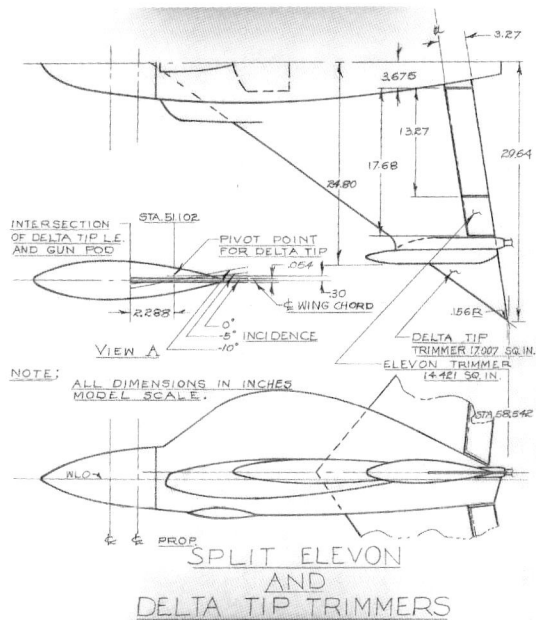

Split elevons and all-moving delta tip trimmers were also evaluated to study trim effects in normal flight conditions.

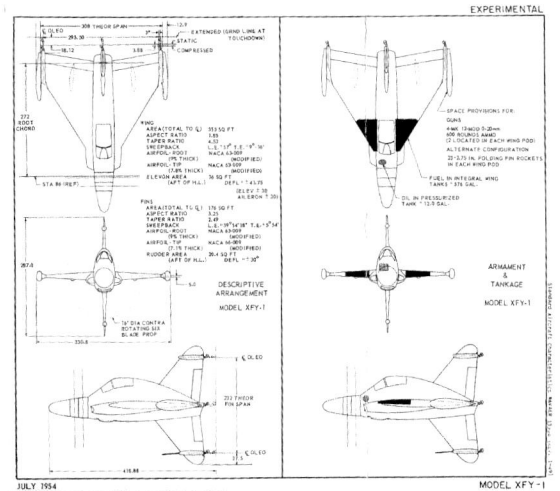

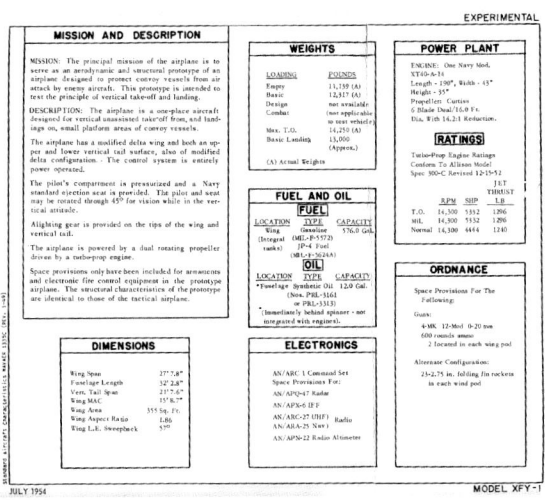

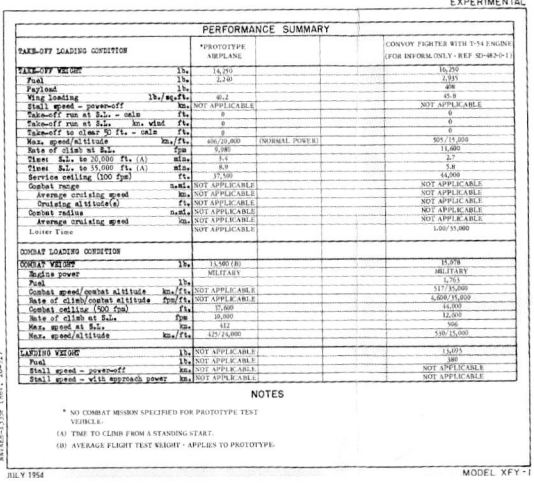

The July 1954 charts basically depict the final configuration of the XFY-1 as it was actually flown. Compared to the 1952 configuration, small castoring wheels replaced the solid feet of the landing gear and the pitot was moved from the tip of the ventral vertical fin upwards to the tip of the landing gear fairing on the same surface.

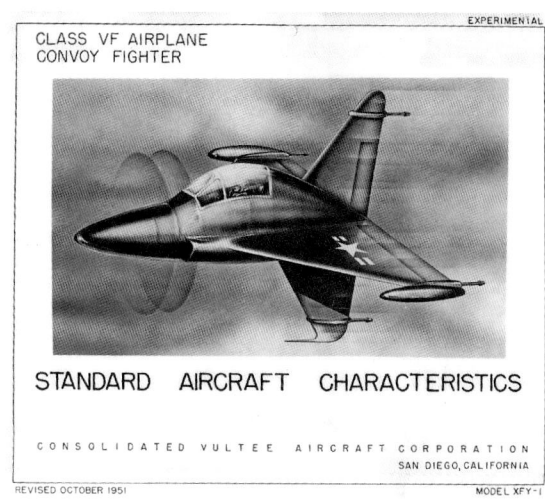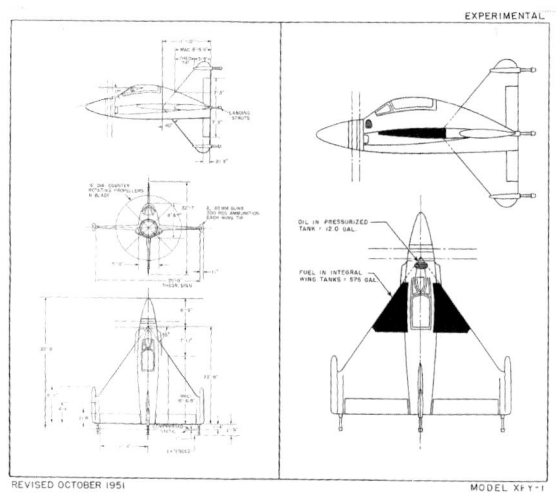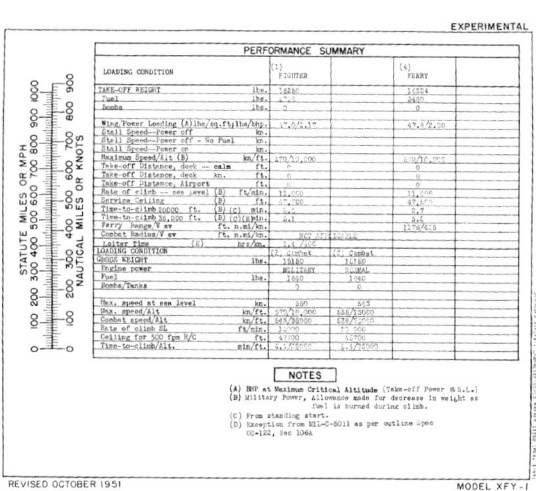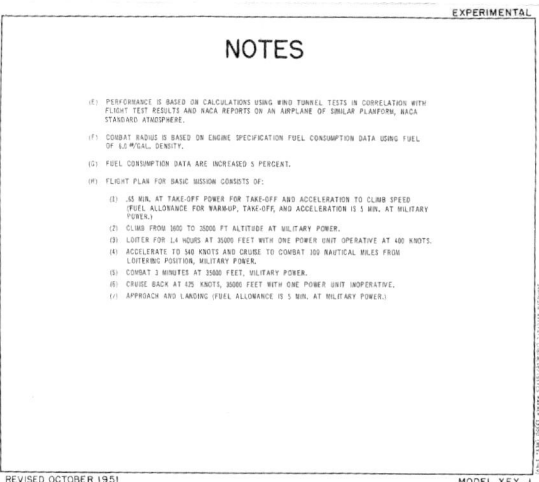

The XFY-1 Standard Aircraft Characteristics charts from October 1951 depict a configuration very similar to the mock-up which was shown to the Navy several months before. Note the pitot mounted on the tip of the ventral fin shown in the artist's impression, which is not present in the general arrangement drawing. Also note that the Pogo is still being depicted in a Glossy Sea Blue scheme at this point.

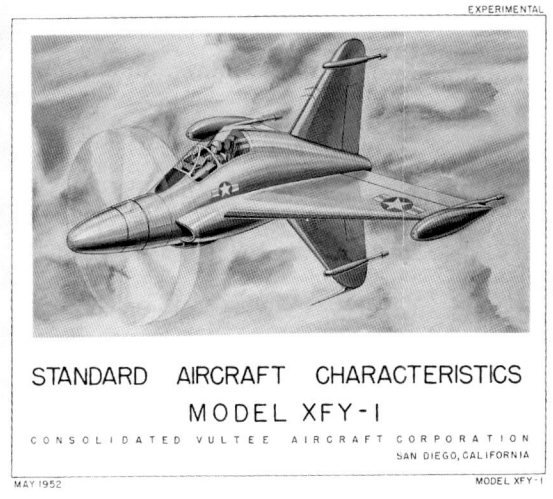
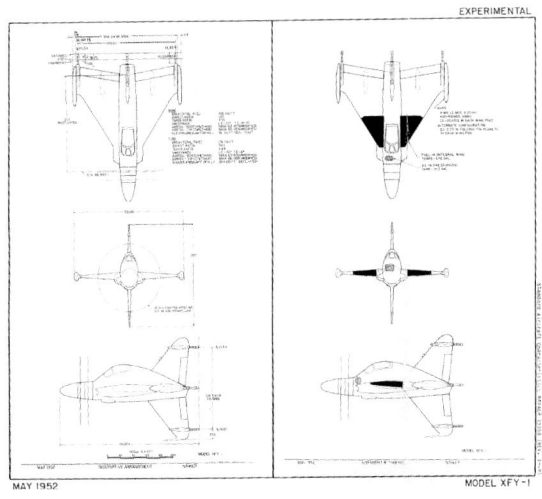
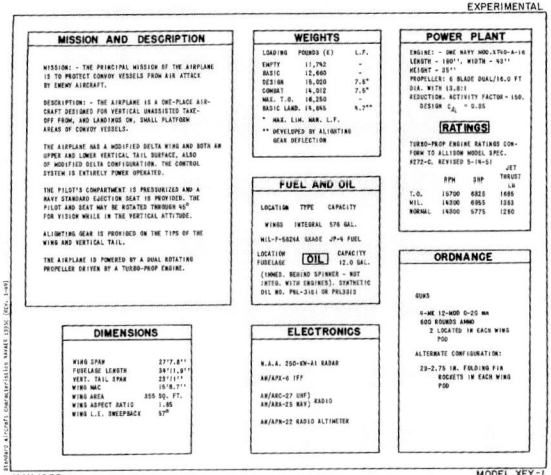
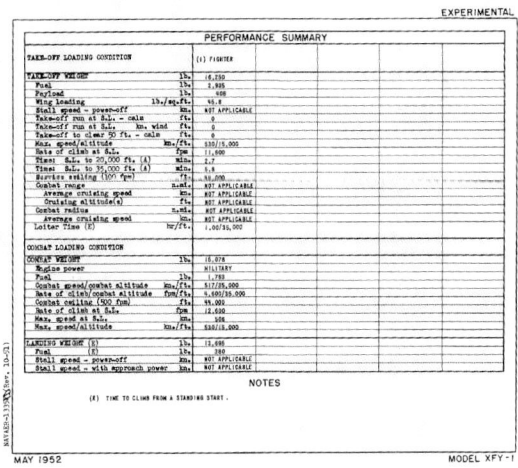
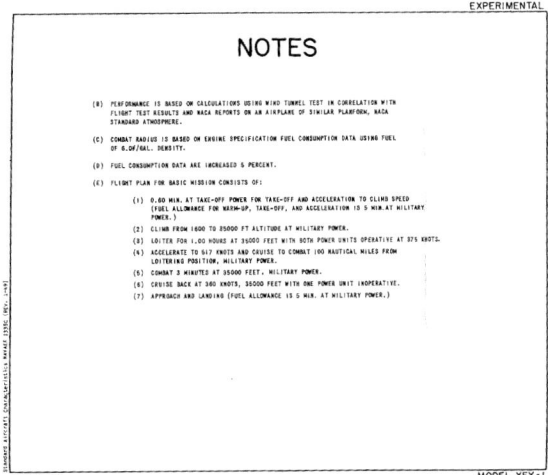

The greatest changes in the configuration of the Pogo occurred between October 1951 and May 1952, when this new set of Standard Aircraft Characteristics charts were produced. The trailing edges of the wings and tail surfaces were modestly swept. The intake shape was changed to a more oval shape and moved slightly aft. The ventral air scoop was made more prominent and the upper fuselage hump became more prominent. The artist's impression now depicts the aircraft in the natural metal scheme which characterized the actual prototype.

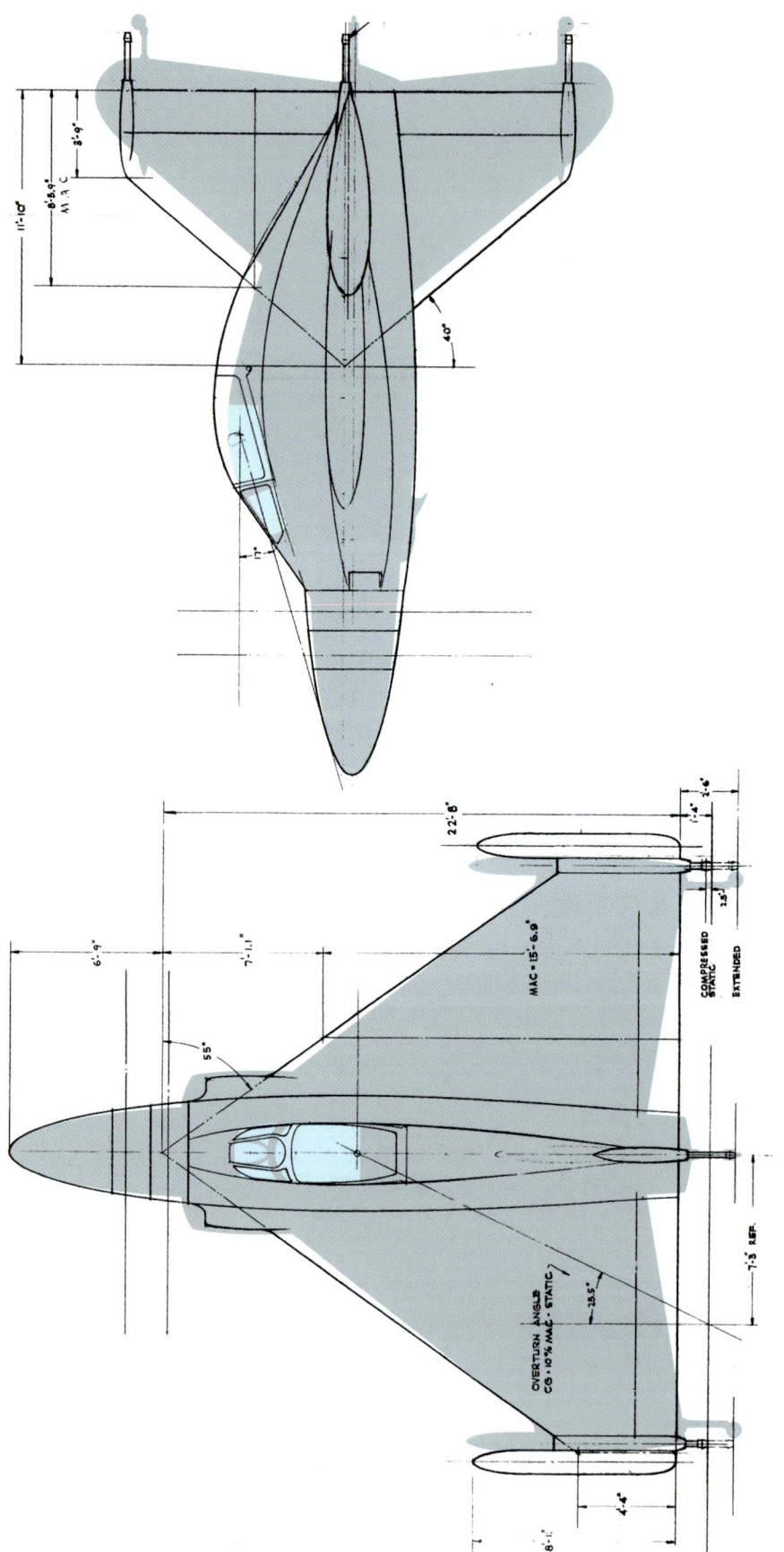

Original plan of the Convair Class VF Convoy Fighter overlaid with a silhouette of the XFY-1 Pogo as built. While the XFY-1 retained the basic configuration of the original proposal, nearly every contour was subtly altered in the type's journey from the drawing board to an actual functioning aircraft.

Lockheed XFO-1/XFV-1

Lockheed was awarded a contract on April 19, 1951 for two full-scale prototypes and a static test article under the designation XFO-1; this would later be changed to XFV-1 when BuAer's code for Lockheed was changed from O to V. The following sections are a fragmentary overview of the early development of the XFO-1/XFV-1 aircraft after the contract award up to the first flight, followed by a summary of its brief flight test career and ultimate cancellation.

NACA wind tunnel testing (1951)

Lockheed representatives visited the NACA Langley Aeronautical Laboratory on March 12, 1951 to discuss testing of the XFO-1, as it was then designated. At the time the tail assembly consisted of three units (essentially an inverted vee tail plus a conventional vertical tail), and it was pointed out to them that such an arrangement would probably give unsatisfactory longitudinal stability because it was likely that the propeller slipstream would miss the V-tail at some angles of attack. The tail was subsequently revised into a four unit X-tail arrangement which would provide much better longitudinal stability. The tail control surfaces (called 'tailerons') provided aileron, rudder, and elevator control and there were no provisions for control surfaces on the wing.

Quarter scale flying model

Lockheed initiated the construction of a ¼ scale flying model of the XFO-1 which would be tested in the Ames 40 x 80ft tunnel in early 1952. It had a span of 7ft 3in and a length of 9ft. This model, which cost an estimated $33,000, had a scaled-down weight of 244lb and was equipped with two 38hp water-cooled electric motors. Each 'taileron' was operated by a single servo and the control mixing was done on the ground.

The cable containing the power and control leads was brought out of the top surface of the model at the centre of gravity and with an operator supporting the cable with a pole so that only a short slack length of cable was supported by the model. Provisions were made for a safety cable attached to the nose but there was no propeller guard to keep this cable out of the propeller. Instead there was a drum inside the propeller spinner which was supposed to automatically reel the cable in to prevent any slack in the cable. Tethering lines attached to each wing tip and led out to the side for holding the model in place while the tunnel speed was run up to the desired test speed. The model would be flown both forward and sideways at steady tunnel speeds from zero to about 80ft per second.

1/10 scale wind tunnel model

Static high-speed stability and control tests of a 1/10-scale powered model of the XFO-1 were performed in the Ames 12ft wind tunnel in the latter half of 1951. This model was equipped with a six-component sting balance furnished by Lockheed. Two 40hp water-cooled motors drove the dual-rotation propellers at 11,000rpm through a hollow-shaft, direct-drive arrangement.

Mock-up conference

The inspection of the mock-up of the XFO-1 airplane was held at Lockheed's Burbank facilities from October 10-12, 1951. The following is a summary of the Mock-Up Conference report and subsequent Mock-Up Board report.

Primary flight control system

The pilot used a conventional control stick and a rudder bar to actuate the four identical surfaces on the cruciform tail. All four surfaces were used together to control the airplane about any one of the three axes (pitch, yaw and roll), either independently or concurrently.

Stick motion was translated by means of bevel gears into motion of two separate cable control systems, so that the two systems operated in the same direction with fore-and-aft stick movement and in opposite directions with lateral stick movement. Each of these two cable systems actuated two of the four control surfaces, assisted by a hydraulic booster on each system.

In order to provide for differential motion of each of these pairs of surfaces as required for rudder control, the elevator aileron boosters were connected to the surfaces by a 'block and tackle' cable system with the dead ends of the cables attached to the rudder control system. This system was actuated by the pilot's rudder bar, connected by a cable system to a hydraulic booster.

The hydraulic boosters also functioned as servo units when the airplane was controlled by the automatic pilot.

Power plant
The power plant was the Allison XT40-A-6 turboprop engine rated at 7,500 Equivalent Shaft Horse Power with a Curtiss 16ft diameter six-bladed dual contra-rotating propeller. The reduction gear ratio was 14.3:1, giving a propeller rpm of 1,000 at military power. The mounting in the fuselage consisted of three vibration isolators, two on the reduction gearbox and one between the power sections located approximately at the end of the compressor.

The engine air intake ducts were located on each side of the fuselage, the entrances being directly behind the aft propeller. The forward portions of the ducts were an integral part of the fuselage, whereas the aft portions were removable to facilitate access to the aircraft accessories which were mounted on the aft face of the reduction gearbox.

The power plant installation was divided into two compartments, zone one and zone two. This was accomplished by the installation of a vertical fire-seal at the aft end of the compressor.

Accessibility to the engine, accessories, etc. was made possible by removal of five access panels which were hinged and latched between the lower fuselage longerons.

The starter was an Airesearch 140hp air turbine starter. The power to the starter was supplied by an Airesearch gas turbine compressor which was ground supplied.

Hydraulic system
The normal hydraulic system, powered by one engine driven variable volume hydraulic pump, had an operating pressure of 3,000psi and was used to operate the aileron-elevator and rudder booster system and the radar scanner drive mechanism. An electric driven emergency stand-by system maintained in a constantly charged condition of 1,500psi was immediately available automatically whenever the normal system pressure dropped below that of the emergency system. This system operated only the aileron-elevator boosters. No common hydraulic lines or fittings existed in the 'normal' and 'emergency' systems.

The reservoir was pressurized to 25psi gauge pressure and incorporated design features that eliminated the possibilities of uncovering the suction port to the pump in any flight attitude. The design of the system was predicated on the use of non-inflammable hydraulic fluid.

Fuel system
A nylon fuel cell having a capacity of 525 US gallons was installed in the fuselage aft of the cockpit. In order to minimize the fuel loss at high altitude high rates of climb, the system was pressurized to 2psi gauge pressure across the tank walls above 18,000ft.

Fuel was supplied to the engine by four submerged electrically driven fuel boost pumps, two of which were located in the forward end of the fuel cell and two in the aft end. The system was designed so that the failure of any one pump did not interfere with the continuous operation of the system in any flight attitude.

The fuel quantity was measured by capacitance type gauges in either the horizontal or vertical flight attitudes.

Oil system
The oil system consisted of one oil tank and one oil cooler suitably manifolded to supply oil to both power sections and to the reduction gearbox. Oil cooling was accomplished by means of exhaust gas augmentation.

Oxygen system
The diluter demand oxygen system incorporated a positive pressure breathing regulator and had an oxygen supply sufficient for approximately a three hour duration. The oxygen was supplied from one 295 cubic inch bottle charged to 1,800psi at 70° F. Charging of the system was accomplished through a high pressure filler valve located on the left hand side of the fuselage. A quick disconnect oxygen breathing tube and radio connection assembly was installed adjacent to the pilot's seat.

Electronics
The electronic system basic power consisted of an engine driven 15 KVA AC generator and a 300 ampere DC generator. The nominal voltage of the

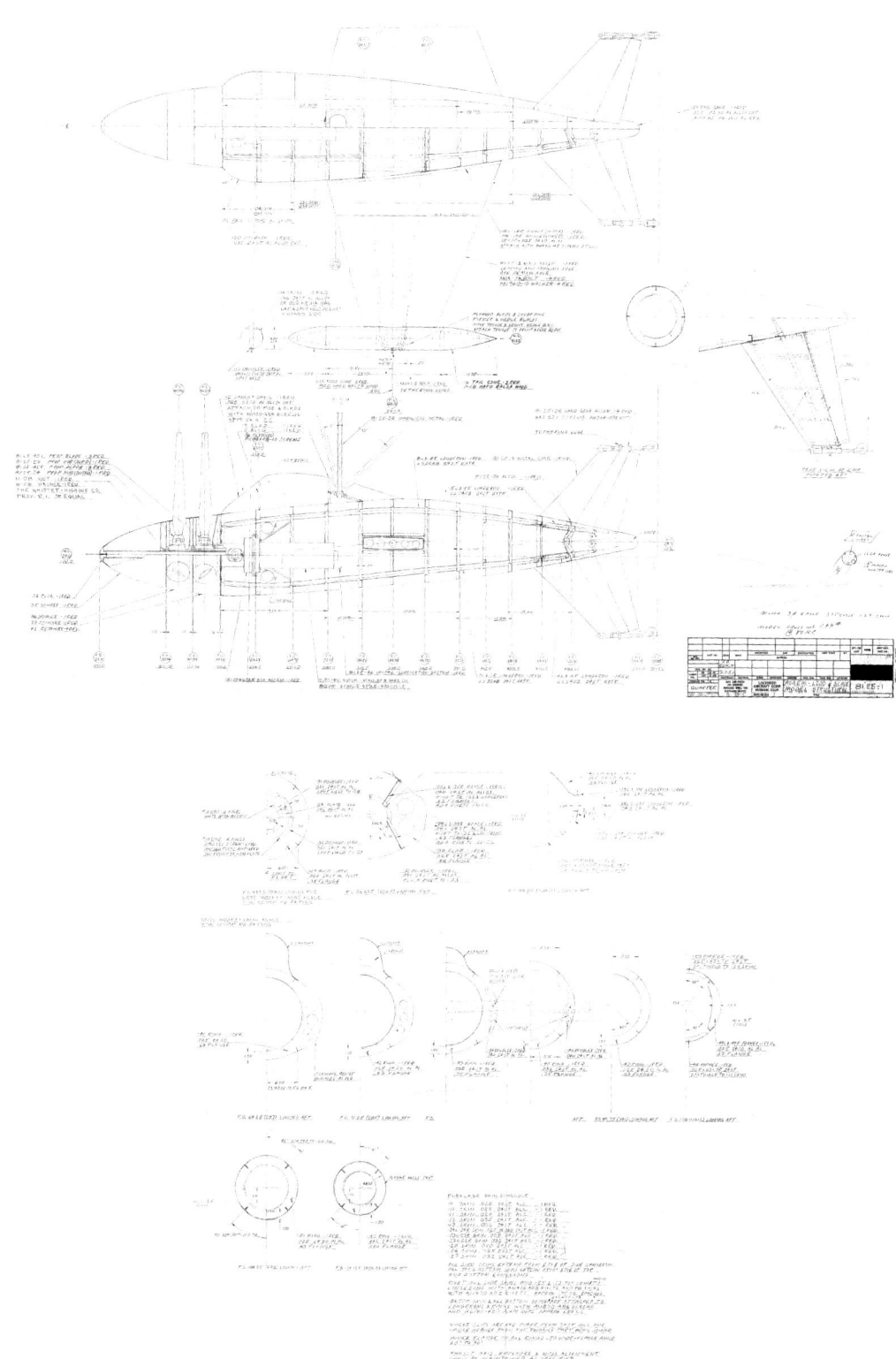

Structural diagram of the ¼ scale flying model of the Lockheed L-200 dated July 7, 1951; by this time the tail had been changed to a four unit X-tail arrangement to provide better longitudinal stability. Also note the switch from an eight-bladed to a six-bladed dual-rotating propeller.

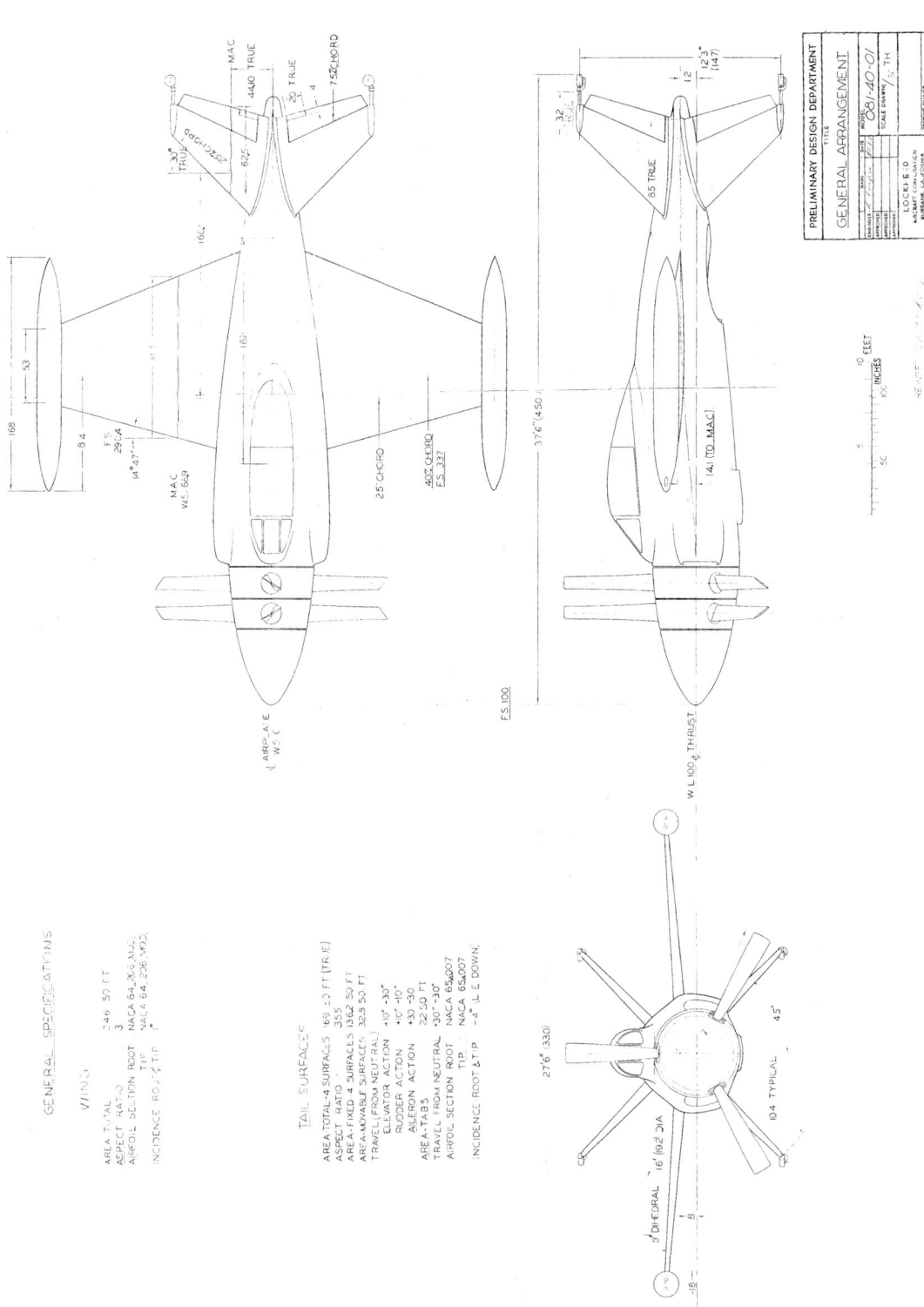

Three-view of the Model 81 (the internal Lockheed designation for the XFO-1) as of July 20, 1951; compared to the aircraft as built, it lacked the aft fins on the wing tip pods, had a different ventral radiator shape, and had smaller wheel fairings on the tips of the tail.

Photos of Lockheed's 1/10 scale powered model of the XFO-1 under test in the NACA Ames 12ft wind tunnel in the latter half of 1951. By this time fins had been added to the wing tip pods and the wheel fairings lengthened.

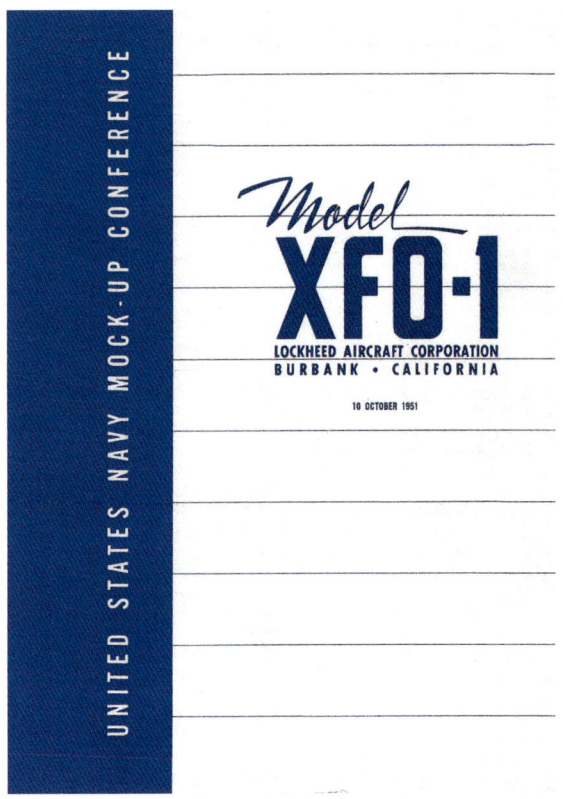

Cover to the Lockheed XFO-1 Mock-Up Conference report dated October 10, 1951.

AC system was 200 volts phase to phase and 115 volts line to neutral at 400 cycles, and the DC system was 28 volts. The standby, or emergency power, was derived from a battery with a 500 VA inverter for the AC system.

The DC and AC contactors were located adjacent to the generators in an accessible area along with the power relays and external receptacles. The voltage regulators were located in the electronics compartment. The feeders from the generators to the AC bus and the DC bus in the electronics compartment had extra mechanical protection other than the integral insulation. All sub-feeders into the cockpit were 'smoke-free' protected except where circuits demanded their being connected directly to the bus.

The main distribution centre was in the electronics compartment along with the electronic equipment. Wherever possible, the electronics were concentrated into one compartment where the maintenance personnel could stand within the compartment and have the major number of adjustments, removals and repairs conveniently available. All wiring runs initiated at this distribution centre and routed to the engine compartment, to the cockpit in short straight runs through ducts accessible from outside the aircraft, and rearward to the major portion of the equipment on the shock mounted rack and on the electronic compartment door.

All shock mounted electronic equipment that was not in itself capable of isolating vibration in both flight attitudes was located on a specially designed rack which, through special type shock mounts, would protect this equipment throughout all flight attitudes. The standard Government Furnished Equipment (GFE) shock mounts would not be used. The units would be solidly captive to the rack with standard type pins and hollow knurled front knobs with guide rails to facilitate installation and removal.

All console panels in the cockpit would plug into the bottom of the wire troughs on either side of the fuselage. This plug connection served two purposes: it incorporated a pressure seal entry into the cockpit and met the requirement of plug connection for each console. In the case of Lockheed fabricated consoles, the plugs would be on the end of a cable in lieu of mounting a receptacle on the console panel. In the case of GFE panels, a jumper cable with plugs on either end would be used.

The wire troughs provided circuitry checks, inspections, repairs and general maintenance on main wire runs without having to gain access to the cockpit (console panels were also accessible and removable from outside the aircraft). In addition, they provided a convenient entry into the pressurized area with the most direct cable run with minimum cable weight. Plug connections in the wire trough were AN moisture proof type with structural precautions provided to drain overboard any entrapped moisture.

Gun pod

Two MK. 12 Mod. 0 20mm guns were mounted in each wing tip pod. The pneumatic system, supplied by a 3,000psi storage bottle, for operating the gun chargers, feed mechanisms, and buffers was completely contained in each pod. Loss of pressure due to temperature drop was prevented by the use of electric heating and insulating covers on the air storage bottles.

Ammunition (150lb per gun) was stored in horizontal ammunition chutes, extending spanwise in the wing ahead of the main beam. Empty cases and links were expended, below and outboard respectively, from the gun pods.

A large access door, comprising roughly the upper half of the pod centre section, opened for servicing the guns, charging the air bottles or loading

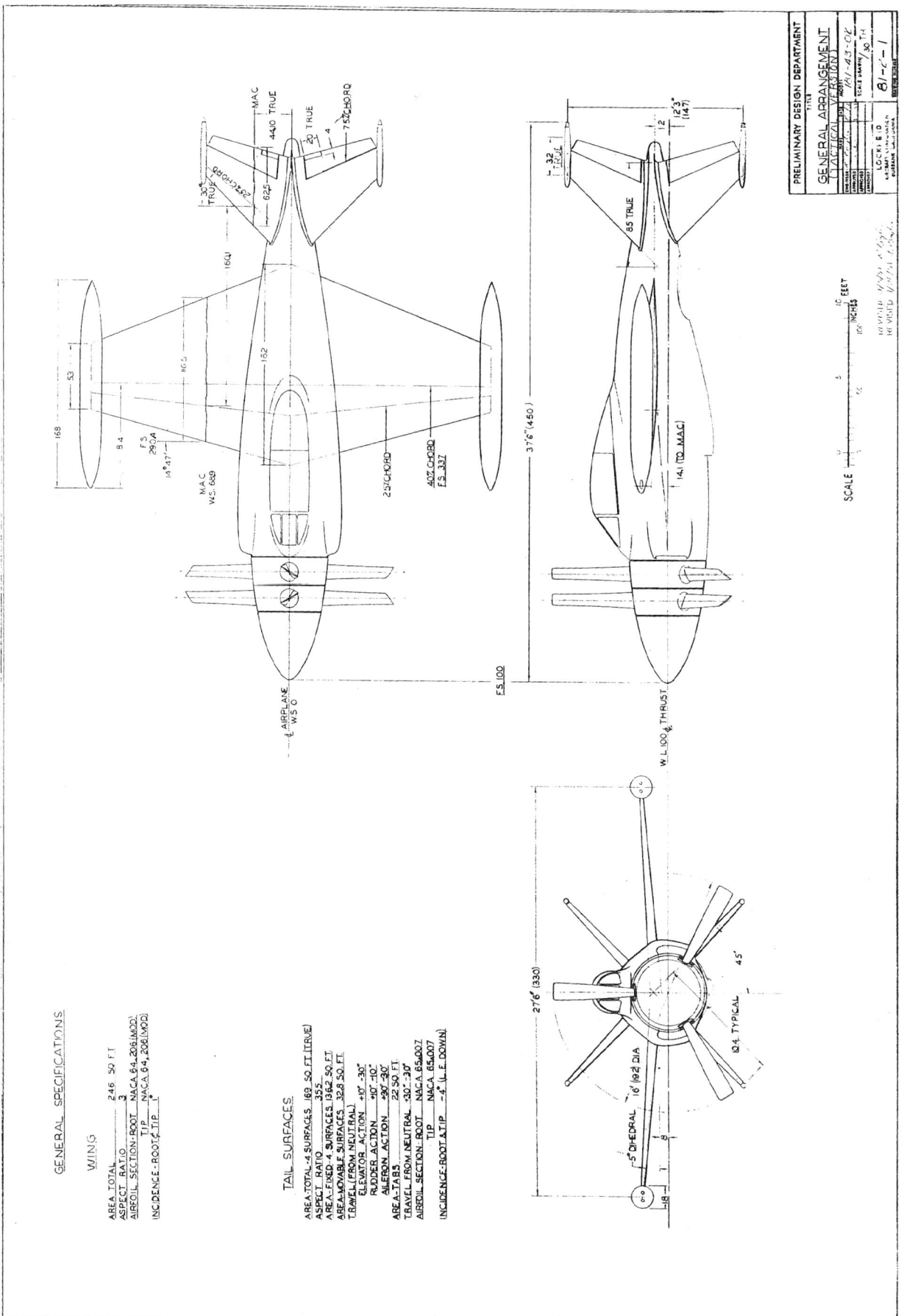

Three-view of the XFO-1 revised September 5, 1951, which was similar to the vehicle as built, except for the lack of fins on the wing pods and the shape of the ventral radiator.

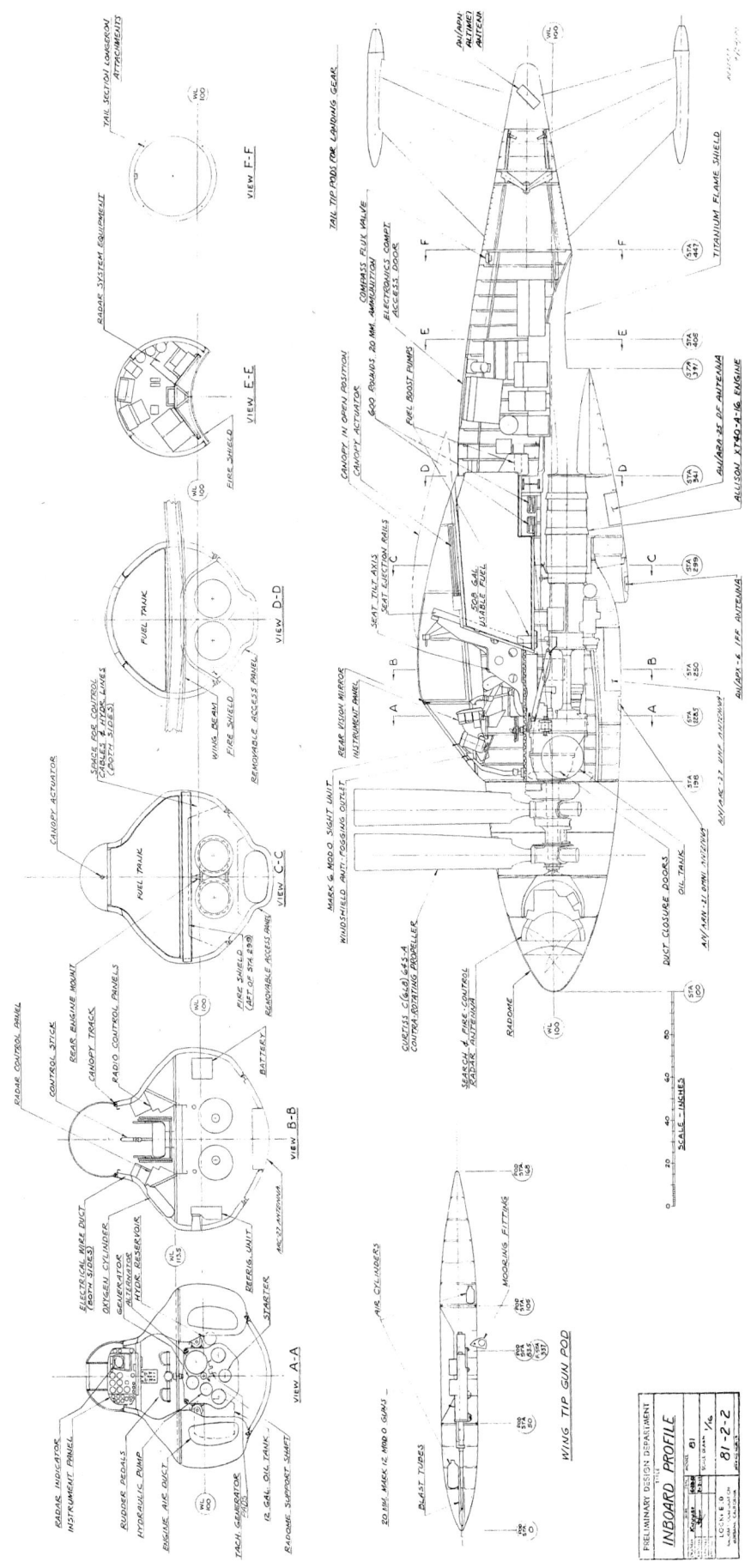

Inboard profile of the Lockheed XFO-1 revised September 24, 1951.

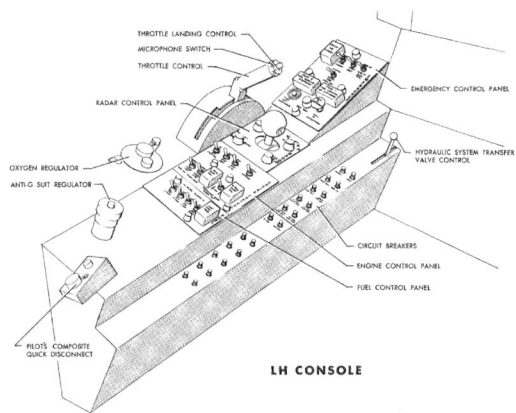

Illustration of the aircraft's left hand console.

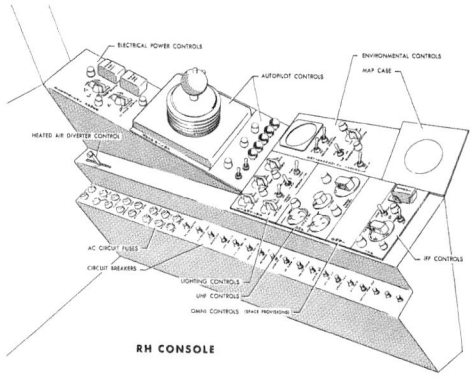

Right hand console of the XFO-1.

ammunition. Thus, the easy and rapid servicing of the guns was readily accomplished from suitable stands with the airplane in either the vertical or horizontal position. Ammunition was readily pulled into the ammunition storage chutes by hand crank operated cables.

Frangible nose rocket pod

An alternate rocket installation was presented with a single bank of 24 rockets and frangible covers for each tube opening. This pod was interchangeable with the gun pod. The blast of the first rocket blew off the tip of the tail cone to open the exhaust port. Any predetermined number of rockets could be fired in a ripple. After firing the rockets, the tip pods had open holes in the nose, and the exhaust port in the tail was also open.

Expendable rocket pod

The expendable rocket pod was presented as another alternate proposal. This pod consisted of a streamlined end plate, interchangeable with the gun pod, and the expendable rocket-carrying pod. The end plate contained the release mechanism for dropping the expendable portion of the pod after firing the rockets. The expendable portion of the pod consisted of:

1. A frangible nose section which was blown off with the firing of the first rocket.
2. A centre cylindrical section consisting of 24 rocket tubes, the insulating and heating blanket, and the skin covering, all mounted on a simple supporting structure.
3. A tail cone which had a frangible tip that opened the exhaust port when blown off.

This pod offered the lightest rocket installation and the least drag penalty on the return trip. Since the pod was designed to be shipped knocked down, it presented no serious space problem in storing a supply of rockets and pods on board the ship. The centre section could serve as the sealed shipping container for the 24 rockets, and a number of nose and tail segments could be nested in crates to minimize stowage space requirements. The sections could be readily assembled by unskilled personnel during the time the rockets were being temperature stabilized prior to mounting on the airplane.

'Normal' landing gear

For the purpose of preliminary flight testing, a normal landing gear was designed. This landing gear could be temporarily bolted on the outside of the airplane structure. The added safety afforded the airplane and pilot by the addition of this temporary gear was in the ability to make a power-off landing, and the opportunity to make the first transitions at safe altitudes. The addition of this gear did not affect the operation of the vertical landing gear; therefore, it was possible to take off on one gear and land on the other.

The landing gear consisted of two main wheels and brakes, suspended on shock absorber struts, and supported by a tube structure, which was bolted to the outside of the fuselage between the main engine mount bulkheads at the hoist fittings, as well as a drag truss attached to the bottom of the main wing

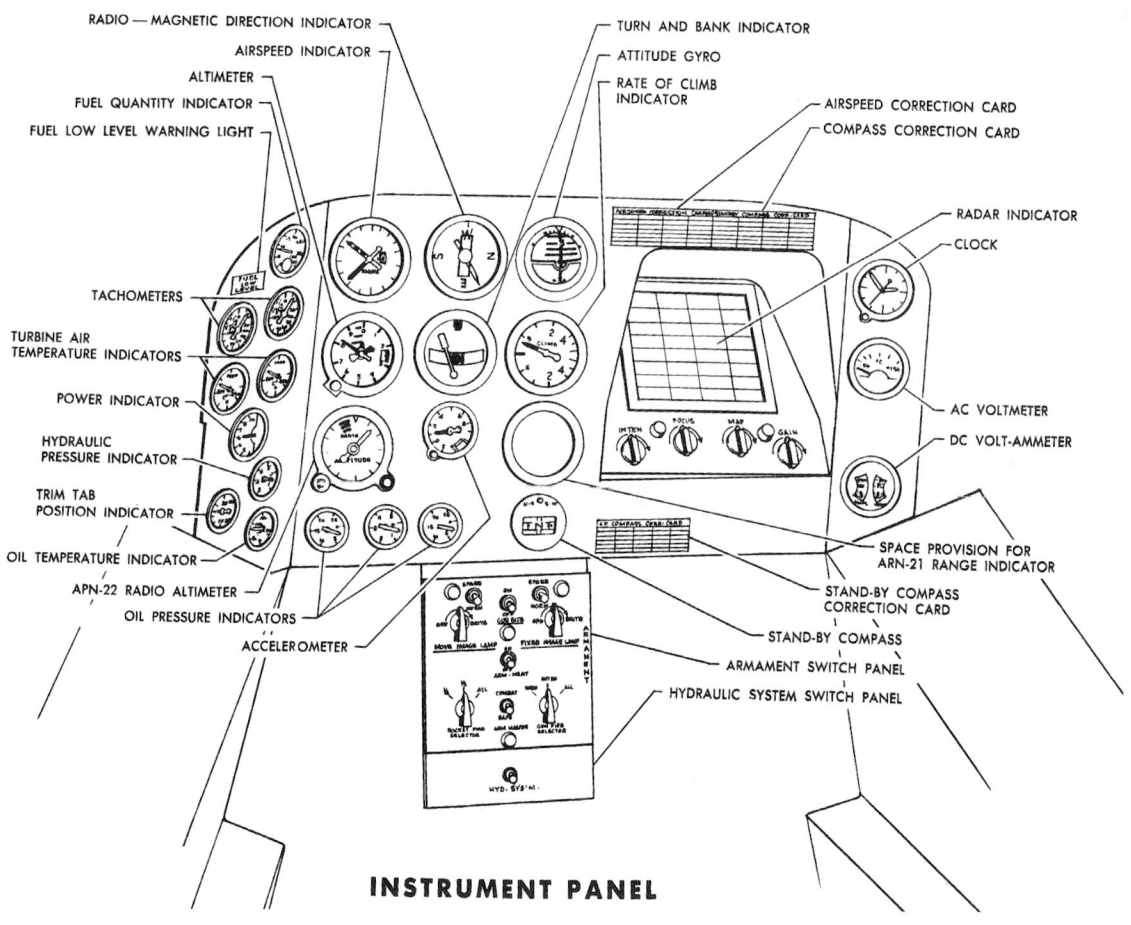

Drawing of the main instrument panel.

beam near the fuselage. Two tail skids were attached to the two lower empennage tips.

It was felt that this type of gear would be a very desirable unit to adapt to tactical articles for pilot training purposes.

Ground handling cart

The accompanying photos (see p324) show a model of a proposed ground handling cart which was used to rotate the airplane from one position to the other. This cart looked so promising that it appeared that horizontal maintenance and servicing could be more practical than in the vertical.

The first photo is of the cart with the airplane horizontal. A platform was shown to indicate access to the cockpit. The smaller cart was for engine removal, and it could be rolled up the ramp under the engine compartment. This smaller cart also had sufficient vertical travel in its linkage so that, with a removable platform, it could be used to service the tip pods; or with suitable cradles attached it could be used to service the radome or propellers.

The second picture showed the cart with the airplane in the vertical position. Ladders extended up the sides of the cart (only one side shown) so that personnel could have access to the adjustable attachment fittings.

MOCK-UP BOARD REPORT

The items of major interest discussed during the Mock-Up Inspection of the XFO-1 are summarized below.

The location and grouping of the circuit breakers was subject to considerable discussion. It was finally determined that the position shown in the mock-up (below the consoles on a narrow shelf) was suitable except that the breakers were moved to project horizontally into the cockpit when open with the labels on the shelf directly above them. They were also rearranged so that the most important functions were at the forward end of the panel since some of the more aft positions on the panel were not too accessible or visible because of obstruction by the pilot's seat.

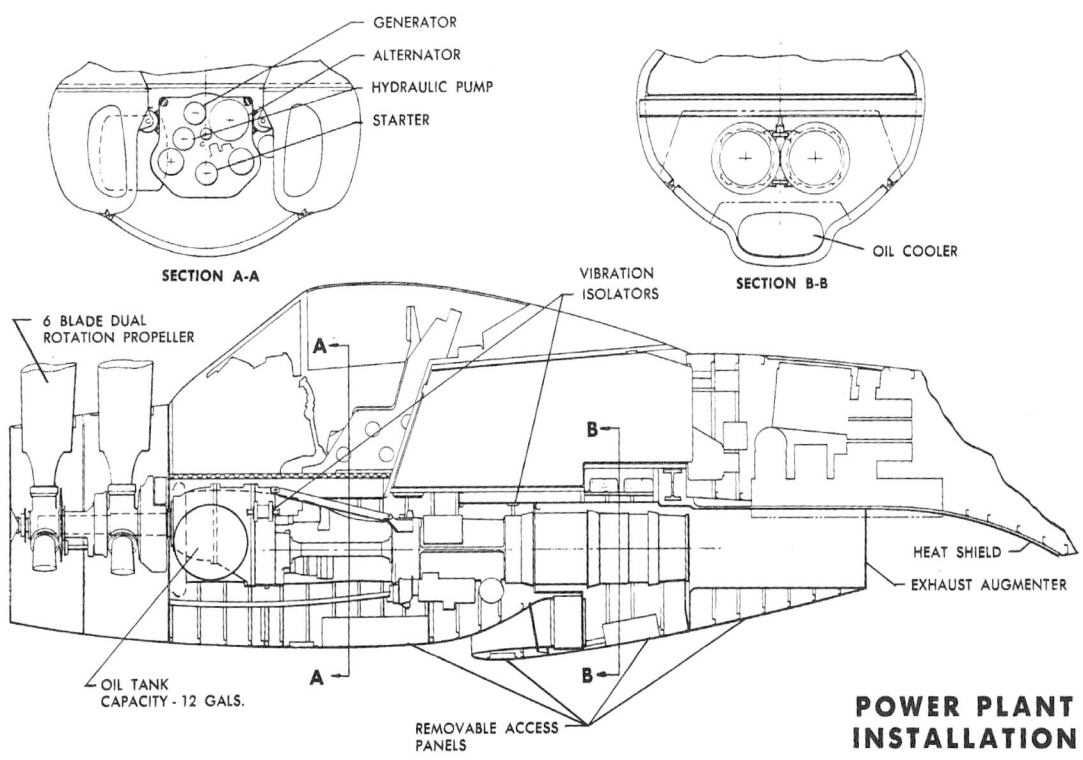

The power plant installation consisted of an Allison XT40-A-6 turboprop engine with a Curtiss 16ft diameter six-bladed dual contra-rotating propeller.

The sequence of operation of the ejection seat was modified to provide a single operation for the primary means of firing. This consisted of the face curtain pull. An alternate means of firing was also requested, consisting of the manual raising of leg braces which then extended a hand grip firing control and also removed the firing pin. It was the opinion of the Mock-Up Board that such an alternate firing means was a necessity since there had been accidents in which it was known that the pilot could not use or had considerable difficulty in using the face curtain.

The board rejected a request for provisions to permit removal of a single power section from the airplane without disturbing the other section because of the considerable airplane redesign involved. It was also pointed out that although there were two power sections there was only one engine control and readjustment of this in the airplane after a section replacement was considered a questionable procedure. In view of the importance of this control function, it was felt that it should be rechecked on a test stand prior to airplane installation of the engine.

Numerous recommendations were made to provide improved maintenance and accessibility of airplane components. The majority of these were directly tied to the problem of handling and servicing the airplane, as a result of its vertical attitude. Lockheed demonstrated a model of a handling cart which would be used for all major servicing of the airplane; normal routine servicing would still be accomplished in the vertical. Other ground handling equipment was also extensively discussed, and it was finally agreed that additional coordination on this would be required with BuAer. Consequently, it was made the subject of conferences with Lockheed and subsequent action by BuAer.

The board requested that Lockheed provide a lighting mock-up of the cockpit for later inspection. In addition the board provided that certain other items would also be reinspected at that time.

The various changes approved by the board resulted in a weight decrease of about 10.5lb. The changes mostly involved the removal/modification of minor items.

NACA flying model tests (1952)
A ¼ scale dynamic model of Lockheed's Convoy

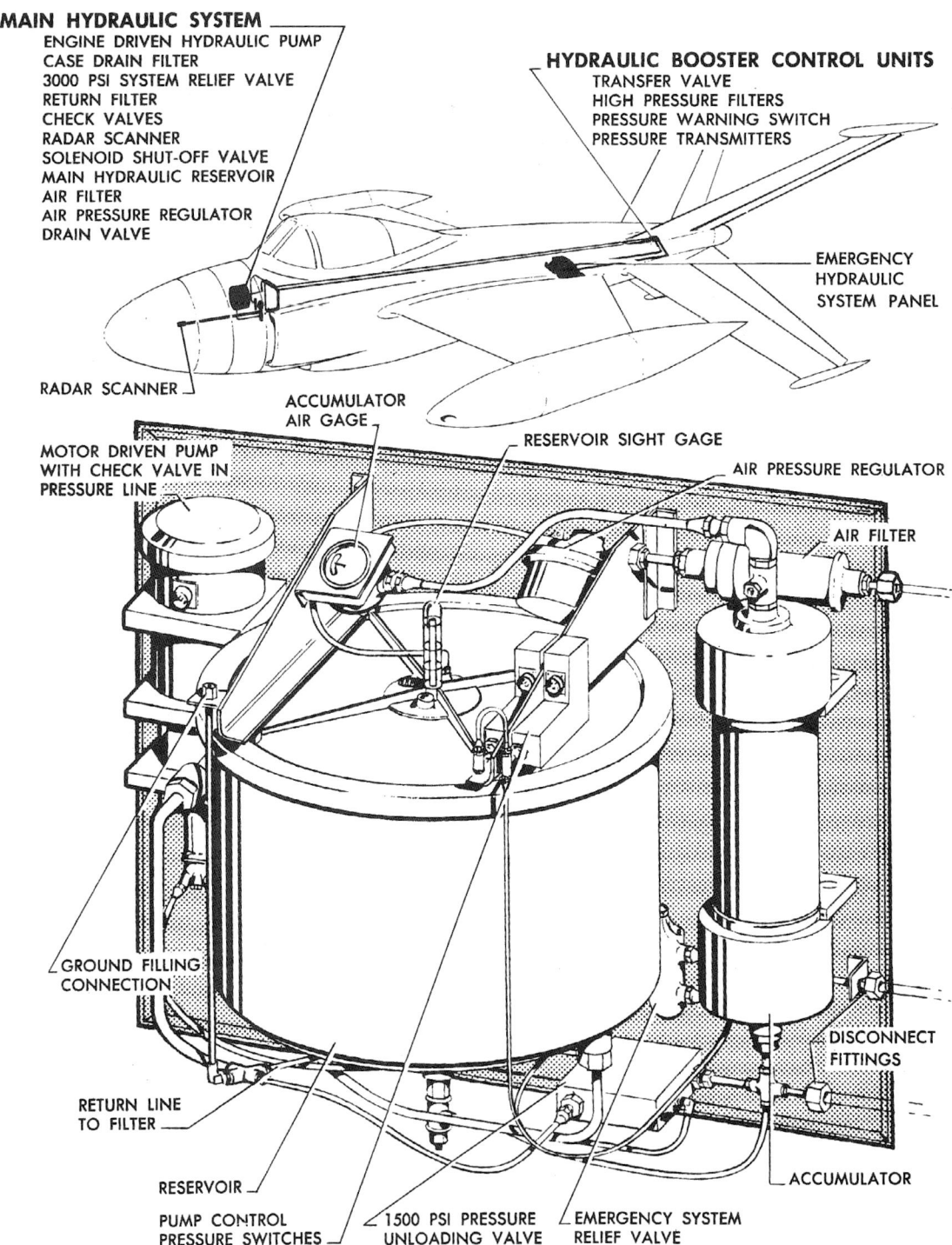

Diagram of the XFO-1 hydraulic system, which was powered by one engine driven variable volume hydraulic pump.

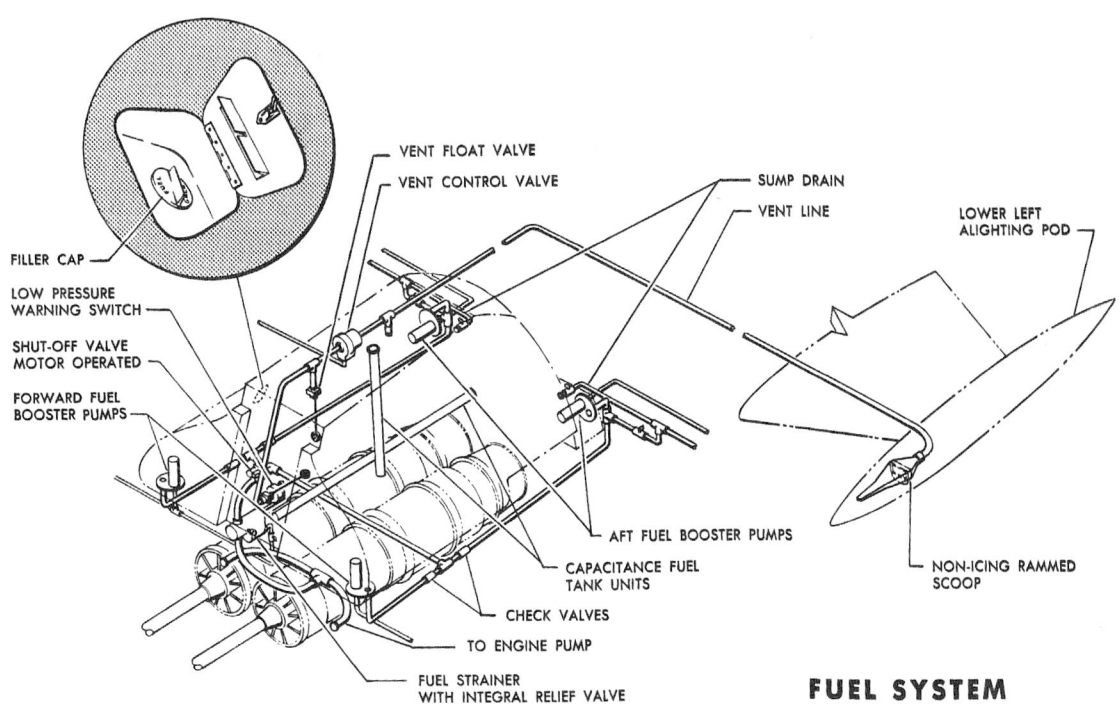

Diagram of the fuel system, the main component of which was a nylon fuel cell having a capacity of 525 US gallons installed in the fuselage aft of the cockpit.

Fighter was constructed in 1951 to determine the handling characteristics and controllability in the completely new flight regimes of hovering, horizontal translation while in the vertical attitude, and transition from normal flight to hovering flight. By early 1952, the aircraft had been redesignated XFV-1. After a checkout period at the Lockheed plant, the model was transported to the Ames Aeronautical Laboratory at Moffett Field and was set up and tested at the 40 x 80ft wind tunnel in the 12 week period from March 24 to June 6, 1952. The tests were conducted by Lockheed personnel with NACA and BuAer representatives observing.

The first series of hovering flights in the tunnel began on April 2, 1952. On these flights the operator at the master control station controlled pitch and yaw only. The roll and power controls were manned independently by two other operators. This was done to restrict the number of modes of motion which any one pilot would have to control until some familiarity with the model had been obtained. In addition to the three flight controllers, one man was used to handle the wing tip tethering lines, and two men were stationed at the winch used to operate the nose tethering line.

The test technique used on these flights was as follows: with the model on the floor, power was applied to give sufficient thrust to enable the men at the winch to lift the model easily. The model was then hoisted to an altitude of about 25ft and restrained by maintaining tension on all tethering lines. The power was then increased to the rated power of the motors and when all operators had signalled their readiness for flight the nose line was slacked off.

Power was then decreased until the wing tip lines were slack and until the desired flight altitude had been reached. On the first flight a large amount of artificial damping about the pitch, roll, and yaw axis was provided by using a large gain on the rate gyros. The motions of the model during this flight were slow and steady, but control was marginal since the signal from the controller was small compared to that from the rate gyros. No difficulty was experienced with the power control and the desired flight altitude was easily maintained.

On subsequent flights the artificial damping was gradually reduced until a balance between stability and controllability satisfactory to the pilot had been obtained. Two attempts were made to fly with no

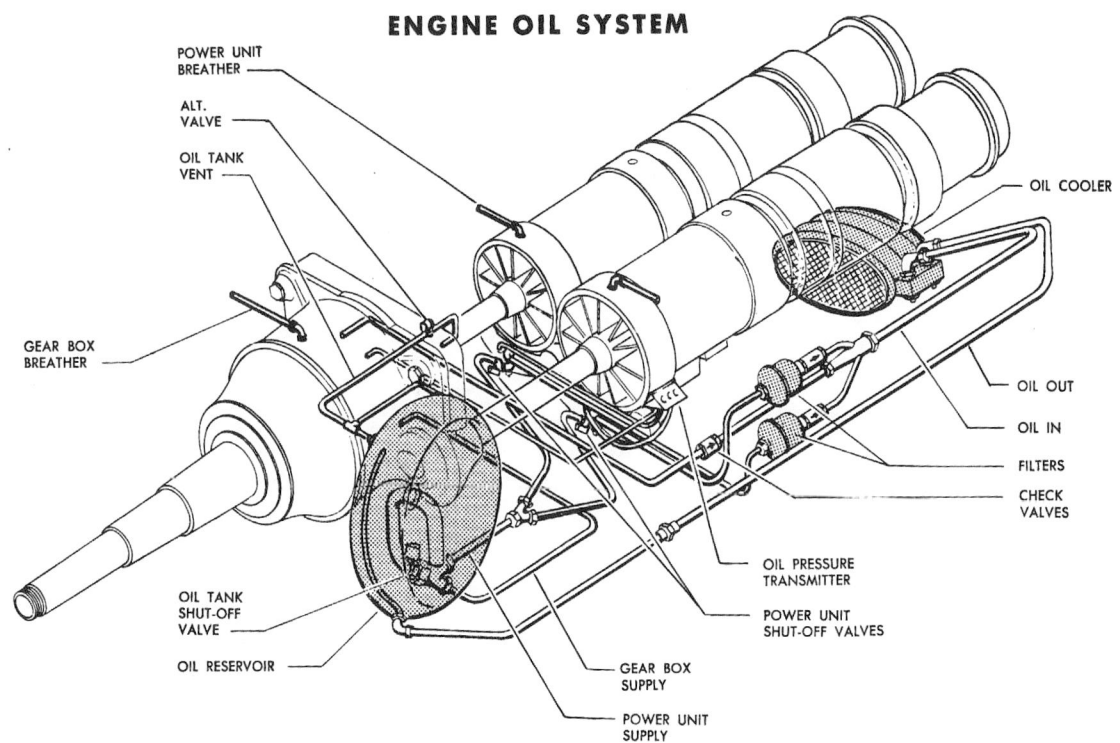

The XFO-1 oil system consisted of one oil tank and one oil cooler suitably manifolded to supply oil to both power sections and to the reduction gearbox.

artificial damping. However, the model motions became rapid and violent and it was necessary to switch the damping on again to retain control. Recovery to steady hovering flight was easily accomplished without tethering when the damping was provided.

After the operators had developed some familiarity with the handling characteristics of the model, flights were made at low altitude in order to determine the height at which reduced control effectiveness due to the proximity of the ground became apparent. It appeared that satisfactory control could be maintained until the tail wheels were about a foot from the floor. When it was attempted to fly at lower altitudes, control was lost and the model moved rapidly in a horizontal direction until restrained by the tethering lines.

On some of these flights, landings were attempted by hovering the model about 1-2ft above the floor and suddenly reducing the power. However, the range of power available from the power-control rheostat was insufficient to reduce the power enough to keep the model securely on the floor. On all of these attempted landings the model descended to the floor smoothly and without any significant change in attitude. However, after contact with the floor the model skipped along the floor out of control until restrained by the tethers.

The next series of flights was performed with the roll, pitch, and yaw controls all handled at the master control station. The power controls were operated independently as before. Flights similar to those described above were made with no appreciable difficulty. On one flight the artificial damping in yaw was completely lost due to a broken connection in the circuit. The model motions in yaw were controlled but erratic, especially so until the pilot became used to the increased effectiveness of the yaw control without damping. As a result of this flight it was decided to attempt to fly the model with no artificial damping by eliminating the gyro signal from one control system at a time.

For the third series of tests the pitch, yaw, roll, and power controls were all handled by the pilot at the master control station. Flights similar to those described above were performed satisfactorily, although at times the pilot had some difficulty coordinating the controls. Successful landings were also accomplished on these flights by opening the main power switch to the motors just before contact with the floor was made. This procedure eliminated all tendency to skip or bounce after contact with the floor.

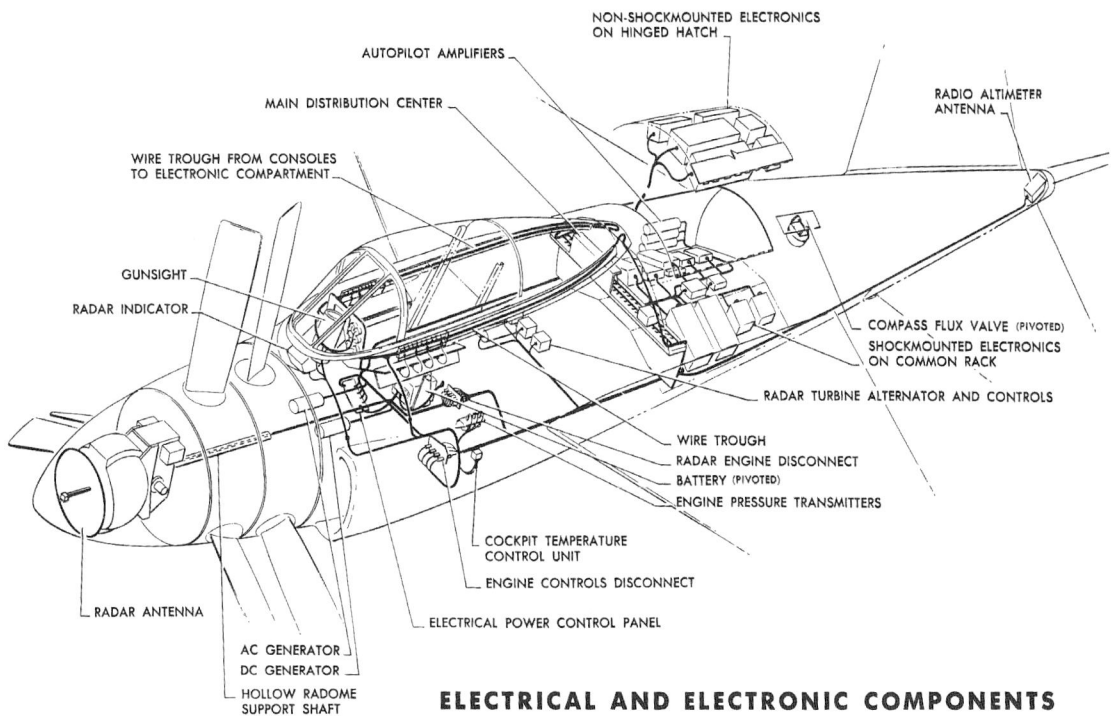

The basic power of the aircraft's electronic system consisted of an engine-driven 15 KVA AC generator and a 300 ampere DC generator.

Takeoffs were also attempted but these were not satisfactorily accomplished in early testing. The main difficulty was in the inability of the 40 x 80ft wind tunnel motor-generator set to change speed quickly enough to lift the model rapidly through the region of low control effectiveness near the ground. Takeoffs were eventually accomplished by holding the model on the ground with the wing tip tethers while the power was applied, and then quickly slacking off on the tethering lines. This difficulty due to the limited speed of power response was not expected to be encountered on the full-scale airplane.

On the last flight performed, the pilot attempted to manoeuvre the model from one predetermined position to another. In general, this was accomplished although some difficulty was encountered by the pilot in starting and stopping the translation from one position to another. At the end of this flight, while manoeuvring for a landing, the nose line was inadvertently allowed to become too slack and was cut by the propellers. The overload breakers opened immediately and the model dropped tail first to the ground from an altitude of about 8ft.

Damage was sustained on the fuselage which buckled in the vicinity of the tail, on two blades of the front propeller, on the extension shaft to the spinner and nose boom, and to the nose boom and cable connections. This damage required two weeks to repair. New parts were brought to Ames by Lockheed personnel. The model was equipped with a longer nose boom, a ring guard around the propellers, and a guard around the tail surfaces to reduce the chance of a similar accident occurring again. The nose boom, one end of which was pivoted at the propeller hub, was now 4ft long. The model propeller radius was 2ft. With the model vertical, the nose boom would be horizontal if it contacted the propeller ring guard. The cable attached to the boom was therefore about 2ft outboard of the propeller tips if the boom contacted the propeller guard. Hydraulic dampers were installed in the control system and eliminated, for all practical purposes, the control system flutter which had been present throughout the first series of flights.

Photos of the ¼ scale model of the XFV-1 subsequent to the accident of April 16, 1952 were taken immediately after the crash landing occurred and, except for the removal of the nose spinner, before the model had been moved or worked on in any way. As indicated in the photographs (see p325), the main damage was restricted to the nose spinner and extension shaft, and to the fuselage in the vicinity of the tail. The damage in the vicinity of the nose was done in flight when the nose cable became fouled

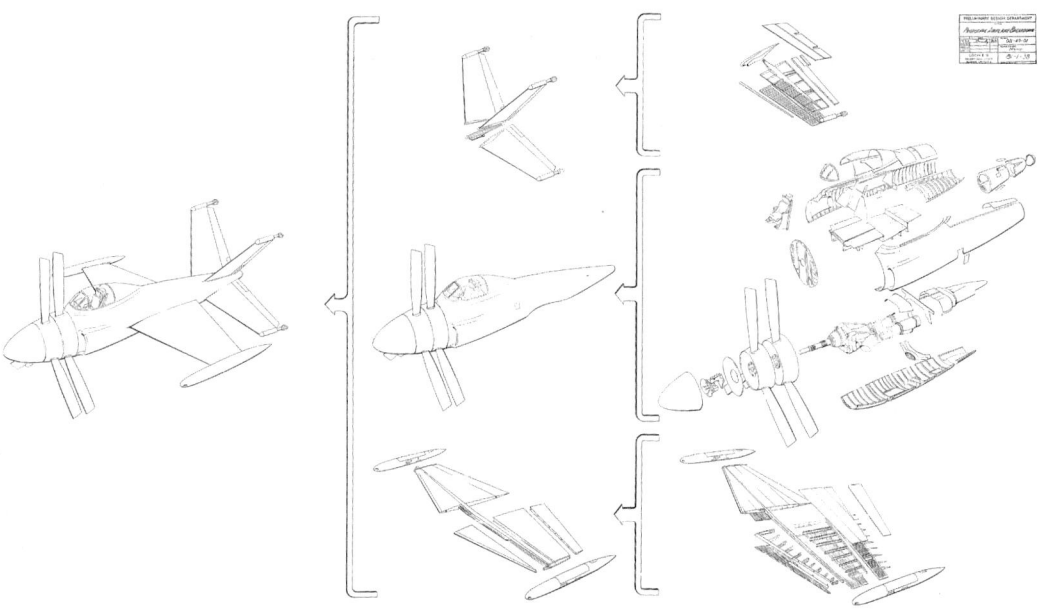

Production breakdown of the Lockheed XFO-1.

in the propeller. The damage due to the impact with the floor following a drop was localized to the fuselage in the vicinity of the tail.

Additional hovering flights of the repaired model were performed until it was felt that translation tests could be attempted. The first of these was a check flight to make sure the equipment was functioning properly and to enable the test personnel to regain familiarity with their tasks. The second flight was primarily a demonstration flight for the benefit of Mr. A. Flock, Lockheed project engineer for the XFV-1, and Mr. G. Danch of BuAer.

Some flights in yaw and pitch translation were made. Controllability was demonstrated in yaw translation at speeds of 15ft/sec and 20ft/sec. However, at a yaw translation of 25ft/sec, the effectiveness of the controls in pitch was so poor that controlled flight was impossible. There still appeared to be some response of the model to yaw control inputs, however. Controllability was demonstrated in pitch translation at speeds of 10, 20 and 30ft/sec. Slight combined rolling and yawing motions were noticed, but these were easily controlled by the pilot. One flight at 35ft/sec was attempted but steady state conditions were never attained.

The difficulty was primarily due to the tethering system which was essentially the same as that used for the hovering tests. When the model was tethered, it was in approximately a hovering attitude and the tethering forces due to the drag of the model were quite large. When the model was pitched to its trimmed flight attitude and moved so as to slacken the tethering lines, the sudden removal of the large tethering forces was too much for the pilot to compensate for. Since apparently a new tethering system was needed and the maximum speed available in the existing location in the 40 x 80ft wind tunnel was being approached, it was decided to move the model into the test section for the remainder of the pitch translation tests.

In a letter to BuAer dated June 23, 1952, Lockheed announced the completion of the scheduled model tests on the hovering and translating flight characteristics of the XFV-1 airplane. In general, the results of tests with the ¼ scale free flight model agreed with earlier estimates, which were based on the force tests of the 1/10-scale model. The only notable exception was that control loss in yaw translated occurred at a lower speed and in a different manner than predicted.

For simplicity, the discussion of the characteristics was separated according to the flight regime:

Hovering

The model was controllable by a single pilot handling pitch, yaw, roll and power remotely through airplane-type controls. Artificial damping in roll, pitch and/or yaw was necessary because of the remote position of the pilot, but could have been dispensed with by sufficient pilot training. It was possible to fly without

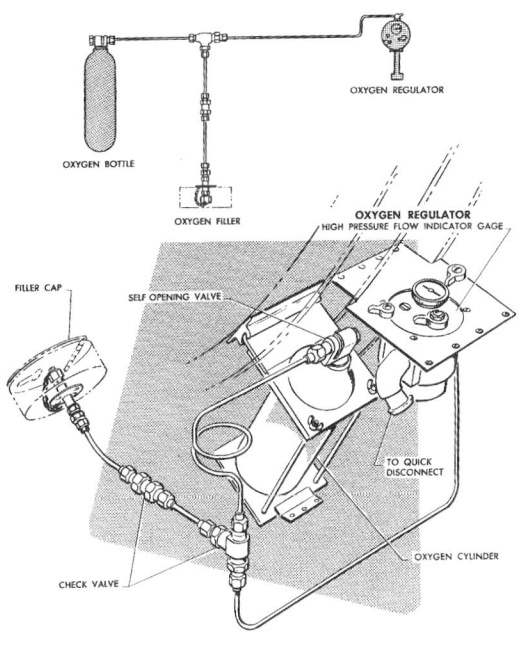

Drawing of the oxygen system, which incorporated a positive pressure breathing regulator and had an oxygen supply of about three hours.

artificial damping about any one, but not all, of the three axes. Hovering flight was found to fatigue the pilot in two to three minutes, however, because the completely neutral stability (expected at zero speed for any aircraft of this type) meant continual movement even when the anticipatory senses had been developed. It was the pilot's opinion that artificial attitude stabilization should normally be used for hovering and landing, although pilots should keep in practice on unstabilized hovering and landing.

Landing

Landings were easily accomplished by cutting the power completely at about 1.5ft (model scale; 6ft full scale) off the floor. The plane settled gently without bouncing. It was found necessary to reduce the thrust below thrust-weight ratio of 0.6 after contact to prevent the model from rising on the belly-side oleos and 'walking'. This action was caused by the wing lift in the slipstream combined with the propeller thrust, giving a resultant force inclined up and back. Pilots were cautioned not to attempt position corrections after retarding the throttle to land, as they could exceed the tip over angle at contact.

Takeoff

The rate of thrust change available in the model was too low for a successful takeoff, even though the movies showed a takeoff sided by the tip tethers to prevent horizontal motion. The tests indicated that it would be necessary for the pilot to keep the thrust below about 60% of the weight until he was ready for takeoff, then to advance the throttle rapidly to takeoff power. For both takeoff and landing he had to pass fairly rapidly through the range within 3ft (full scale) of the ground where the control effectiveness was markedly reduced by the ground effect. With the model, control became inadequate at 8-12 inches from the floor.

Yaw translation

The model was flown successfully (full control, no apparent adverse dihedral effect) with wing tip into winds as high as 24 kts full scale, but had no control of pitch and/or roll at 30 kts full scale. The exact speed at which control was lost was not determined because of time limitations. The mechanism of the control loss was believed to be that the slipstream was blown off the two upstream tail surfaces, so that either a pitch or roll input signal produced a diluted combination of pitch and roll moments from the remaining two downstream surfaces. In other words, two control surfaces were inadequate for three degrees of freedom.

Pitch translation and transition

These tests were carried up to 83 kts speed full scale, or an angle of attack of 23° to 25°. They were completely successful both with and without leading edge slats. Above about 10 kts full scale speed there was a favourable dihedral effect, and all controls were effective through the speed range. The model was flown through a continuous slowdown transition from 70 kts full scale to zero speed in about 2.5 minutes, the time being determined by the maximum retardation rate of the tunnel. The model was subsequently stored at Lockheed for use later in the programme.

In subsequent correspondence with BuAer dated January 13, 1953, Lockheed addressed BuAer's concerns about automatic stabilization. The company noted that it had always realized the desirability of automatic stabilization for hovering flight as well as for the takeoff and landing manoeuvres. However, it believed that the XFV-1 could be satisfactorily hovered without autopilot. The difference between a pilot sitting in the airplane and a model operator trying to fly the model from a distance was

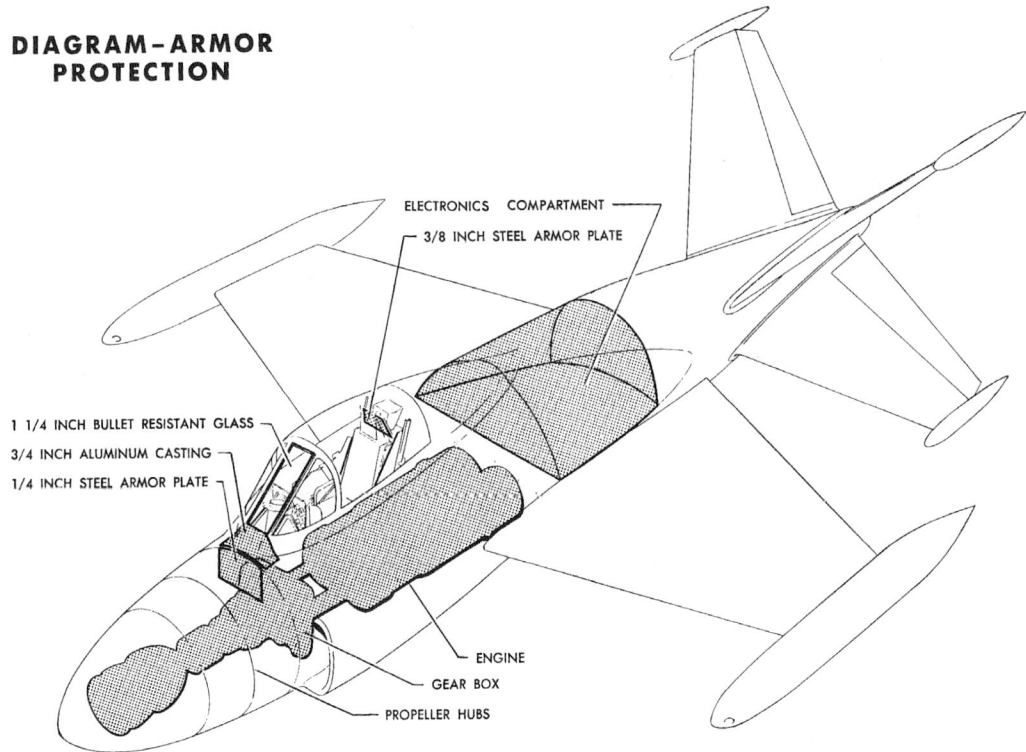

Diagram of the armour protection provided in the operational XFO-1.

probably similar to the difference in flying the model with and without the model rate gyro stabilization.

The flying model only had rate stabilization. The automatic pilot constructed for the actual airplane would have complete stabilization and position trim. The autopilot would be installed in the airplane prior to the first attempts at vertical flight.

Lockheed made no effort to improve the yaw translation control, as there was considerable uncertainty as to whether the airplane would not be quite superior to the model in maximum possible yaw translation speeds. The airplane change required to increase this limiting speed would be the addition of an inboard aileron type surface on one of the airplane wings. This surface would be actuated with the roll control and thus allow the cancellation of the roll which resulted from pitch moment requirements at high translational speeds. Such a device, however, increased the weight and complexity of the airplane and therefore was not considered further until after flight test of the aircraft.

Subsequent history of the XFV-1

The XFV-1, powered by the 5,332 hp Allison YT40-A-6 turboprop engine, began flight testing in November 1953 with a temporary non-retractable undercarriage; this was similar to the design shown in the original proposal. Ground testing and taxiing were held at Edwards AFB with Lockheed chief test pilot Herman 'Fish' Salmon at the controls; his surname ended up becoming associated with the aircraft, though it was never officially adopted by the Navy. The XFV-1 accidentally made a brief hop on December 22, 1953, when Salmon taxied the aircraft past liftoff speed. At the time, it lacked the rear portion of its spinner. The official first flight wouldn't take place till June 16, 1954.

Delays in the production of the 7,100shp Allison YT54 limited the XFV-1's VTOL capabilities, which were never fully realized. The aircraft made a total of 32 flights at Edwards AFB, none of which included vertical takeoffs or landings. A few transitions were made in flight from the horizontal to the vertical attitude, with one instance of the aircraft hovering at altitude. BuAer eventually realized that even with the more powerful YT54 engine, the XFV-1 would be outclassed by contemporary jet fighters. Furthermore, landing the aircraft was a hazardous operation requiring the most experienced pilots, which were in limited supply. These and other factors led to its cancellation in June 1955.

Final verdict

By any measure, the XFV-1 was a disappointing

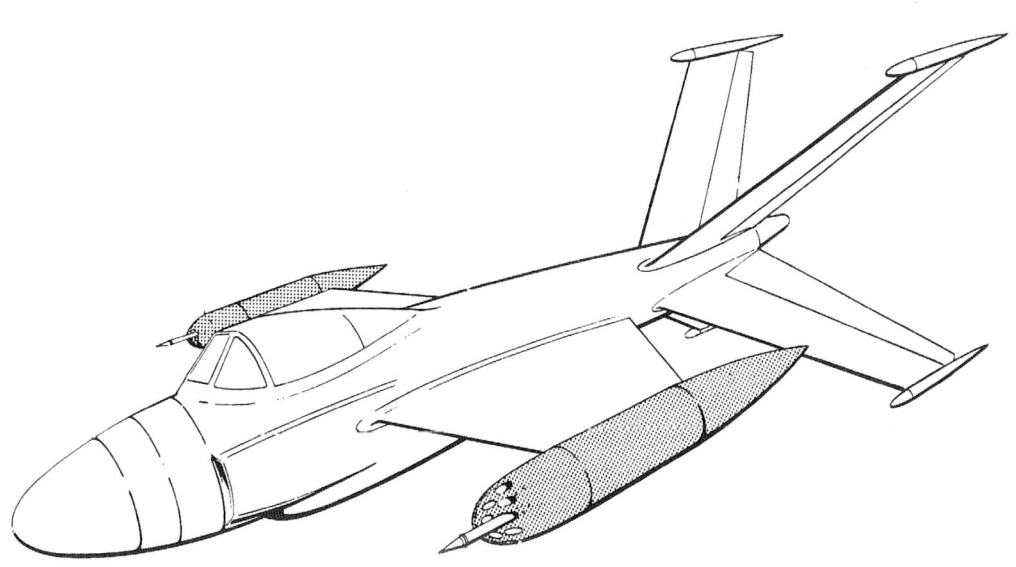

FRANGIBLE ROCKET TUBE COVERS
(ALTERNATE PROPOSAL)

As an alternative to the 20mm guns, Lockheed proposed a rocket installation with each wing pod having a single bank of 24 rockets and frangible covers for each tube opening.

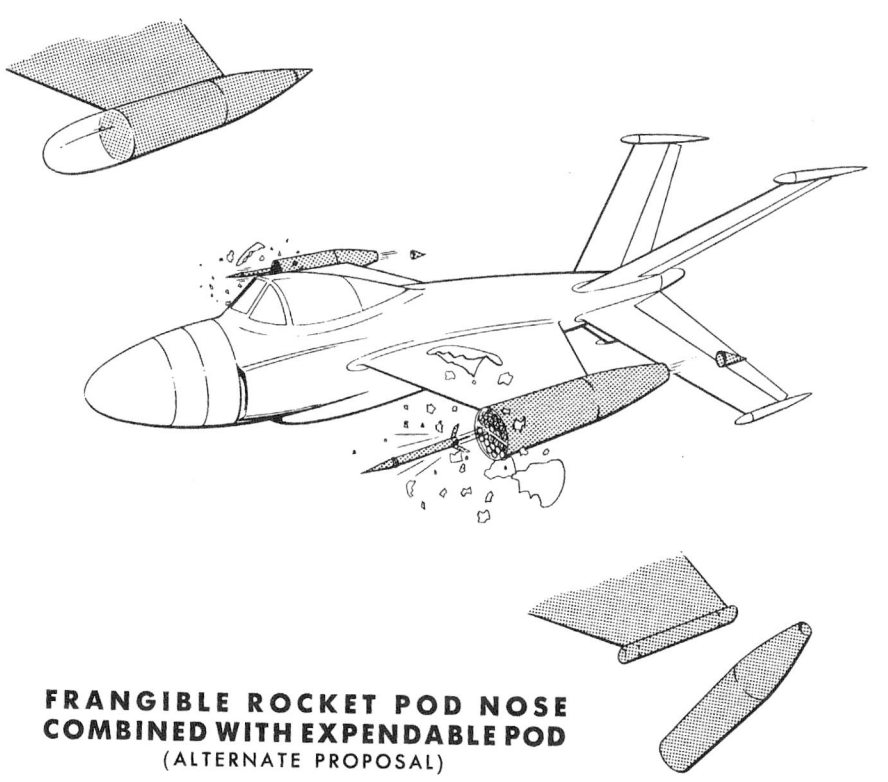

**FRANGIBLE ROCKET POD NOSE
COMBINED WITH EXPENDABLE POD**
(ALTERNATE PROPOSAL)

A second armament alternative was an expendable rocket pod, which offered the lightest rocket installation and the least drag penalty on the return trip.

aircraft, notably inferior to the Convair XFY-1, which at least managed to take off and land vertically on multiple occasions. The US Navy is often faulted for being too conservative in its procurement policies, but the Convoy Fighter programme is a case where it was overly ambitious and confident, greatly underestimating the challenges of developing such an unconventional aircraft. These included the great difficulty of landing a tailsitter aircraft vertically, which required a sophisticated autopilot device that was probably beyond the capabilities of the aerospace industry of the early 1950s.

BuAer also underestimated the time and money it would take to develop the necessary turboprop engine-gearbox combination to power an operational Convoy Fighter; concurrently, it did not foresee how rapidly the jet fighter would overtake its turboprop-powered rivals.

Hindsight is 20/20, and the obvious conclusion from the test programmes of the XFY-1 and XFV-1 is that the VTOL turboprop tailsitter concept just was not viable given the limitations of the early 1950s aerospace technology. Of course, BuAer failed to grasp just how difficult it would be to operate this type of aircraft when it first started studying the concept immediately after the Second World War.

Giving it the benefit of the doubt, I believe BuAer should have procured the less expensive 0.766 reduced scale prototype and rigorously tested it before cutting any metal on a full-scale aircraft; preferably, it should have only bought the Convair design, as it appears to have had the best configuration and lowest takeoff gross weight. This could have been a joint X-plane type programme with the other services and NACA, as the majority of these organizations had an interest in the VTOL turboprop tailsitter concept for various roles. Such an approach would have been a less expensive way of testing the viability of the concept and gathering useful research data. However, there is little doubt that the same problems encountered with the XFV-1 and XFY-1 programmes would have been revealed, especially with regards to safely landing and recovering the machine, and the VTOL turboprop tailsitter concept would have died just the same – just at less cost to the taxpayer.

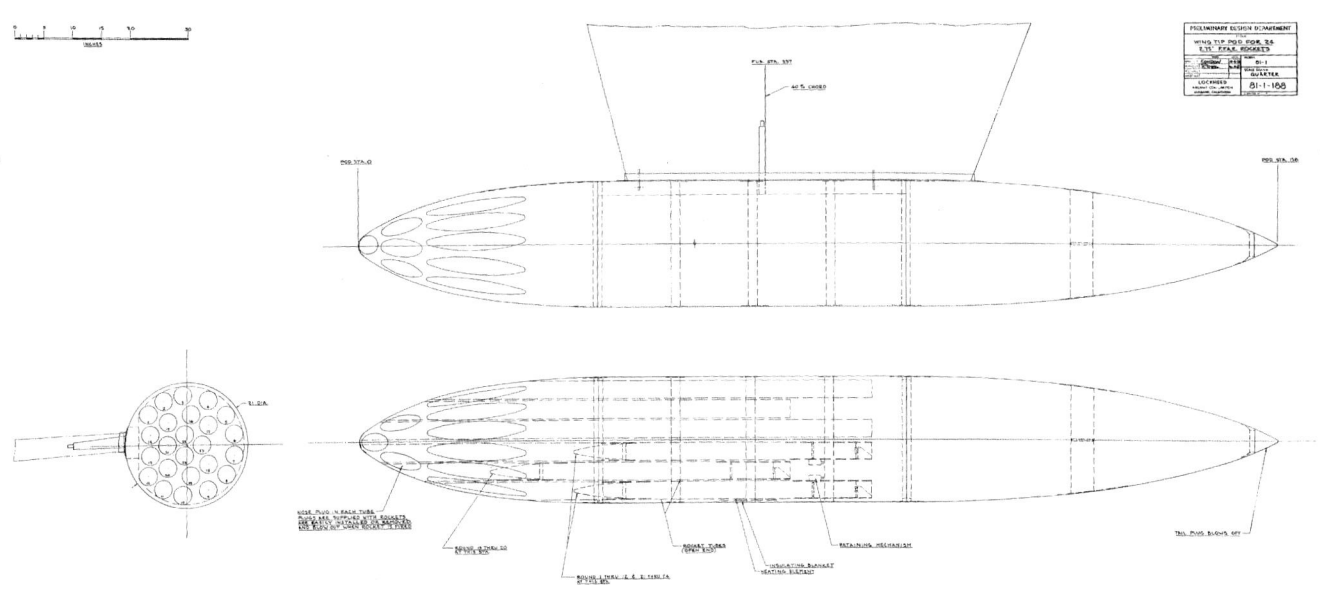

Diagram of the fixed rocket pod with frangible covers for each tube opening.

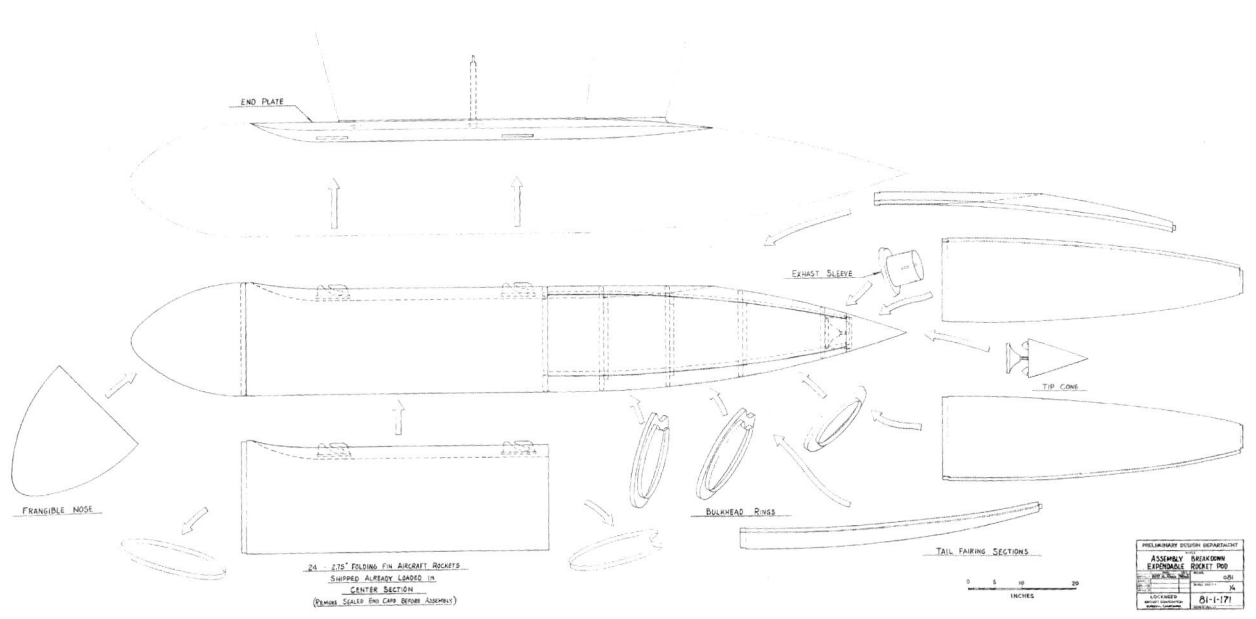

Assembly diagram of the expendable rocket pod. Neither design would be built or tested on the actual XFV-1.

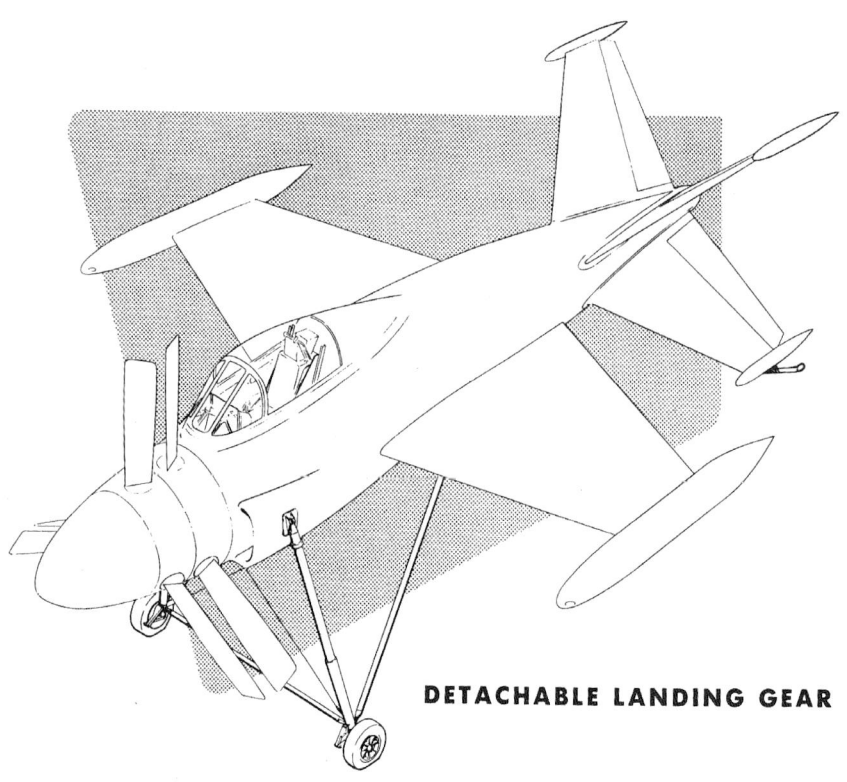

A detachable landing gear was designed for the purpose of preliminary flight testing; this is generally similar to what was used on the actual XFV-1 aircraft.

Model of the XFO-1 on its proposed ground handling cart, which was used to rotate the airplane from one position to the other; this photo shows the cart with the airplane horizontal. Also shown is a smaller cart used for engine removal.

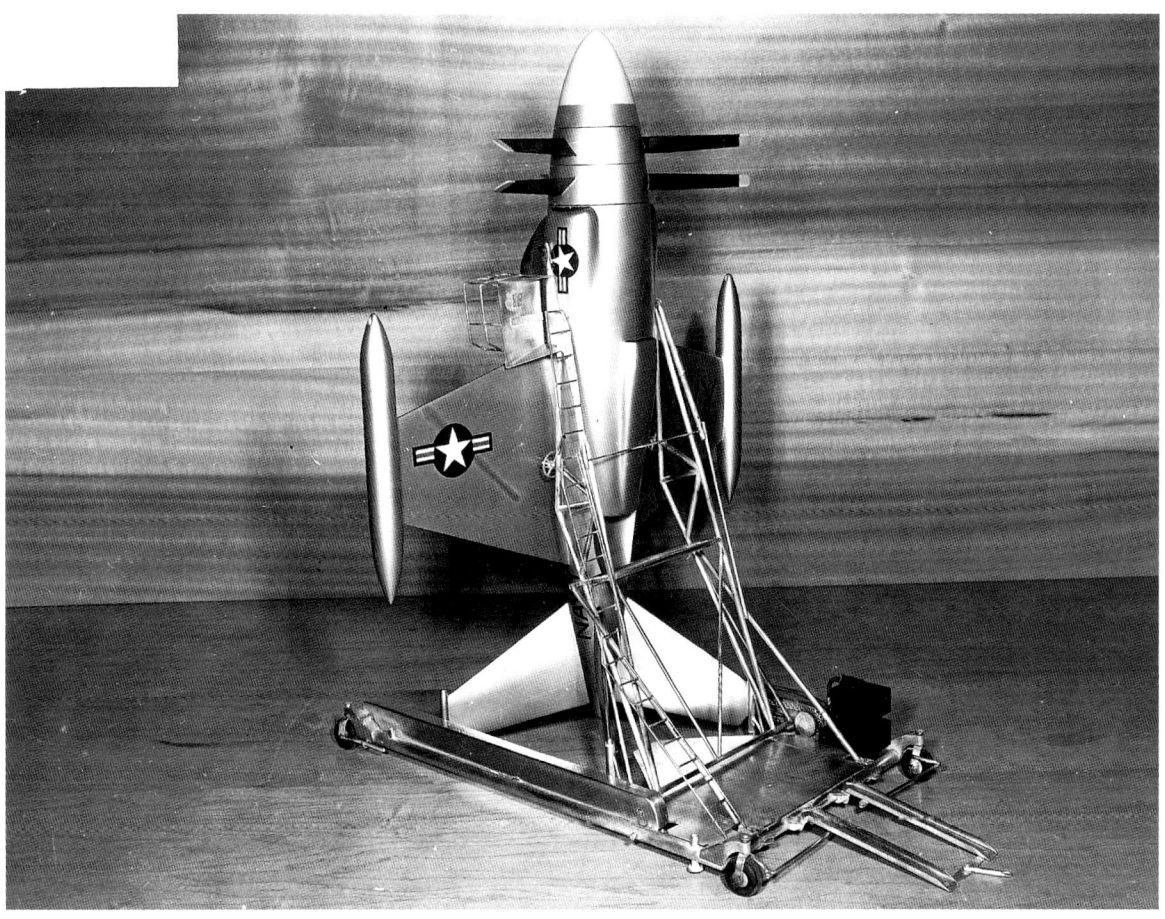

Photo of the cart with the airplane in the vertical position; ladders extended up the sides of the cart to permit access to the adjustable attachment fittings.

The ¼ scale model of the XFV-1 was damaged in an accident on April 16, 1952; this photo shows the crumpled tail section caused by the impact with the floor following a drop of approximately 8ft.

Detail view of the damaged nose spinner.

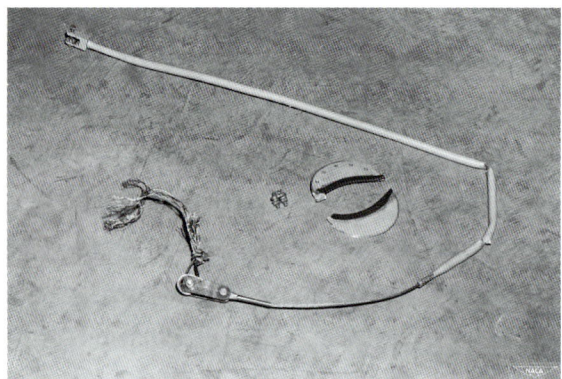
Close-up of the nose cable which was accidentally severed by the propellers.

Rare colour photo of the ¼ scale flying model of the XFV-1 which was tested in the Ames 40 x 80ft tunnel in early 1952. It weighed 244lb and was equipped with two 38-horsepower water-cooled electric motors.

Photo of the repaired and modified ¼ scale flying model of the XFV-1, which was now equipped with a longer nose boom, a ring guard around the propellers, and a guard around the tail surfaces.

General view of the flying scale model immediately after the crash landing. The model required two weeks to repair, receiving several modifications to prevent the nose line from being caught in the propellers again.

LOCKHEED XFV-1 (STRIPPED VERSION) MODEL 081-40-01
7 February 1952

AREA TOTAL FOUR SURFACES	169 ft²
ASPECT RATIO	3.55
AIRFOIL SECTION — ROOT & TIP	NACA 65A007
WHEELS — SPECIAL	8.5" x 4 SOLID TYRES

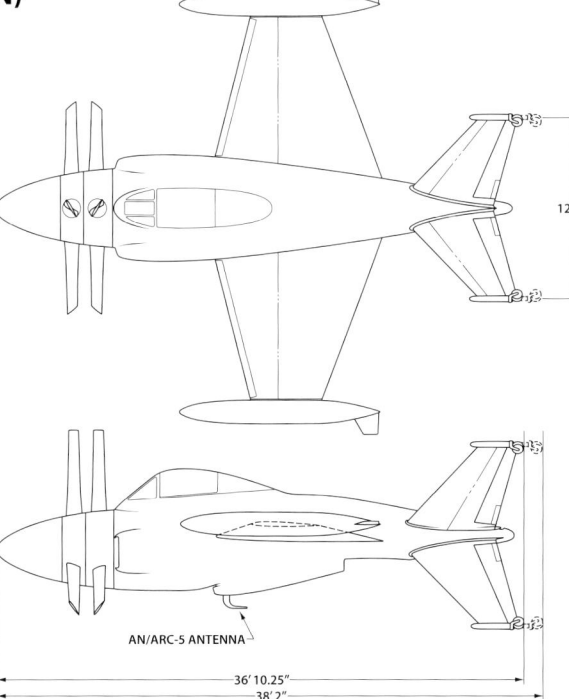
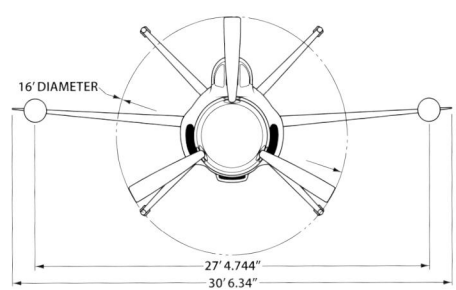

Three-view of the XFV-1 revised February 7, 1952; this configuration basically represents the aircraft as built, though the ailerons are not depicted.

8

Related Projects & Studies

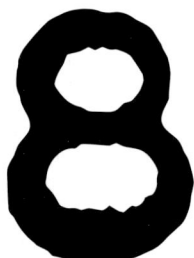

BOTH CONVAIR and Lockheed produced a variety of turboprop tailsitter studies in the mid-1950s inspired by their work on the XFY-1 and XFV-1, respectively. These were, however, aimed at the US Air Force and Army, which did not have direct experience with the difficulties of operating this type of aircraft. The following represents the unbuilt turboprop tailsitter studies by these companies thus far discovered; hopefully more will emerge with additional research.

CONVAIR STUDIES

T54 Powered Tactical VTO Airplane

On April 6, 1955, Convair (now a division of General Dynamics Corporation) proposed a substantially modified tactical version of the XFY-1 powered by a more powerful Allison T54 turboprop engine. The blueprint of this study accompanied a report titled A Comparison of Three Power Plant Types for Tactical VTOL Aircraft, which is summarized below. It is not clear if this proposal was submitted to the US Navy or Air Force (or both); given that shipboard operations are not mentioned in the report, it seems likely that the latter was the intended customer.

Introduction

A study was made of the relative merits of vertical takeoff and landing (VTOL) tactical aircraft powered by three types of propulsion systems. These systems were turbojet, turbofan and turboprop. At the outset of the study it was decided to consider the best engines for which performance specifications were available regardless of developmental status or estimated delivery dates. It was also decided that a delta wing airframe type designed to alight in a nose up attitude would be used for all three engines. This decision made it possible to use extensive available data on this airframe type and permitted a direct comparison of the engines. Design of the airframes was carried out to the point where weights and drag could be calculated and feasibility of the arrangements could be established.

Aircraft considered

The turboprop airplane was powered by an Allison T54 engine. The airplane was designed to operate at low altitudes resulting in a small, compact and light airframe. This airplane was similar in concept to the XFY-1 airplane and would not have required extensive additional development.

The turbojet airplane was designed for an engine with the performance specifications of the X61. Since this engine was not being developed the airplane represented what could be done with an advanced turbojet having adequate thrust to give desired radius performance. The airframe used a delta wing and was designed to cruise at altitudes

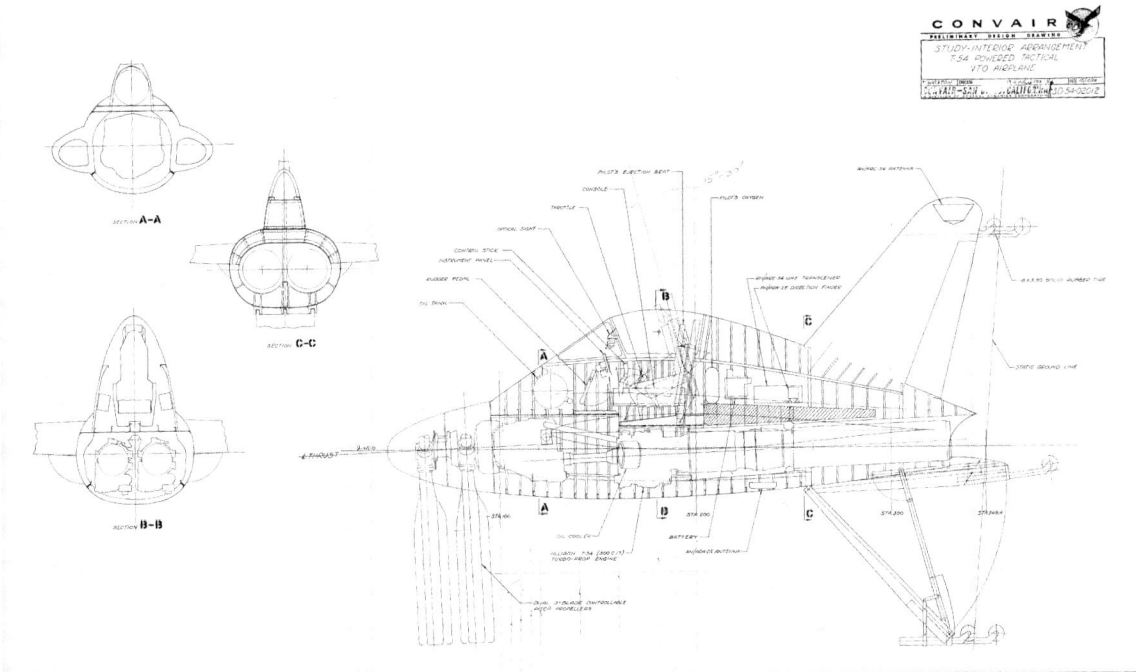

An inboard profile of a tactical version of the XFY-1 Pogo powered by a more powerful Allison T54 turboprop engine. It was intended to carry an atomic bomb in a large external store underneath the fuselage; note the replacement of the ventral fin with a prominent aftward-folding landing gear unit, likely to save weight and provide better clearance for the carriage and release of ordnance.

best suited to the engine characteristics.

The turbofan-powered airplane had essentially the same airframe as the turbojet with modifications to permit installation of the General Electric X84 engine. The engine was selected because its performance represented the immediate future potential of the type and it was under development at the time the study was undertaken.

Mission

The aircraft considered in Convair's report were intended for use by a land based theatre tactical air force. Their principal armament was tactical atomic bombs. These tactics differed considerably from those used when only conventional weapons were available.

In the initial phases of a tactical air war where atomic bombs were used, the principal emphasis would probably have been on destruction of enemy aircraft on the ground by bombing. There appeared to be little value in either ground defences or in extensive interceptor activity. Ground defences operated best when they could be concentrated around valuable targets. Against bombers using atomic bombs, such tactics would have eventually failed since only one successful bomb drop was required to destroy the target and possibly the ground defence installations also.

Air interception required large numbers of interceptors and an extensive control and early warning net. The effectiveness of even a modern well-designed system was poor against small high-performance aircraft intent on bombing rather than aerial combat. Furthermore, the interceptor system was also vulnerable to bombers so it probably could not have even protected itself against a concerted attack. This situation called for tactics where passive defence had to be emphasized and the principal flying effort devoted to reconnaissance and bombing missions.

The bombing effort had to be directed at the enemy aircraft on the ground until such tactics began to yield diminishing returns. Since the enemy was expected to adopt the same tactics, such a war would have been characterized by very high aircraft attrition with the 'winning' side finally left with more aircraft operating than could be employed fully for attacks against the remaining enemy air potential.

These tactics required aircraft for bombing and reconnaissance missions with operating properties suited to passive defence. The ability of VTOL aircraft to operate in small numbers from concealed and dispersed mobile sites made them particularly well suited to these requirements. Payload requirements were determined by the ability to evade defensive measures. These were the principal factors indicating the effectiveness of a tactical airplane.

Interceptor capabilities appeared to be of secondary importance. The aircraft which were investigated were examined for their effectiveness in performing the bombing and reconnaissance missions for which the most pressing need appeared to exist.

In the study, a payload of 1,000lb was assumed for the standard bombing mission. This appeared to be a reasonable figure for a streamlined tactical atomic bomb of moderate yield. The bombing equipment assumed consisted of the LABS AERO 18A system and an optical gunsight. This equipment was applicable to both the 'loft' bombing technique of delivery from low altitude and the 'glide' bombing technique of delivery from medium altitude. These two methods of bomb delivery had proven to be the simplest and most effective for attacks with small atomic bombs by tactical aircraft against ground targets.

Performance

The three aircraft were compared directly by the radius to which they could perform a bombing mission. Comparable assumptions were made concerning the fuel required for takeoff, landing and reserves. Cruise conditions depended on the specific airplane and engine. While the top speeds of the aircraft were calculated, the importance of maximum speed was hard to determine because the required mission did not call for aerial combat. Furthermore, the use of high speed reduced the radius available for the bombing mission and might have made the completion of the mission impossible.

The turboprop airplane could cruise efficiently on one power section with varying fractions of normal rated power at altitudes from sea level to over 20,000ft and to higher altitudes with two power sections. The different conditions gave different operating radii. The airplane was definitely subsonic with a maximum Mach number of .90 at 35,000ft and .81 at sea level. The most economical cruise condition was with one burner operating well below normal related power.

The best radius could be obtained by flying at an altitude near 20,000ft and a speed near 33mph. For combat missions, cruising at medium altitudes might have been inadvisable so a mission profile with medium altitude cruise over friendly territory and low level cruise over enemy territory would be possible. The limits of the radius performance are shown in a chart. Mission radii were shown for

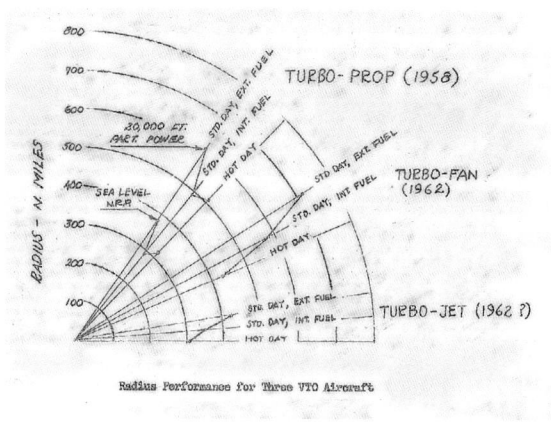

Radius Performance for Three VTO Aircraft

standard day with full internal fuel load, standard day with external fuel, and hot day for normal rated power at sea level and partial power at 20,000ft.

The turbojet airplane cruised most economically at altitudes near 50,000ft. The high speed of this airplane at altitude was over Mach 2, but this performance could not be used effectively because of the greatly reduced radius at high speed. The radius performance characteristics are shown in the above figure.

The turbofan airplane also cruised best at near 50,000ft. Its high speed was supersonic but accompanied by very high fuel consumption. Radius performance is also shown in the figure.

Comparison of radius performance from the figure shows that the turboprop airplane had a combat radius which varied considerably with the flight condition selected. It varied generally between that of the turbofan which had the best standard day radius and the turbojet which had generally the smallest radius. The turbofan lost the most performance on a hot day while the turboprop actually gained radius over the internal fuel standard day mission due to limited internal fuel tank capacity.

Operating properties

The turboprop powered airplane was best suited to low altitude attack and loft bombing. Navigation would have probably been largely by ground contact. This airplane could have probably landed and taken off on any hard ground without special preparation. The low altitude cruise would have been valuable for avoiding detection and interception.

The turbojet- and turbofan-powered airplanes had to cruise at altitudes close to 50,000ft for best mission radius. Operating at such altitudes posed problems in navigation and target location as well as the usual pilot survival needs. The high altitude, speed and small size of the airplane would have made interception difficult. Glide bombing was best adopted to the high altitude cruise.

Specially prepared alighting sites might have been required in order to prevent the jet reflection from the ground from damaging the airframe. The equipment required could have been a heat-resistant grid which supported the airplane while letting the jet gases pass through. Further development of the jet VTOL type was required to determine the landing site requirements and whether landing at unprepared sites was possible.

In comparing the operating properties of the three aircraft under consideration it appeared that the two jet types had roughly equal characteristics while the turboprop type differed from them in several important ways. From the point of view of vulnerability to enemy active defences there was no obvious advantage to either high or low altitude cruise. Operating at low altitude had the advantage of maintaining ground contact in overcast weather. This was an important factor in navigation and target location.

Bombing accuracy was about the same for the loft and glide techniques but the visual sighting requirements of both indicated an advantage for the low altitude loft technique in overcast weather. A retro-rocket bombing technique might have improved low altitude target location and delivery accuracy. The ability of the turboprop airplane to fly slowly and search for a target at low altitudes without exorbitant fuel consumption was a definite advantage over the turbojet types. The fuel load of the turboprop type was about half that of the jet types so that logistic support of the former was easier.

Technology and availability

A major factor in the evaluation of VTOL aircraft was the availability of engines. The three aircraft compared here were powered by engines with different availability and development dates. The T54 engine was predicted to be available in 1958 if development was continued. At the time it was being considered, the future of the engine was not certain. The X84 engine was under development but the programme was not being pushed. An availability date of 1962 appeared to be reasonable given the progress at the time the report was written. The X61 engine had been cancelled and no development work was being done on a comparable engine. If development work was restarted, the engines might have been available in 1962.

According to Convair, most of the problems peculiar to VTOL aircraft were solved for the turboprop type in the course of the XFY-1 programme. A T54-powered airplane could have been built without extensive additional development. The situation with the turbojet models was quite different. The problems of stabilization and control in hovering still required further development. Special alighting equipment likely would have been needed because of the high jet temperature and velocities. Conditions requiring partial afterburning also needed investigation and special afterburner development might have been required. The initial flight test programme would have been complicated considerably by the exhaust products of the downward-directed jet.

Conclusions

Convair's study indicated the following points of comparison between the three aircraft types considered:

1. The most critical factor in VTO airplane selection was the power plant. A suitable turboprop power plant was not available until 1958. A turbofan power plant was probably available by 1962. A suitable turbojet did not appear to be in active development.

2. The radius performance of the turbofan was generally superior to the turboprop and definitely much better than the turbojet. The turboprop airplane had adequate radius for an operational tactical airplane while the turbojet was marginal, further turboprop development by 1962 would have certainly improved the radius performance of the turboprop type.

3. All of the airplanes appeared to be reasonably capable of avoiding destruction by enemy defences. The turboprop airplane could operate effectively at low altitudes while the turbofan and turbojet designs operated at high altitudes and showed small radar returns. The usefulness of the high speed for the jet aircraft was questionable because of the sacrifice in radius which resulted.

4. The low altitude operation of the turboprop

airplane appeared to favour navigation, target location and attack particularly in overcast weather.

5. Convair was certain that the turboprop airplane could use unprepared landing sites at least in emergencies while the temperatures and velocities in the engine exhaust of the jet aircraft might have made special alighting equipment essential. The ability to use unprepared sites for operations or emergency landings contributed appreciably to the versatility of a tactical VTOL airplane.

6. The turboprop airplane used about half the fuel needed by the jet airplanes for comparable missions. This made the logistic support of turboprop equipped units significantly easier than jet-equipped units with the same sortie potential. Simplification of logistic requirements were an important advantage to a tactical air force in an atomic air war.

Convair's proposal did not find favour with its intended customer and nothing came of this Allison T54-powered variant of the XFY-1 intended for tactical nuclear bombing. Drawings of the turbofan- and turbojet-powered Pogo variants described in Convair's report have yet to be located; one can only speculate how these unusual aircraft studies might have appeared.

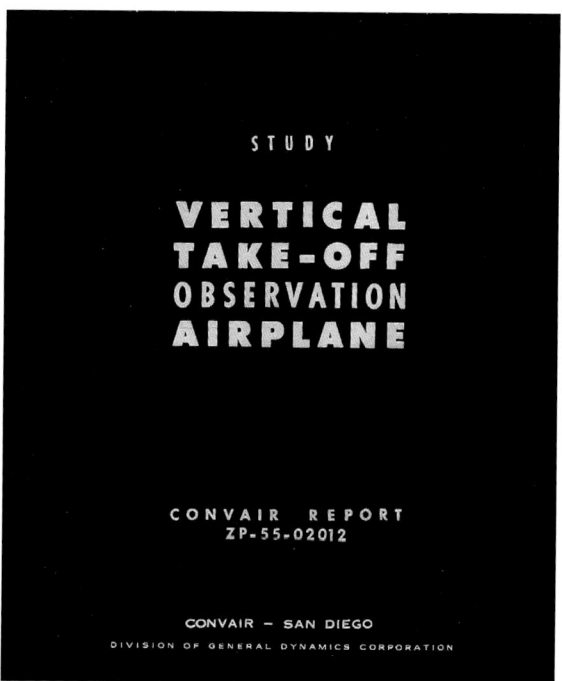

Cover to Convair's Vertical Take-Off (VTO) Observation Airplane proposal to the US Army dated April 28, 1955. This document presented three tailsitter configurations derived from the company's experience with the XFY-1.

Artist's impression of the Convair VTO Observation Airplane study of 1955. The unusual double bubble canopies are noteworthy.

Concept of operations illustration showing the potential versatility of Convair's VTO design.

Army Vertical Takeoff Observation Airplane Proposal

Though not directly related to the XFY-1, Convair pitched a similar tailsitter concept to the Army in April 1955 for the observation mission. According to the proposal document, during the previous 4½ years Convair had been engaged in intensive research and engineering development of various types of aircraft designed to take off and land vertically. These studies were based on the philosophy that vertical takeoff aircraft should not require special takeoff or landing facilities; i.e., they should be capable of landing practically anywhere.

The feasibility of the vertical takeoff aircraft was demonstrated by extensive wind tunnel tests, by flight tests of remotely controlled models and many actual flights of the Navy XFY-1 Pogo. Transition from vertical to horizontal flight had been demonstrated for both the takeoff and return to complete the actual vertical descent to a landing.

TABULATED PHYSICAL CHARACTERISTICS

	WING	VERTICAL
S	150 ft²	60 ft²
b	18'-0"	8'-0"
AR	2.17	1.07
TR	7.64:1	2:1
C_R	14'-7-1/2"	5'-0"
C_T	1'-11"	2'-6"
MAC	9'-10"	3'-11"
Λ @ L.E.	55°	30°
NACA Section	63-009 (mod)	63-009 (mod)

Engine: Allison T-56 (501-D7) Turbo-prop (alt. T-38)
Propeller: (for T-56) 13'-6" dia. 6 blade dual rotation
R.P.M. = 977 Activity factor = 140
Design lift coefficient = 0.5

ESTIMATED WEIGHT

Wing group	542 lb	Fuel system	51 lb
Tail group	200	Instruments	47
Fuselage	361	Surface Controls	146
Landing Gear	237	Hydraulic System	150
Nacelle	86	Electrical	155
Engine Installation	1610	Communicating	62
Engine Accessories	152	Furnishings	323
Power Plant Controls	19	Weight Empty	5340
Propellers	1090	Useful Load	1273
Starting System	25	(incl. 800 lb fuel)	
Lubricating System	80	T.O. Gross Weight	6613 lb

Physical characteristics and weights of Convair's VTO Observation Aircraft study for the Army.

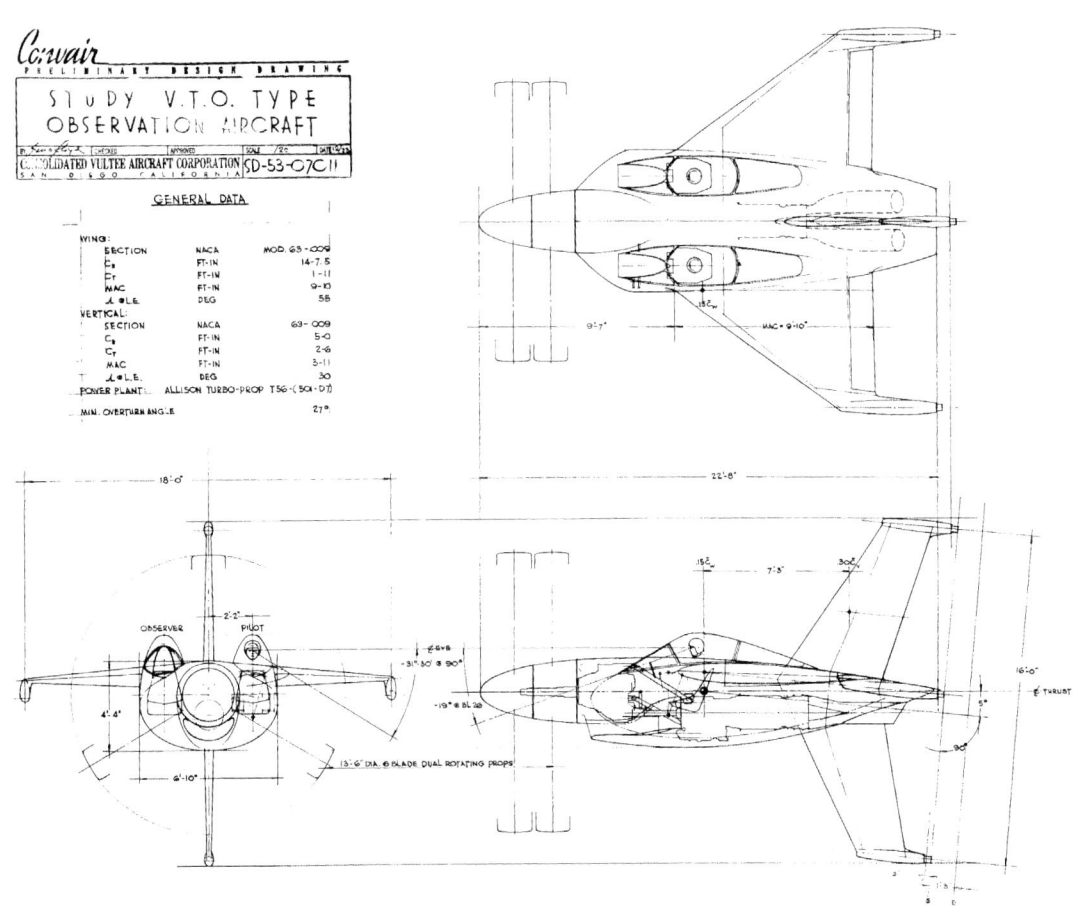

General arrangement drawing of the Convair VTO Observation Aircraft study. At 22ft 8in long, it was substantially smaller than the XFY-1, which was 32ft in length.

333

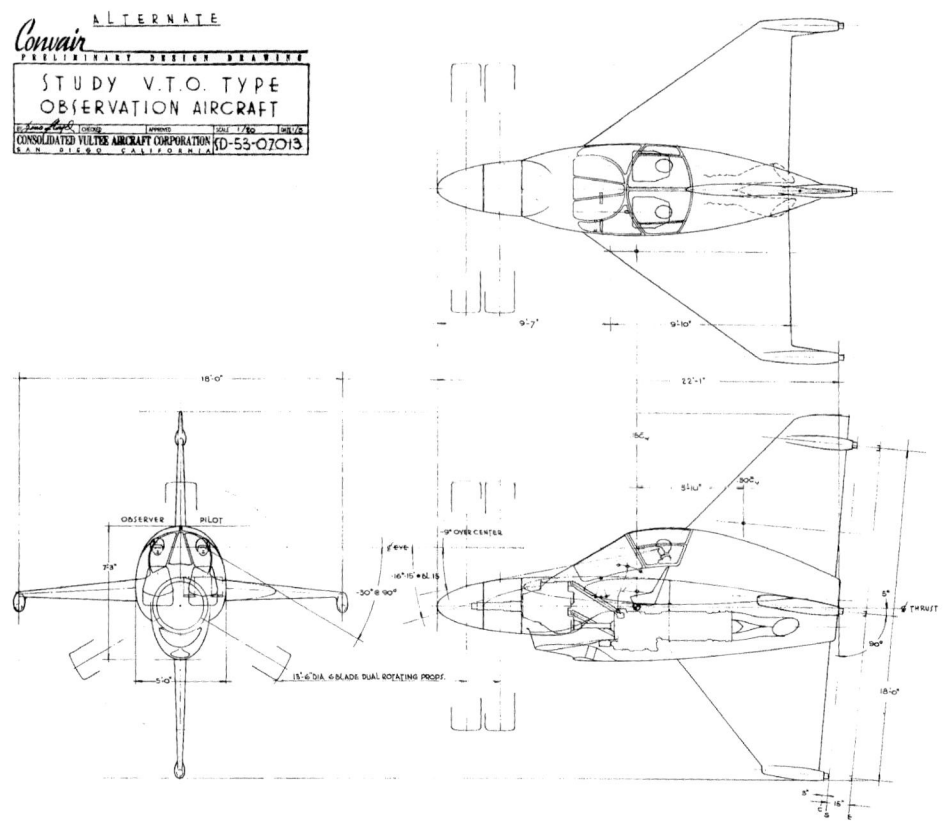

Convair also offered this alternate configuration of its VTO Observation Aircraft study with side-by-side seating under a single large canopy.

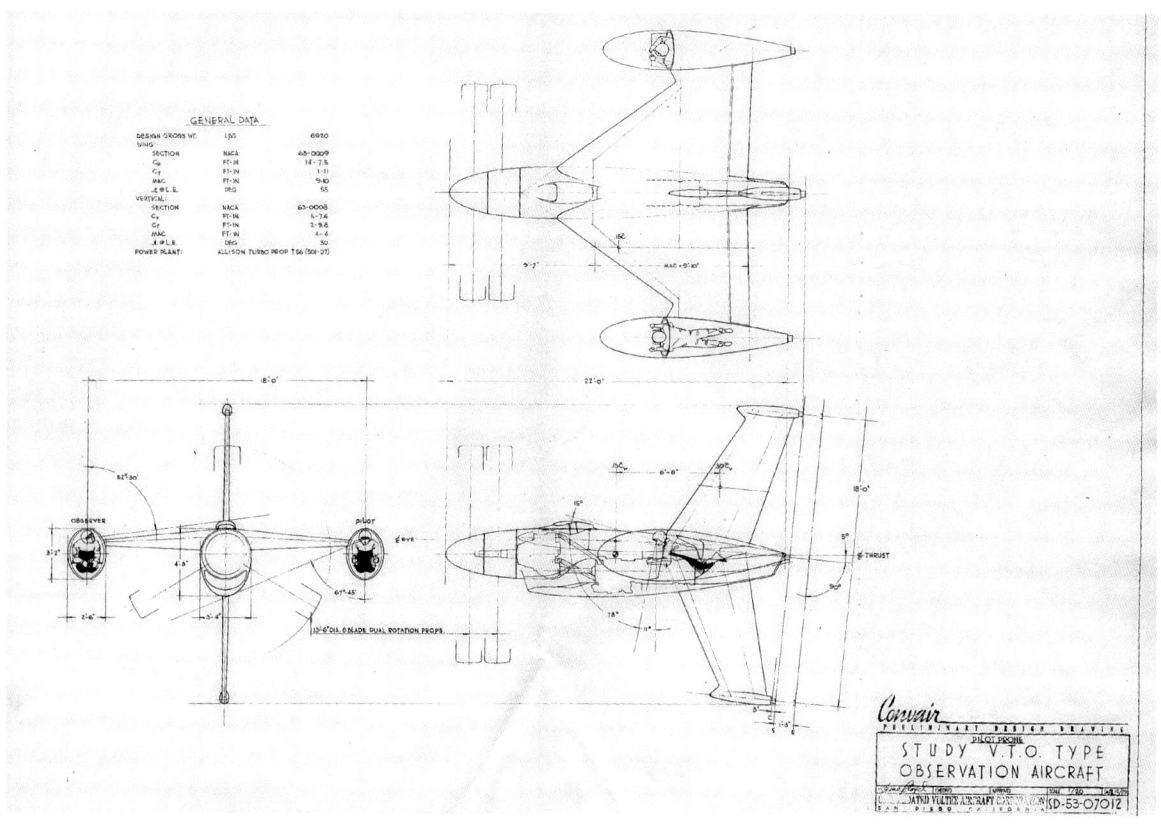

The third and final known Convair VTO Observation study placed the pilot and observer in prone positions in separate wingtip-mounted pods. Ultimately, the Army rejected the concept entirely, perhaps due to the Navy's negative experience with the XFY-1.

RELATED PROJECTS & STUDIES

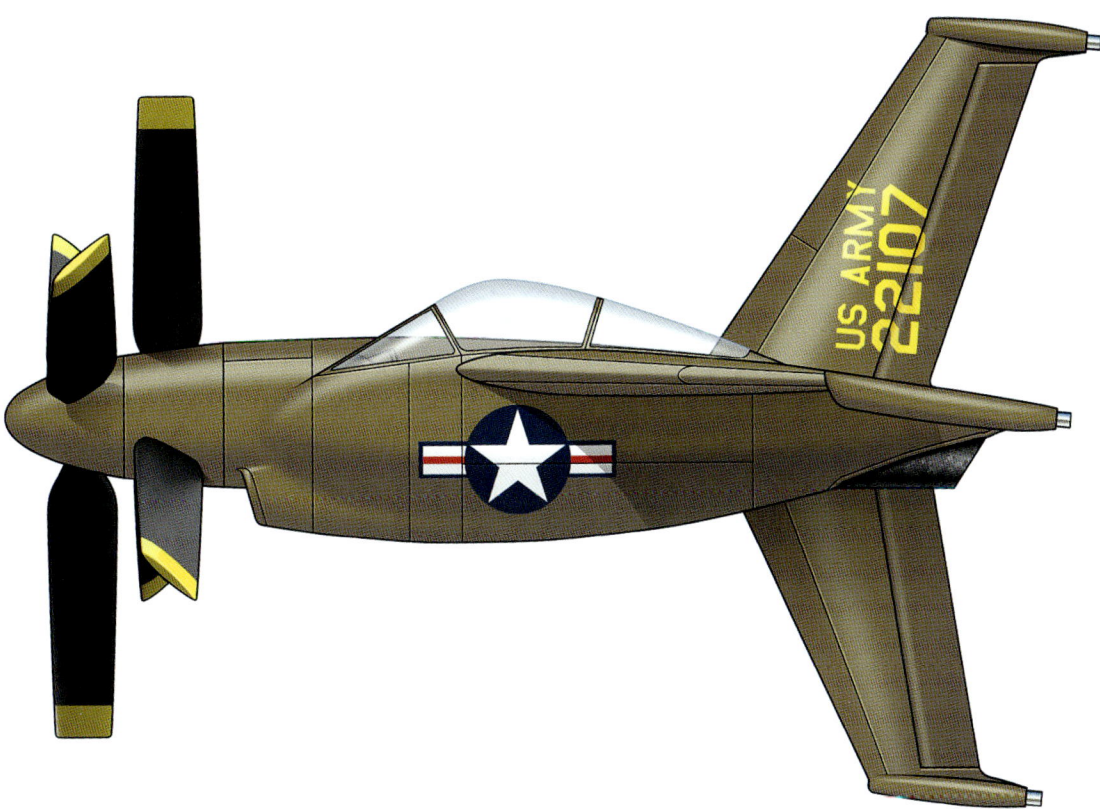

In April 1955 Convair pitched a vertical takeoff observation airplane concept to the US Army derived from its experience with the XFY-1 Pogo. This is the first of two configurations proposed, which featured a double bubble canopy arrangement and a central turboprop engine. The speculative Olive Drab scheme is inspired by contemporary Army observation airplanes.

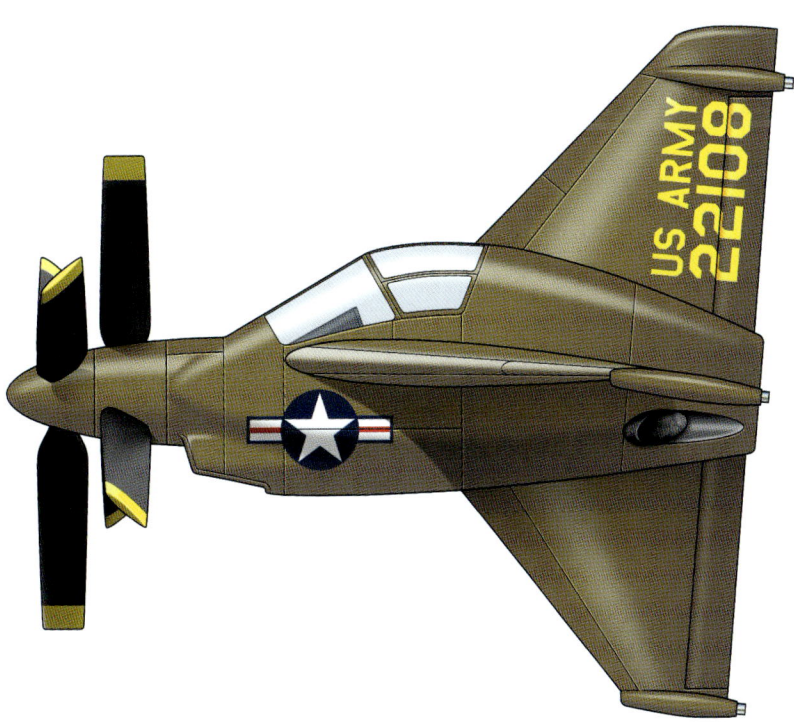

The second configuration of the vertical takeoff observation airplane had both the pilot and observer in a side-by-side seating arrangement under a single large canopy. The Army ultimately rejected both versions of the design, likely due to the poor performance of the XFY-1 during Navy flight tests, which reflected poorly on the turboprop tailsitter concept.

The following is Convair's engineering review of the application of a vertical takeoff airplane design to a two-place high performance Observation and Reconnaissance airplane capable of:

1. Operating from forward area bases.
2. Landing and taking off vertically from small, cleared areas.
3. High rate of climb and acceleration.
4. High speed cruise to and from target area.
5. Slow speed of less than 50 kts for reconnaissance.
6. Loiter time over target of three hours or more.

VTO observation mission
For the purpose of establishing design criteria, it was assumed that this VTO airplane would be employed with artillery batteries and field forces as described below and would have certain general characteristics also described herein.

Convair assumed that it was desirable to operate the aircraft in small groups of two to four airplanes directly from various field command posts about ten miles behind the MLR (main line of resistance). The airplane had to be capable of takeoff from a concealed, unprepared area and had to be able to penetrate up to 200 miles from enemy territory. However, an average penetration of 30 miles was considered to be required for a typical mission.

The airplane had to be as small as possible as an aid to ground handling and camouflage. Provisions were required for a pilot and an observer with a wide range of visibility. Provisions for tilting the pilot and observer seats forward and backward were also provided. Rotation of the observer's seat was considered for horizontal flight. The observer was replaceable with photographic equipment when desired.

General design considerations
In order to determine the feasibility of a small vertical takeoff airplane designed to perform the mission outlined previously, the power output per pound of weight of a number of engines in the small and medium power class was examined. The horsepower per pound with dual rotation gearbox for reciprocating engines investigated was found to vary between 0.52 and 0.88, which was considered insufficient for efficient VTO design. The horsepower per pound for the turboprop engines varied between 1.57 and 2.22. Turbojet engines were not considered for this application because of their excessive fuel requirements.

Based on the use of a turboprop power plan, a preliminary design chart was prepared. The chart indicated a minimum attainable value for several design parameters necessary to meet takeoff conditions as follows:
- Minimum takeoff weight: 6,000lb
- Minimum horsepower: 1,400
- Desirable propeller diameter: 13-14ft

A slightly higher weight and power were required if a good margin of takeoff performance was desired to permit operation on hot days or from high altitude terrain. The increased power was also required to meet the 500 kts maximum speed.

Description
The T38 and T56 engines were selected as the only domestic power plant in the desired power range which was already in flight status. The T56 engine offered greater power and superior specific fuel consumption with negligible penalty in overall installed weight. Design studies were made based on the use of the T38 or T56 interchangeability.

The delta wing planform was selected because of its aerodynamic superiority in the transition range at high angles of attack. Wind tunnel and flight tests had shown it was possible to slow to zero velocity with this configuration smoothly without changing altitude. The use of suitable airfoil camber and wing twist permitted good cruising performance to be obtained. Since the airplane mission did not require Mach numbers in excess of 0.9, no high speed design problems would occur with this type of airplane configuration.

The upper and lower vertical tail arrangement ensured a stable ground platform and provided adequate aerodynamic directional stability for all phases of flight. The ground attitude of the airplane was tilted forward slightly by the shock strut arrangement of the vertical tails. This was done so that the ground attitude coincided with the estimated hovering trim attitude.

The forward location of the pilot and observer ensured a wide range of visibility for both pilot and observer.

Provision for up to 2,400lb of fuel was included for extended range and endurance although this amount of fuel may seldom have been required for normal operation.

Performance characteristics
The performance was based on use of the T56 engine with ESHP rating of 3,750 and propellers at 977rpm.

Takeoff. The static thrust at sea level under standard atmospheric conditions was estimated conservatively to be 10,250lb. This gave sufficient excess thrust at any loading condition to allow takeoff up to altitudes of 10,000ft or takeoff under hot day conditions. Time from standstill to level flight at best climb velocity was approximately 30 seconds.

Landing. The time required to decelerate from a level flight approach speed of 120 kts to vertical flight at zero speed at 200ft altitude was approximately 20 seconds and a let down from 200ft required approximately 20 additional seconds. Flight tests had demonstrated the ability of similar VTO configurations to make satisfactory spot landings, even in cross winds.

Climb and velocity. The aircraft had an estimated performance of 500 kts at sea level with a slightly higher maximum velocity available at altitude.

Cruise. An average mission which was initiated in friendly territory from ten miles behind the MLR and penetrated 30 miles beyond the MLR with fuel reserve for three minutes at maximum velocity at sea level and standard reserves for landing was assumed. The ferry range of the airplane was estimated to be 1,000 miles. If desired, this could have been substantially increased with additional fuel and assisted takeoff.

Acceleration. An observation airplane of this type had to depend on its small size, speed and manoeuvrability for defence. In order to effectively use the speed potentialities, the airplane required very good acceleration characteristics. Preliminary calculations indicated that the airplane studied here was capable of accelerating from cruise at 250 kts at sea level to a maximum velocity of 500 kts in less than one minute.

Rejection
The Army passed on Convair's VTO Observation Airplane study and the type was never built. It should be noted that the proposal was submitted in late April 1955, about three months before the XFY-1 was cancelled. The Army likely inquired with the Navy and found out about the problems encountered with the Pogo in flight testing, concluding that Convair's tailsitter observation airplane would have similar flaws, making it a less useful and reliable platform than the company claimed.

LOCKHEED STUDIES

In addition to the L-200, Lockheed investigated the turboprop tailsitter concept for the Air Force; while these studies are not directly related to the Convoy Fighter programme, they are included here as bonus content for enthusiasts of secret aircraft projects. The accompanying drawings have been digitally traced from the originals to improve clarity. Due to the poor quality of the source material, some educated guesswork was required in depicting the finer details.

L-203 Liaison and Transport Studies
The L-203 designation encompassed a series of studies for a VTOL liaison aircraft produced around the same time as the L-200 at the request of the Air Materiel Command (AMC) of the US Air Force. The contract also included a limited investigation of VTOL transports. Numerous studies were produced but none were finalized; three of these are shown here. Performance requirements for the liaison aircraft included a rate of climb at sea level of 6,900ft/min; a cruise speed of 310 kts; and a maximum altitude of 16,400ft. The requirements for the transport studies were less defined, except for a maximum speed of approximately 300 kts for a tilt wing design and 400 kts the others. Details of each design are discussed below:

L-203-3—A simple design for a small liaison aircraft with large diameter contra-rotating propellers and a parachute in the spinner, possibly used to slow the descent of the aircraft while landing or in the event of engine failure. During landings, the pilot's seat rotated forward 45° while the observer's seat rotated backwards 90°, with the latter looking outwards.

L-203-5—A more unorthodox liaison aircraft design in which the pilot and observer were placed in separate pods mounted on the wing tips, which rotated 90° during takeoffs and landings. Two types of pods are shown; a teardrop-shaped pod with the crew laying in a prone position, and a pod with a stepped canopy and conventional seating positions.

L-203-7—A large assault transport capable of carrying 24 troops. The seats for the troops rotated 90° during takeoffs and landings; large doors and retractable ladders were provided underneath the aircraft for boarding and deployment. The seats of the pilot and co-pilot appear to have rotated 45°. The

L-203-7 featured a Burnelli-type lifting fuselage and two widely-spaced cockpits. It may have also had an asymmetric cockpit layout, as only the left side of the aircraft is shown in the original top view drawing.

L-210 Ground Attack Aircraft Studies
Lockheed prepared this series of studies in the early 1950s, also likely for the AMC. These were heavily armed turboprop-powered aircraft with STOL/VTOL capabilities; at least 19 configurations were produced. The engine for several may have been the Allison T40 which powered the XFV-1, though this is not confirmed. Little is known about this contract and no performance data is available. While only one of these studies is an actual tailsitter, all known variants are reproduced here for sake of completeness:

L-210-1—This design study is reminiscent of a slimmed-down North American XA2J-1 Super Savage in general layout. The aircraft was powered by a pair of Allison T38-C turboprop engines. It had a wing area of 400sq ft and a basic span of 49.2ft; including the wing tip tanks, the span increased to 57.2ft. The aircraft had a bicycle landing gear with the tandem main gear retracting into the fuselage and the outrigger struts retracting into the outer nacelles. It carried a crew of two and featured a basic armament of two 1,000lb bombs and four T118 guns with 300 rounds of ammunition per unit. Alternate armament configurations are listed in the table below. The surviving drawing indicates the possibility of a bulged forward fuselage for more efficient pressurization.

L-210-2—This configuration appears to utilize an XFV-1-style forward fuselage with an eight-bladed contra-rotating propeller, mating it with a low unswept wing and a swept tail. This was confirmed by overlaying a side view of the L-200-1 plan over this one. While no dimensions are shown on the original drawing, the forward fuselage match between the L-200-1 and L-210-2 indicate that they likely had the same propeller diameter—15ft 6in. The other dimensions shown on the tracing are derived from this assumption. The L-210-2 featured two large pods mounted mid-wing, each containing six guns, probably the T118 or 50 cal. guns of the L-210-1. It took off and landed conventionally and likely had STOL capability.

L-210-1 Armament Alternates				
	A	B	C	D (long-range, clean)
Napalm tanks	4 x 275 gallons	4 x 275 gallons	0	0
T118 or .50 cal. guns	10	4	4	4
HVAR rockets	32	56	24	0
External fuel	0	0	1,000 gallons	1,000 gallons
1,000lb bombs	0	0	6	2

L-210-4—This was a twin boom pusher armed with ten guns (T118 or 50 cal.), six in the nose and four in the forward part of each boom. It was a low wing monoplane with unswept flying surfaces. A notable feature of the engine installation was the inward cant of the turbojets, which were connected by long extension shafts to the gearbox; this drove an eight-bladed contra-rotating propeller. The exact dimensions of the study are unknown, though I believe it had the same propeller diameter as the earlier L-210-2.

L-210-10—A tailless, swept wing VTOL aircraft with a takeoff gross weight of 16,700lb, putting it in a similar class as the L-200, though carrying heavier armament. An overlay of the L-200-1 side view shows a close match in the forward fuselage and propeller diameter. Two basic payloads were proposed:

Useful Load	
Option A	Option B
12 x .50 cal. guns	8 x .50 cal. guns
3,000 rounds ammo	2,000 rounds ammo
350 gals. fuel	447 gals. fuel
Overload condition—two standard 110 gal. napalm tanks with four JATO units of 500lb thrust for 30 secs.	

It appears that a droppable takeoff cart was provided for the aircraft to take off conventionally when in the overload condition. The dimensions shown in the accompanying drawing assume a propeller diameter of 15ft 6in.

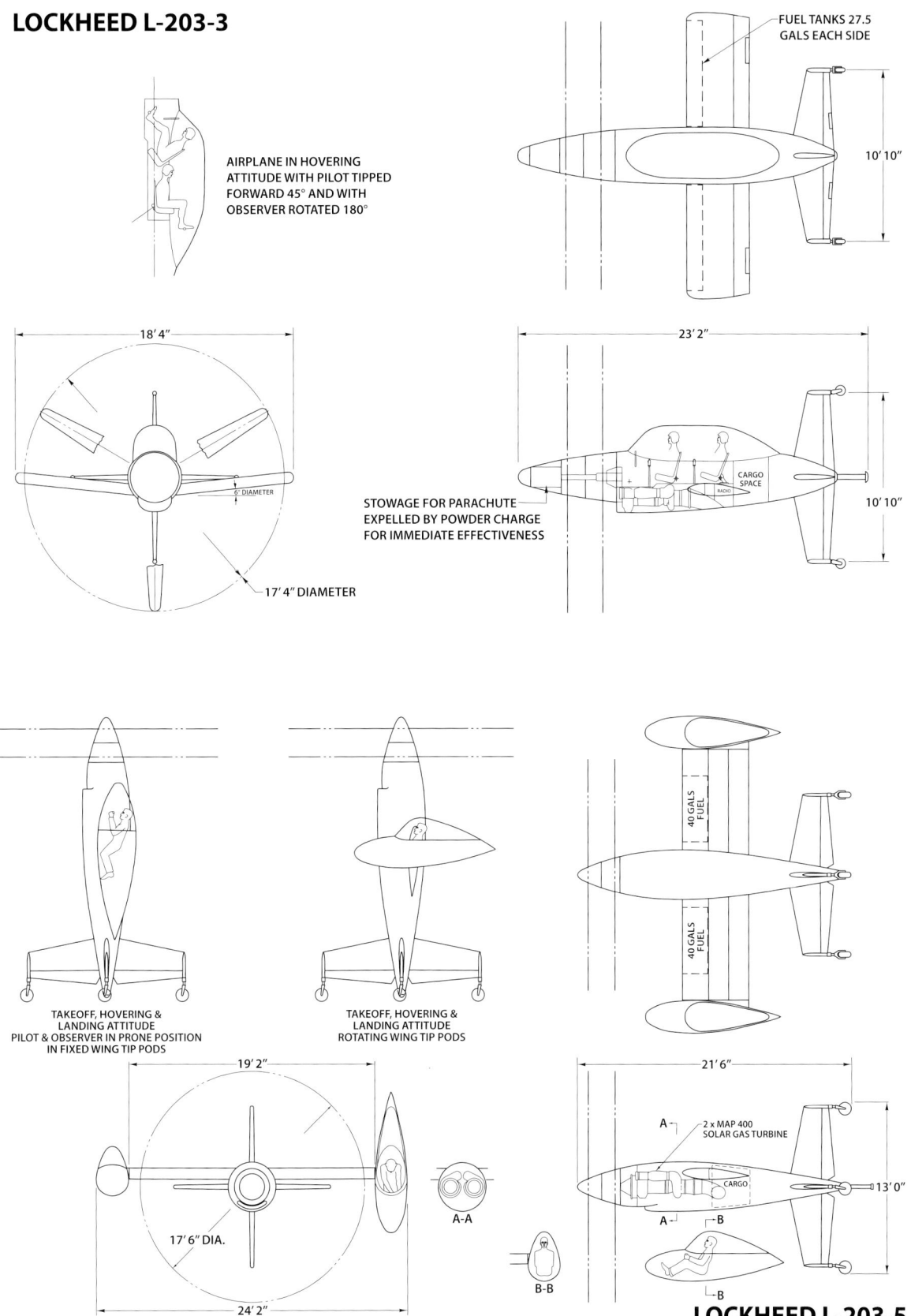

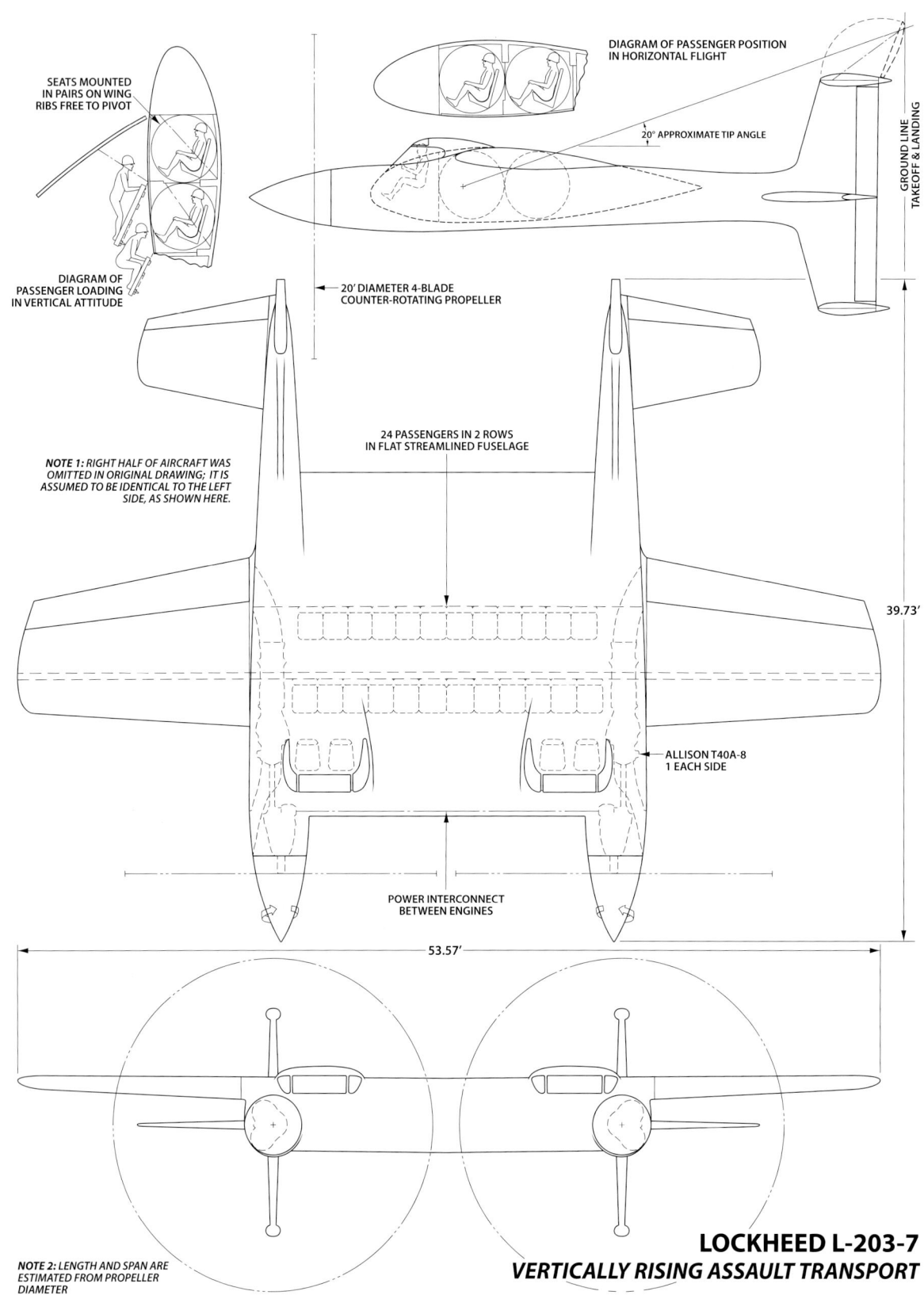

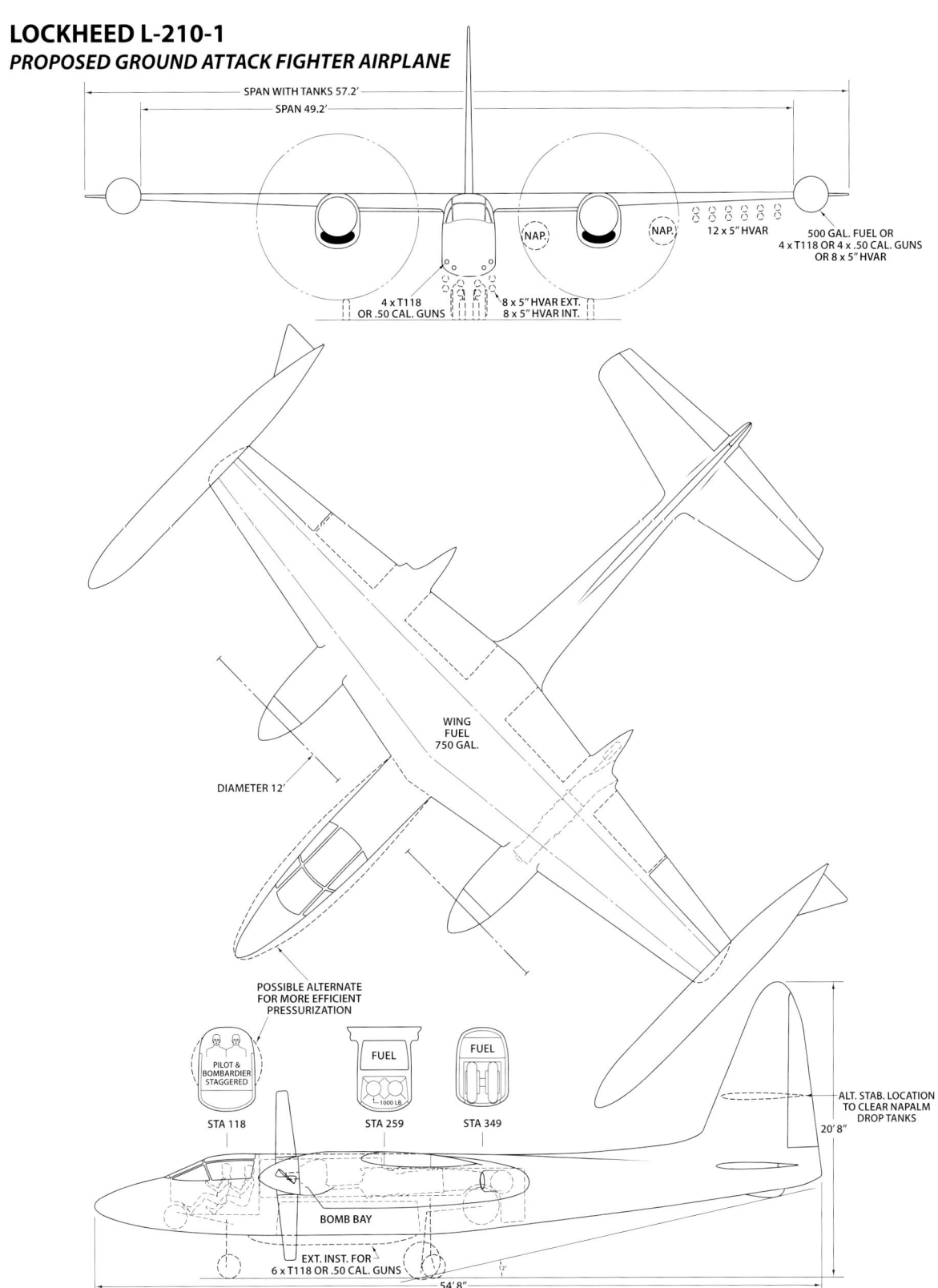

LOCKHEED L-210-10

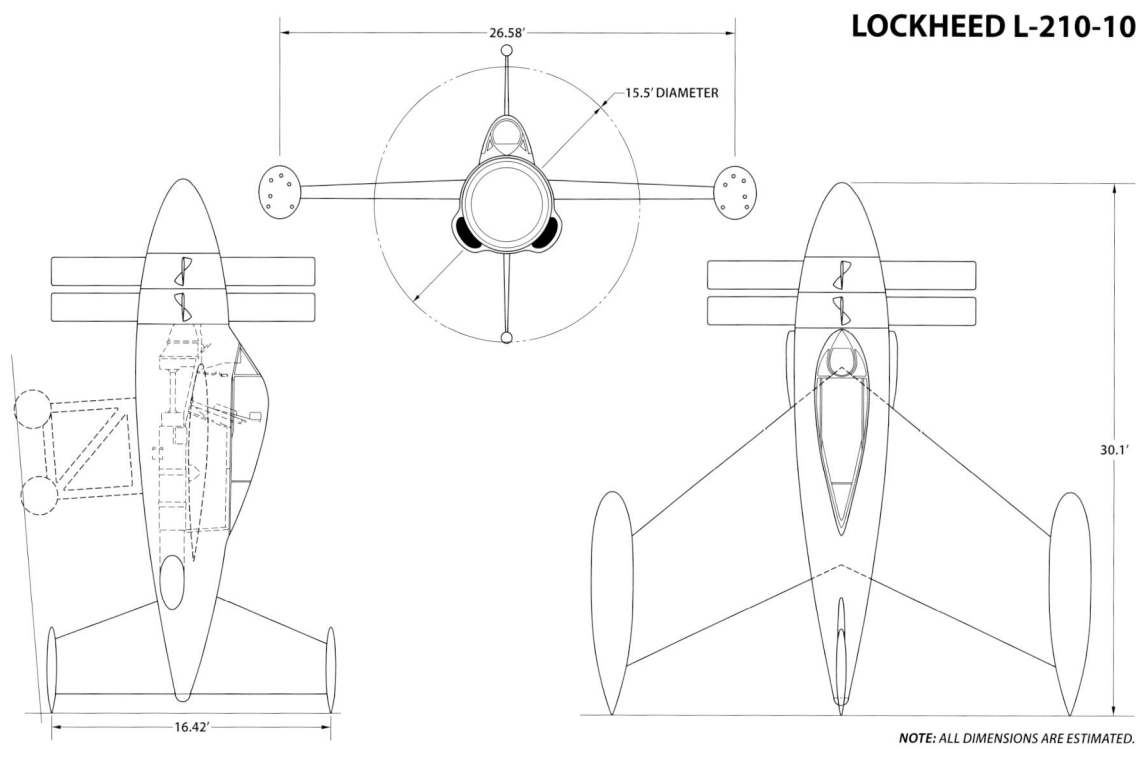

LOCKHEED L-210-2

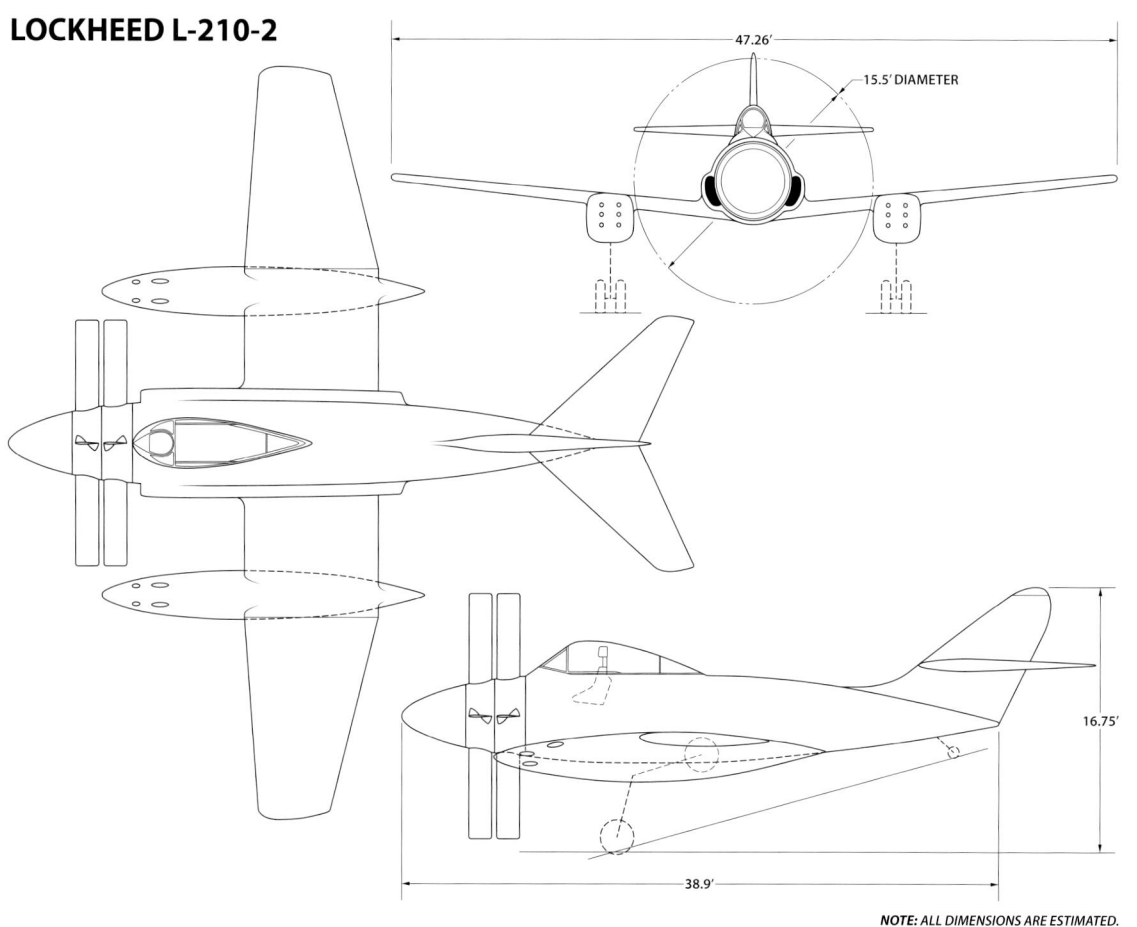

LOCKHEED L-210-4

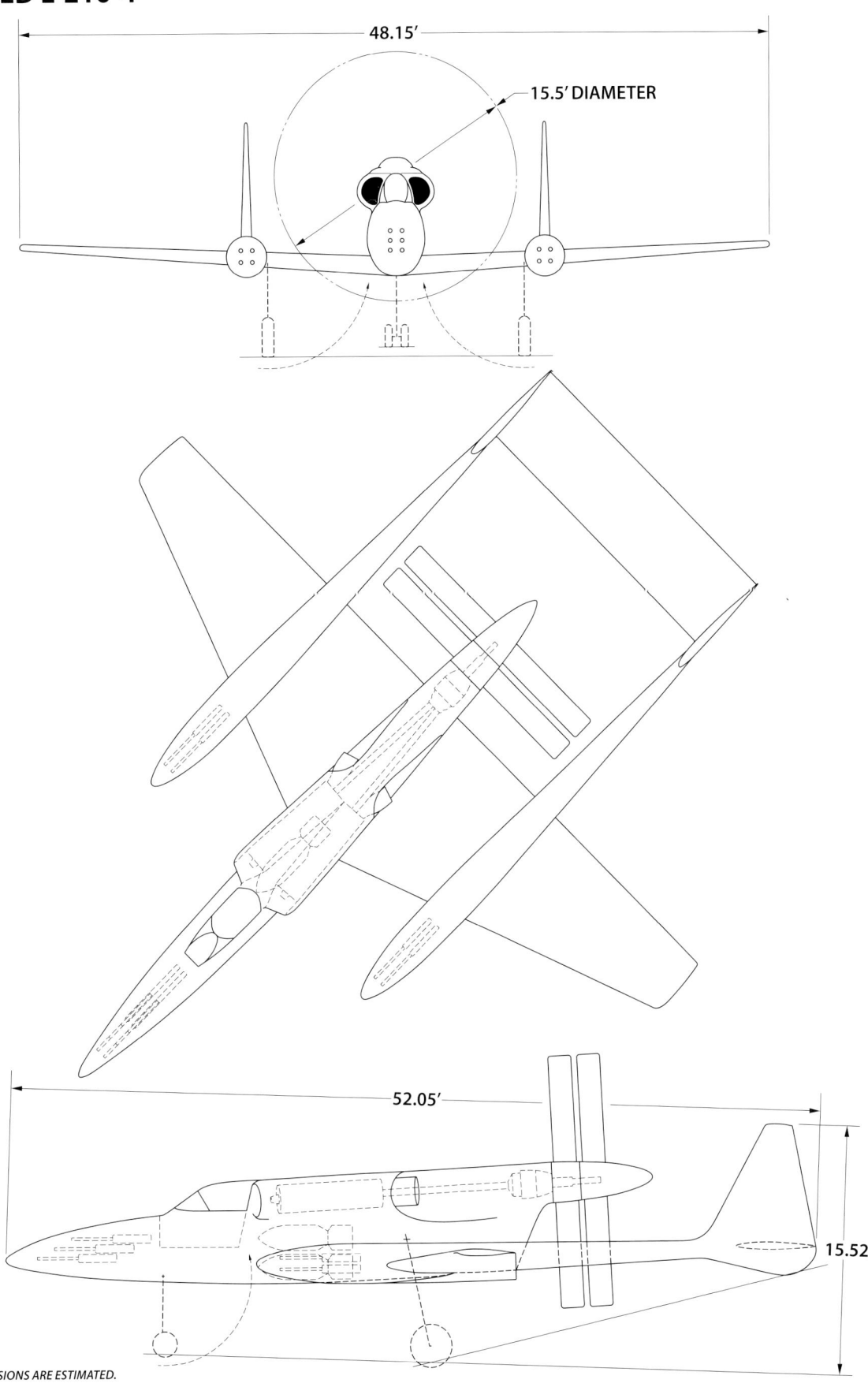

NOTE: ALL DIMENSIONS ARE ESTIMATED.

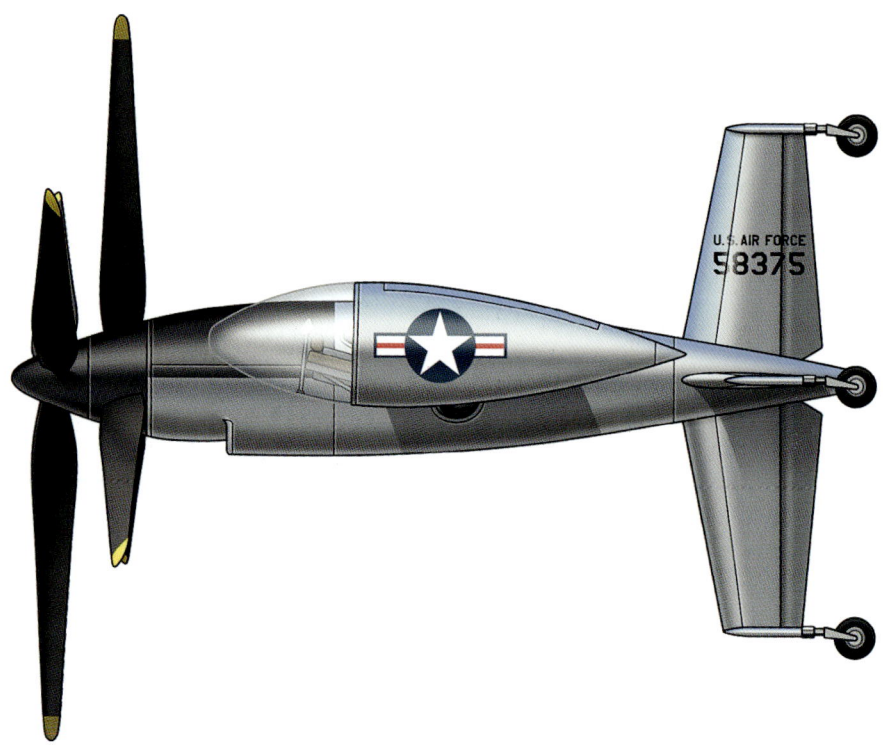

Before it lost favour with the military, the VTOL turboprop tailsitter concept was applied to other contracts by Lockheed. This artwork depicts the L-203-5, one of several studies for a liaison aircraft done for the USAF in the early 1950s. Note the swiveling pods on the wing tips for the pilot and observer, who were seated in a prone position. A more conventional upright seating position was also considered.

The L-203-7 was a study for a large assault transport capable of carrying 24 troops; the seats rotated 90° for takeoffs and landings. The aircraft featured a Burnelli-type lifting fuselage and was powered by two Allison T40-A-8 engines.

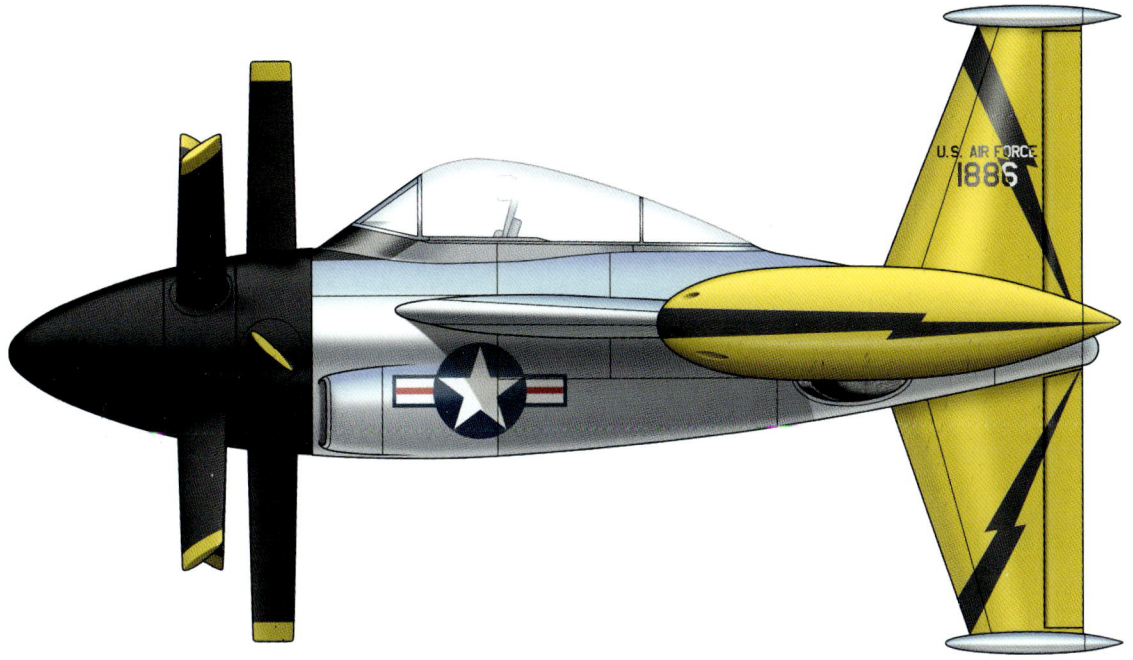

The L-210-10 was a tailless, swept wing VTOL ground attack aircraft with a formidable armament of 8-12 .50 cal. guns in its wing tip pods. It could also take off conventionally with a droppable dolly when in overload condition.

Sources and Bibliography

Sources

Additional Low-Speed Wind Tunnel Tests of a .15 Scale Powered Model of the XFY-1 Airplane with 57° Wing Leading Edge Sweepback, Report CVAL 107A. Consolidated Vultee Aircraft Corporation, March 31, 1953.

Additional Low-Speed Wind Tunnel Tests of a .15 Scale Powered Model of the XFY-1 Airplane with 57° Wing Leading Edge Sweepback, Report CVAL 107B. Convair–San Diego, A Division of General Dynamics Corporation, May 14, 1954.

Barling, Walter H. Aircraft. US 2,308,802, United States Patent and Trademark Office, 19 January 1943.

Class VF Airplane Convoy Fighter Standard Aircraft Characteristics. Consolidated Vultee Aircraft Corporation, October 1951.

Class VF Airplane Prototype for Convoy Fighter. Consolidated Vultee Aircraft Corporation, November 1950.

Clousing, L. Memo to Director, National Advisory Committee for Aeronautics. 1 August 1951.

Cockpit Mock-up Report Model L-200-1 Convoy Fighter, Report No. 7720. Lockheed Aircraft Corporation, 9 November 1950.

A Comparison of Three Power Plant Types for Tactical VTOL Aircraft. Convair–San Diego, A Division of General Dynamics Corporation, April 6, 1955.

Consolidated-Vultee Corporation. "Model Convoy Fighter– Proposal For." Received by Chief, Bureau of Aeronautics, 29 November 1950. National Archives II, RG 72.

Convoy Fighter VF Model N-63, Brochure No. 32. Northrop Aircraft, Inc., 6 November 1950.

DeFrance, Smith J. Letter to Lockheed Aircraft Corporation. 25 April 1952. National Archives II, RG 255.

Design Features of the lockheed L-200 Convoy Fighter, Report No. 7711. Lockheed Aircraft Corporation, 10 November 1950.

Electronic Equipage, Report No. 7721. Lockheed Aircraft Corporation, 14 November 1950.

Engineering Analysis of Proposed 1000 h.p. Helicopter-Type Aircraft. Lloyd H. Leonard, April 1939.

FY-1 (Convair Model "5") Mock-up. Consolidated Vultee Aircraft Corporation, June 29, 1951.

Free Flight Model Tests, Report No. 7715. Lockheed Aircraft Corporation, 10 November 1950.

Gamby, Georges Henri. Letter to Aeronautical Patents and Design Board. 12 March 1934. National Archives II, RG 255.

Glenn L. Martin Company. "Convoy Fighter–Cost Proposal For." Received by Chief, Bureau of Aeronautics, 1 November 1950. National Archives II, RG 72.

Goodyear Aircraft Corporation. "Informal Cost Proposal for Convoy Fighter Airplane Project." Received by Navy Department, Bureau of Aeronautics, 1 November 1950. National Archives II, RG 72.

Lockheed Aircraft Corporation. "Model Convoy Fighter – Price and Proposal." Received by Chief, Bureau of Aeronautics, 28 November 1950. National Archives II, RG 72.

Low-Speed Wind Tunnel Tests of a .15 Scale Powered Model of the XFY-1 Airplane with 57° Wing Leading Edge Sweepback, Report CVAL 107. Consolidated Vultee Aircraft Corporation, November 1, 1952.

The Martin Model 262 Convoy Fighter for the United States Navy, Engineering Report No. 4120. Glenn L. Martin Company, 17 November 1950.

Martin Model 262 P, Standard Aircraft Characteristics. Glenn L. Martin Company, 15 November 1950.

Mock-up Board Report for Model XFO-1 Aircraft. US Navy, Bureau of Aeronautics, 11 December 1951.

Mock-up Board Report for Model XFY-1 Aircraft. US Navy, Bureau of Aeronautics, 24 September 1951.

Model 262 P (Convoy Fighter Prototype), Engineering Report No. 4128. Glenn L. Martin Company, 15 November 1950.

Model N-63A Convoy Fighter VF Prototype, Brochure No. 33. Northrop Aircraft Inc., 6 November 1950.

Model XFO-1 United States Navy Mock-up Conference. Lockheed Aircraft Corporation, 10 October 1951.

National Advisory Committee for Aeronautics. Letter to

Aeronautical Patents and Design Board. 11 March 1929, National Archives II, RG 255.

National Advisory Committee for Aeronautics. Letter to Georges Henri Gamby. 28 March 1934. National Archives II, RG 255.

Navy Convoy Fighter Class VF, ZP-50-15002. Consolidated Vultee Aircraft Corporation, November 1950.

Northrop Aircraft, Inc. "Convoy Fighter–Cost Proposal." Received by Department of the Navy, Bureau of Aeronautics. 24 November 1950. National Archives II, RG 72.

Northrop Progress Report Project N-63–Convoy Fighter (VF) No. 1. Northrop Aircraft, Inc., 22 December 1950.

Proposal GA-28A Transition Training and Research Airplane, Report No. GER 2401. Goodyear Aircraft Corporation, 20 November 1950.

Proposal GA-28B Convoy Fighter Airplane, Report No. GER 2402. Goodyear Aircraft Corporation, 20 November 1950.

Prototype Proposal for Lockheed L-200 Convoy Fighter, Report No. 7724. Lockheed Aircraft Corporation, 10 November 1950.

Remmen, J. Letter to Aeronautical Patent and Design Board. 26 February 1929. National Archives II, RG 255.

Standard Aircraft Characteristics, Goodyear Aircraft Model A 28A. Goodyear Aircraft Corporation, 20 November 1950.

Standard Aircraft Characteristics, Goodyear Aircraft Model A 28B. Goodyear Aircraft Corporation, 20 November 1950.

Standard Aircraft Characteristics, Model N-63A. Northrop Aircraft, Inc., 6 November 1950.

Standard Aircraft Characteristics, Model XFY-1. Consolidated Vultee Aircraft Corporation, May 1952.

Standard Aircraft Characteristics, Model XFY-1. Consolidated Vultee Aircraft Corporation, July 1954.

Stripped Down Convoy Fighter with Present 5525 Equiv. H.P. T-40 Engine (Alternate for Double Mamba Prototype), ZP-50-15003. Consolidated Vultee Aircraft Corporation, November 1, 1950.

Study: Vertical Take-off Observation Airplane, Convair Report ZP-55-02012. Convair–San Diego, A Division of General Dynamics Corporation, April 28, 1955.

USAAC, Materiel Division. Letter to Lloyd H. Leonard. 31 July 1939. National Archives II, RG 342.

Various Methods for Takeoff and Alighting Model L-200 Convoy Fighter, Report No. 7714. Lockheed Aircraft Corporation, 10 November 1950.

Young, Arthur Middleton. Aircraft. US 2,382,460, United States Patent and Trademark Office, 14 August 1945.

Bibliography

Barker, Ralph. The Hurricats. London, Pelham Books, 1978.

Bradley, Robert E. Convair Advanced Designs II: Secret Fighters, Attack Aircraft, and Unique Concepts 1929-1973. Manchester, Crecy Publishing, 2014.

Coleman, Skeets, and Steve Ginter. Naval Fighters Number Twenty-Seven: Convair XFY-1 Pogo. Simi Valley, Ginter Publications, 1994.

Ginter, Steve. Naval Fighters Number Thirty-Two: Lockheed XFV-1 VTOL Fighter. Simi Valley, Ginter Publications, 1996.

Gunston, Bill. The Development of Jet and Turbine Aero Engines, 4th Edition. Sparkford, Haynes Publishing, 2006.

Hughes, Terry and Costello, John. The Battle of the Atlantic. New York, Dial Press, 1977.

Paust, Gilbert. "Convertible Jet Helicopter." Modern Mechanix, March 1948.

Slayton, Bill. A History of Lockheed Preliminary Designs. TS, Peter Clukey Collection, 1999.

Zichek, Jared. Convair Class VF Convoy Fighter: The Original Proposal for the XFY-1 Pogo. Hauser, Retromechanix Publications, 2017.

Zichek, Jared. Goodyear GA-28A/B Convoy Fighter: The Naval VTOL Turboprop Tailsitter Project of 1950. Hauser, Idaho, Retromechanix Publications, 2015.

Zichek, Jared. Lockheed Model L-200 Convoy Fighter: The Original Proposal and Early Development of the XFV-1 Salmon–Part 1. Hauser, Retromechanix Publications, 2017.

Zichek, Jared. Lockheed Model L-200 Convoy Fighter: The Original Proposal and Early Development of the XFV-1 Salmon–Part 2. Hauser, Retromechanix Publications, 2017.

Zichek, Jared. Martin Model 262 Convoy Fighter: The Naval VTOL Turboprop Project of 1950. Hauser, Retromechanix Publications, 2015.

Zichek, Jared. Northrop N-63 Convoy Fighter: The Naval VTOL Turboprop Tailsitter Project of 1950. Hauser, Retromechanix Publications, 2015.

Image credits

Chapter 1 Convoy Fighter Origins
p9 top and bottom: National Archives II, RG 255
p11 top: © 2017 Jared A. Zichek
p11 bottom: National Archives II, RG 255
p12-15: National Archives II, RG 72
p16, p18: United States Patent and Trademark Office
p19: Fawcett Publications

Chapter 2 Convair
p25 © 2021 Jared A. Zichek
p26 © 2022 Jared A. Zichek
p48 © 2022 Jared A. Zichek
All other images: National Archives II, RG 72

Chapter 3 Goodyear
p60 © 2022 Jared A. Zichek
p80 © 2022 Jared A. Zichek
All other images: National Archives II, RG 72

Chapter 4 Lockheed
p95 © 2017 Jared A. Zichek
p141 © 2017 Jared A. Zichek
p142-144: John Aldaz Collection
p172, p173: © 2022 Jared A. Zichek
All other images: National Archives II, RG 72

Chapter 5 Martin
p217 bottom: © 2022 Jared A. Zichek
p218, p219: © 2022 Jared A. Zichek
All other images: National Archives II, RG 72

Chapter 6 Northrop
p259 © 2022 Jared A. Zichek
p260, p261 © 2022 Jared A. Zichek
All other images: National Archives II, RG 72

Chapter 7 Winners and Losers
p302 © 2022 Jared A. Zichek
p305-307: National Archives II, RG 255
p325 bottom, p326, p327 top: National Archives II, RG 255
p327 bottom: © 2017 Jared A. Zichek
All other images: National Archives II, RG 72

Chapter 8 Related Projects and Studies
p329-330: National Archives II, RG 342
p335: © 2017 Jared A. Zichek
p339-345: © 2022 Jared A. Zichek
All other images: National Archives II
All other images: National Archives IImons Collection

Index

Aeronautical and Patents Design Board 9
Aeroproducts Propellers 22, 220
Airesearch 304
Allison Engine Company 110, 117, 119, 160
Allison T38-A-1 22, 336, 338
Allison T40 22, 25, 26, 30, 34, 109, 110, 119, 130, 158-163, 167, 206, 338
Allison T40-A-6 113, 131, 167, 206
Allison T40-A-8 113, 117, 131, 157, 167
Allison T54 328, 331, 332
Allison XT40-A-6 65, 107, 110, 234
Allison XT40-A-8 22, 25, 26, 30, 174, 189, 196, 200, 220
Allison YT40-A-14 294
Allison YT54 294, 320
Ames, Joseph S. 11
Ames Research Center 303, 315, 317
Armstrong Siddeley Double Mamba III 23, 25, 30, 53, 107, 109, 110, 130, 157-163, 167, 196, 200, 205, 251

Barling, Walter H. 15, 17
Barling Aircraft 16
Bell Aircraft Corporation 17, 19, 20
Bell P-39 Airacobra 17, 20
Bell P-63 Kingcobra 20
Boeing B-29 Superfortress 88, 89
BuAer 8, 14, 22, 24, 25, 30, 34, 49, 59, 65, 81, 110, 117, 119, 134, 157, 158, 162, 167, 168, 200, 203, 205, 206, 232-234, 238, 262, 263, 313, 315, 318-320, 322
BuAer DR-72/DR-72A 22, 51, 59, 107
Burbank, California 303

Carroll, F. O. 15
Coleman, James F. 294
Consolidated Vultee XP-81 21
Convair Aircraft Corporation 23-25, 28, 30, 33-35, 49, 171, 262, 263, 290, 291, 294, 322, 328, 331-333, 336
Convair Vertical Take-Off Observation Plane 332-333
Convair XF-92A 25, 27, 28, 33

Convair XFY-1 Pogo 7, 17, 24, 35, 263, 268, 290, 291, 294, 322, 328, 331, 332, 333, 337
Curtiss Propellers 304

Danch, G. 318
David Taylor Model Basin 222
Douglas Aircraft Company 11

Edwards Air Force Base, California 320

Flock, A. 318
Focke-Wulf Fw 200 Condor 20
Focke-Wulf Triebflügel 15, 21

Gamby, Georges Henri 11-13
General Electric X84 329, 331
General Electric XT31 21
Goodyear Aerospace Corporation 22, 23, 49-51, 53-55, 58, 59, 63, 65, 262, 263
Goodyear GA-28A 49, 50, 51, 53-55, 58, 59, 65
Goodyear GA-28B 49, 50, 51, 53-55, 58, 59, 61, 63, 65, 262
Grumman Martlet 20

Hawker Hurricane 20, 21
Hawthorne, California 220
Heinkel Lerche 21
Heinkel Wespe 21
Hibbard, Hal 168

Johnsville, Pennsylvania 22, 50, 81, 134, 138, 140

Langley Memorial Aeronautical Laboratory 13, 303
Leonard, Lloyd H. 13-15
Leonard Helicopter-Type Aircraft 13-15
Lewis, G. W. 10-12
Lindbergh Field, California 290
Lockheed Constellation 158
Lockheed Corporation 11, 23, 35, 81, 84, 86, 88-91, 93, 109, 110, 113, 117, 119-127, 130-132, 134, 135, 142, 147, 151, 157-160, 162, 165, 167, 168, 171, 262, 263, 303, 308, 313, 315, 317-320, 328, 337, 338

Lockheed F-80 Shooting Star 158
Lockheed F-90 158
Lockheed F-94 Starfire 121, 126, 129
Lockheed L-200 81, 88, 103, 105, 107, 109, 110, 113, 121, 122, 124, 126-128, 131, 134-136, 142, 144, 146, 162, 262, 337, 338
Lockheed L-203 337, 338
Lockheed L-210 338
Lockheed P2V Neptune 158
Lockheed XFV-1 (aka XFO-1 aka Model 81) 7, 35, 81, 171, 263, 303, 312, 315, 317-320, 322, 328, 338

Martin Company 23, 35, 174-176, 178, 179, 182, 184, 186, 187, 189, 193, 195, 198, 200, 201, 203, 205, 206, 262, 263
Martin Model 262 174, 175, 198, 200, 203, 262
Martin Model 262P 174, 200
Martin XB-51 186, 189
Modern Mechanix 20
Moffett Field, California 294, 315
Multhopp, H. 184

National Advisory Committee for Aeronautics (NACA) 8, 10, 12, 13, 28, 61, 107, 142, 220, 303, 313, 315, 322
Naval Air Development Center (NADC) 22, 50, 81, 134
Naval Auxiliary Air Station, California 294
North American XA2J-1 Super Savage 338
Northrop Corporation 23, 220-223, 232-234, 238, 240, 263
Northrop F-89 Scorpion 238
Northrop N-63 220, 221, 223, 232-234, 240, 262
Northrop N-63A 220, 223, 232-234
Northrop P-61 Black Widow 238

Operational Specification 121 (OS-121) 107, 158
Operational Specification 122 (OS-122) 22, 30, 49, 62, 81, 120, 122, 167, 187, 200, 262, 263

Pagé, Victor W. 10
Pearl Harbor, Hawaii 20
Pittsburgh, Pennsylvania 9
Pratt & Whitney PT2E 119
Pratt & Whitney R-2800 Double Wasp 14
Project Hummingbird 21

Remmen, J. 9-11
Remmen Helicopter Plane 9-11
Rolfe, Douglas 20
Ryan X-13 Vertijet 22

Salmon, Herman 320
Santa Monica, California 11
Semipalatinsk, Kazakhstan 21

US Army Air Corps Materiel Command 8, 337
US Navy Bureau of Aeronautics (see BuAer)

Vertigo Plane 11, 12
Vought FG-1 Corsair 49

Warner Aircraft Corporation 12
Witteman-Lewis XNBL-1 'Barling Bomber' 15
Wright Field, Ohio 15

Young, Arthur M. 17, 20